Guidelines for Mastering the Properties of Molecular Sieves

the Properties of
Molecular Sieves

Relationship between the
Physicochemical Properties of
Zeolitic Systems
and Their Low Dimensionality

NATO ASI Series

Advanced Science Institutes Series

A series presenting the results of activities sponsored by the NATO Science Committee,
which aims at the dissemination of advanced scientific and technological knowledge,
with a view to strengthening links between scientific communities.

The series is published by an international board of publishers in conjunction with the
NATO Scientific Affairs Division

A	**Life Sciences**	Plenum Publishing Corporation
B	**Physics**	New York and London
C	**Mathematical**	Kluwer Academic Publishers
	and Physical Sciences	Dordrecht, Boston, and London
D	**Behavioral and Social Sciences**	
E	**Applied Sciences**	
F	**Computer and Systems Sciences**	Springer-Verlag
G	**Ecological Sciences**	Berlin, Heidelberg, New York, London,
H	**Cell Biology**	Paris, and Tokyo

Recent Volumes in this Series

Series B: Physics

Guidelines for Mastering the Properties of Molecular Sieves

Relationship between the Physicochemical Properties of Zeolitic Systems and Their Low Dimensionality

Edited by

Denise Barthomeuf

Université Paris VI
Paris, France

Eric G. Derouane

Université Notre Dame de la Paix
Namur, Belgium

and

Wolfgang Hölderich

BASF
Ludwigshafen, Federal Republic of Germany

Plenum Press
New York and London
Published in cooperation with NATO Scientific Affairs Division

AIR PRODUCTS

AIR PRODUCTS AND CHEMICALS, INC.
ALLENTOWN, PA 18195
ATTN: INFO. SERVICES (R&D #1)

Proceedings of a NATO Advanced Research Workshop on
Physicochemical Properties of Zeolitic Systems
and Their Low Dimensionality,
held April 24–28, 1989,
in Chantilly, France

Library of Congress Cataloging in Publication Data

NATO Advanced Research Workshop on Physicochemical Properties of Zeolitic
Systems and Their Low Dimensionality (1989: Chantilly, France)
  Guidelines for mastering the properties of molecular sieves: relationship be-
tween the physicochemical properties of zeolitic systems and their low dimen-
sionality / edited by Denise Barthomeuf, Eric G. Derouane, and Wolfgang
Hölderich.
     p.   cm.—(NATO ASI series. Series B, Physics; vol. 221)
  "Published in cooperation with NATO Scientific Affairs Division."
  "Proceedings of a NATO Advanced Research Workshop on Physicochemical
Properties of Zeolitic Systems and Their Low Dimensionality, held April 24–28,
1989, in Chantilly, France"—T.p. verso.
  Includes bibliographical references and index.
  ISBN-13: 978-1-4684-5789-6   e-ISBN-13: 978-1-4684-5787-2
  DOI: 10.1007/978-1-4684-5787-2
  1. Molecular sieves—Congresses. 2. Zeolites—Congresses. I. Barthomeuf,
Denise. II. Derouane, E. G. III. Hölderich, Wolfgang. IV. North Atlantic Treaty
Organization. Scientific Affairs Division. V. Title. VI. Series: NATO ASI series.
Series B., Physics; v. 221.
TP159.M6N38   1989   90-7497
660′.284245-dc20   CIP

© 1990 Plenum Press, New York
Softcover reprint of the hardcover 1st edition 1990

A Division of Plenum Publishing Corporation
233 Spring Street, New York, N.Y. 10013

SPECIAL PROGRAM ON CONDENSED SYSTEMS OF LOW DIMENSIONALITY

This book contains the proceedings of a NATO Advanced Research Workshop held within the program of activities of the NATO Special Program on Condensed Systems of Low Dimensionality, running from 1983 to 1988 as part of the activities of the NATO Science Committee.

Other books previously published as a result of the activities of the Special Program are:

SPECIAL PROGRAM ON CONDENSED SYSTEMS OF LOW DIMENSIONALITY

PREFACE

Low dimensionality is a multifarious concept which applies to very diversified materials. Thus, examples of low-dimensional systems are structures with one or several layers, single lines or patterns of lines, and small clusters isolated or dispersed in solid systems. Such low-dimensional features can be produced in a wide variety of materials systems with a broad spectrum of scientific and practical interests. These features, in turn, induce specific properties and, particularly, specific transport properties.

In the case of zeolites, low dimensionality appears in the network of small-diameter pores of molecular size, extending in one, two or three dimensions, that these solids exhibit as a characteristic feature and which explains the term of "molecular sieves" currently used to name these materials. Indeed, a large number of industrial processes for separation of gases and liquids, and for catalysis are based upon the use of this low-dimensional feature in zeolites. For instance, zeolites constitute the first class of catalysts employed all over the world.

Because of the peculiarity and flexibility of their structure (and composition), zeolites can be adapted to suit many specific and diversified applications. For this reason, zeolites are presently the object of a large and fast-growing interest among chemists and chemical engineers.

The usual forms of zeolites consist of AlO_4 and SiO_4 tetrahedra linked together in such a way that cages, from 0.3 to 1.3 nm in size, are formed which, in turn, are associated to create a continuous network of pores extending in one, two, or three dimensions. The presence of such micropores in a three-dimensional crystal generates fascinating properties. Molecules of gases or liquids, with the proper size, adsorb in the cages and they reach a state which is different from that observed either in free molecules or upon adsorption on other porous solids. In the zeolites micropores, molecules are immersed in electrostatic field and electrostatic field gradients since the framework charges distributed in the walls of the cages surround them at molecular distances. In addition, the high curvature of the cavities modify the interactions caused by Van der Waals forces. It follows that very unexpected selectivities are observed for adsorption or for catalysis.

The aim of the Workshop was, first, to describe the present state of the art in order to, then, find the guidelines that may be used to depict and synthesize tailor-made zeolites for specific applications in catalysis or adsorbtion.

This approach requires progress in the fields of synthesis and techniques of characterization, as well as studies of the influence of the charged cages on the adsorbed molecules and of the resulting selectivities in adsorption and catalysis. These points constitute the various chapters of the Workshop.

The organization of the Workshop was made possible with the financial support of NATO and of the Catalysis Division of the Société Française de Chimie and with the help of the Laboratory of Surface Reactivity and Structure at the University of Paris. The kind assistance of Miss F. Armanini and Mr. A. Shikholeslami are particularly acknowledged. Moreover the laboratory of Catalysis at the University of Namur was of great help for the preparation of the book especially Dr. Z. Gabelica.

Many thanks are presented to all of those who contributed to the Workshop and to the book of Proceedings.

P.C. GRAVELLE   D. BARTHOMEUF

CONTENTS

LOCALIZED AND OVERALL PROPERTIES RELATED TO THE NATURE AND STRUCTURAL
ORGANIZATION OF THE FRAMEWORK ATOMS

ORIENTATION OF THE PATH OF REACTIONS (CATALYSIS, ADSORPTION) BY
CHEMICAL OR OTHER GEOMETRIC OR NON GEOMETRIC EFFECTS

OVERVIEW OF THE WORKSHOP

DISCUSSION REPORTS

EFFECTS OF SUBSTITUTION IN SAPO-n FRAMEWORKS ON THEIR PROPERTIES AS ACID CATALYSTS

Machteld Mertens, Johan A. Martens, Piet J. Grobet and Peter A. Jacobs

Laboratorium voor Oppervlaktechemie
K.U. Leuven, Kardinaal Mercierlaan 92
B-3030 Heverlee, Belgium

PHOSPHORUS SUBSTITUTION IN ALUMINOSILICATES

Various similarities exist between the oxygen chemistry of phosphorus and silicon. The synthetic aluminophosphates are numerous, and may sometimes resemble aluminosilicates. Besides, mineral silicates exist in which (SiO_4) is isomorphously replaced by (PO_4)[1]. Barrer and Marshall[2,3] found that aluminophosphates often co-precipitate with aluminosilicates but could not find any evidence for the crystallization of silicoaluminophosphate frameworks. Phosphate-rich hydrous aluminosilicate gels crystallize as separate aluminosilicates and aluminophosphates, with no observable isomorphous substitution[2]. Kühl[4,5] also never encountered phosphate substitution in the zeolite framework when using phosphate as a complexing agent or as a buffer in the synthesis mixture. The aim of such research was primarily to synthesize zeolite structures with increased SiO_2/Al_2O_3 ratios by using complexing agents for Al. No evidence was obtained for the substitution of P for Si.

Flanigen, on the contrary, succeeded to synthesize different types of zeolites with phosphorus in tetrahedral sites in the framework, as evidenced by electron microprobe analysis and infrared spectroscopy[6]. Remarkable phosphate contents were reported, in contrast with the results of Kühl[4,5] and Barrer and Marshall[3]. The compositional area in which the Al-rich P-containing zeolites described by

Guidelines for Mastering the Properties of Molecular Sieves
Edited by D. Barthomeuf *et al.*
Plenum Press, New York, 1990

Flanigen can be obtained, are given in Fig.1 and is denoted as the 'first generation of P-zeolites'. Figure 1 further summarizes most of the literature data on the proportions of Si, Al and P in the oxide frameworks reported for zeolites and molecular sieves.

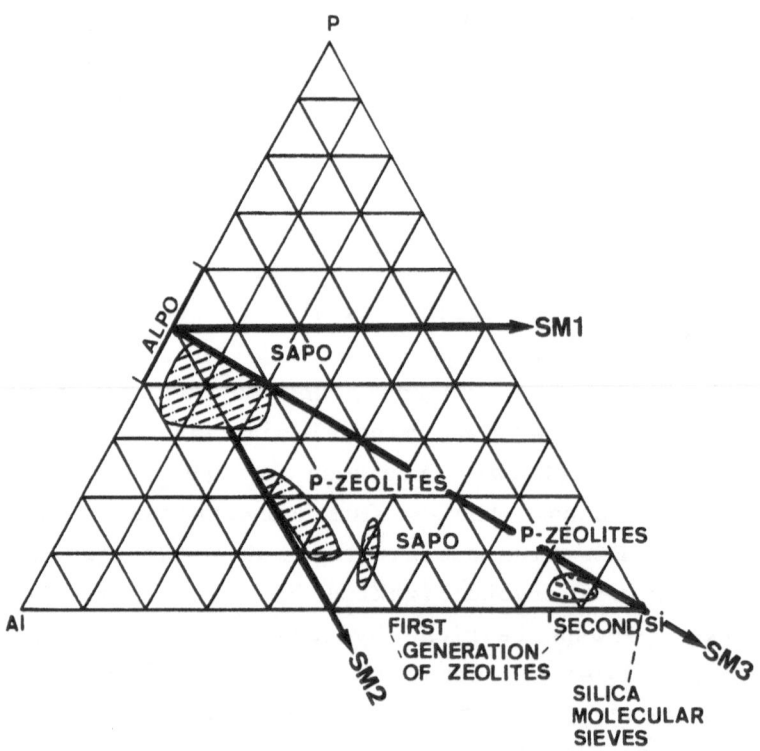

Figure 1. Fraction of Si, Al ,P in the framework of zeolites and molecular sieves from refs. 6,8,9,10.

Further studies on the P substitution for Si during the zeolite synthesis were performed by Barrer and Liquornik[7]. Their observations are in accordance with those of Barrer and Marshall[2,3] and Kühl[4,5] rather than with those of Flanigen and Grose[6]. The P fractions, varying in the range from 10^{-4} to 10^{-3}, were extremely low. Barrer et.al.[7]

emphasized even the fact that these P fractions only represent upper limits for the extent of substitution because some or all of the P could instead be intercalated as discrete phosphate anions in the framework cavities. Rieck et.al.[8] synthesized P-substituted-ZSM-5. The chemical composition of their samples corresponds to Si-rich structures and can, therefore, be classified as a 'second generation of P-containing zeolites (Fig.1).

SILICON SUBSTITUTION IN ALUMINOPHOSPHATES

The breakthrough of phosphorus containing crystalline molecular sieves occurred with the advent of the synthesis of microporous crystalline aluminophosphates (AlPO-n) and derived materials[9]. Microporous crystalline silicoalumino-phosphates are tridimensional tetrahedral oxide frameworks in which the T-atom positions are occupied by Al, P or Si atoms. Most of the obtained framework compositions are included in Fig. 1. The materials patented by Lok et al. are denoted with the acronym SAPO-n, in which n stands for the structure type[10]. According to recent patents and publications, the SAPO-n materials have potential catalytic applications in petroleum refining and petrochemical conversion reactions[11]. For these applications to be successful and accessible to catalyst optimization, it is indispensable to understand and controll the generation of acidity in these SAPO-n catalysts.

It is known that samples of aluminosilicate zeolites of the same structure type can exhibit largely different catalytic properties. Variations in catalytic behaviour may be caused by differences in chemical composition, crystal symmetry, crystal size and morphology, in the frequency and the nature of stacking faults and accompanying pore occlusions or even in the nature of the compositional gradients across the crystals[12]. It can be expected that the presence of a third element (phosphorus) in T-atom positions next to silicon and aluminium will further increase this complexity. Flanigen et al. have stated that 'to date, there does not appear to be any single unifying concept for the development of acidity and catalytic activity in $AlPO_4$-based molecular sieves, and that acidity appears to be a complex function of framework composition, framework charge, the

nature of the element incorporated and structure type.'
In this work it will be shown that the phenomena observed with aluminosilicate zeolites, and in particular, the occurrence of compositional variations across the crystals run indeed into extremes in SAPO-n crystals.

THEORETICAL ISOMORPHOUS SUBSTITUTION MECHANISMS IN SAPO-n MATERIALS

The framework of an AlPO-n molecular sieve is built from strictly alternating PO_2^+ and AlO_2^- tetrahedra and has no net lattice charge. Due to these restrictions on the Al and P ordering, the generation of silicon containing microporous aluminophosphate frameworks is most conveniently simulated by replacing some of the Al and/or P atoms by Si atoms in an isostructural hypothetical $AlPO_4$ framework[13,14]. The theoretically possible isomorphous substitution mechanisms (SM) are illustrated in Fig.2. The substitution mechanisms are Si for Al (SM 1), Si for P (SM 2) and two Si atoms for one Al and one P simultaneously (SM 3). SM 1 generates net positive framework charges which have to be neutralized by anions. When according to SM 2, a PO_2^+ tetrahedron is replaced with SiO_2, the framework becomes negatively charged and cation exchange capacity is generated. The simultaneous replacement of one AlO_2^- and one adjacent PO_2^+ tetrahedron by two SiO_2 tetrahedra (SM 3) doesnot involve changes in the lattice charge.

Fig.3 shows different possible Si T-atom configurations that arise after substitution of Si in $AlPO_4$ frameworks according to the different mechanisms. In the two-dimensional representations of Fig.3 it is only allowed to consider the first T-atom neighbours, as the identification of the second nearest neighbours requires the specification of the exact loop configurations, which depend on the structure type. Fig.3 shows that SM 1 generates Si(4P) environments and that SM 2 results in Si(4Al) configurations. In principle, SM 1 and SM 2 may occur simultaneously, giving rise to isolated Si(4P) and Si(4Al) sites. The occurence of SM 3 in an homogeneous way in the crystals (SM 3 Ho) generates isolated pairs of Si atoms with Si in Si(1Si3Al) and Si(1Si3P) configurations.

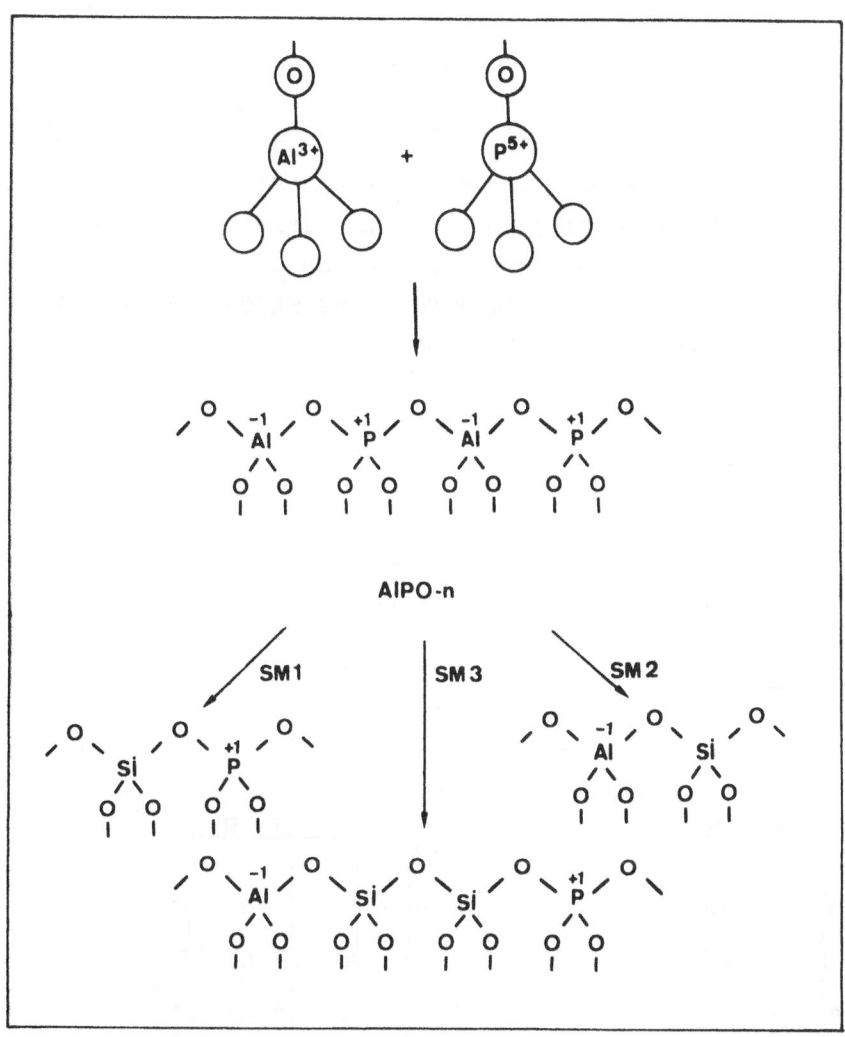

Figure 2. Possible isomorphous substitution mechanisms of Si in AlPO$_4$ frameworks.

The occurrence of SM 1 and SM 3 Ho is unlikely in SAPO-n, since it involves the generation of Si-O-P linkages[13] (Fig.3). However, the formation of Si-O-P linkages by SM 3 Ho can be avoided by its simultaneous occurrence with SM 2, as shown in Fig.3. A silicon patch composed of five Si atoms is generated in this way inside the AlPO-n framework. When SM 3 is applied in an heterogeneous way (SM 3 He) by replacing P and Al atom pairs in a certain domain of the crystal as shown in Fig.3, no P-O-Si linkages are formed. In the siliceous crystal domains thus generated, eventually part of the Si in its turn can be replaced with Al giving rise to an 'aluminosilicate domain' in the SAPO-n crystal. The generation of such silicate and aluminosilicate domains in SAPO-5 and SAPO-11 materials will be illustrated in the following paragraphs.

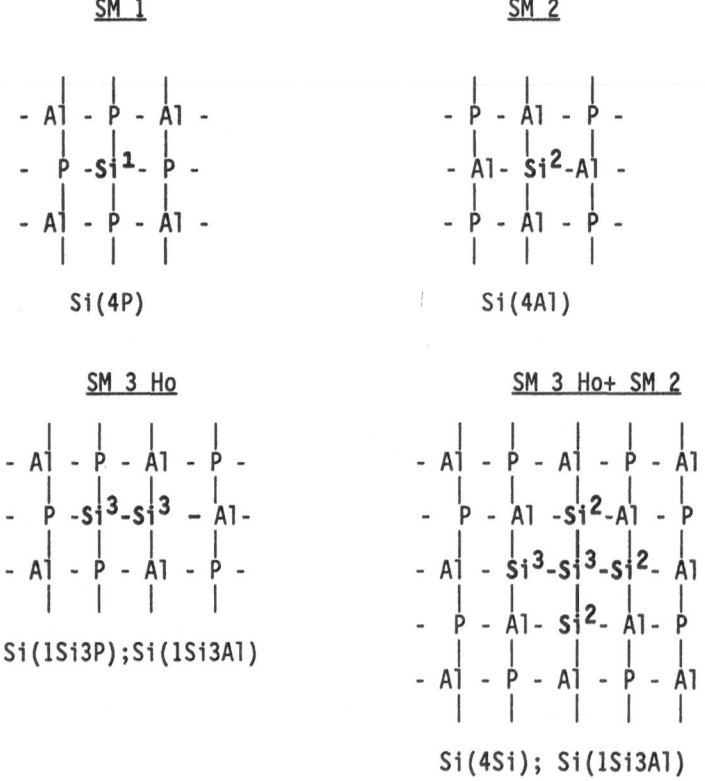

Figure 3. Si T-atom configurations generated by the different isomorphous substitution mechanisms. The exponent of Si denotes the isomorphous substitution mechanism number.

Literature data on the synthesis of AFI-type materials

SAPO-5 is a molecular sieve with a 12-membered ring (12-MR) monodimensional micropore system[10]. Its structure type is denoted as AFI[15].

An overview of the synthesis methods and properties of SAPO-5 materials from literature is given in Table 1. Appleyard et al.[16] observed one single ^{29}Si MAS NMR signal at -92 ppm in a SAPO-5 material with a Si T-atom fraction equal to 0.05. This signal is due to Si(4Al). The sample represents a SAPO-5 material in which the Si is substituted according to SM 2. Except for the nature of the template, which was tripropylamine, no further details regarding the

SM 3 He

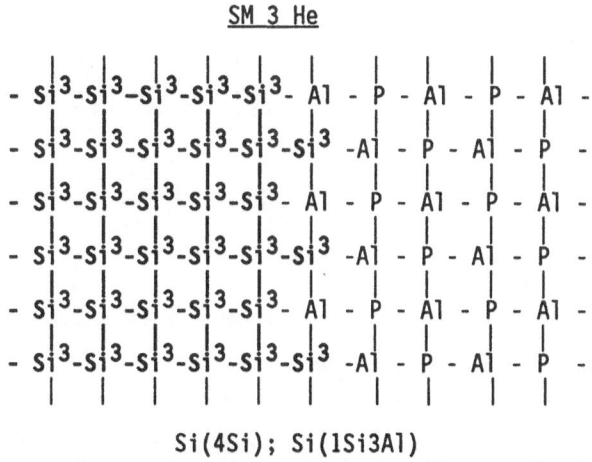

Si(4Si); Si(1Si3Al)

SM 3 He + Si, Al substitution in the Si domain

Si(4Si); Si(3Si1Al); Si(2Si2Al); Si(1Si3Al)

Table 1. Literature data on the synthesis and characterisation of SAPO-5

Source of Al	Source of Si	R[a]	Si T-atom fraction in product	Brønsted sites/ T-atom	Chem. shift of 29Si MASNMR signals (ppm)	SM	Remarks	Ref.
x	x	Pr$_3$N	0.05	x	-92	2	—	16
Pseudo-boehmite	Colloidal Sol	Pr$_3$N	0.02 0.033	0.014 0.017	-92,-111	2+3	Chemical homogeneity of crystals checked by EDAX	17
Pseudo-boehmite	Fumed Silica	Et$_3$N	0.01 0.025 0.045 0.07	0.010 0.020 0.030 0.035		2+3b 2+3b 2+3b	'Condensed phase impurities' formed for Si contents > 10%	18
x	x	Et$_3$N	0.021	0.017± 0.004	-92,-111	2+3b	'Crystalline SiO$_2$' present	19
x	x	x	0.071 0.093 0.043	x x x	-92,-97,-102,-108 -98 -93	2+3 2+3 2	phase purity checked by XRD and micropore adsorption 'SM depends on synthesis method'	20
Pseudo-boehmite	'Fuzed Silica'	Et$_3$N	0.008 0.113 0.361	x x x	-82,-93,-110 -82,-92,-102,-110	2 2	'Non-framework silicon phase' present	21
Aluminum isopropoxide	Colloidal Sol	CHA	0.01 0.28 0.36	x x x	-92 -92,-110 -92,-110	2 2+3 2+3	'Silica patches' in the SAPO-5 crystals	23
x	x	x	0.14	x	x	2+3	T-atom fraction of Si + P > 0.5	14
Pseudo-Boehmite	Colloidal Sol	Pr$_4$NOH	0.90	x	x	3	Hexagonal tablets	10

a, template; Pr$_3$N=tripropylamine; Et$_3$N=triethylamine; CHA=cyclohexylamine; Pr$_4$NOH=tetrapropylammonium hydroxide; b, own interpretation; x, no details reported.

synthesis were given. Chao and Leu synthesized SAPO-5 samples using orthophosphoric acid, pseudoboehmite, Ludox and tripropylamine. Two samples were obtained with a Si T-atom fraction of 0.02 and 0.033, respectively. In these SAPO-5 materials only 70% and 50%, of the Si atoms gave rise to the generation of strong Brønsted acid sites, respectively. SM 2 is likely to occur. The major ^{29}Si MAS NMR resonance was found at -92 ppm. A less intense ^{29}Si MAS NMR signal at -111 ppm was also observed and ascribed to the presence of 'clusters of silica in SAPO-5'[17] and could be due to SM 3 He. Tapp et al.[18] synthesized SAPO-5 samples with Si T-atom fractions up to 0.07 and measured the Brønsted acidity. The number of strong Brønsted acid sites was found to correspond to the number of Si atoms up to a Si T-atom fraction of 0.01. In the sample with the highest Si content only 50 % of the Si atoms gave rise to Brønsted acidity. The Si T-atom fraction of a SAPO-5 material reported by Freude et al. was 0.021[19]. It was synthesized using triethylamine. The number of bridging Si-OH-Al groups could be determined using ^{1}H MAS NMR spectroscopy and was estimated at 1.7±0.4 sites per 100 T-atoms. The presence of a small amount of 'crystalline SiO_2' in the sample was derived from the presence of a ^{29}Si MAS NMR resonance at -112 ppm next to a signal at -92 ppm. The nature of this crystalline SiO_2 was not further discussed.

Blackwell et al.[20] reported ^{29}Si MAS NMR spectra of SAPO-5 materials with Si T-atom fractions of 0.043, 0.071 and 0.093, respectively. ^{29}Si MAS NMR signals were observed at -92, -97, -102 and -108 ppm, the relative intensities depending on the synthesis method of the sample. Flanigen et al. reported on a SAPO-5 sample with Si, Al and P T-atom fractions equal to 0.14, 0.45 and 0.41, respectively and concluded from the sample composition that SM 2 and SM 3 had occured simultaneously[14]. Wang et al.[21] reported on the synthesis of SAPO-5 samples with Si T-atom fractions of 0.008, 0.113 and 0.361 using pseudoboehmite, 'fuzed silica' and triethylamine. Next to a signal at -93 ppm, a multiplicity of lines in the -80 to -120 ppm region as observed in the ^{29}Si MAS NMR spectrum of the sample with the highest Si content. The occurrence of these signals was ascribed to a 'non-framework silicon phase'. The maxima in the reported Si MAS NMR spectrum were at -82, -93, -102 and -110 ppm, respectively[21]. The -82 ppm signal should be ascribed to Si(1Si3OH). The -93 and -102 ppm signals can be due to Si(2Si2OH) and Si(3Si1OH), or alternatively to Si(4Al) and Si(2Al2Si), respectively. The observed relative decrease of the intensity of the -82, -93 and -102 ppm signals with respect to that at -112 ppm after calcination[21] can confirm the assignment to defect sites.

In example 2 of the original patent[10] the synthesis of SAPO-5 crystals with hexagonal shape was disclosed, the chemical composition of which was determined with EDAX. These SAPO-5 crystals contained 90% of Si, 5% of P and 5% of Al, suggesting the occurence of pure SM 3. The synthesis gel was made from orthophosphoric acid, pseudoboehmite, colloidal silica and Pr_4NOH. Van Nordstrand et al. succeeded in preparing an all-silica molecular sieve, isostructural with AlPO-5 from an alkaline gel using quaternary adamantammonium cations, which was denoted as SSZ-24[22].

The literature data of Table 1 illustrate that SAPO-5 is certainly not a unique material and that its physico-chemical properties largely depend on the synthesis conditions. SM 2 seems to dominate for low Si contents, while SM 3 sets in from a certain Si loading on. Extensive contributions of SM 3 seem to be possible only under specific synthesis conditions. We, therefore, denote SAPO-5 materials with pure SM 2 as 'SAPO-5(II)', and SAPO-5 with SM 3 together with SM 2 as 'SAPO-5 (II+III)'. In a previous publication[23] we reported a synthesis method according to which large amounts of silicon can be incorporated in SAPO-5. The incorporation of Si is mainly according to heterogeneous SM 3, with an additional small contribution by SM 2. In the SAPO-5 (II+III) crystals thus obtained the Si T-atom fraction can be as high as 0.36[23].

Appropriate synthesis conditions for SAPO-5 (II+III)

The Si T-atom configurations generated by the different substitution mechanisms may differ wildly (Fig.3). It can therefore be expected that the nature of the substitution mechanism and the extent of Si incorporation depend on the synthesis conditions. SM 2 is likely to occur via silicate monomers, while SM 3 He most likely involves (alumino)silicate oligomers. By analogy to the crystallization of aluminosilicate zeolites[1], the degree of polymerization of silicon in the synthesis mixture could be critical for a successful synthesis of SAPO-n with aluminosilicate domains. It is primarily determined by the pH of the system. The pH conditions for succesful synthesis of SAPO-n(II+III) materials should be comprised between the ones for AlPO-n or SAPO-n(II), which grow from slightly acidic media[9,10], and those for aluminosilicate zeolites, which crystallize from alkaline gels[1].

Literature data show, indeed, that the efficiency of the incorporation of silicon depends on the synthesis recipe. The Si T-atom fraction of SAPO-5 obtained by Freude[19] et al. was 0.02, although in the gel this value was 0.20. Tapp et al.[18] found dense-phase-impurities in their SAPO-5 samples when the Si T-atom fractions in the synthesis mixtures exceeded 0.10. Wang et al.[21] found that the major fraction of silicon was not incorporated in the SAPO-5 crystals in a sample with a Si T-atom fraction of 0.36, the pH of the synthesis gel being between 5.5 and 7.0.

Synthesis of SAPO-5 (II+III) from monophasic systems

In monophasic synthesis systems, water has been the only solvent reported. In biphasic systems, a second solvent such as hexanol is present. We performed monophasic synthesis experiments using orthophosphoric acid (85%) (Janssen Chimica), aluminium isopropoxide (Janssen Chimica) or pseudoboehmite (Catapal B), an ammonia stabilized colloidal silica sol (Ludox AS-40 from Dupont), water and cyclohexylamine (CHA) (Janssen Chimica). First the aluminium source was stirred into water diluted phosphoric acid. Subsequently, the silicate solution and CHA were added. The composition of the synthesis mixtures corresponded to $((Si_xAl_yP_z)O_2\ 0.15CHA\ 13H_2O)$. For the mixture prepared with pseudoboehmite the value of x was 0.03 while the y/z ratio was equal to 1.0. When aluminium isopropoxide was used, the pH of the initial synthesis mixtures was between 10 and 11. The pH of the synthesis mixtures with pseudoboehmite was between 5 and 6. The hydrothermal treatment was performed in stainless steel autoclaves which were rotated at 50 rpm. The synthesis temperature was 473 K and the synthesis time 50 hours.

In the synthesis experiments with pseudoboehmite, only dense phases and amorphous phases were formed. The outcome of the synthesis experiments with aluminium isopropoxide depended on whether the mixtures were agitated or not. Under static conditions, no crystalline phase could be obtained. The outcome of the crystallization from rotated mixtures depended on the values of x, y and z, as shown in Fig.4. Phase-pure SAPO-5 was obtained when $y = z$ and $x < 0.40$. Under other conditions SAPO-44 co-crystallized or pure SAPO-44 was formed.

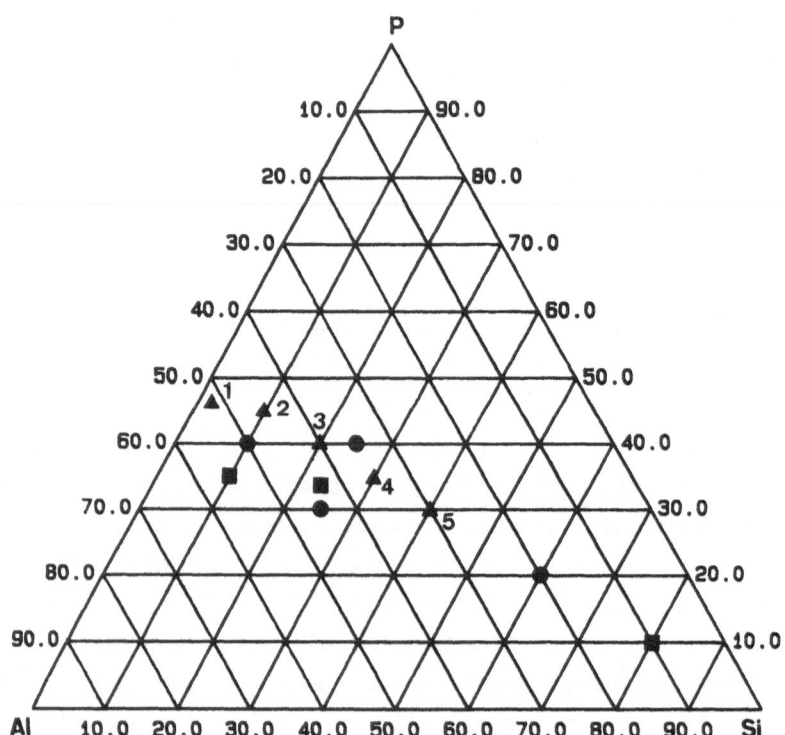

Figure 4. Crystalline phases obtained from synthesis mixtures with composition $((Si_xAl_yP_z)O_2)$ 0.15CHA 13H_2O prepared with aluminium isoproxide: ▲: SAPO-5, ■: SAPO-44, ●: mixture of SAPO-5 and SAPO-44

The Si, Al and P composition found in the SAPO-5 products from Fig.4. is given in Table 2. The Si content of the SAPO-5 materials increases with increasing Si content of the synthesis mixture. The compositions of SAPO-5 presented in Table 2 suggest the occurrence of SM 3 since the T-atom fraction of Al and P both are lower than 0.50. The T-atom fraction of Al is systematically higher than that of P, suggesting the occurence of SM 2. According to the chemical compositions given in Table 2, SAPO-5/1 could be a SAPO-5 (II) material, and the SAPO-5/2 to SAPO-5/5 samples could represent examples of SAPO-5 (II+III).

Table 2. Si, Al and P fractions $(Si_xAl_yP_z)O_2$ of SAPO-5 products.

No	sample	x	y	z	y/z
1	SAPO5/1	0.01	0.50	0.49	1.02
2	SAPO5/2	0.15	0.44	0.41	1.07
3	SAPO5/3	0.15	0.47	0.38	1.24
4	SAPO5/4	0.28	0.39	0.33	1.18
5	SAPO5/5	0.36	0.37	0.27	1.37

A SEM photograph of sample SAPO-5/5 is shown in Fig.5. The size of the crystals ranges from 2 to 10 μm. Some of the crystals exhibit an hexagonal shape.

Properties of SAPO-5 samples from monophasic systems
1.Location of Si. ^{29}Si MAS NMR spectra of the as-synthesized SAPO-5/1 to /5 samples are shown in Fig.6. The ^{29}Si MAS NMR spectra were recorded on a Bruker 400 MSL spectrometer at 79.5 MHz and a spinning rate of 3 kHz. The ^{29}Si MAS NMR spectra consist primarily of two lines at -92 and -112 ppm, respectively. SAPO-5/1 exhibits only the -92 ppm signal. The relative intensity of the -112 ppm line increases with respect to the -92 ppm line as the silicon content of the product increases. The -92 ppm signal represents the Si(4Al) coordination and reflects the occurence of SM 2, the -112 ppm signal the Si(4Si) coordination and SM 3 He. The ^{29}Si MAS NMR spectra of Fig.6 confirm the classification of SAPO-5/1 as SAPO-5(II) and of the samples SAPO-5/2 to SAPO-5/5 as SAPO-5(II+III). The ^{29}Si resonance envelop suggests the

Figure 5. Scanning electron micrograph of SAPO-5/5.

presence of other ill-defined lines in the chemical shift range from -92 to -112 ppm. It was, however, not possible to resolve these signals in a decisive manner, but the presence of signals at ca. -104, -100 and -97 ppm is not questionable.

The ^{29}Si MAS NMR and ^{29}Si MAS NMR CP spectra of as-synthesized and calcined SAPO-5/5 are compared in Fig.7. Calcination at a temperature of 823 K does not alter the ^{29}Si MAS NMR spectra, indicating that the -92 ppm signal represents Si(4Al) rather than Si(2Si2OH) sites; the latter are expected to be unstable under the calcination conditions applied. Under CP conditions the -112 ppm signal is invisible, confirming its assignment to Si(4Si) in a proton-free environment. The -92 ppm signal is observed under CP conditions, probably due to the presence of charge compensating protons or protonated template molecules. Signals at higher field, e.g. at -82 ppm, reported by Wang et al.[21] for samples containing amorphous silica, are not observed.

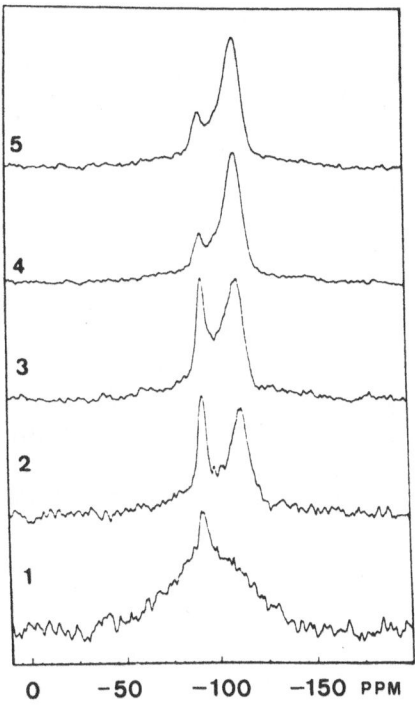

Figure 6. ^{29}Si MAS NMR spectra of the five as-synthesized SAPO-5 samples (pulse length = 3 μs, pulse interval = 3 s).

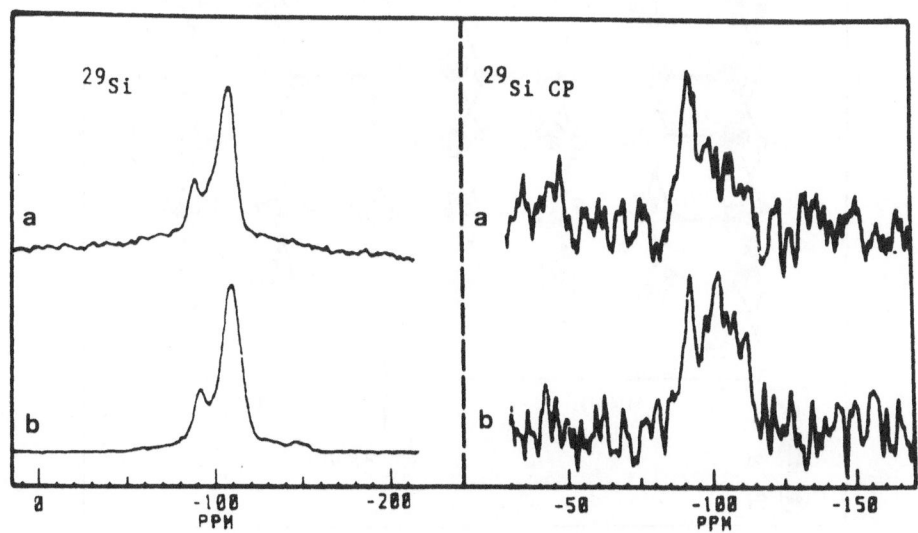

Figure 7. ^{29}Si and ^{29}Si CP MAS NMR spectra of SAPO-5/5 before (a) and after (b) calcination at 823 K (^{29}Si: pulse length = 3 μs, pulse interval = 3 s; ^{29}Si CP: pulse length = 5.2 μs, pulse interval = 5 s, cantact time = 5 s).

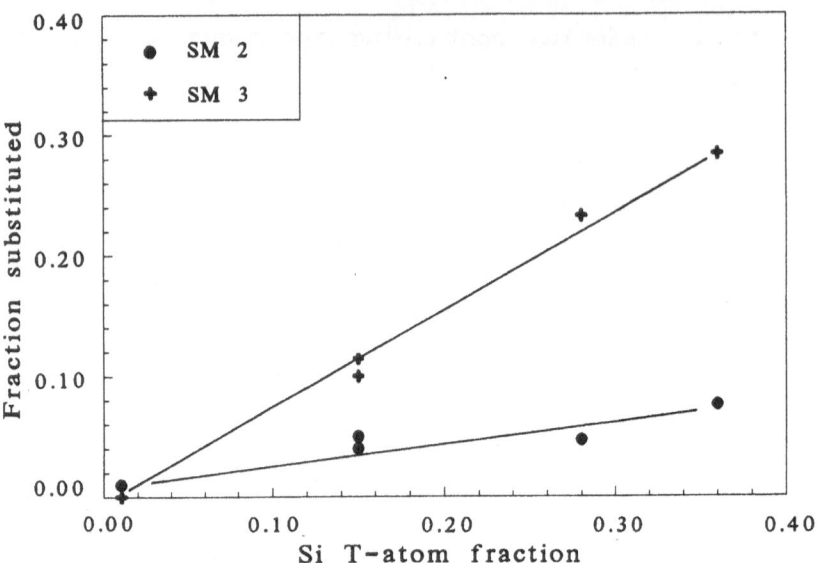

Figure 8. Contribution of SM 2 and SM 3 He in samples SAPO-5/1 to SAPO-5/5.

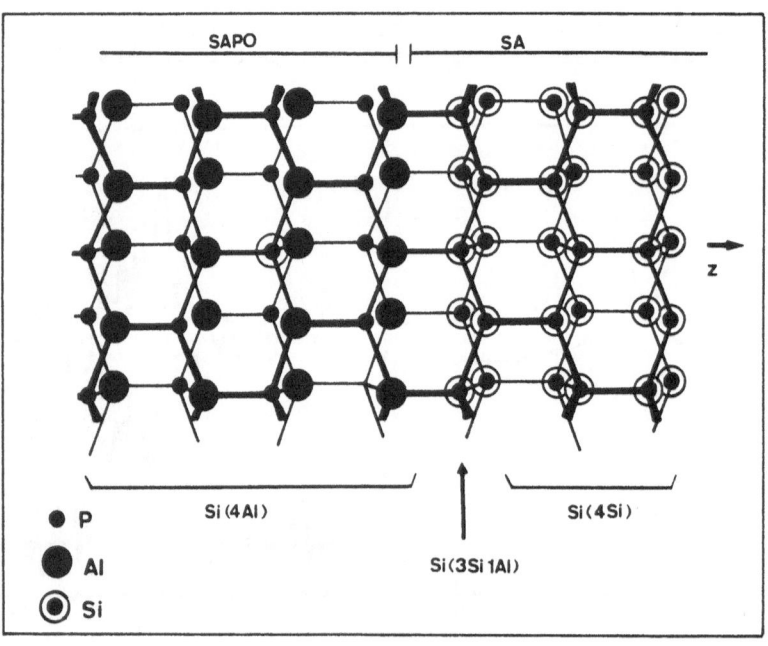

Figure 9. Model for the generation of silicoalumino-phosphate (SAPO) and aluminosilicate (SA) domains in the AFI framework.

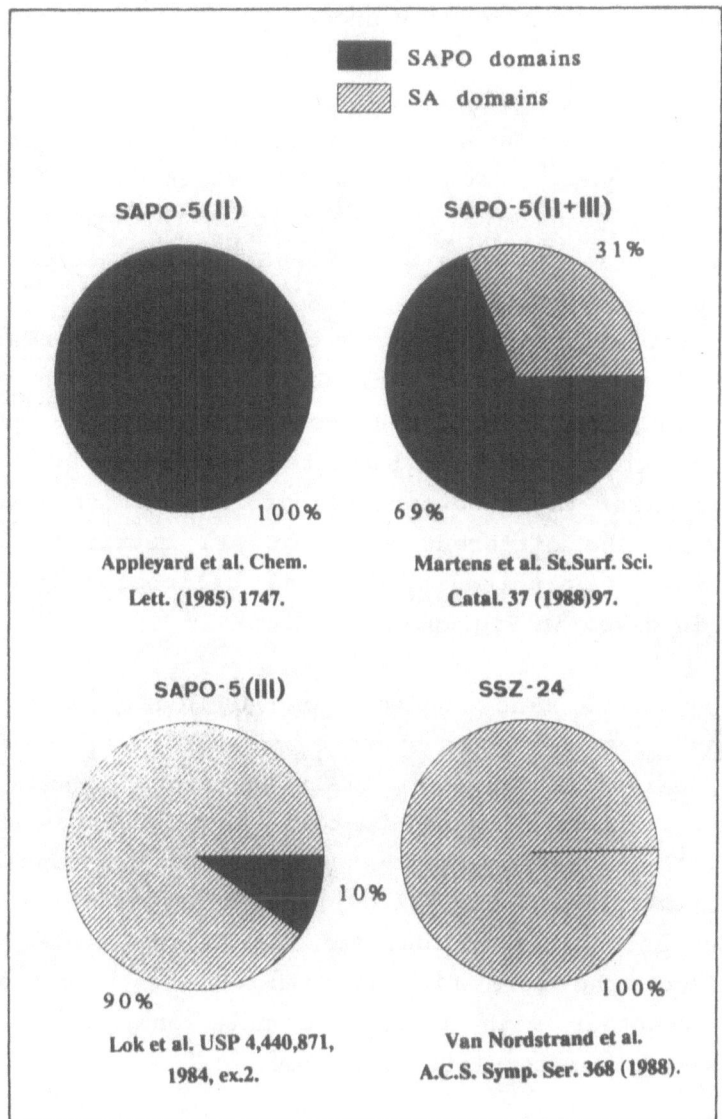

Figure 10. Repartition of silicoaluminophosphate (SAPO) and aluminosilicate (SA) domains in different types of AFI materials.

The contribution of SM 2 and SM 3 He to the total amount of Si substituted in the framework was calculated from the relative intensity of the -92 and -112 ppm signal in the ^{29}Si MAS NMR spectra of Fig.6 and is shown in Fig.8. In the present series of SAPO-5/1 to /5 samples, the Si substitution by SM 2 doesnot exceed 7% of the T-atoms.

The AFI structure seems to be particularly suited for the generation of siliceous crystal domains. In the c direction, the structure corresponds to a stacking of alternating layers of PO_2^+ and AlO_2^- tetrahedra. A siliceous domain can be linked to an $AlPO_4$ region through a layer of SiO_2 tetrahedra, in which Si is in Si(1Al3Si) coordination. This is shown in Fig.9.

In conclusion, the data of this work together with the literature data indicate that Si substitution in the AFI topology is possible over the whole compositional range. At low Si fractions, SM 2 predominates. Heterogeneous SM 3 is responsible for the incorporation of large Si fractions. An overview of the different types of AFI materials and the repartition of aluminosilicate and silicoaluminophosphate domains is given in Fig.10.

2.<u>Microporosity</u>. The number of cyclohexylamine (CHA) molecules present in as-synthesized SAPO-5/1, -4 and -5 samples, allows to fill completely the micropores of an AFI framework, assuming head-to-tail arrangements in the pores[23]. It indicates also that there is no substantial part of the T-atoms (e.g. of silicon) present external to the AFI framework as occluded oxide. For the set of SAPO-5 samples synthesized, the pore filling with CHA is illustrated in Fig.11, together with literature data on SAPO-5 samples filled with other template molecules.

The decane test is a method to characterize the micropore architecture of bifunctional zeolite catalysts[24]. Information on the micropore architecture of an unknown zeolite is derived from the composition of the reaction products from decane, obtained over the zeolite under specific standardized experimental conditions, using as reference the same values obtained over known zeolites. Eight criteria based on the composition of the isomerization

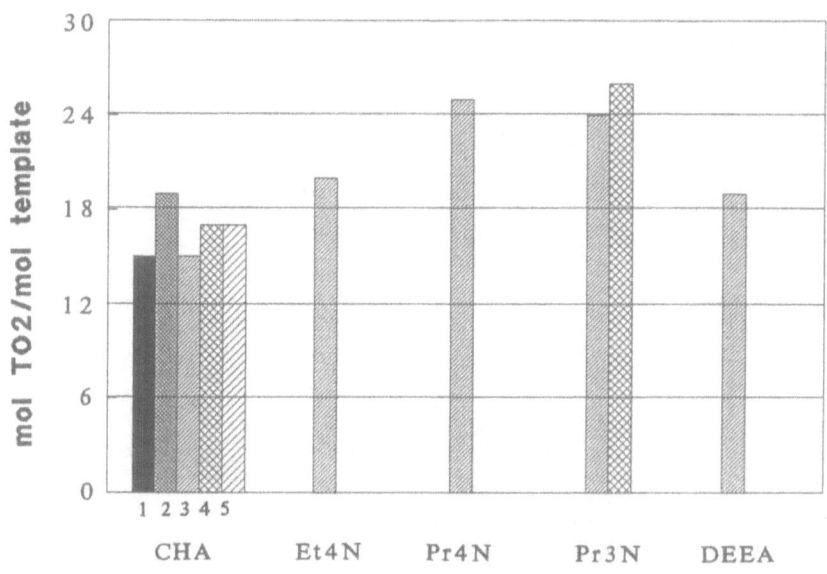

Figure 11. The TO_2 to template molar ratio of as-synthesized SAPO-5/1 to SAPO-/5; Data on SAPO-5 with Et_4N, Pr_4N, Pr_3N and DEEA (diethylethanolamine) are from ref.10.

and hydrocracking products are handled to characterize a sample[24].

The results of the decane test on the SAPO-5/1 to SAPO/5 samples are given in Fig.12. The first criterion of the decane test uses the refined constraint index, CI*, which is the 2-methylnonane/5-methylnonane product ratio at 5% isomerization of the feed. CI* correlates with the effective size of the apertures to the cages and channels of the catalyst[25]. The SAPO-5/1 to SAPO-5/5 samples exhibit a CI* value ranging from 1.3 to 1.5, thus confirming the presence of 12-MR pores with pore openings larger than 0.7 nm. Criterion 2 is the ethyloctane selectivity in the monobranched isomers at 5% isomerization of the feed. From Fig.12 it can be seen that over the different SAPO-5 samples, comparable amounts of ethyloctanes are formed. The second criterion confirms the classification of SAPO-5 samples as 12-MR structures. 4-Propylheptane is the bulkiest isomer of decane. It can only be formed in 12-MR zeolites. The extent of formation of 4-propylheptane correlates well

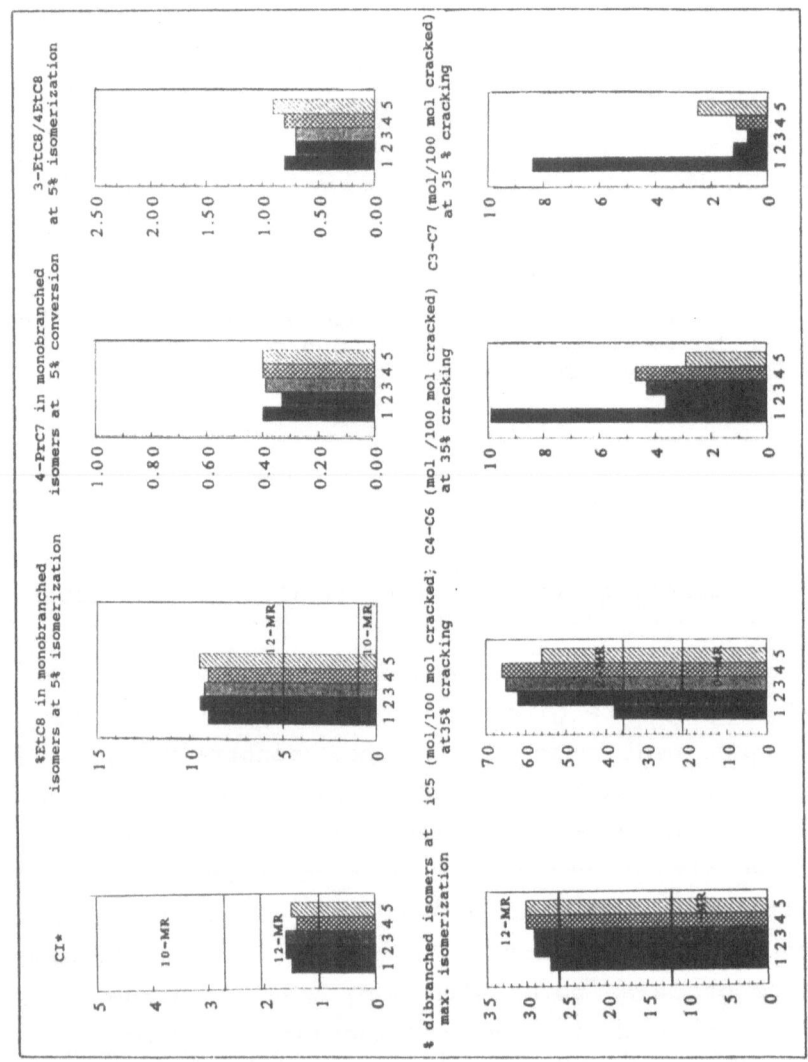

Figure 12. Decane test on SAPO-5 samples

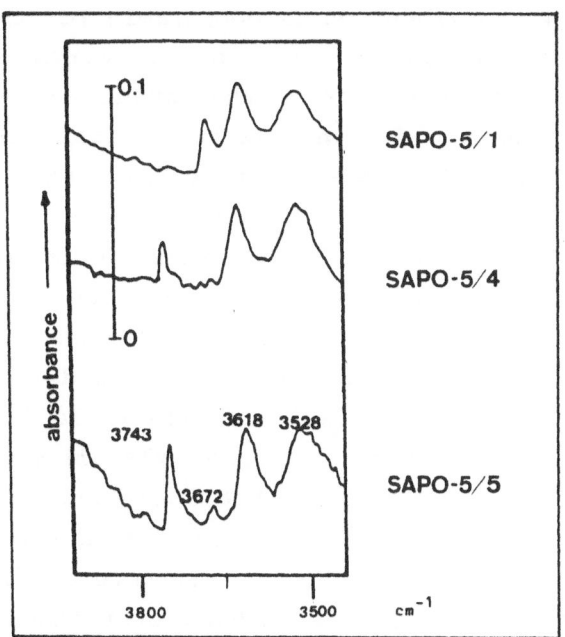

Figure 13. Hydroxyl spectra of SAPO-5/1, -/4 and -/5 calcined at 823 K.

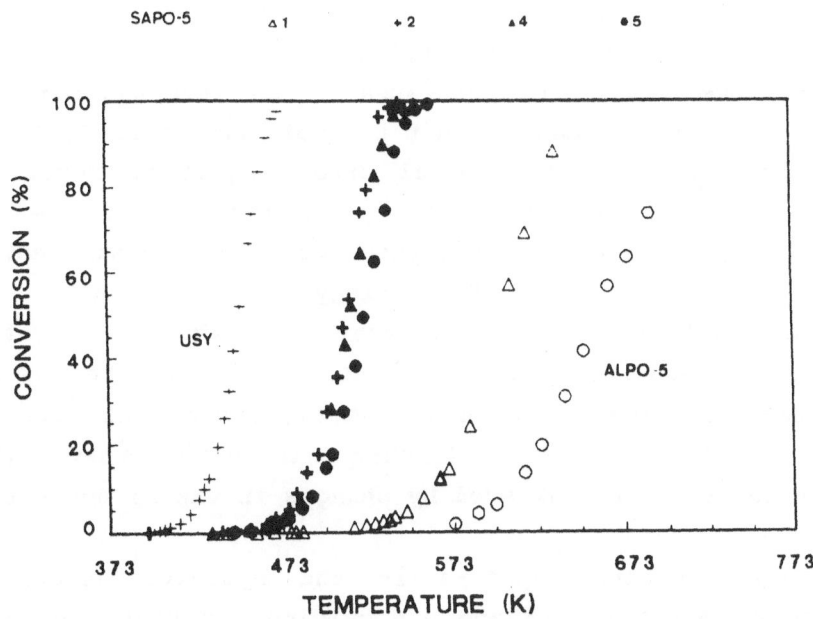

Figure 14. Conversion of decane against reaction temperature over SAPO-5/1, -/2, -/4 and -/5 loaded with 0.5% of Pt. W/F_o=0.5 kg s mmole-1, P_{H2} = 0.35 MPa and P_{decane} = kPa.

with the space available in the intracrystalline voids. The yield of 4-propylheptane is comparable for the different SAPO-5 samples, indicating that the intracrystalline voids are identical. The ratio of 3-ethyloctane to 4-ethyloctane at 5% isomerization is indicative of the size of the microcavities. For the SAPO-5 samples this ratio ranges from 0.7 to 0.9. According to this criterion, SAPO-5 has smaller pores than faujasite, for which this ratio is equal to 0.6[24]. Consecutive branching of decane gives rise to mono-, di-, and tribranched isodecanes, which become progressively more bulky in nature. The yield of dibranched isomers in the isodecanes at the maximum isomerization yield is handled as the fifth criterion. This criterion classifies all SAPO-5 samples as 12-MR zeolites. The yield of isopentane per 100 mole decane cracked at 35% cracking is another parameter which is sensitive to the zeolite structure. The more the feed hydrocarbon undergoes branching rearrangements before cleavage, the higher the probability will be for central scission; the latter will result in the formation of branched, viz. isopentane fragments. All SAPO-5 samples show 12-MR behaviour. The lower isopentane yield found for SAPO-5/1 is due to hydrogenolysis of part of the feed over the metal function. 10-MR and 12-MR zeolites with intersecting pores generally show a symmetrical distribution of cracked products. A zeolite is permeated by an intersecting pore system when it is observed that the mole fractions C4-C6 and C3-C7 are, within experimental error, equal to zero. The cracked product distributions found for SAPO-5 are all asymmetric, confirming the presence of non-intersecting pores as required by the AFI topology.

In conclusion, the decane test shows that the crystals all contain non-intersecting channels, circumscribed by 12-membered ring openings with a diameter larger than 0.7 nm. These catalytic data further indicate that this microporosity is not altered by changes in the Si content.

3.<u>Acidity</u>. Infrared spectra in the hydroxyl stretching region of SAPO-5/1, -/4 and -/5 calcined at 823K are shown in Fig.13. Hydroxyl vibrations are observed at 3743, 3672, 3618 and 3528 cm^{-1}. The 3743 cm^{-1} band is due to silanol groups. This band is absent in SAPO-5/1. Its intensity is low for SAPO-5/4 and -/5. These silanol groups can be accounted for by crystal termination. The 3672 cm^{-1} band can be ascribed to P-OH groups[26] and the 3618 and 3528 cm^{-1} signals to bridging Si-OH-Al[27]. The bathochromic shift of the 3528 cm^{-1} band is due to additional electrostatic interaction of the protons in six membered rings with the nearest oxygen atoms, a phenomenon well-known in faujasite zeolites[25]. From Fig.13 it appears that the number of P-OH groups is lower and that of Si-OH-Al higher in SAPO-5/4 and /5 compared to SAPO-5/1. The acid strength of the bridging hydroxyl groups was measured from the wavenumber shift upon interaction with adsorbed benzene molecules[23] (Table 3). The shift of the 3618 cm^{-1} vibration was estimated to be between 250 and 390 cm^{-1} for SAPO-5/1, and between 270 and 410 cm^{-1} for the SAPO/4 and /5 samples. This is in the range of the frequency shifts reported for the bridging hydroxyl groups of aluminosilicate zeolites (Table 3).

Table 3. Wavenumber shifts (cm^{-1}) of the OH-bands of molecular sieves upon interaction with benzene.

sample	wavenumber shift (cm^{-1})	
	3740-3720 cm^{-1}	3610-3600 cm^{-1}
zeolite Y[a]	40	326
HZSM-5[a]	80	300
		348
HZSM-11[a]	72	220
		300
		340

sample	3618 cm^{-1}		
		min	max
SAPO-5/1	-	250	390
SAPO-5/3	-	270	410
SAPO-5/4	-	270	410

a. data from ref 35

The conversion of decane over the different SAPO-5 samples at increasing temperatures is shown in Fig.14 and compared with AlPO-5 and USY zeolite. The activity of the SAPO-5 samples is situated between that of AlPO-5 and USY zeolite. SAPO-5/1 is substantially less active compared to the other SAPO-5/ samples, which have comparable activities. The catalytic activities reflect the extent of Si substitution by SM 2 (Fig.8) and the number of bridging Si-OH-Al groups (Fig.13).

All these data confirm the conclusions on the substitution mechanisms derived in earlier sections.

ISOMORPHOUS SUBSTITUTION IN SAPO-11
Literature data on SAPO-11

The structure type of SAPO-11 is AEL and characterized by having non-intersecting 10-MR ring channels[29].

The SAPO-11 materials described in the original patent[10] and subsequently in literature[18,30,31] are characterized by having a Si T-atom fraction equal to or lower than 0.14, depending on the specific synthesis conditions. Khoudzami et al.[30] synthesized a series of SAPO-11 samples using Pr_2NH, hexanol, aluminium isopropoxide, silica gel and orthophosphoric acid. XPS data showed that the surface of the SAPO-11 crystals was enriched with silicon[30]. It was concluded that this silicon is probably present as amorphous silica layers since it disappeared partially after calcination[30]. Furthermore, SAPO-11 samples with Si T-atom fractions between 0.01 and 0.05 showed the highest crystallinity[30]. Weyda et al.[31] synthesized SAPO-11 using pseudoboehmite, fumed silica and Pr_2NH. The highest crystallinity of SAPO-11 was obtained from a synthesis mixture with composition $((Si_{0.07}Al_{0.465}P_{0.465})O_2)$ $0.12Pr_2NH$ $11.6H_2O$.

From the literature data, it can be derived that SAPO-11(II) as well as SAPO-11(II+III) type samples exist. Blackwell et al.[20] recorded a ^{29}Si MAS NMR spectrum of a SAPO-11 material with Si, Al and P T-atom fractions of 0.09, 0.50 and 0.43, respectively. Only a single ^{29}Si signal at -93 ppm was found, indicative of pure SAPO-11(II). Flanigen et al.[14] reported a SAPO-11 material with Si, Al and P T-atom fractions of 0.14, 0.44 and 0.42, respectively, to illustrate the simultaneous occurence of SM 2 and SM 3 in SAPO-11. The synthesis method of this SAPO-11(II+III) sample was not mentioned, however.

Monophasic synthesis of SAPO-11(II+III) using aluminium isopropoxide

The synthesis recipe in monophasic conditions was based on example 18 of the original patent[10]. Aluminium isopropoxide (from Janssen Chimica) was added to a mixture of orthophosphoric acid (85%, from Janssen Chimica) and water. Subsequently, an ammonia stabilized silicate solution containing 40 wt-% of silica (Ludox AS-40, from DuPont) and dipropylamine (Pr_2NH) (from Janssen Chimica) were added under stirring. The composition of the reaction mixtures corresponded to $((Si_xAl_yP_z)O_2)$ $0.20Pr_2NH$ $3.7H_2O$.

The Si, Al and P fractions $((Si_xAl_yP_z)O_2)$ of the synthesis mixtures are shown in Fig.15. The pH of the reaction mixtures was between 9 and 10. The reaction mixtures were transferred into 120 ml autoclaves, rotated at 50 rpm and heated at 473 K under autogenous pressure for 48 h. At the end of the crystallisation it was verified that for the mixtures with a P/Al ratio of 1, the pH was unchanged with respect to the initial value. The solid reaction products were recovered by filtration, washed with water, and dried in air at 333 K.

SAPO-11 phases were obtained from the synthesis mixtures with an Al/P ratio of 1.0 and a Si T-atom fraction of 0.60 or lower. SAPO-34 was crystallized from the mixtures with a Si T-atom fraction of 0.7. When the Al/P ratio of the mixture was lower than 1.0, tridymite was crystallized. On the other hand, when the Al/P ratio was higher than 1.0, the SAPO-11 phase was obtained. The phase purity of the crystallization products was verified by means of the XRD. In order to illustrate the phase-purity, the XRD pattern of SAPO-11/1 and SAPO-11/3 are compared in Fig.16 with that of SAPO-11 from the patent[10].

The T-atom compositions of the SAPO-11 products crystallized from synthesis mixtures 6 to 10 (see in Fig.15) are given in Table 4. For all the samples the Al/P ratio is

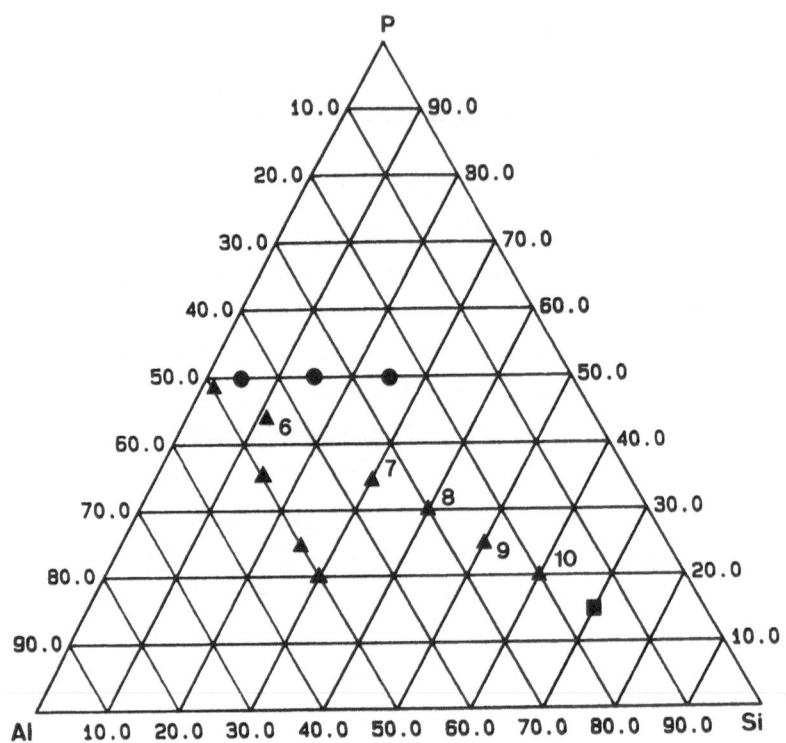

Figure 15. Crystalline phases obtained from synthesis mixtures with composition $((Si_xAl_yP_z)O_2 \ 0.2Pr_2NH \ 3.7H_2O$ prepared with aluminium isoproxide: ▲ : SAPO-11; ■ : SAPO-34; ● : tridymite.

Table 4. The Si, Al and P fractions $(Si_xAl_yP_z)O_2$ of SAPO-11 products.

No	sample	x	y	z	y/z
6	SAPO11/1	0.04	0.52	0.44	1.18
7	SAPO11/2	0.21	0.43	0.36	1.20
8	SAPO11/3	0.39	0.33	0.28	1.18
9	SAPO11/4	0.42	0.31	0.27	1.15
10	SAPO11/5	0.46	0.30	0.24	1.25

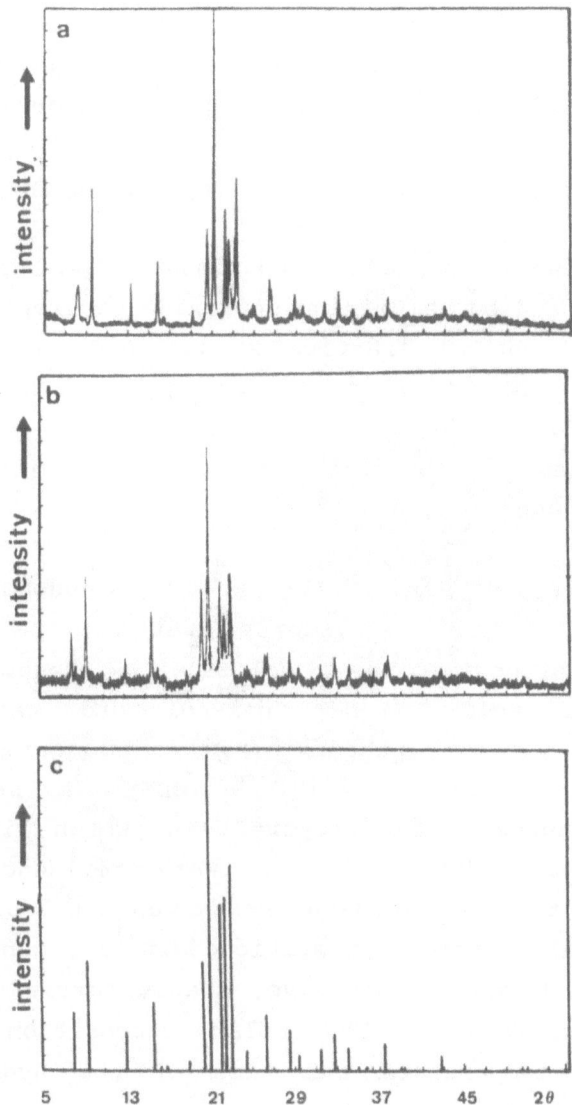

Figure 16. XRD pattern of SAPO-11/1 (a), SAPO-11:3 (b) and patent SAPO-11 sample[10] (c).

larger than 1.0, suggesting the occurence of SM 2. Furthermore, with SAPO-11/2 to -/5, the Al T-atom fraction is lower than 0.50, pointing to SM 3. It is seen from Fig.15 and Table 4 that the Si T-atom fraction in the product is systematically lower than in the synthesis mixture.

A SEM photograph of SAPO-11/1 and SAPO-11/5 is shown in Fig.17. The SAPO-11/1 sample consists of aggregates of spherical particles with a diameter of 5 to 10 μm. The SAPO-11/5 sample consists of irregularly aggregated crystallites.

Monophasic synthesis of SAPO-11(II+III) using pseudo-boehmite

In another synthesis experiment the monophasic procedure was used but aluminiumisoproxide was replaced with pseudoboehmite. The Si fraction x, was 0.03 and the y/z ratio was 1.0. The initial pH of the synthesis mixture was between 5 and 6.

The synthesis experiment with pseudoboehmite yielded only dense and amorphous phases.

Biphasic synthesis of SAPO-11(II+III) with pseudoboehmite

Biphasic synthesis mixtures were prepared from tetraethyl orthosilicate (TEOSi), pseudoboehmite, orthophosphoric acid, Pr_2NH, hexanol and water. The composition of the synthesis mixture corresponded to $((Si_xAl_yP_z)O_2 \ 0.19Pr_2NH \ 6.5.H_2O \ 1.1HEX)$. A series of synthesis mixtures were prepared in which increasing equiatomic amounts of Al and P were replaced by Si. Initially, the pH in the aqueous phase was between 4 and 5. The hydrothermal treatment in static coditions, consisted of one, two or three consecutive steps carried out at increasing temperatures. The molar composition of the mixtures, the temperature and time of the hydrothermal treatment and the phases obtained are summarized in Table 5. The crystalline phases obtained were SAPO-11, SAPO-41, VPI-5 and tridymite. In none of the syntheses a pure SAPO-11 phase was obtained (Table 5). When the Si fraction, x, was equal to 0.10, and when a one step heating procedure at 473 K was applied, VPI-5 was the co-crystallizing phase. Prolongation of the heating at 473 K didnot influence significantly the proportions of VPI-5 and SAPO-11 in the product (Table 5).

Table 5. Crystallization products from biphasic synthesis mixture with composition $((Si_xAl_yP_z)O_2 \cdot 0.19Pr_2NH \cdot 6.5H_2O \cdot 1.1HEX$.

No.	sample	Si_x	T_1 (K)	t_1 (h)	T_2 (K)	t_2 (h)	T_3 (K)	t_3 (h)	product[a]
		0.03	473	48					11+41
		0.10	473	24					VPI5+11[b]
			473	48					VPI5+11[b]
			473	96					VPI5+11[b]
11	SAPO-11*/1		363	24	473	24			11(+TRID)[c]
			363	24	473	48			11+TRID[d]
		0.20	473	2					11+TRID[d]
				2.5					11+TRID[d]
				3.5					11+TRID[d]
				24					11+TRID[d]
				65					11+TRID[d]
				72					11+TRID[d]
			363	4	403	20	473	24	11+TRID[d]
			363	4	403	20	473	38	11+TRID[d]
12	SAPO-11*/2		363	24	473	24			11(+TRID)[c]
		0.30	473	24					11+TRID[d]
				48					11+TRID[d]
				65					11+TRID[d]
				72					11+TRID[d]
13	SAPO-11*/3		363	24	473	24			11(+TRID)[c]
		0.40	473	24					11+TRID[d]
				72					11+TRID[d]
				96					11+TRID[d]
			363	4	403	20	473	24	11+TRID[d]
			363	4	403	20	473	38	11+TRID[d]
			363	4	403	20	473	96	11+TRID[d]
14	SAPO-11*/4		363	24	473	24			11(+TRID)[c]
		0.50	473	24					11+TRID[d]
				39					11+TRID[d]
				48					11+TRID[d]
				72					TRID(+11)[e]
				96					TRID(+11)[e]
15	SAPO-11*/5		363	24	473	24			11(+TRID)[c]

a, 11: SAPO-11; 41: SAPO-41; TRID: tridymite.
b, ca. 50% SAPO-11; c, less than 5% TRID; d, less than 20% TRID; e, less than 20% SAPO-11.
T: temperature; t: time of hydrothermal treatment.

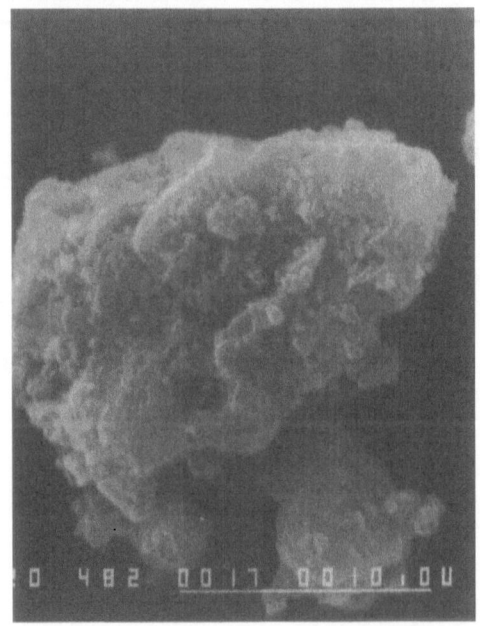

Figure 17. Scanning electron micrograph of SAPO-11/1(A) and SAPO-11/5 (B).

SAPO-11 phases with only minor tridymite impurities were obtained when the mixtures were heated at a temperature of 363 K for 24 h and subsequently, at 473 K for another 24 h. The samples thus obtained will be denoted as SAPO-11*. The XRD pattern shown in Fig.18 illustrates the contamination of SAPO-11*/3 with tridymite. At the end of the crystallization of the SAPO-11* samples, the pH was between 9 and 10, except for sample SAPO-11/5*, where the final pH was 8.

The T-atom compositions of the SAPO-11* samples are given in Table 6. The Si T-atom fraction in SAPO-11/1* is 0.02 and increases to 0.11 in SAPO-11/5*. Because of the presence of tridymite impurities, it is not possible to derive the type of substitution mechanism from the chemical composition. The relation between the Si T-atom fraction in the products and in the synthesis mixtures is given in Fig.19. This figure shows that the incorporation of silicon in the SAPO-11* samples is much more difficult than in the SAPO-11 and SAPO-5 samples obtained from monophasic systems using aluminium isopropoxide.

The morphology of SAPO-11*/2 is shown in Fig.20. The sample consists of large aggregates of individual crystals of ca. 5 μm.

Table 6. Si, Al and P fractions $(Si_xAl_yP_z)O_2$ of SAPO11* products.

No	sample	x	y	z	y/z
11	SAPO11*/1	0.02	0.48	0.50	0.96
12	SAPO11*/2	0.06	0.45	0.49	0.96
13	SAPO11*/3	0.08	0.46	0.46	1.00
14	SAPO11*/4	0.07	0.45	0.48	0.94
15	SAPO11*/5	0.11	0.52	0.37	1.41

Biphasic synthesis of SAPO-11(II+III) with aluminium isopropoxide

Biphasic synthesis mixtures were prepared using tetraethyl orthosilicate (TEOSi), aluminium isopropoxide, orthophosphoric acid, Pr_2NH, hexanol and water. The

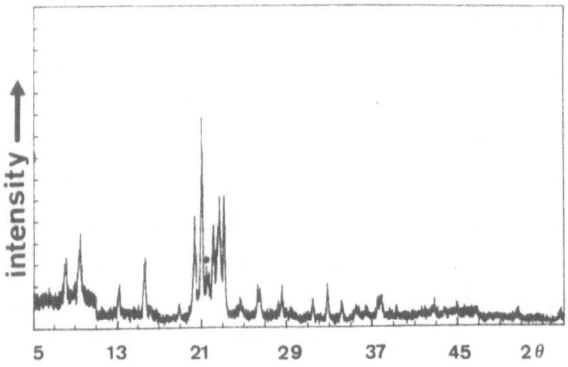

Figure 18. XRD pattern of SAPO-11*/3

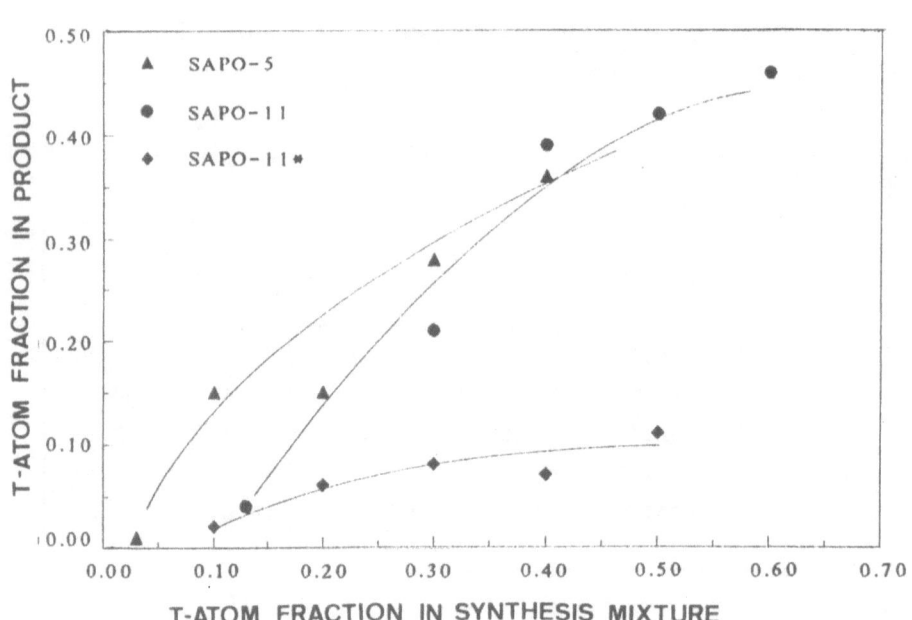

Figure 19. Si T-atom fraction in products against Si T-atom
fraction in synthesis mixtures of SAPO-11*, SAPO-11 and
SAPO-5.

composition of the synthesis mixture corresponded to $((Si_xAl_yP_z)O_2)$ $0.19Pr_2NH$ $6.5H_2O$ $1.1HEX$ with $y/z=1$. The pH of

Figure 20. Scanning electron micrograph of SAPO-11/2*.

Table 7. Si fraction, x, of the synthesis mixtures, the temperature and time of the hydrothermal treatment and the obtained phases from biphasic mixtures with composition $((Si_xAl_yP_z)O_2)$ $0.19Pr_2NH$ $6.5H_2O$ $1.1HEX)$ with $y=z$.

x	T_1 (K)	t_1 (h)	T_2 (K)	t_2 (h)	T_3 (K)	t_3 (h)	product
0.07	473	48					11+am[a]
	473	72					11+am
	403	24	473	48			11
0.12	403	24	473	48			11
0.30	473	48					TRID
	363	24	473	24			TRID

a, am =amorphous; 11 = SAPO-11; TRID = tridymite.

the initial mixture was between 5 and 6. The molar compositions of the mixtures, the temperature and time of the hydrothermal treatment and the phases obtained are given

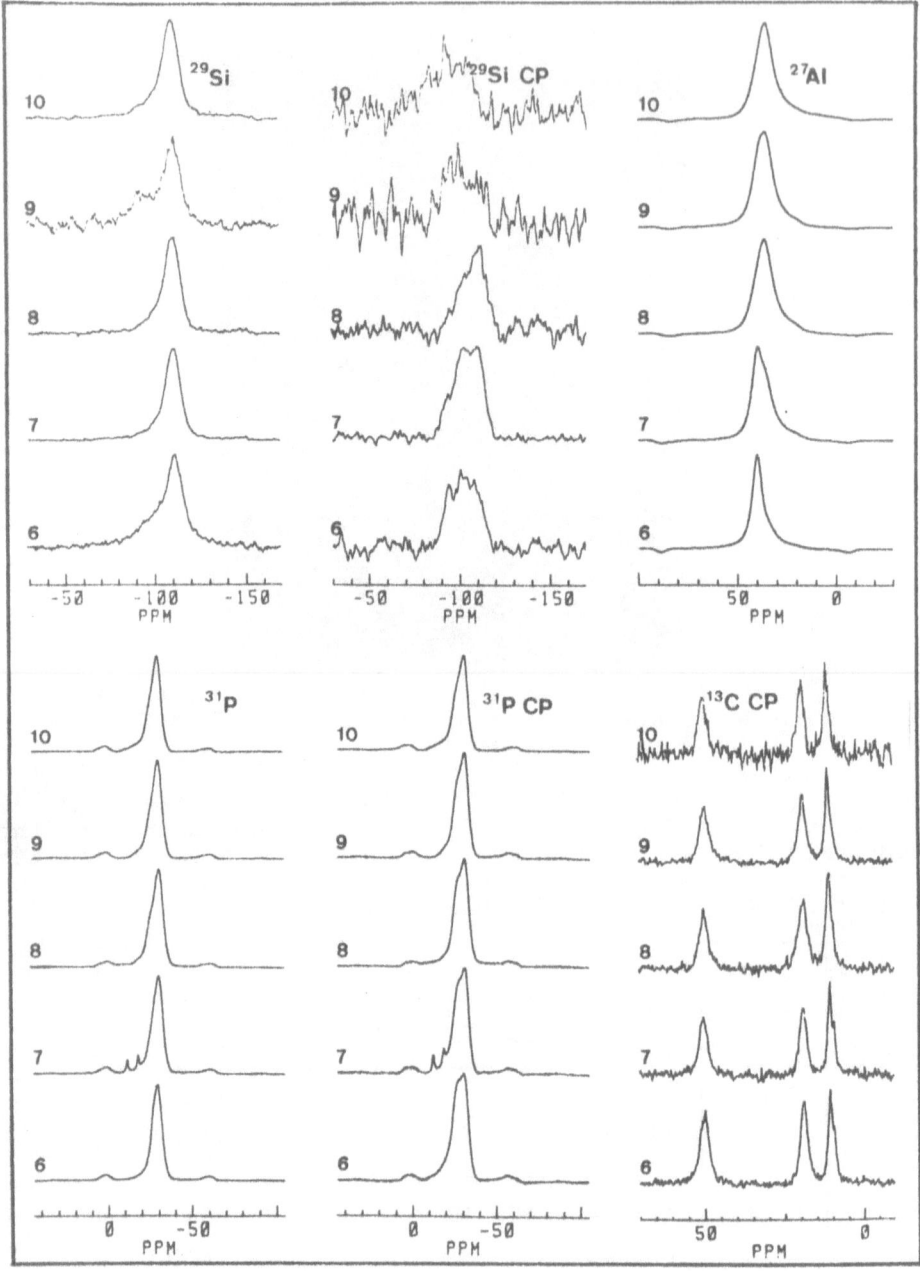

Figure 21. ^{29}Si, ^{29}Si CP, ^{27}Al, ^{31}P, ^{31}P CP and ^{13}C CP MAS NMR spectra of as-synthesized SAPO-11 samples (numbers correspond to sample numbers of Table 4)

in Table 7. Pure SAPO-11 phases were obtained from mixtures with Si T-atom fractions of 0.07 and 0.12 using a two step hydrothermal treatment, which was performed under static conditions, at temperatures of 403 and 473 K during 24 and 48 h, respectively. SAPO-11 of lower crystallinity was obtained when the low temperature step was omitted. Higher Si contents of the synthesis mixture resulted in tridymite formation.

Properties of the SAPO-11 synthesis products

1.<u>Space filling with Pr$_2$NH.</u> According to XRD no other crystalline phase is present in the synthesized SAPO-11 samples besides the AEL phase (Fig.16). In the SAPO-11* samples a small amount of tridymite could be detected (Fig.18). The ^{13}C CP MAS NMR spectra of the as-synthesized SAPO-11/1 to -/5 samples, shown in Fig. 21, indicate that the samples contain residual template molecules. The infrared spectra of SAPO-11/1 recorded at increasing calcination temperatures (Fig.22) show that the template is removed after calcination at 673 K. The same observations are made for SAPO-11* samples. Flanigen et al.[13] have shown that total space filling of the micropores of the AEL framework with Pr$_2$NH template requires one Pr$_2$NH molecule for every 20 TO$_2$ tetrahedra. The molar TO$_2$/ Pr$_2$NH ratio of the as- synthezised SAPO-11 and SAPO-11* samples was derived from the N, Si, Al and P content and is shown in Fig.23 together with published data on the template content of low-silica SAPO-11 and AlPO-11 phases. Fig.23 shows that the molar ratio TO$_2$/ Pr$_2$NH of the SAPO-11/1 to -/4 samples varies between 16 and 22 and is comparable to the values found in literature. **If present, the share of the Si atoms in non-template containing amorphous phase should be small.** The somewhat higher TO$_2$/ Pr$_2$NH values found for SAPO-11* samples can be explained by the presence of the dense phase tridymite.

2.<u>Micropore structure</u>. The results from the decane test on the SAPO-11 samples are shown in Fig.24. According to the CI*, which has a value of 3.5, the SAPO-11/1 to -/5 samples are 10-MR zeolites with very similar pore dimensions. The CI* value of 3.5 further indicates that SAPO-11 is among the

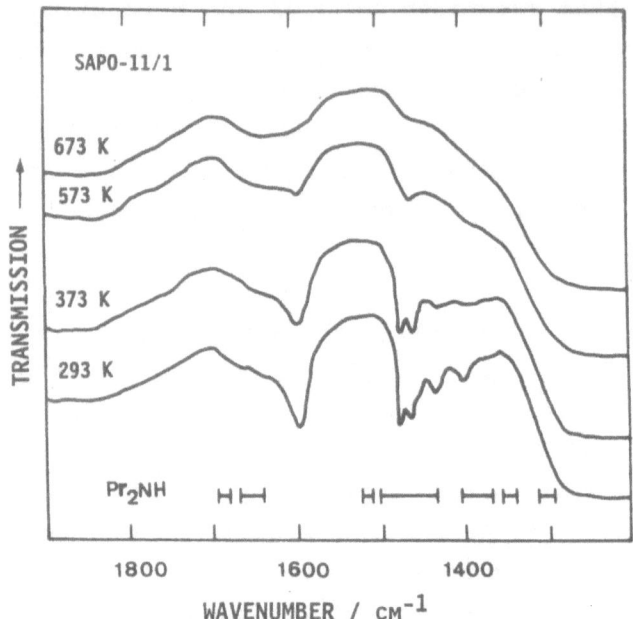

Figure 22. Infrared spectrum of the organic material in as-synthesized SAPO-11/1 sample heated at different temperatures under vacuum.

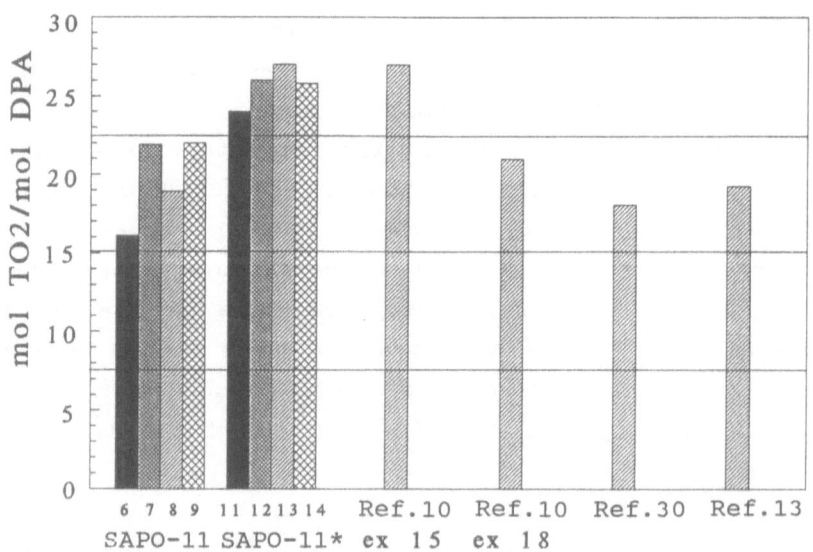

Figure 23. Molar TO_2/ Pr_2NH ratio of as-synthesized SAPO-11/1 to -/4 , SAPO-11*/1 to -/4 (sample numbers of Table 4 and Table 6) and comparison with literature data.

most open 10-MR structures and less shape selective compared, for instance, to ZSM-5 which has a CI* ranging from 6.8 to 9.4 depending on the crystal size[24].

10-MR zeolites generally do not yield ethylbranched isodecanes. With the SAPO-11/1 to -/5 samples from 0.8 to 2.4% of ethylbranching is obtained (Fig.24). It indicates that ethylbranching in these structures is less suppressed than in common 10-MR zeolites. The 10-membered rings in the AEL structure have dimensions of 0.63 x 0.39 nm. The largest dimension is larger than usually found for 10-MR structures with more circular windows.

The selectivity for dibranching and the isopentane selectivity obtained with the SAPO-11/1 to -/5 catalysts are larger than expected for common 10-MR zeolites (Fig.24). This is another manifestation of the somewhat larger pores in this AEL-type of material, in line with the selectivity of monobranching (CI* and ethylbranching criterion).

The carbon number distributions of the hydrocracked products show an excess of C3 and C4 products obtained over SAPO-11/1 to /5 with respect to C7 and C6, respectively, confirming the presence of non-intersecting tubular pores.

The results of the decane test on the SAPO-11*/1 to -/5 samples are also shown in Fig.24. In general, the SAPO-11* samples show a similar behaviour as the SAPO-11/1 - /5 samples and can be classified among the most open 10-MR structures. All SAPO-11* samples have 10-MR openings according to the CI* criterion. CI* of the SAPO-11* samples shows a larger variation than found for the SAPO-11 samples. The selectivity for ethyloctanes is in the range of what is expected for a 10-MR structure.

All SAPO-11* samples form dibranched isomers in larger amounts than found for common 10-MR zeolites. The results of the C3-C7 and C4-C6 criterion show that the SAPO-11* crystals are permeated with non-intersecting pores.

The ^{29}Si MAS NMR spectra of the SAPO-11/1 to /5 samples consist mainly of one intense signal at -112 ppm (Fig.21). In the spectra of some of the samples less intense signals appear as shoulders upfield of this signal till a chemical shift of -92 ppm (Fig.21). These resonances can be

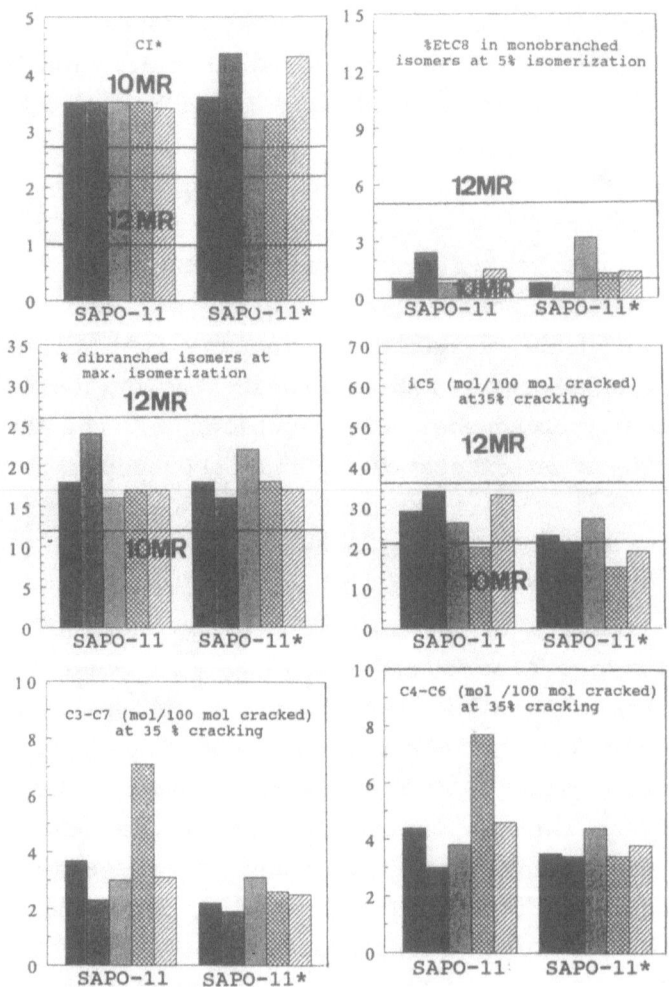

Figure 24. Results of the decane test on SAPO-11/1 to -/5 and SAPO-11*/1 to-/5.

3. <u>Location of silicon, aluminium and phosphorus</u>. The experimental conditions used in the NMR experiments were as follows:

| Parameter | ^{29}Si | | ^{27}Al | ^{31}P | | ^{13}C |
	DEC	CP	DEC	DEC	CP	CP
MAS frequency (MHz)	79.5	79.5	104.2	161.9	161.9	100.6
Pulse length (μs)	3	5.2	0.6	4.5	4.9	4.9
Repetition time (s)	3	5	0.1	5	10	5
Spinning rate (kHz)	3	3	3	5	5	5
CP MAS contact time (ms)	–	5	–	–	1	1
Decoupling (ms)	1	–	–	1	–	–
Number of scans	18000	10000	3000	48	48	1500

assigned to Si with different numbers of Al and Si neighbours. The low intensity of the -92 ppm signal indicates that the contribution of SM 2 in SAPO-11/1 to /5 is lower than in the SAPO-5 samples (Fig.8). However it was not possible to derive the relative contribution of SM2 and SM3 from the ^{29}Si MAS NMR spectra.

The ^{29}Si CP MAS NMR spectra show only very weak signals with maxima positioned at -94, -101 and -111 ppm (Fig.21). The -112 ppm signal under CP conditions is less intense when more silicon is present in the sample (Fig.21). As the -112 ppm signal represents silicon in the siliceous domains in the crystals, this observation can be explained by the assumption that the distance between these Si atoms and OH groups (Si-OH, P-OH and bridging Si-OH-Al) becomes larger due to the increasing size of the siliceous domains.

Fig.21 shows that SAPO-11/1 is characterized by a ^{27}Al MAS NMR resonance at 40 ppm. In SAPO-11/2 a shoulder at 36 ppm appears on the 40 ppm signal. The 36 ppm signal dominates the 40 ppm signal in SAPO-11/3 to -/5 (fig.21). This ^{27}Al signal is typical for Al(4P) coordinations[32]. The ^{31}P NMR spectra consist mainly of a line at -30 ppm with a shoulder at -27 ppm. The intensity of the -27 ppm signal is enhanced under CP conditions (Fig.21). Significant changes of the ^{31}P MAS NMR spectrum of ALPO-11 depending on the degree of hydration of the sample have been reported by Goepper et al.[28] and can account for the differences found among the samples.

The ^{29}Si and ^{29}Si CP MAS NMR spectra of SAPO-11* samples are shown in Fig.25. The ^{29}Si MAS NMR spectra of SAPO-11* samples are similar to those found with SAPO-11/1 to -/5 (Fig.21). For all the SAPO-11* samples, the ^{29}Si signals were found in the chemical shift range from -80 to -120 ppm (Fig.25). SAPO-11/1* shows a broad signal with maximum intensity at ca.-100 ppm. In the other samples, the main ^{29}Si resonance is positioned at around -110 ppm, indicative of Si(4Si) environments and of the presence of siliceous crystal domains. The main Si signal at -110 ppm is accompanied by signals at higher field due to presence of Si with Al atoms as nearest neighbours. It was, however, not possible to deconvolute the spectra and to quantify the amount of Si(nAl) environments. The ^{29}Si spectra of SAPO-11* samples with and without CP are similar (Fig.25). Cross polarization between the Si atoms of the siliceous parts of the crystals with protons is efficient, suggesting that the siliceous domains could be smaller than e. g. in the SAPO-11/4 and /5 samples (Fig.21). The smaller size of the siliceous domains could also be from the lower Si content of the SAPO-11* samples (Table 4 and 6). The ^{29}Si and ^{29}Si CP spectra of the SAPO-11* samples suggest the presence of siliceous domains in the crystals. The predominance of SM3 over SM2 in the SAPO-11* samples is shown by the relative intensity of the signal at -110 ppm compared to one at the -92 ppm.

The ^{31}P, ^{31}P CP and ^{27}Al MAS NMR spectra of the SAPO-11* samples are shown in Fig.25. All samples show a ^{31}P resonance at -30 ppm and a shoulder at -27 ppm. The intensity of the -27 ppm signal is enhanced under CP conditions. These ^{31}P resonances denote the presence of P(4Al) environments[32]. The main ^{27}Al signal is at 40 ppm and is due to tetrahedrally coordinated aluminium in AlPO$_4$

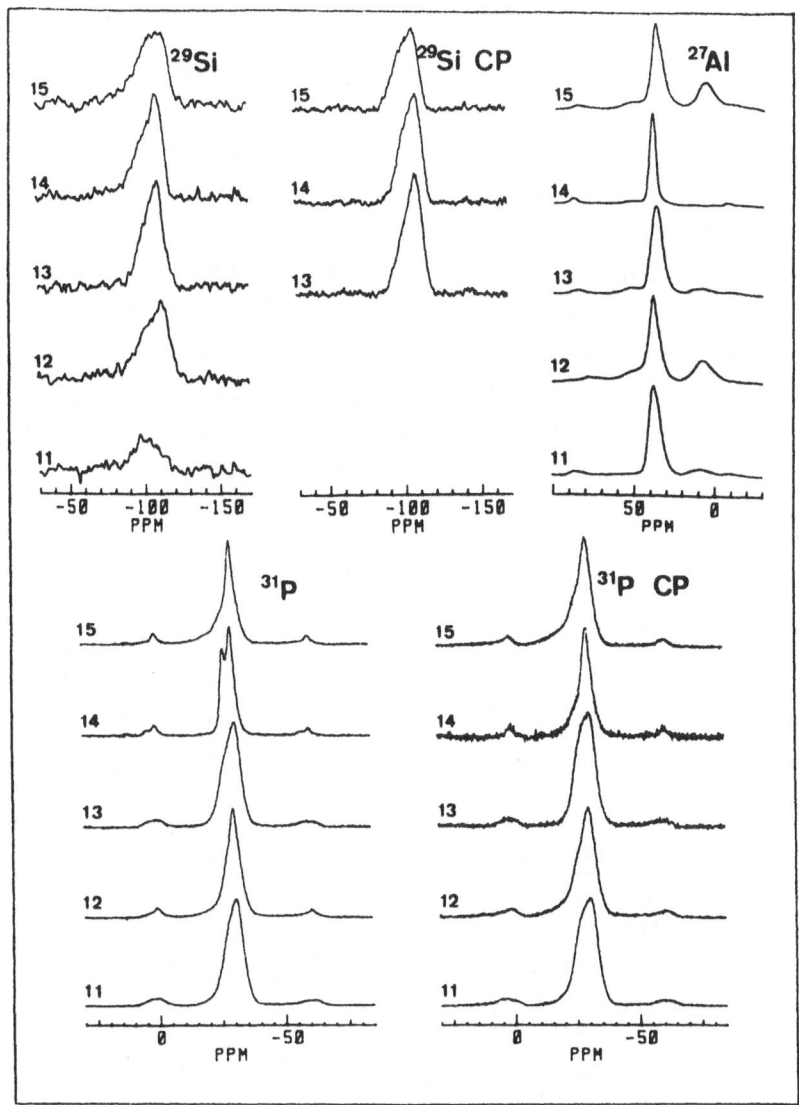

Figure 25. ^{29}Si, ^{29}Si CP, ^{27}Al, ^{31}P and ^{31}P CP MAS NMR spectra of as-synthesized SAPO-11* samples the number corresponds to the sample number of Table 6.

environment[32]. The ^{27}Al resonance at 8 ppm is due to amorphous alumina present in the sample[33]. In some of the samples a signal at ca. 50 ppm is also present. This chemical shift could be due to aluminosilicate[34].

4. <u>Brønsted acidity</u>. The hydroxyl spectra of SAPO-11/1 to -/5 are shown in Fig.26. The samples contain at least three different types of OH groups. Silanol groups are found at 3740 cm^{-1}, P-OH vibrations at 3670 cm^{-1} and bridging Si-OH-Al are vibrating at 3620 cm^{-1}. For all samples except for SAPO-11/4, a low frequency bridging Si-OH-Al vibration is observed at 3530 cm^{-1} . The intensity of the hydroxyl vibrations is low compared to the intensity of the hydroxyl vibrations of zeolite H-Y (Fig.26). The same observation was made for SAPO-5/1 to -/5 samples (Fig.13). The silanol groups can account for crystal termination. The amount of

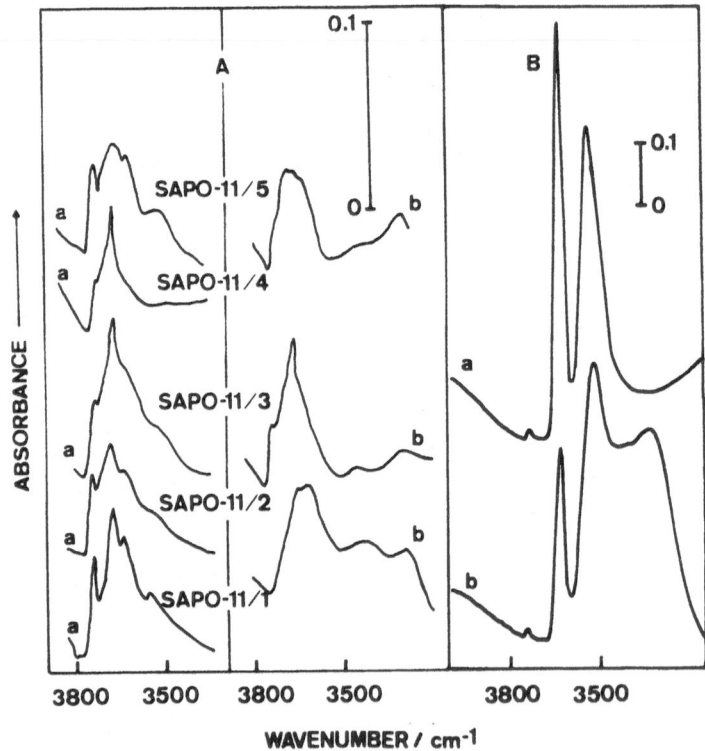

Figure 26. Hydroxyl spectra of SAPO-11 samples (A) and zeolite H-Y (B) pretreated at 673 K (a) and after adsorption of benzene (b).

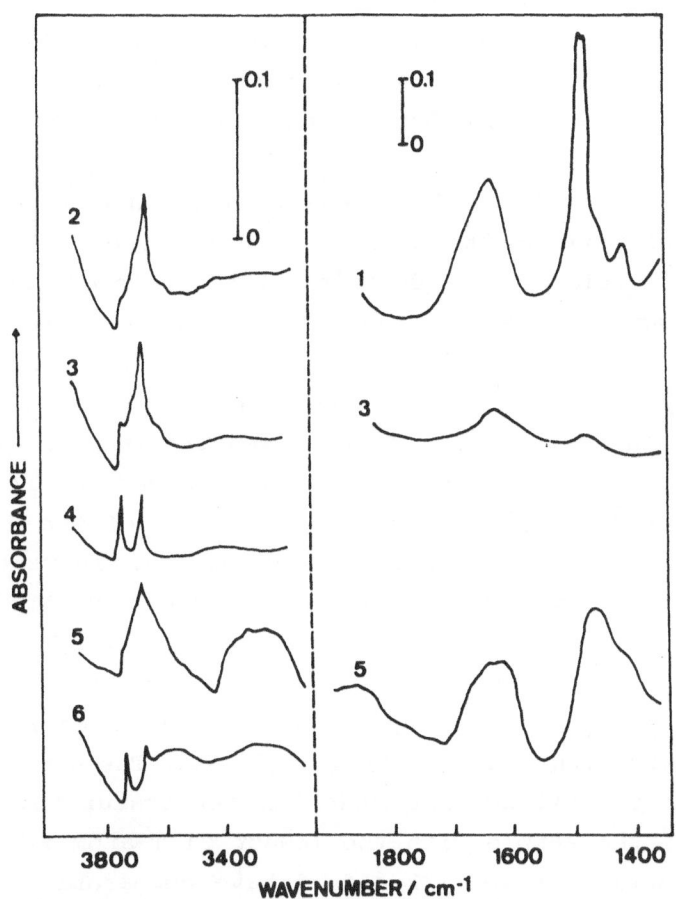

Figure 27. Hydroxyl spectra of SAPO-11/4 after different pre-treatments: (1) as-synthesized; after calcination at 673 K (2); after calcination at 823 K (3); after calcination at 1123 K (4); after calcination at 823 K and adsorption of NH_3 (5); after calcination at 1123 K and adsorption of NH_3 (6).

bridging Si-OH-Al groups is systematically lower in SAPO-11/1 to /5 compared to SAPO-5/1 to -/5 samples (Fig.13). This is in agreement with the ^{29}Si MAS NMR results, showing that the -92 ppm signal is less intense in the SAPO-11 samples. The P-OH vibration is intense in the SAPO-11 samples.

A heat treatment of SAPO-11/4 at 823 K during 24 h does not change the hydroxyl spectrum (Fig.27). After calcination of the sample at 1123 K during 24 h, the intensity of the P-OH vibration has decreased substantially.

After adsorption of benzene on the SAPO-11 samples, a substantial part of the hydroxyl bands shifts from 3450 to 3300 cm^{-1} (Fig. 26), denoting the presence of strong Brønsted acid sites. The Brønsted acid character of the hydroxyl groups is confirmed by the formation of ammonium cations upon adsorption of ammonia on the samples (Fig.27).

The hydroxyl spectra of the SAPO-11* samples are shown in Fig.28. Hydroxyl vibrations are observed at 3740, 3680 and 3630 cm^{-1}, respectively. The intensity of the 3740 cm^{-1} band, which is due to the silanol groups, increases with increasing Si content of the sample. The SAPO-11* samples contain more silanol groups than SAPO-11/1 to -/5 and SAPO-5/1 to /5 (Fig. 13 and 26).

The catalytic activity of the SAPO-11/1 to /5 samples was measured with the coversion of decane. The conversion of decane is plotted against reaction temperature in fig.29. SAPO-11/1, the sample with the lowest Si T-atom fraction is the most active. SAPO-11/2 and /5 have comparable catalytic activities. The catalytic activity of SAPO-11/1 is comparable to that of SAPO-5/5, which is the least active SAPO-5 sample (Fig.14). The lower activity of the SAPO-11 compared to SAPO-5 is in agreement with the lower contribution of SM 2.

The conversion of decane at increasing reaction temperatures over the SAPO-11* samples is given in Fig.30. The activity of the SAPO-11* samples decreases slightly with increasing Si content. The activity of the SAPO-11* samples is comparable to that of the most active SAPO-11/1 to /5 samples (Fig.29).

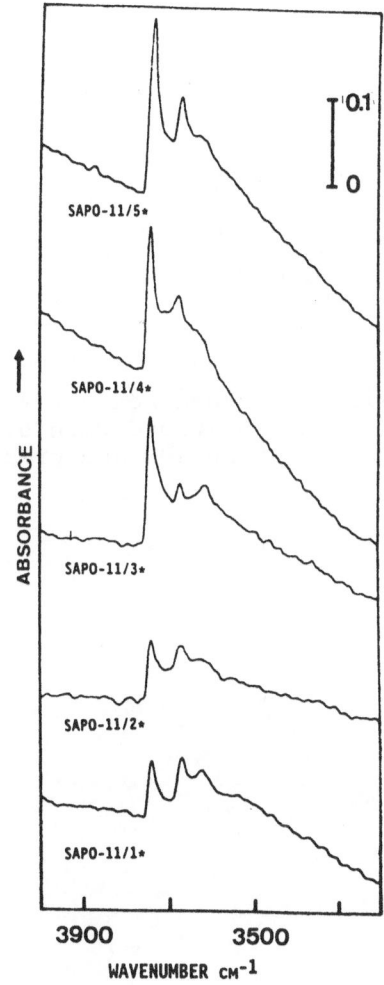

Figure 28. Hydroxyl spectra of SAPO-11* samples calcined at a temperature of 823 K.

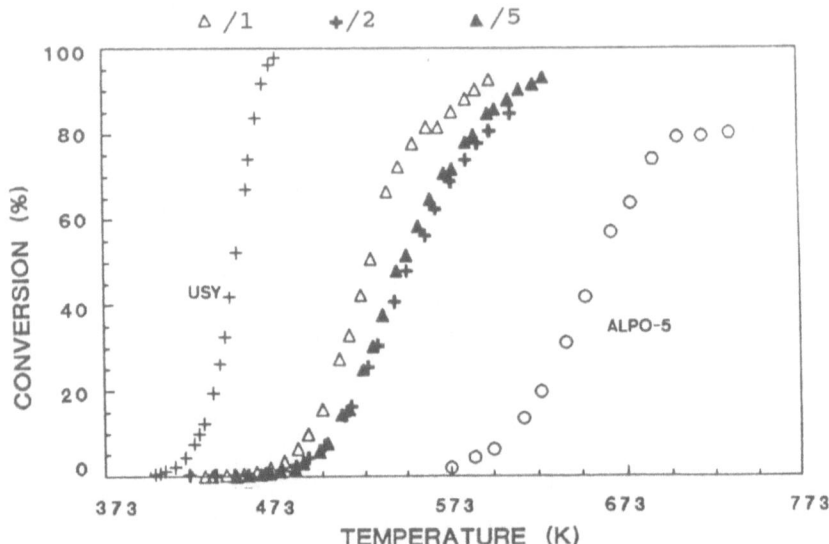

Figure 29. Conversion of decane against reaction temperature over SAPO-11/1, -/2 and -/5 loaded with 0.5% of Pt. $W/F_o=0.5$ kg s mmole^{-1}, hydrogen/decane=100 and pressure=0.35 MPa.

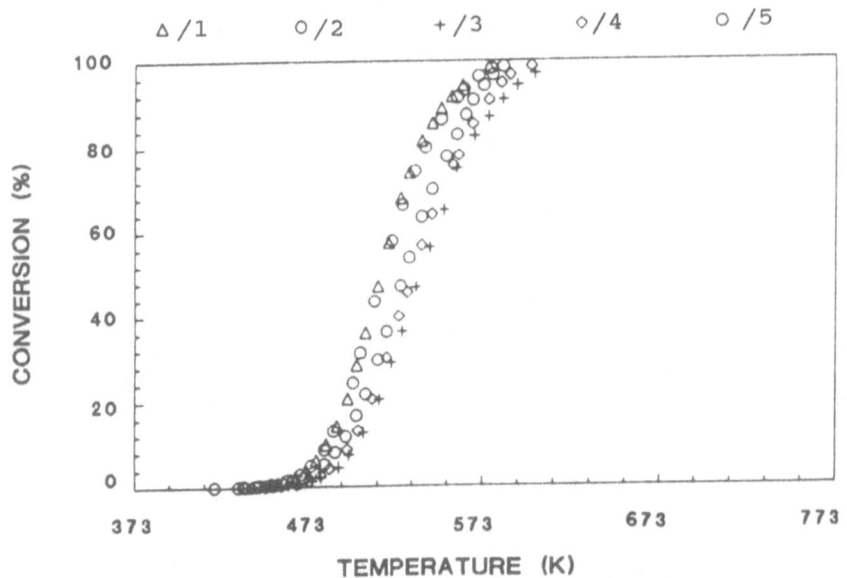

Figure 30. Conversion of decane against reaction temperature over SAPO-11* samples loaded with 0.5% of Pt. $W/F_o=0.5$ kg s mmole-1, hydrogen/decane = 100 and pressure = 0.35 MPa.

Table 8. Summary of the results of the synthesis experiments using DPA and CHA and equiatomic amounts of Al and P.

type of synthesis	R^a	solvent	Si source	Al	pH^b	agita-tion	phase obtained
I_{MR}	CHA	water	Ludox	Aluminium-isopropoxide	10-11	-	amorphous
I_{MQ}						+	SAPO-5(II) SAPO-5(II+III)
I_{PR}				pseudoboehmite	5-6	+	dense + amorphous
I_{MQ}	DPA	water	Ludox	Aluminium-isopropoxide	9-10	-	amorphous
I_{MR}						+	SAPO-11(II) SAPO-11(II+III)
I_{PR}				pseudoboehmite	5-6	+	dense + amorphous
II_{MQ}		water+ hexanol	$TEOS^e$	Aluminium-isopropoxide	$6-7^c$	-	SAPO-11(ND)^d
II_{PQ}				pseudoboehmite	$5-6^c$	-	SAPO-11(II+III) +TRID (low Si content)

a, templating agent; b, pH immediatly after adding the R; c, pH of the water phase; d, SM not determined; e, tetraetyl orthosilicate

It is important to emphasize that in this work all the synthesis mixtures were prepared according to the same procedure. The mixing sequence and the mode of mixing was always the following. First, the aluminium source was added diluted phosphoric acid. Subsequently, the Si source (dissolved in hexanol if applicable) and the organic were added. The mixtures were transferred into autoclaves and immediately introduced in a preheated furnace.

Table 8 summarizes the results of the synthesis experiments. The synthesis procedures can be classified into two different types. Type I corresponds to mixtures containing water as the solvent. Type II refers to mixtures with two solvents, viz. water and hexanol. In the type I synthesis a polymeric source of Si (Ludox) was used. It is an ammonium stabilized silica sol which is stable at a pH of 9.6. In the syntheses of type II, the source of Si was tetraethyl orthosilicate, which is a monomeric source of Si. The two types of synthesis can be subdivided in subtypes depending on whether a polymeric (P) source of Al was used (pseudoboehmite) or a monomeric (M) type (aluminium isopropoxide). A further subdivision depends on whether the synthesis mixture was rotated (R) during hydrothermal treatment, or not (Q).

I_M versus I_P synthesis

Aluminium isopropoxide is a reactive source of Al. It immediately starts to hydrolyse when added to the diluted phosphorus acid, resulting in an increase of the pH. Pseudoboehmite is much less reactive and its addition to the phosphoric acid solution doesnot increase the pH immediately. **This pH effect is the major difference between I_M and I_P syntheses.** Table 8 shows the pH of the synthesis mixture after the addition of the organic. The pH is always substantially higher when aluminium isopropoxide is used. The pH will determine whether and to what extent the silica of the ammonium stabilized silica sol precipitates. It can be expected that silica precipitation is less pronounced in type I_M compared to type I_P syntheses. Table 8 shows that I_P syntheses always result in dense and amorphous products.

Table 8 further shows that the use of aluminium isopropoxide is not sufficient to obtain SAPO-5 and SAPO-11 materials, but that, in addition, the mixture has to be homogenized by rotating the autoclave during the hydrothermal treatment. Only the I_{MR} method successfully yields pure SAPO-5 and SAPO-11 crystals.

The amount of Si that can be incorporated via SM2 using this I_{MR} method is limited. The remaining Si is incorporated in the framework according to SM 3 and generates siliceous crystal domains.

Type II synthesis

In type II syntheses the silicon source is dissolved in hexanol. It can be expected that the silicon is transferred only slowly into the water phase, especially under static conditions and when the hydrothermal treatment starts with a low temperature step. Initially, the formation of amorphous or dense silica phases under the acidic conditions of the water phase can thus be avoided. Table 8 shows that the nature of the aluminium source is much less critical in type II compared to type I syntheses and that microporous crystals can be obtained with both systems. The incorporation of Si in SAPO-11 synthesized with method II is mainly according to SM3. With respect to the amount of silicon that can be incorporated, syntheses of type II are less efficient than those of type I.

CONCLUSIONS

Microporous silicoaluminophosphates of the type SAPO-5 and SAPO-11 crystallize from monophasic as well as from biphasic synthesis mixtures. In monophasic synthesis systems and when a silica sol is used as source of Si, aluminium isopropoxide is the preferred source of aluminium. The source of aluminium is less relevant in biphasic systems when using tetraethylorthosilicate as source of Si.

The amount of Si incorporated in the crystallisation product depends on the synthesis method. The monophasic method allows to obtain SAPO-5 and SAPO-11 materials with up to 36% and 46%, respectively, of the T atom positions occupied by Si. The maximum Si loading of SAPO-11 synthesized from biphasic mixtures is 10%. The crystallisation products from the biphasic synthesis method are often contaminated with dense and amorphous phases.

When the amount of silicon in the synthesis mixture is low, all silicon is incorporated via SM2. The amount of Si that can be incorporated via SM 2 is always limited. The contribution of SM2 and SM3 could be evaluated from the relative intensities of the -92 and -110 ppm ^{29}Si MAS NMR signals. In the most siliceous SAPO-5 sample, SM2 involved only 7 % of the T-atoms. In the SAPO-11 samples, SM2 was active but its contribution was too low to be measured with ^{29}Si MAS NMR. In SAPO-5 and SAPO-11 containing appreciable amounts of Si, the major part of the Si is incorporated in the framework according to SM 3, generating siliceous crystal domains. ^{29}Si CP MAS NMR spectra indicate that in the most siliceous SAPO-5 and SAPO-11 samples these siliceous domains should be large.

The micropores of SAPO-5(II+III) and SAPO-11(II+III) materials are totally filled with template molecules after synthesis. After calcination the samples exhibit Brønsted acidity and catalytic activity. SAPO-5(II+III) and SAPO-11(II+III) materials contain bridging Si-OH-Al, Si-OH and P-OH groups, vibrating at 3618 and 3528, 3670 and 3740 cm-1, respectively. The decane test indicates that the micropores are not altered substantially when the extent of Si incorporation via SM3 is changed.

ACKNOWLEDGMENTS

The authors acknowledge the Flemish NFWO and the Belgian Ministery for Science Policy for a research grant. M.M. is grateful to IWONL for a fellowship. J.A.M., P.J.G and P.A.J. acknowledge the Flemish NFWO for a research position as Research Associate, Senior Research Associate and Research Director, respectively.

REFERENCES

1. R.M. Barrer, "Hydrothermal Chemistry of zeolites", Acad. Press, New York (1982).
2. R.M. Barrer, D.J. Marshall, Chemistry of Soil Minerals. Part I, Hydrothermal Crystallisation of some Alkaline Al_2O_3-SiO_2-P_2O_5 Compositions, J. Chem. Soc., 6616 (1965).
3. R.M. Barrer, D.J. Marshall, Chemistry of Soil Minerals, Part II, Reactions of Phosphates with Kaolin and Faujasite, J. Chem. Soc., 6621 (1965).
4. G.H. Kühl, Infuence of phosphate and other complexing agents on the crystallisation of zeolites, in: "Molecular Sieves", Soc. Chem. Ind., London (1968).

5. G.H. Kühl, Crustallization of Zeolites in the Presence of a Complexing Agent, in: "Molecular Sieve Zeolites, - I", E.M. Flanigen, L.B. Sand, eds., Am.Chem.Soc., Adv.Chem.,Ser.,101:76, (1971).
6. E. M. Flanigen and R.W. Grose, Phosphorus Substitution in Zeolite Frameworks, in: "Molecular Sieve Zeolites-I", Am.Chem.Soc., Adv.Chem.Ser. 101:76 (1971).
7. R.M. Barrer, D.J. Marshall, Hydrothermal Chemistry of Silicates. Part XX. The Question of Phosphorus Substitution for Silicon during Zeolite Synthesis, J.Chem.Soc., Dalton Trans., 2126 (1974).
8. H-P. Rieck, H-J. Kalz, Europäiche.P.A. 0 111 748, (1984).
9. S.T. Wilson, B.M. Lok and E.M. Flanigen, U.S. Patent 4,310,440 (1982).
10. a, B.M. Lok, C.A. Messina, R.L. Patton, R.T. Gajek, T.R. Cannan, E.M. Flanigen, Crystalline Silicoaluminophosphates, U.S.Patent 4,440, 871, 1984; b, J. Am. Chem. Soc. 106:6092 (1984).
11. J.A. Rabo, R.J. Pellet, P.K. Coughlin, E.S. Shamshoum, Skeletal Rearrangement Reactions of olefins, Parafins and Aromatics over Aluminophophate based Molecular Sieves, in: "Zeolites as Catalysts, Sorbents and Detergent builders", H.G. Karge and J. Weitkamp, eds., Elsevier, Stud. Surf. Sci. Catal., Elsevier, Amsterdam, Oxford, New York, Tokyo 46:1-18 (1989) and references therein.
12. P.A. Jacobs and J.A. Martens, "Synthesis of High-Silica Aluminosilicate Zeolites", Stud. Surf. Sci. Catal. 33, Elsevier, Amsterdam, Oxford, New York, Tokyo (1987).
13. E.M. Flanigen, R.L. Patton, S.T. Wilson, Stuctural, Synthetic and Physicochemical Concepts in Aluminophosphate-based Molecular Sieves, Stud. Surf. Sci. Catal. 37, P.J. Grobet, et. al. (eds). Elsevier, Amsterdam, pp.13-27 (1988).
14. E.M. Flanigen, B.M. Lok, R.L.Patton, S.T. Wilson, Aluminophosphate Molecular Sieves and the Periodic Table, Proc. 7th Int. Zeol. Conf., Y. Murakami et. al. (Eds.), Kodanska, Tokyo 1986, 103.
15. W.M. Meier, D.H. Olson, "Atlas of Zeolite Structure Types", Butterworths,1987.
16. I.P, Appleyard, R.K. Harris, F.R. Fitch, 27-Aluminium, 31-phosphorus and 29-Silicium MAS NMR Studies of the silicoaluminophosphate SAPO-5, Chem. Lett. 1985, pp.1747-1750.
17. K.J. Chao, L.J. Leu, Catalytic and Physical Properties of Silicon substituted AlPO-5 Molecular Sieves, Stud. Surf. Sci. Catal. 46, 1989, pp. 19-27.
18. N.J. Tapp, N.B. Milestone, D.M. Bibby, Generation of acid sites in substituted aluminophosphate molecular sieves, Stud. Surf. Sci. Catal. 37, P.J. Grobet, et.al. (Eds.), Elsevier, Amsterdam, 1988, pp. 393-402.
19. D. Freude, H. Ernst, M. Hunger, H. Pfeifer, E. Jahn, Magic-Angle-Spinning NMR studies of zeolite SAPO-5, Phys. hem. Lett. 143, 1988, pp.477-481.
20. C.S. Blackwell, R.L. Patton, Solid-State NMR of Silicoaluminophosphate Molecular sieves and Aluminophosphate Materials, J. Phys. Chem., 92, 3970.
21. X. Wang, X. Liu, T. Song, J. Hu, Qiu, Substitution of Si in SAPO-5, Chem. Phys. Lett., 1989, pp. 87-91.

22. R.A. Van Norstrand, D.S. Santilli, S.I. Zones, "An All-Silica Molecular Sieve That is Iso-Structural with AlPO-5", in:"Perspectives in Molecular Sieve Science", A.C.S. Symp. Ser. 368, 1988, pp. 236-245.

23. J.A. Martens, M. Mertens, P.J. Grobet, P.A. Jacobs, Synthesis and Characterization of Silicon-rich SAPO-5., Stud. Surf. Sci. Catal., 37, P.J. Grobet, et. al. (Eds.), Elsevier, Amsterdam, 1988., pp. 97-105.

24. J.A. Martens, P.A. Jacobs, The Potential and Limitations of the n-decane hydroconversion as a test reaction for characterization of the void space of molecular sieve zeolites, Zeolites, 6, 1986, 334.

25. P.A. Jacobs, J.A. Martens, Exploration of the Void Size and Structure of Zeolites and Molecular Sieves Using Chemical Reactions, Proc. 7th Int. Zeol. Conf., Y. Murakami et. al. (Eds.), Kodanska, Tokyo, 1986, 23.

26. J.B. Peri, Surface Chemistry of $AlPO_4$ - A mixed Oxide of Al and P, Discuss. Faraday Soc., 52, 1971, pp. 55-65.

27. P.A. Jacobs, J.B. Uytterhoeven, Assignment of the hydroxyl bands in the infrared spectra of zeolites X and Y. 1. Na-H zeolites, J. Chem. Soc., Faraday Trans. I, 1973, 69, 359.

28. M. Goepper, F. Guth, L. Delmotte, J.L. Guth, H. Kessler, Effect of template removal and rehydratation on the structure of $AlPO_4$ and AlPO4based microporous crystalline solids, Stud. Surf. Sci. Catal., 49, P.A; Jacobs et. al. (Eds.), Elsevier Amsterdam, 1989, pp. 857-866.

29. J.M. Bennet, J.V. Smith, Enumeration of 4-connected 3-dimensional nets and classification of framework silicates. 3D nets based on the 4.6.12 and $(4.6.10)_4$ (6.6.10):2D nets, Zeitschrift für Kristallographie, 171, 1985, pp. 65-68.

30. R. Khoudzami, G. Coudurier, B.F. Mentzen, J.C. Vedrine, Structural, Acidic and Catalytic Propoerties of SAPO-11 Molecular Sieves, Stud. Surf. Sci. Catal., 37, P.J. Grobet et. al (Eds.), Elsevier Amsterdam, 1988, pp. 355-363.

31. H. Weyda, H. Lerchert, Kinetic Studies of Alumino-phosphate- and Silicoaluminophosphate Molecular Sieves, Stud. Surf. Sci. Catal., 49, P.A; Jacobs et. al. (Eds.), Elsevier Amsterdam, 1989, pp. 169-178.

32. D. Müller, E. Jahn, G. Ladwig, U. Haubenreisser, High-Resolution Solid-State ^{27}Al and ^{31}P NMR: Correlation between chemical shift and mean Al-O-P angle in $AlPO_4$ polymorphs, Chem. Phys. Lett., 109,4,1984, pp. 332-336.

33. P;J; Grobet, H. Geerts, M. Tielen, J.A. Martens, P.A. Jacobs, Framework and non-framework Al species in dealuminated zeolite Y, in: "Zeolites as Catalysts, Sorbents and Detergent builders", H.G. Karge and J. Weitkamp, eds., Elsevier, Stud. Surf. Sci. Catal., Elsevier, Amsterdam, Oxford, New York, Tokyo 46, 1989, pp. 721-734.

34. J.A. Martens, C. Janssens, P.J. Grobet, H.K. Beyer, P.A. Jacobs, Isomorphous substitution of silicon in SAPO-37, Stud. Surf. Sci. Catal., 49, P.A; Jacobs et. al. (Eds.), Elsevier Amsterdam, 1989, pp. 215-225.

35. P.A. Jacobs, J.A. Martens, J. Weitkamp, H.K. Beyer, Shape-seletivity Changes in High-silica Zeolites, Faraday Disc. Chem. Soc., 72, 1981, pp.353-369.

ZEOLITE SYNTHESIS AND CRYSTAL TAILORING

François Fajula

URA 418 CNRS
ENSCM, 8, Rue de l'Ecole Normale
34 075 Montpellier Cedex France

I INTRODUCTION

In the past thirty years zeolites have stimulated a huge amount of academic and industrial research and impacted several adsorption and catalytic processes. The commercial success of zeolites was due primarily to their unique structure and composition controlled properties which permitted the improvement of known processes and the development of new ones. On the other hand, the microporous zeolite frameworks have provided the degree of perfection needed for the study of fundamental events in sorption, catalysis, spectroscopy and surface science in general.

All along this period of time, the discovery of new concepts has been closely related to the discovery of new structure types and compositions. The number of known and topologically distinct frameworks approaches, today, seventy (including the AlPO and SAPO's families) and is expected to continue to increase at a rate of at least three per year[1].The research in synthesis has been -and will certainly continue to be- thus very active and productive, at least at a laboratory scale. A detailed review of all this work may be found in the excellent books by Breck (2), Barrer (3) and Jacobs and Martens(4).

This paper will focus on two aspects of zeolite synthesis, namely the control of the morphology of the crystals and the generation of composition gradients in the crystallites. The molecular shape selective properties of a zeolite are mainly due to the size, the shape and the tortuosity of its cavities or pores (5). These topological characteristics are intrinsic properties of the three-dimensional framework under consideration. At variance, the size, the habit and the homogeneity of the composition of the crystals depend on the synthesis conditions. These parameters also can decisively affect the sorption and catalytic behaviours and their control is as important as the control of the crystallographic purity for the production of solids with optimized properties. The discussion below summarizes some current and potential ideas which are put forward to reach these goals.

Guidelines for Mastering the Properties of Molecular Sieves
Edited by D. Barthomeuf *et al.*
Plenum Press, New York, 1990

II CONTROL OF THE MORPHOLOGY OF THE CRYSTALS

A complete control of the morphology of crystals of zeolites implies that the size and the habit of the individual crystallites be adjusted independently. In fact, both characteristics are determined by the supersaturation level of the synthesis medium - a parameter ill-defined in the case of zeolites - and show twinned evolutions : high supersaturation levels cause the appearance of numerous nuclei which develop into small-sized-crystals exhibiting high-index faces and curved surfaces whereas at low values of supersaturation, a smaller amount of nuclei is produced ; their growth leads to large particles exposing faces with low Miller indexes and high lattice density.

For the sake of clarity and despite the above limitation, the control of the size and of the morphology of zeolite crystals will be created separately.

Table 1. Major factors influencing the crystallization of the MFI structure (4).

Increase of Al content	Decreases crystallization rate.
Increase of TPA/SiO$_2$ ratio	Favours nucleation.
H$_2$O/SiO$_2$ ratio	Little or no effect.
Nature of cation	Nucleation rate increases with increasing electrostatic potential. No effect on growth rate.
Increase of the OH/SiO$_2$ ratio	Increases rate of nucleation.
Increase of the polymerization degree of silica	Decreases nucleation rate.

II-1 Control of the size of zeolite crystals

The formation of zeolites is a nucleation controlled process occurring from inhomogeneous alkaline hydrogels. The size of the crystals is determined by the experimental factors which alter the relative rates of nucleation and of crystal growth. Basically, high nucleation rates combined with small crystallization ones will tend to produce the smaller crystals. The reverse is true for producing large particles.

The vast domain of stability of the zeolites of the MFI familly provides an useful -and rather unique- model showing how this can be achieved by adjusting the composition of the initial reaction medium. A critical evaluation of the various parameters influencing the crystallization kinetics has been made recently by Jacobs and Martens (4) and is summarized in Table 1.

All other factors remaining constant, the two parameters which have the most drastic effect on the crystal size are the nature of the inorganic cation and the alkalinity of the medium. Figure 1a shows the effect of the addition of ammonium ions to a tetrapropylammonium (TPA) silicoaluminate hydrogel on the rate of nucleation (expressed by the reciprocal time of the induction period). As the ammonium fraction was increased from 0.2 to 0.8, the nucleation rate was decreased by nearly one order of magnitude and the size of the crystals increased from 1-2 μm to 40-50 μm. When small amounts of sodium or potassium ions were added to the ammonium rich gel, the nucleation rate increased again (Figure 1b).

These evolutions are in line with the structure -forming ability of the three cations (6). Ammonium ions exert a structure- breaking effect (disrupting the water structure through breaking of the hydrogen bonds) on the gel which would decrease the rate of nucleation whereas sodium, and to a lesser extent potassium, ions would favour nucleation due to their templating or structure -directing properties.

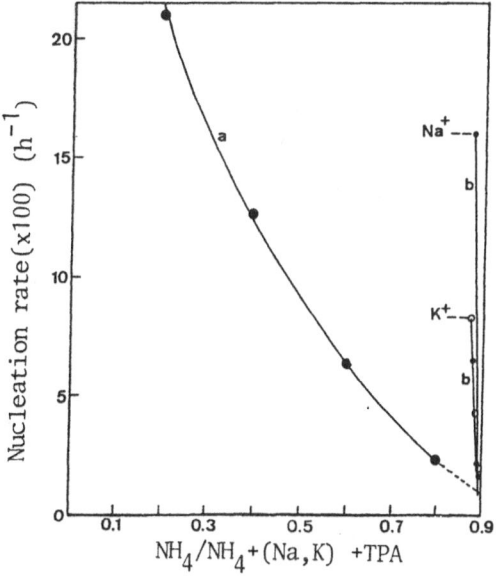

Fig. 1. Change of the nucleation rate of ZSM-5. a/ as a function of the NH$_4$ molar fraction, b/ addition of Na or K to the ammonium-rich gel (Ref. 4, Fig. II-9 page 69).

The influence of the alkalinity on kinetics is well recognized (3). As a general rule, higher alkalinities produce smaller crystals. The influence of this parameter must be however analyzed very cautiously. Increasing the alkalinity increases the degree of supersaturation and of dissolution of the crystals -which both tend to decrease final crystal size- but also increases the rate of growth. An optimum alkalinity will often exist. Moreover the alkalinity of a gel is a complex function of the total dilution, of the nature and content of alkali, and of the degree of polymerization of the silica.

The definition of the alkalinity of a medium by its sole OH/SiO_2 ratio is usually not sufficient. The effect of hydroxide content on crystal sizes is nevertheless clear, as illustrated in the work by Hayhurst et al. for example (7). The authors prepared silicalite samples by reacting two series of TPA containing silica-gels in which increasing amounts of hydroxide ions were introduced by using either sodium hydroxide (S-series) or TPAOH (T-series) as alkali source. As expected, the size of the crystals changed substantially with the hydroxide content (Figure 2) on both series of experiments. The crystals formed at higher alkalinity were however smaller in the S-series (12 x 13 μm compared to 20 x 24 μm for the T-series) and the optimum size was attained at a value of the OH/SiO_2 ratio of 0.02 (instead of 0.01 for the T-series) confirming then the decisive influence of the cations on the crystallization kinetics.

In reaction media of extremely low alkalinity (prepared by using salts instead of hydroxides as the only source of alkali ions (8,9) or eventually neutral or acidic (prepared with fluoride ions (10) very large crystals reaching 400 to 500 μm in size are commonly produced.

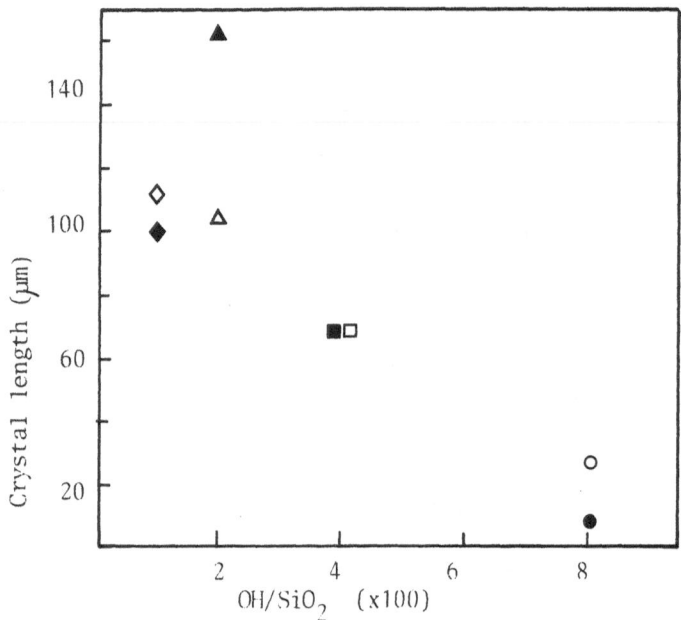

Fig. 2. Influence of the hydroxide concentration on the size of silicalite crystals. Hydroxide source : full symbols, NaOH ; open symbols, TPAOH. (Ref. 7, data taken from Table 2).

As said above, the case of the MFI structure must be regarded as an exception. Usually, zeolite crystals are metastable in their synthesis medium and the composition field in which they can be isolated as pure products is narrow. Little possibility is then left to gain significant variations in sizes through changes of the stoichiometry of the initial reaction mixtures. In those situations seeding may be used. In the absence of secondary nucleation, the volume of the final crystals is roughly determined by the amount of seeds added as demonstrated by Kacirek and Lechert (11) for the growth of zeolite NaY on seeds of NaX. Initial fractions of seeds (crystals of NaX with a mean radius of 0.23μm) of 0.044, 0,44 and 4.4%, resulted in crystals with mean radii of 2.7, 1.25 and 0.6 μm respectively. The amount of new crystal deposited on each seed was therefore proportional to the total surface of the seeds.

Poorly crystallized solids, fragments or dust of crystals show a better efficiency for seeding than large well-formed ones. Seeding may also be achieved by using an (X-ray) amorphous mixture previously aged or a portion of the mother liquor from a previous synthesis.

Pre-ageing the gels at low temperature before crystallization is a related method which can lead to a decrease of the final crystal sizes. While the gel is maintained at low temperature, crystal growth is prevented, allowing for a larger number of viable nuclei to form. Figure 3 shows data taken from the work of Zdhanov and Samulevich (12) in the case of NaX. By room temperature ageing of the starting mixtures before their crystallization at 90°C the authors were able to decrease the size of the crystals by a factor of ten. If the volume of the crystals is considered, it can be calculated that the number of nuclei which developed after 70 days of ageing was three orders of magnitude greater than in the fresh gel.

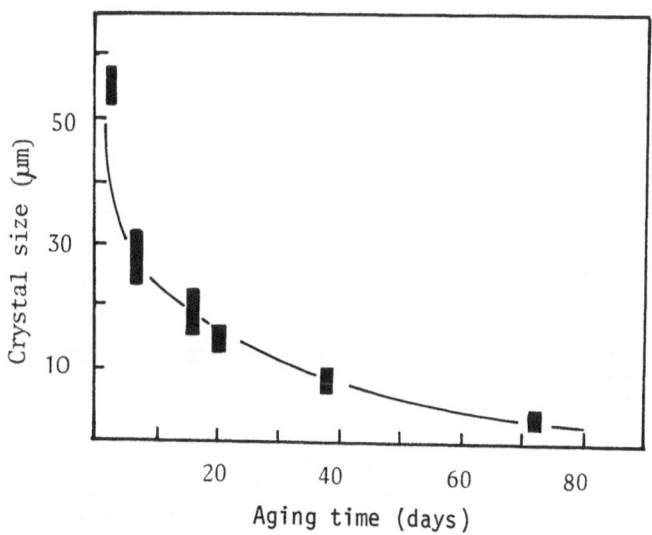

Fig. 3. Influence of the time of aging at room temperature on the size of NaX crystals grown at 90°C (ref. 12).

In summary, this brief survey shows that various methods exist for modifying zeolite particle sizes. It is difficult to say, however, that the synthesis science has reached the point were crystal sizes can be really controlled. Such a control would require a better knowledge of the phenomena associated to the nucleation step. As long as this knowledge will be missing, particle size distributions will be better explained than controlled.

II-2 Control of crystal habit

The high supersaturations needed for zeolites to form make that the first crystals which appear are spherical. Only at the end of the crystallization the consumption of nutrient allows low-index faces to develop. Crystals with plane faces and an overall shape correlated to the symmetry group appear therefore during the final period of growth, under low supersaturation. The use of starting materials with low solubilities provides favourable conditions for the formation of crystals exposing low-index faces. Such an approach has been followed in our laboratory for the synthesis of hexagonal crystals of zeolite omega (13,14). Crystallization under low supersaturation was achieved by using kaolinite as aluminium source. In order to avoid secondary nucleation, the mixtures were seeded by a portion of a reactive gel previously aged to contain nuclei of omega (15). In all experiments, the growth of the crystals resulted in a slow decrease of the amount of aluminium dissolved (Figure 4).

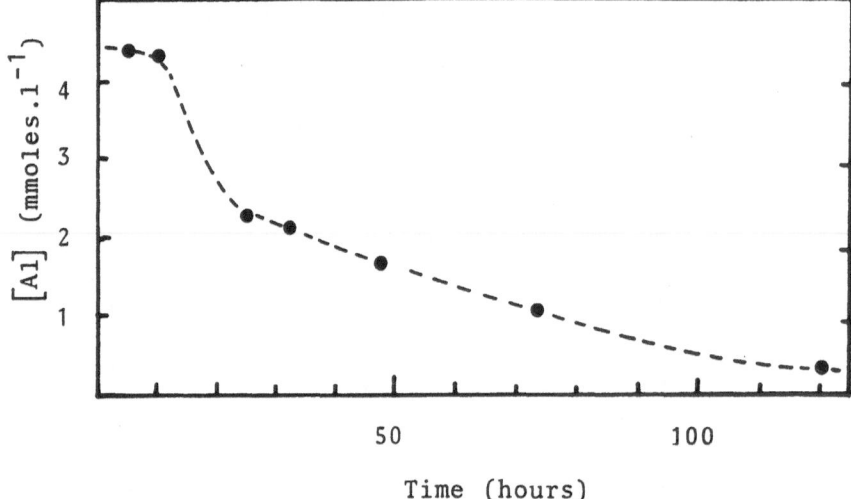

Time (hours)

Fig. 4. Change in the concentration of dissolved aluminium during the hydrothermal conversion of kaolinite into zeolite omega at 115°C (ref. 14, Fig. 5).

As shown by Figure 5, a clear correlation was established in the temperature range 100-135°C between the aluminium concentration in the solution and the variation of crystal habit. Spherulites featuring high-index faces existed at the higher supersaturations. When the aluminium content became lower than 4-5 mmol/l, (001) faces appeared and the spheres evolved into barrel-shaped particles bounded by (hk0) faces. Below 2-2.5 mmol/l, (100) faces formed and the habit evolved to euhedral hexagonal. No further growth in the <100> direction was noticed when the crystals were maintained in the medium up to the complete dissolution of the kaolinite. Crystal growth occurred solely in the <001> direction, leading ultimately to elongated prisms with aspect ratios (length/width) greater than 3. The observed evolution in morphology is consistent with the overlapping principle : the crystal habit is determined by the faces with the lowest growing rate (16).

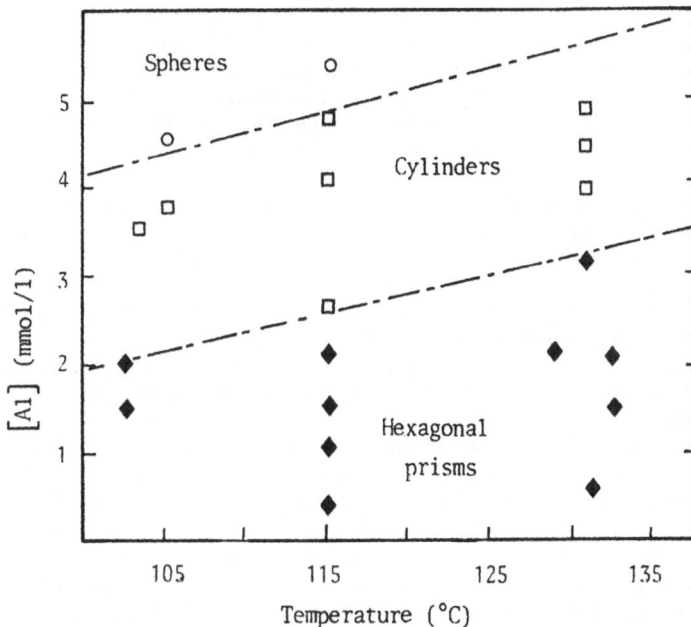

Fig. 5. Evolution of crystal habit of omega as a function of the amount of aluminium in the liquid phase. Hydrothermal conversion of kaolinite (ref. 14, Fig. 6).

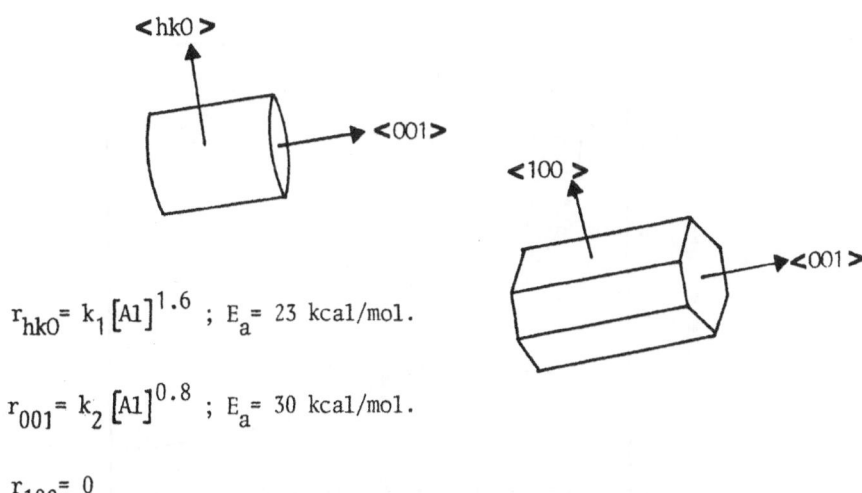

$$r_{hk0} = k_1 [Al]^{1.6} \; ; \; E_a = 23 \text{ kcal/mol.}$$

$$r_{001} = k_2 [Al]^{0.8} \; ; \; E_a = 30 \text{ kcal/mol.}$$

$$r_{100} = 0$$

Fig. 6. Kinetic laws for the development of the various faces of crystals of zeolite omega. (ref. 14)

The growth rates (in nm/h) were expressed as a function of the aluminium concentration and the rate equations summarized in Figure 6 were found. From these data one can see that the aspect ratio of the crystals (which correspond to the $<001>/<hk0>$ size ratio) can be controlled by adjusting a single parameter of the system : the concentration of aluminium. In effect, elongated crystals (high aspect ratios) will be produced if they are grown at an aluminium concentration lower than c.a. 2 mmol/l, whereas low aspect ratios will require that growth be conducted in an optimal range of concentration between 2.5 and 4 mmol/l. The practical way in which we realized these objectives was the use of initial synthesis media containing only the aluminium present in the seeding mixture. Once crystallization started, an aluminate solution was injected in the autoclave by a volumetric pump in order to maintain the aluminium concentration at the desired level. The expected increase in growth surface was calculated using the above equations and the flow rate of the injected solution was continuously adjusted in order to balance the incorporation of nutrient by the crystals.

Samples of hexagonal crystals of zeolite omega with aspect ratios ranging from 8 to 1.3 were thus obtained in quantities of 30-50 g per preparation.

It is out of the scope of this lecture to draw mechanistic conclusions out of these data. However, the different activation energies and dependences on aluminium concentration for the growth rates on the (001) and (hk0) faces support the view that different growth mechanism apply to the various faces or, at least, that different aluminosilicate species are participating in the growth along the different directions.

A similar conclusion has been reached in the work on silicalite crystallization by Hayhurst et al. (7) already cited. The dependence of growth rate on hydroxide concentration was found greater for the a and b directions than for the c direction. The practical result of this, in relation with crystal habit modification, is the existence of a direct and unique relationship between the hydroxide content and the aspect ratio of the particles (Figure 7).

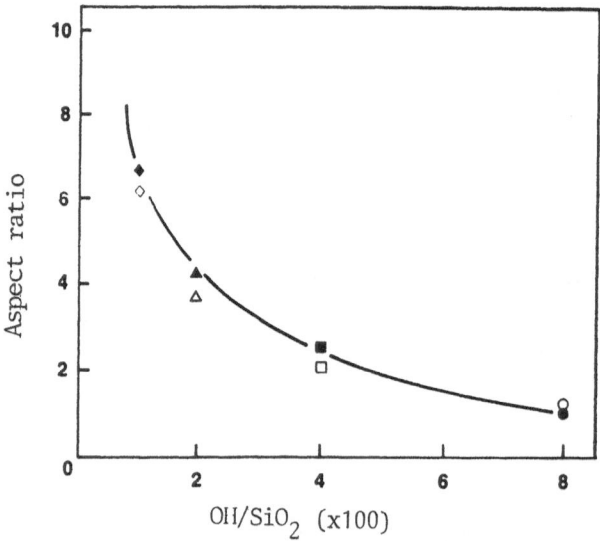

Fig. 7. Aspect ratios of silicalite crystals as a function of hydroxide content. Same meaning of symbols than Fig. 2 (ref. 7, Fig.4).

The situation presented in these investigations, showing that non-uniform crystallization kinetics operate on a same crystal, is probably common to most zeolites. Kinetic studies aiming at the determination of the parameters governing crystal growth should permit to achieve controlled (predictible) changes in crystal habit. These measurements represent indeed quite an important investment. It should be kept in mind however that, since reaction rates must correlate directly with the concentration of the basic units involved in crystal growth, kinetic studies are expected to be also of great significance for the elucidation of the mechanisms of zeolite formation.

III COMPOSITION GRADIENTS IN ZEOLITES

III-1 Composition gradients

The first experimental evidence for the presence of composition gradients (zoning) in zeolite crystals has been provided by von Ballmoos and Meier (17). Electron microprobe scanning of large ZSM-5 crystals revealed that crystal cores contained ten times less aluminium than the outer parts in both longitudinal and lateral directions (type I gradient). Three additional aluminium profiles have been proven now (7) : enrichment of the core of the crystals (type II), homogeneous distribution within the particle (type IV) and a combination of types IV and I in which aluminium is present in all the crystal but with a net enrichment near the rim (type III). Type I gradients have been attributed to a solution phase mechanism where all the aluminium of the mixture is trapped in the gel and nearly Al-free ZSM-5 nucleates in solution. Silica is then consumed preferentially. As the silica content decreases, the gel dissolves and more and more aluminium is progressively incorporated by the crystal. The gradients of type IV and II have been associated with a crystallization process via gel nucleation in which Al-rich species are incorporated or occluded initially. Finally, the type III gradient has been interpreted by the epitaxial growth of an Al-rich phase on the surface of preformed crystals.

The relevant parameters which can be used for the design of a given Al profile in a given ZSM-5 morphology have been discussed by Jacobs and Martens (4). Among them the nature of the silica source, the basity and the aluminium content of the medium play a dominant role. One main conclusion of the authors is that large crystals without an aluminium gradient should be very difficult to obtain.

Composition gradients can also be introduced into crystals when seeds with a different composition than the crystallization mixture are used. NaX seeds are frequently used to initiate the crystallization of zeolite Y (11,18) and they, of course, introduce a zoning of aluminium. Monocrystals of ZSM-5 with a layered composition have been similarly prepared by growing siliceous external shells on the surface of aluminium containing crystals (19).

III-2 Distribution gradients

A zoning of a different type has been evidenced recently (20). It concerns the distribution of aluminium among the crystallographic sites of zeolite omega. The framework of omega is composed of tetrahedra located in two non-equivalent crystallographic sites corresponding to the four-membered rings (site A) and the six-membered rings (site B) of the structure. There are 24 sites A and 12 sites B per unit cell. The occupancy of both sites by aluminium atoms in intermediate samples of products collected during the hydrothermal conversion of kaolinite (see section II-1) has been examined by high field ^{27}Al MAS NMR spectroscopy.

Figure 8 presents the spectra of samples collected at 50% (Fig. 8a) and 90% (Fig. 8b) conversion. The deconvolution of the two peaks of the

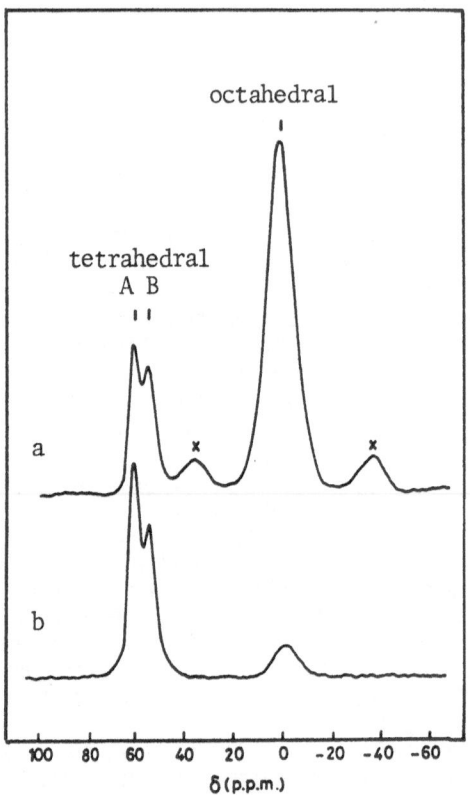

Fig. 8 ^{27}Al MAS NMR spectra at 104.27 MHz, crosses indicate spinning side bands of the 0 ppm signal of kaolinite (ref. 20, Fig. 2).

tetrahedral aluminium at 50-65 ppm (the peak at 0 ppm corresponds to the octahedral aluminium of kaolinite) gave values of 1.10 in the first sample and 1.25 in the second indicating that the distribution of aluminium between sites A and B was not homogeneous in the crystal. The preferential incorporation of Al in the six membered rings (site B) in the earlier stages of the crystallization was attributed to growth under more supersaturated conditions. This hypothesis was confirmed by a systematic investigation of the Al distributions in samples of pure omega grown under various conditions (21). The six membered rings of the structure were definitely the more aluminous the more reactive the silicon and aluminium sources.

Since both the strength of the acid sites and the thermal and hydrothermal stabilities of the zeolites are influenced by the location of the aluminium atoms (22-24), the detection of composition or distribution gradients in crystals is of particular interest. Crystals which are large enough to permit direct analysis give generally clear answers. It is nevertheless likely that gradients also exist in smaller crystals, which are materials of more practical interest for sorption and catalysis but much less easy to characterize.

IV - INFLUENCE OF MORPHOLOGY AND GRADIENTS ON PHYSICOCHEMICAL AND CATALYTIC PROPERTIES

IV-1 Influence on catalysis

It is a widespread belief that, in reactions catalyzed by zeolites the volume of the crystals is not fully utilized. The two reasons which can explain that the reactions occur preferentially in the regions close to the surface of the crystallites are an inhomogeneous distribution of the active sites due to the presence of gradients and a control of the kinetics by diffusion.

Studies dealing with the effect of the crystal size on catalytic activity are numerous and often contradictory conclusions have been reached. No indications for inhibition by diffusion and accordingly no influence of the crystal size was reported for ZSM-5 catalysts in the cracking of n-hexane (25,26) and 3-methylpentane (26) or the conversion of dimethylether (27). By contrast Ratnasamy et al. (28) for instance reported a decrease in activity in xylene isomerization with increasing crystal size. A similar observation was made by Herrmann et al. (29) for the isomerization of n-hexane, the conversion of methanol to olefins and the amination of methanol but the authors could not attribute the decrease of activity to the increasing size of the crystals or to a surface depletion in active sites. This kind of difficulty arises from the fact that activities and selectivities in zeolite catalyzed reactions are highly influenced by the Si/Al ratio (30) (31).

The influence of crystal size on reaction kinetics has been analyzed by Chutoransky and Dwyer (32) in the isomerization of o-xylene on two Mobil aromatic processing catalysts with particle sizes of 0.2-0.4 μm and 2-4 μm. In the absence of diffusion limitations (small crystals) the data were consistent with a linear reaction scheme :

$$\text{o-xylene} \rightleftharpoons \text{m-xylene} \rightleftharpoons \text{p-xylene}$$

When diffusion-resistance existed, with the larger crystals, the m-xylene molecules remained in the crystallite long enough for several reaction steps to occur and a more complex reaction scheme resulted in accordance with the diffusion theory developed by Wei (33).

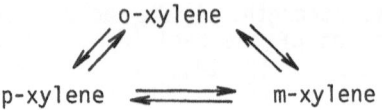

A change in the apparent kinetics -i.e., those determined from the product distributions in the bulk phase- constitutes a major argument to prove a crystal size effect.

Many other manifestations of crystal size effects, such as increased para-selectivities in the conversion of aromatics, increased deactivation rates, changes in the degree of branching of hydrocarbons, increase of hydrogen transfer rate, are expectable and often observed. They are all in line with an enhancement of the molecular shape selective behaviour of this type of catalysts (5,34).

Differences in the aluminium distribution have been put forward to explain differences in selectivity between ZSM-5 catalysts prepared in the presence or absence of template (35). The ethylbenzene disproportionation reaction was used as a test. Microprobe analyses of the crystals synthesized in a template free system (a) and in the presence of TPA (b) are shown in Figure 9.

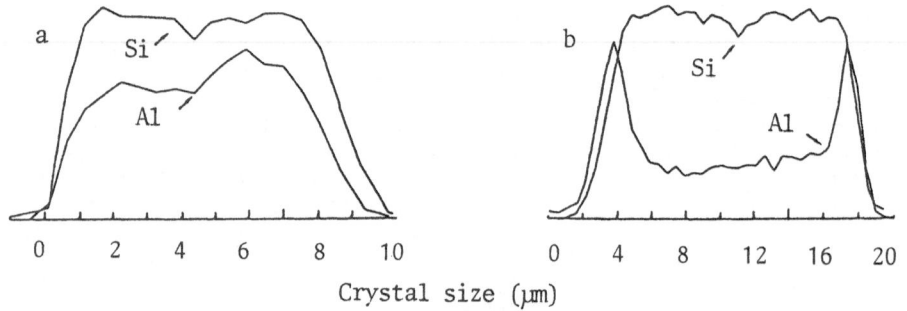

Fig. 9. Aluminium and silicon distributions in ZSM-5 crystals synthesized in the absence (a) or presence (b) of TPA. (ref. 35, Fig. 6).

The catalytic activity of the solids varied linearly with the number of aluminium atoms located in the zeolite framework over a broad range of Si/Al ratio between 20 and 50 suggesting a full utilization of the crystal during the reaction. When the selectivities were compared, the solids synthesized with templates yielded a product composition close to 60/40 (w/w of meta-diethylbenzene to para-diethylbenzene) whereas the ratio reached 30/70 for the template-free catalysts. To rationalize these data it was assumed that, due to their more homogeneous aluminium distribution across the crystal, the samples synthesized without template would contain less catalytically active sites at the external surface. Surface reactions, which are known to decrease para-selectivities, would be then minimized in these solids.

Morphology and gradient effects in catalysis are receiving more concern in the recent literature. Both are difficult to evidence clearly because they intervene essentially in processes which are already controlled by diffusion. Moreover the activation treatments required to develop the activity may severely influence the aluminium distribution in the crystals as well as the topology of the frameworks. The size limitation in actual catalysts outlined in the last paragraph of section III-2 and the possible presence of defects call for cautious interpretations.

IV-2 Influence of crystal habit on thermal stability

Among the physical properties which determine the commercial acceptability and success of a zeolite, thermal stability is of primary importance. The main parameters which alter the thermal stability of a zeolite are the silica content, the size of the particles, the type of compensating cation and the degree of exchange (36). The influence of the habit of the crystals has been studied with offretite (37). Single crystals appearing as rods (rice-like morphology) or hexagonal prisms have been synthesized in the systems Na-K-TMA (TMA=tetramethylammonium) and K-TMA respectively. The crystals had comparable sizes (1.5 x 0.8 μm) and silica contents (Si/Al $\approx$ 3) and the absence of faulting was ascertained by sorption measurements and transmission electron microscopy. The thermal stability in air of homoionic (K+) forms of the two zeolites is compared in Figure 10.

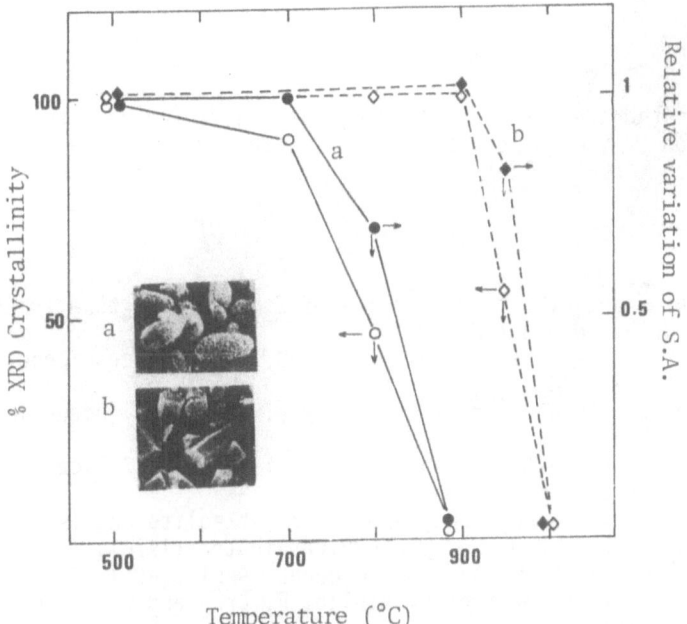

Fig. 10. Thermal stability of offretite.
a/ rod-like crystals, b/ hexagonal prisms
(ref. 37, Fig. 5).

Both X-ray crystallinities and surface area measurements revealed that the hexagonal crystals exhibited a much higher stability than the rod-like particles. In the ammonium form, structural degradation started below 600°C in the latter whereas the former did not suffer any damage up to 750°C. The practical application of this difference is obvious for the preparation of acidic catalysts.

The higher thermal stability of the hexagonal crystals most probably reflects the higher thermodynamic stability of flat (low index) surfaces. The surface topography of the particles would be then a significant factor in stability. In that respect, zeolites would therefore behave like amorphous silica-alumina or metals for which the ability of sintering is directly related to the radius of curvature of the grains.

V CONCLUDING REMARK

This report has emphasized some experimental approaches and means aiming at a better control of the morphology and composition of zeolite crystals by synthesis in order to tailor solids with predetermined properties.

Examples have been given showing how the above parameters can alter the catalytic behaviour of zeolites. The morphology of the crystals is thought to play a major role for many other applications based on the sorption and ion-exchange properties of zeolites.

The fundamental concepts which must be applied to achieve morphology changes are known. They correspond to the general laws of crystallization from solutions. Considerable progresses have been made for the crystal growth step- which determines the crystal habit- but much remains to be done for nucleation.

ACKNOWLEDGEMENT

The author thanks the following publishers for permission to use a reproduction of the following figures : Elsevier (Fig. 1,9), Butterworth (Fig. 7), the Royal Society of Chemistry (Fig. 8) and Gauthier Villars (Fig. 10).

REFERENCES

1. W.M. Meier and D.H. Olson, "Atlas of Zeolite Structure Types", Butterworth, London (1987).
2. D.W. Breck, "Zeolite Molecular Sieves", Wiley Interscience, New York, (1974).
3. R.M. Barrer, "Hydrothermal Chemistry of Zeolites", Academic Press, London (1976).
4. P.A. Jacobs and J.A. Martens, "Synthesis of High-Silica Aluminosilicate Zeolites" Elsevier, Amsterdam (1987).
5. S.M. Csicsery, in:ACS Monograph N° 171, "Zeolite Chemistry and Catalysis", J.A. Rabo, ed., p.680,Washington (1976).
6. Z. Gabelica, N. Blom and E.G. Derouane, Appl. Catal. 5:227 (1983).
7. D.T. Hayhurst, A. Nastro, R. Aiello, F. Crea and G. Giordano, Zeolites 8:416(1988).
8. R. Mostowicz and L.B. Sand, Zeolites 3:219 (1983).
9. J. Kornatowski, Zeolites 8:77 (1988).
10. J.L. Guth, H. Kessler and R. Wey, in: "New Developments in Zeolite Science and Technology", Proceedings 7th Intern. Zeolite Conf., Y. Murakami, A. Iijima and J.W. Ward, eds.,p. 121, Kodansha, Elsevier, Tokyo, Amsterdam, (1986).

11. H. Kacirek and H. Lechert, J. Phys. Chem. 79:1589(1975) ; 80:1291 (1976).
12. S.P. Zdhanov and N.N. Samulevich, in: "Proceedings 5th International Conference on Zeolites", L.C.V. Rees, ed., p. 75, Heyden, London (1980).
13. S. Nicolas, PhD Thesis Montpellier, France (1988).
14. F. Fajula, S. Nicolas, F. Di Renzo, C. Gueguen and F. Figueras, in: ACS Symp. Ser. N°398, "Zeolite Synthesis" M.L. Occelli and H.E. Robson, eds., ch. 34, Washington (1989).
15. S. Nicolas, P. Massiani, M. Vera Pacheco, F. Fajula and F. Figueras. Stud. Surf. Sci. Catal. 37:115 (1988).
16. R.F. Strickland - Constable in: "Kinetics and Mechanisms of crystallization", p. 7, Academic Press, London (1968).
17. R. von Ballmoos and W.M. Meier, Nature 289:782(1981).
18. E.F. Freund ; J. Crystal Growth 34:11 (1976).
19. L.D. Rollmann, US. Patent, 4 203 869 (1980).
20. P. Massiani, F. Fajula and F. Di Renzo, J. Chem.Soc., Chem. Commun. 814 (1988).
21. P. Massiani, F. Fajula, F. Figueras and J. Sanz, Zeolites 8:332(1988).
22. D. Barthomeuf, Mater. Chem. Phys. 17:49 (1987).
23. J. Dwyer, Stud. Surf. Sci. Catal. 37:333 (1987).
24. A. Goursot, F. Fajula, C. Daul and J. Weber, J. Phys. Chem. 92:4456 (1988).
25. W.O. Haag, R.H. Lago and P.B. Weisz, J. Chem. Soc., Faraday Discuss. 72:317(1981).
26. W. Holderich and L. Rickert, Chem. Ing. Tech. 58:412 (1986).
27. H.J. Doelle, J. Heering, L. Rickert and L. Marosi, J. Catal. 71:27 (1981).
28. P. Ratnasamy, G.P. Babu, A.J. Chandwadkar and S.B. Kulkarni, Zeolites 6:98 (1986).
29. C. Herrmann, J. Haas and F. Fetting, Appl. Catal. 35:299 (1987)
30. H.G. Karge, Y. Wada, J. Weitkamp, S. Ernst, U. Girrbach and H.K. Beyer, Stud. Surf. Sci. Catal. 19:101 (1984).
31. W.W. Kaeding, J. Catal. 95:512 (1985).
32. P. Chutoransky, Jr. and F.G. Dwyer, Adv.Chem. Ser. 121:540 (1973)
33. J. Wei, J. Catal. 1:526 (1962).
34. E.G. Derouane in: "Zeolites Science and Technology" NATO ASI Series, N°80, F.R. Ribeiro, A.E. Rodrigues, L.D. Rollmann and C. Naccache, eds.,p. 347, Martinus Nijhoff, The Hague (1984).
35. A. Tissler, P. Polanek, U. Girrbach, U. Müller and K.K. Unger, Stud. Surf. Sci. Catal. 46:399(1989).
36. C.V. Mc. Daniel and P.K. Maher, in: ACS Monograph N°171, "Zeolite Chemistry and Catalysis", J.A. Rabo, ed.,p.285, Washington (1976).
37. F. Fajula, L. Moudafi, R. Dutartre and F. Figueras, Nouv. J. Chim. 8:207(1984).

NEW MOBILIZING AND TEMPLATING AGENTS IN

THE SYNTHESIS OF CRYSTALLINE MICROPOROUS SOLIDS

J. L. Guth, P. Caullet, A. Seive, J. Patarin, F. Delprato

Laboratoire de Matériaux Minéraux URA-CNRS-428
ENSCMu, 3, rue A. Werner, 68093 MULHOUSE CEDEX (France)

INTRODUCTION

The structure of crystalline microporous solids, such as zeolites[1] and molecular sieves[2], consists of :
- a framework TO_2 (T = Si, Al, P...), each TO_4 tetrahedron being linked to four others through T-O-T bridges
- regular micropores of molecular size (containing species such as cations, water, ionic pairs) which communicate with the environment.
These ionocovalent macromolecular solids are sparingly soluble in the aqueous solutions corresponding to their formation media (Table 1).

If the different constituents, and especially the framework-forming elements, are mixed in their soluble form (e. g., concentrated silicate and aluminate solutions) no precipitation of crystalline material takes place generally. The fast condensation rate between the species in the solution cannot lead to a complex organized structure. An amorphous solid, forming a "gel" with the aqueous phase comes into existence. This gel will be the source of reactants, leading mainly to the framework-forming elements, involved in the crystallization reaction.

There is general agreement that the crystallization of zeolites and related materials occurs in the solution[5-8]. The latter is supplied by continuous dissolution of the solid phase of the gel as the crystallization proceeds. The presence of a mobilizing agent is required in order to increase the dissolution rate and to bring the solution into the supersaturated state needed for the nucleation and the crystal growth. Thus, if the framework is based on silicon associated with aluminum, the hydroxide anions are generally used as the mobilizing agent. They bring about the dissolution of silica and alumina thus producing silicate and aluminate anions.

Table 1. Solubility of zeolite A and analcite in aqueous NaOH (0.5 mol. 1^{-1}) solution at 60°C[3, 4]

Name	Composition	SiO_2(g. 1^{-1})	Al_2O_3(g. 1^{-1})
Zeolite A	Na_2O, Al_2O_3, $2SiO_2$, $4.5H_2O$	0.41	0.34
Analcite	Na_2O, Al_2O_3, $4SiO_2$, $2.0H_2O$	0.14	0.06

Guidelines for Mastering the Properties of Molecular Sieves
Edited by D. Barthomeuf *et al.*
Plenum Press, New York, 1990

During crystallization, the micropores are filled with species present in the solution and called templating agents which control the build-up of the framework and contribute to the stabilization of the structure[9].

In the former part of this paper the effects of conventionally used OH^- and of new F^- mobilizing agents are compared. The latter part displays results concerned with a new family of templates which provides a means for preparing silica-rich zeolites of the faujasite type.

MOBILIZING AGENTS : OH^- and F^- ANIONS

The role of the mobilizing agent is to provide the solution with the T elements required to build the framework, in the form of "useful species" and at an "adequate rate". "Useful species" refers to species which may be incorporated into the framework, i.e. with condensable ligands ($\equiv T-OH$) and with the T element in tetrahedral coordination. The achieved concentration levels of the "useful species" must be high enough to bring the solution out of equilibrium, in order to allow nucleation and crystal growth. These conditions must be kept constant throughout the crystallization by an "adequate" dissolution rate. Table 2 shows the compositions of the synthesis mixtures and the reaction conditions used when OH^- or F^- is the mobilizing agent.

Chemistry of Silica and Alumina Solutions in the Presence of OH^- Anions. Some Consequences in the Synthesis of Aluminosilicate Zeolites

At the neutral pH the only soluble form of silica is monosilicic acid $Si(OH)_4$. As the pH increases, the solubility of silica increases through the ionization of the silanol groups and the hydrolysis of oxygen bridges according to the following reactions

$$\equiv Si - OH + OH^- \rightleftharpoons \equiv Si - O^- + H_2O$$

$$\equiv Si - O - Si \equiv + HO^- \rightleftharpoons \equiv Si - O^- + HO - Si \equiv$$

Silicate solutions, thus, contain a whole series of anions with different degrees of polymerization and ionization[9].

Table 2. Compared synthesis conditions (OH^- or F^-)

Parameters	OH^- medium	F^- medium
pH	> 8	< 12
OH^- or F^- sources	- Bases (e.g., NaOH) - Salts and molecules yielding OH^- in water (Na_2CO_3, amines...)	- Acids (e.g., HF) - Salts and molecules yielding F^- in water (NH_4F, BF_3...)
T element sources	Oxides, hydroxides, alkoxides, salts (amorphous or crystalline solids)	
Templates	Ionic compounds : e.g., Pr_4NBr	
Temperature	> 20°C	> 40°C
Duration	several hours to several days	

The behaviour of aluminum in alkaline media is quite different, because of the non-ionizable Al-OH groups. That is why, in such solutions, aluminum occurs mainly in the form of tetrahedral $Al(OH)_4^-$ anions[9]. In basic solutions containing both silica and alumina, aluminosilicate anions are formed in addition to silicate and aluminate ions[10].

Finally, as the pH increases, the silanol functions are more and more ionized (the $[\equiv Si-OH/\equiv Si-O^-]$ ratio decreases), while the four Al-OH groups of the $Al(OH)_4^-$ ion remain unchanged. For instance, if one compares an alkaline silicate solution with an aluminate solution, both having the same concentration of monomeric species (i.e. $[$monomeric silica$]$ = $[Al(OH)_4^-]$), calculations show that the monosilicic acid concentration level is much lower than that of the $Al(OH)_4^-$ ion, and decreases quickly with increasing pH[11]. As the silicate and aluminate anions undergo condensation reactions, mainly through the hydroxide functions, the ability to condensation decreases with increasing pH in the case of silicate species, and remains constant for the $Al(OH)_4^-$ ions. Thus, the Si/Al average ratio of the supposedly formed aluminosilicate species is expected to decrease with increasing pH, all other chemical parameters remaining unchanged. This dependence was confirmed by means of a computation model. This model makes it possible to calculate the concentration of any oligomer species in alkaline aluminosilicate solutions, provided the concentrations of monomeric silica and $Al(OH)_4^-$ (and pH) are known[11]. As an example the distribution of cyclic tetramers is given against pH in figure 1.

As aluminosilicate species are probably involved in the building of the zeolite framework which takes place during the synthesis, the same dependence similar to the above-mentioned one is expected between the Si/Al ratio of the resulting material and the pH of the crystallization medium. Figure 2 shows, indeed, that the incorporation of Al atoms in the framework, here in the case of chabazite, is increasingly favoured with increasing alkalinity of the reaction mixture[12].

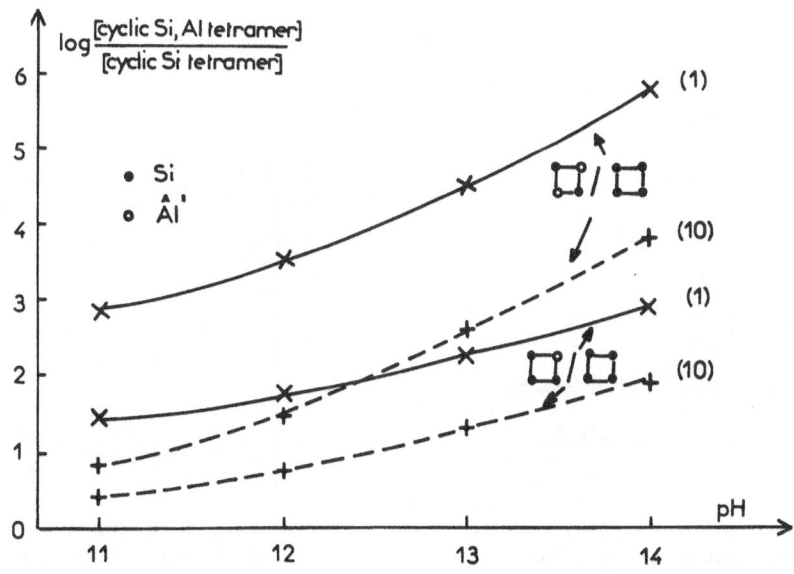

Fig. 1 . Calculated distribution of cyclic tetramers ($[$monomeric silica$]$ + $[Al(OH)_4^-]$ = 0.002 mol.1.$^{-1}$) (numbers in brackets refer to the concentration ratios monomeric silica / $Al(OH)_4^-$)[11].

It is worth noting that, in some cases, the pH of the synthesis medium is not constant and actually increases during crystallization. For instance, the pH changes during the synthesis of the EU-1 zeolite are shown in figure 3[13]. This phenomenon can be explained as follows : the many $\equiv$T-OH and$\equiv$T-O$^-$ groups still present in the gel must condense during the transformation of the latter into a crystalline product, thus giving rise to the release of the hydroxide anions. Figure 4 displays schematically the proposed interpretation. This pH increase induces an heterogeneity (inter- and (or) intracrystalline ?) of the resulting crystals, just as for the Si/Al ratio[14].

The pH of the reaction mixture must be high enough to allow crystallization and, therefore, zeolites with a very high content of silica cannot be easily prepared in alkaline media. Two parameters, the pH and the nature of template may be used to increase the Si/Al ratio of the synthesized products :
- as the pH decreases, mordenite could, for instance, be prepared with Si/Al ratio up to 10, in sodium containing media[15]
- the use of "specific" templates able to interact strongly with the zeolite framework makes it possible to prepare silica-rich zeolites, so that a purely silicic product, silicalite, has been obtained in the presence of tetrapropylammonium hydroxide[16]. In that case, no coulombian interaction arises between compensation cations and the negative charges of the framework, and other types of interaction (e. g. Van der Waals forces) are instrumental in the stabilization of the structure.

As a general rule, it appears that higher synthesis temperatures are required to prepare silica-richer zeolites. The hydrolysis of Si-O$^-$ groups might be favoured by an increased temperature.

High OH$^-$ concentrations prevent complete $\equiv$Si-O-Si$\equiv$ bridging in silica-rich zeolites such as silicalite I[17]. $\equiv$Si-O$^-$M$^+$ defects (M = H, alkaline or organic cations) are formed. Fig. 5a shows the ^{29}Si NMR MAS of an as-synthesized MFI-type zeolite (Si/Al>1000) prepared at pH 12 and 200°C. The molar composition of the reaction mixture was as follows: 1 silica (Aerosil) : 0.15 Pr$_4$NBr ; 0.2 NaOH ; 20 H$_2$O .

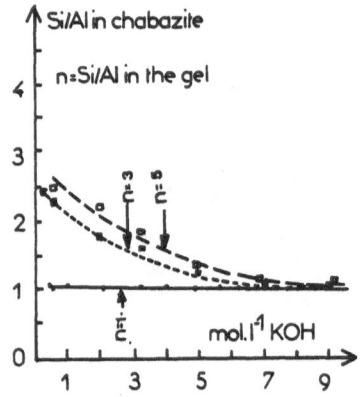

Fig. 2. Dependence of Si/Al ratio of chabazite vs. pH of synthesis medium[12]. (From R. M. Barrer, ref. 12, p. 156)

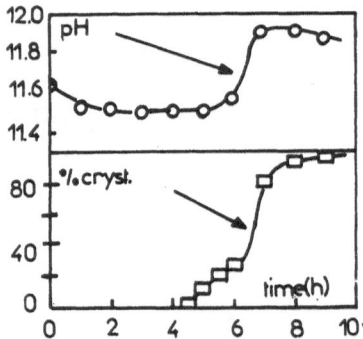

Fig. 3. Time-dependence of pH and % of cryst. during synthesis of EU-1 zeolite[13]. (From J. L. Casci and B. M. Lowe, ref. 13, p186)

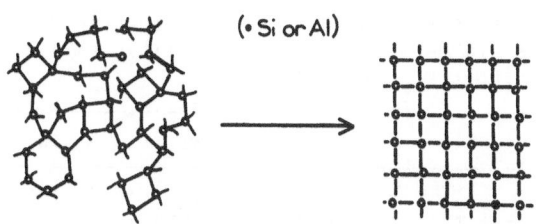

(• Si or Al)

Incompletely condensed gel   crystal
(with many ≡Si-OH and
≡Si-O⁻ groups)

Fig. 4. Schematic interpretation of pH increase during zeolite synthesis

The presence of defects is responsible for the non-resolved Si(4Si) signal centered at -112 ppm (vs. T.M.S.) and for the signal around -103 ppm which corresponds to the incompletely bridged silicons. Calcination of the sample at 550°C transforms the organic cations into protons. In the absence of alkaline cations (e.g., sample prepared with Pr₄NOH and without NaOH), this treatment allows the healing of the defects by condensation of two vicinal ≡Si-OH groups. The ²⁹Si NMR MAS spectrum (Fig. 5b) is then well resolved. If M⁺ is mainly an alkaline cation, as in the above mentioned sample, which contains 2.5 Na⁺/u.c., only partial healing takes place. The ²⁹Si NMR MAS spectrum is somewhat better resolved than before calcination, but the CP-MAS spectrum still shows the presence of non-bridging groups. After a cation exchange against H⁺ and a second calcination the ²⁹Si NMR MAS spectrum is then almost the same as presented in Fig. 5b.

The Use of F⁻ Anions in Zeolite Synthesis

The nature and concentration in solution of fluorosilicate[18], fluoroaluminate[19] ions and of other fluorocomplexes of trivalent (e.g., Fe[III], Ga[III])[20] or tetravalent (e.g., Ge[IV], Ti[IV])[21] elements are not well-known. In the presence of F⁻ anions, the solubility of these elements is increased through the formation of complexes, F⁻ replacing OH⁻ or H₂O. At high pH values, such complexes hydrolyse and at low pH values, water can also act as ligand. As a rule, octahedral coordination of these complexes takes place in the case of high F⁻ contents and in neutral to acidic solutions. We are in the opinion that tetrahedral coordination in hydroxide complexes of elements such as silicon or aluminum changes progressively to octahedral coordination with incrasing F⁻ concentration.

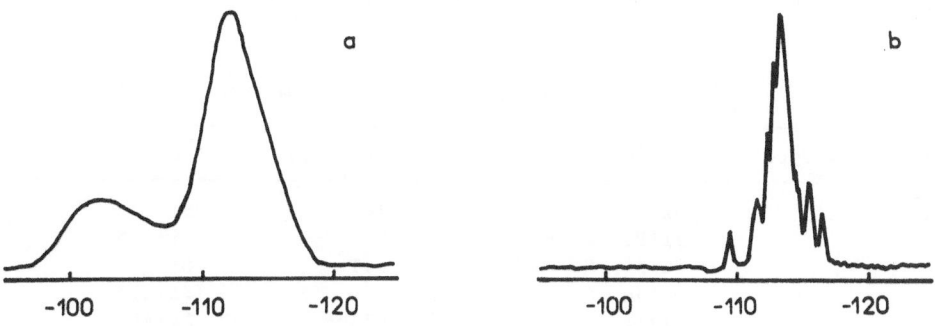

Fig. 5. ²⁹Si NMR spectra of silicalite a) as-synthesized b) calcined after synthesis in absence of alkaline cations (ppm/TMS)

Table 3. Mean number N of F⁻ bonded to one T element (pH = 5, overall F⁻/T =6, T =10^{-2}mol.l⁻¹) and solubility (mol.l⁻¹) of T in solution without F⁻, at the same pH

T	Si	Al	Fe	Ga
N	1,5	4,3	2,5	1,9
Solubility of T without F⁻	2.10^{-3}	1.10^{-5}	1.10^{-11}	1.10^{-8}

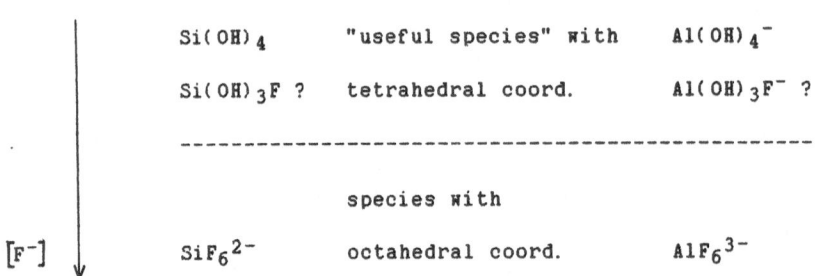

The free fluoride concentration was measured in our laboratory, by using a specific fluoride anion electrode, for solutions with an overall F⁻/T ratio of 6 (T = Si, Al, Fe, Ga) and a T element concentration of 10^{-2} mol.l⁻¹ at pH = 5. The mean number N of fluoride bonded to one T element was then computed for the above mentioned T elements (Table 3).

In conclusion F⁻ is a possible mobilizing agent, especially for silica-rich zeolites synthesis. The substitution of Si^{IV} by T^{III} or T^{IV} elements, will be controlled by the values of the complexation constants of their fluoride complexes. Table 4 displays the zeolites obtained in F⁻ media.

A large number of MFI-type zeolites were crystallized with different templates ($Pr_xNH_{4-x}^{+}$ with x = 4 to 2 and $NH_2(CH_2)_yNH_2$ with y = 6 to 8) and by substituting different T elements for Si. $NH_2(CH_2)_7$ or $_8 NH_2$ and Pr_4N^{+} are however the best templates. They can be used to synthesize phases with or without Si substitution. The poorer efficiency of the other templates (poorer pore volume fitting ?) makes the synthesis more difficult, especially in the absence of trivalent T elements, such as Al^{III} (absence of stabilizing coulombian interaction with framework charges). Seeding is, then, necessary and mixtures of two phases are reported in some cases[22].

Table 4. Zeolites obtained in F⁻ media.

Structure type	T elements	Structure type	T elements
MFI	Si	TON	Si
	Si+Be, Si+B		Si+Al*
	Si+Al, Si+Ga	MTT	Si
	Si+Fe, Si+Ge		Si+Al*
	Si+Ti	LTA**	Si+Al
FER	Si*	NU1**	Si+Al
	Si+Al, Si+Fe...	MTN	Si, Si+Al

* Difficult to synthesize
** Non reproducible synthesis.

Table 5. pK_n as related to $TF_n^{3-n} \rightleftharpoons T^{3+} + nF^-$ (T = Ga, Fe, Al)

	Ga^{III}	Fe^{III}	Al^{III}
pK_1	4.4 - 4.9	5.2	6.1
pK_2	3.8	3.9	5

The purely siliceous MFI type zeolites, prepared with the three propylammonium templates, contain respectively $4Pr_4NF$, $4Pr_3NHF.8H_2O$ and $4.8\ Pr_2NH_2F\ 6.5H_2O$ per unit cell. Water molecules occupy the volume of the channels which are left empty by the templates carrying less than four propyl groups. The presence of water and defects, resulting from the channels being filled incompletely by the templates, is brought to light with ^{13}C and ^{29}Si NMR MAS spectra, who show lesser resolution[22].

Substitution of Si by T^{III} elements is controlled by the complexation constants[21] of the fluoride complexes of T^{III} (Table 5).

The degree of substitution can be adjusted using three parameters. Substitution increases with the
- decrease of the initial Si/T^{III} and $F^-/(Si+T^{III})$ ratios
- increase of the temperature of the synthesis.
These trends are illustrated by the examples shown in Table 6 and corresponding to the incorporation of Al^{III} in MFI type zeolites.

The homogeneity for crystals with Si/T^{III} ratios approaching 24 is quite perfect. It decreases with increasing Si/T^{III} ratio : the core of the crystal is richer in T^{III} than the outer-shell. This gradient becomes weaker from Ga^{III} to Fe^{III} and Al^{III} : a more stable fluoride complex retains the T^{III} element in the solution and, thus, favours the homogeneity of the crystals.

The concentration gradient of T^{III} in the crystal can be strongly reduced when Si and T^{III} are available in the same reactive, solid source. Fig. 6 shows SEM and X-ray emission mapping of Fe^{III} for two Fe^{III}-containing, MFI-type crystals (Si/Fe=70 prepared from two different reaction mixtures, having both the same Si/Fe ratio, equal to 70). The first mixture was a (Si, Fe)-hydrogel prepared by hydrolysis of tetraethoxysilane in the presence of $FeCl_3$, and distillation of the ethanol. The resulting crystals (Fig. 6a, b) display a homogeneous iron distribution : the dissolution of the solid gel phase during crystallization, keeps the activity of the T element species at a nearly constant level. The second mixture was a Aerosil silica mixed with a $FeCl_3$ solution. In this case, (Fig. 6c, d) the iron is distributed in the core of the crys-

Table 6. Si/Al ratios obtained in MFI-type zeolites by varying synthesis variables

Si/Al (gel)	F/Si (gel)	$\theta °C$	Si/Al (zeolite)
12	1	80	110
12	1	150	43
12	1	200	30
12	0,25	150	24
7	1	150	30
7	1	200	19
7	0,25	190	15

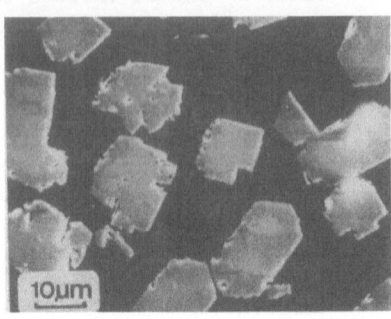
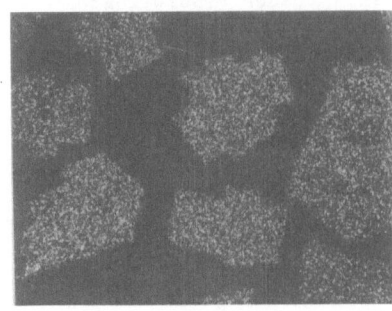
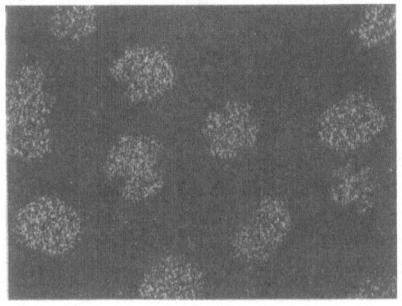

Fig. 6. SEM and X-ray emission mapping of iron
.(a) and (b) (Fe, Si)- MFI homogeneous crystals prepared from
a (Si, Fe)-hydrogel
.(c) and (d) (Fe, Si)- MFI heterogeneous crystals prepared from
a mixture of silica(Aerosil) and a ferric solution.
(From J.L. Guth et al, ref.22)

tals ; all the required iron was available in the solution and built in
right from the onset of the crystallisation.

When several T^{III} elements are competing, the T^{III} having the lowest
stability constant (Table 5) in the fluoride complexes, is incorporated
first. Fig. 7 illustrates the distribution of Ga^{III} and Al^{III} in MFI-type
crystals synthesized with the following reaction mixture :
100 SiO_2 ; 1 Al_2O_3 ; 1 Ga_2O_3 ; 25 NH_4F ; 3000 H_2O (200°C, 6 days)
Gallium is in the core of the crystals and aluminum in the outer-shell.

It is also possible to use the fluoride method with the object of
substituting tetravalent elements for silicon. With Ti^{IV}, the substitution degrees achieved ($Si/Ti \geq 20$) compare with those observed with T^{III}
elements. High substitution degrees however could be reached with Ge^{IV}
(up to $Ge/Si = 2/3$), especially for pH values between 9 and 10, the composition of the reaction mixture being as follows :
a $GeCl_4$; b SiO_2 ; HF ; 0.5 Pr_4Br ; 8 CH_3NH_2 ; 35 H_2O (a+b=2,
150°C, 15 hours)[23]. The monoclinic-orthorhombic transition temperature
of the calcined samples increases linearly from 80°C (0 Ge/u.c.) to
$\approx$ 250°C (32.8 Ge/u.c.).

Synthesis in F^- media allows direct incorporation of new compensating cations, for instance, of cations which are unstable (e.g., NH_4^+) or
sparingly soluble (e.g. Co^{2+}) at high pH. In this way, zeolites can be
prepared without any alkaline cations, and their protonic form can be
obtained after a simple calcination without ion exchange.

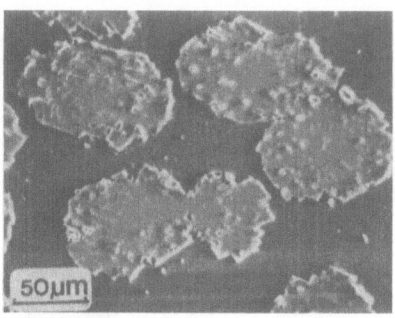

a

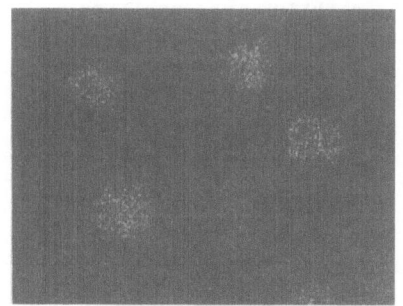

b

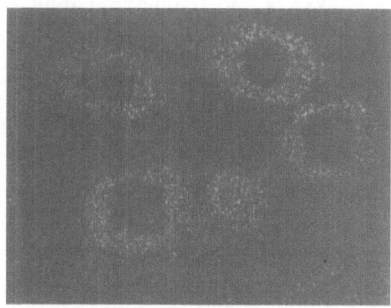

c

Fig. 7. Section across (Ga, Al, Si)-MFI crystals
. Scanning electron micrographs (a)
. X-ray emission mapping of (b) Ga and (c) Al.
(From J. L. Guth et al, ref. 22

As shown in Table 4, zeolites other than MFI-type zeolites were prepared with this new method. With the exception of one phase, they are silica-rich materials. The conditions of the synthesis and their characteristics were reported elsewhere[22].

The supersaturation achieved in the solution with respect to framework-forming species, is generally lower and more stable during the synthesis (no pH change) with F^- than with OH^- anions. This allows a better control of the crystallization :
- the number of metastable phases is lower
- the size and morphology of the crystals is easier to control through different parameters (static, agitated or seeded reaction mixtures). Fig. 8 shows microphotographs of crystals corresponding to different phases and compositions
- the number of defects is lower.
    But this requires :
- generally longer crystallization times
- optimized templates.

The factors affecting the formation of defects are high crystallization rates but also the presence of non-condensable groups on the T element species in the solution. The neutral or acidic pH prevents, for instance, the formation of $\equiv Si-O^-$ groups which are responsible of nonbridging defects in a silicon-rich framework. Fig. 9 shows the ^{29}Si MAS NMR spectrum corresponding to an as-synthesized purely siliceous orthorhombic MFI-type sample. The spectrum displays well-resolved Q^4 signals.

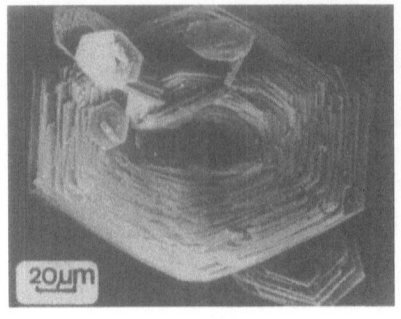
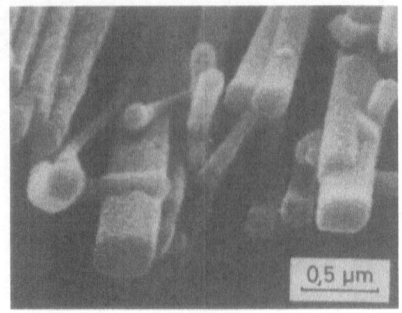

a   b

Fig. 8. Microphotographs of a) FER-type and b) TON-type zeolites

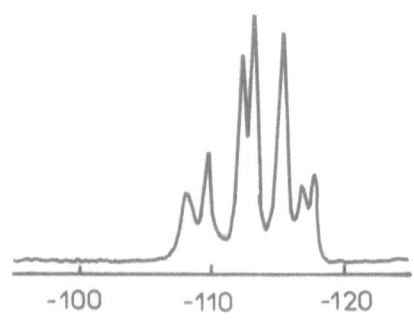

Fig. 9. ^{29}Si NMR MAS spectrum of purely siliceous MFI-type zeolites
           synthesized at pH=7, 200°C, 2 days, F/Si=0.1, Pr_4N^+/Si = 0.1,
           H_2O/Si = 30 (ppm/TMS)

It is very different from those recorded with as-synthesized silicalite-I
samples, prepared in alkaline media (Fig. 5a). After calcination, the
crystal structure becomes monoclinic, and the new ^{29}Si MAS NMR spectrum
characterizing the material is similar to that of Fig. 5b.

THE TEMPLATING AGENT

The incorporation of a templating agent in the crystalline microporous solid during synthesis
- controls the formation of the framework by a template effect involving
  the geometry (form and size) of the template
- contributes to the stability of the material by the bonds that are
  created, and makes, thus, the crystallization possible.

It should be borne in mind that the removal of the template after
synthesis generates the microporous space.

Different types of templates are in current use :
- cations balancing the framework charges : inorganic (e. g., Na^+, Ca^{2+})
  or organic (e. g. alkylammonium)
- molecules : inorganic (e. g., H_2O) or organic (e. g., amines)
- ionic pairs : inorganic (e. g., Na^+Cl^-) or organic (e. g., $Pr_4N^+F^-$).

To estimate the efficiency of a chemical compound as a template, one still has to proceed by a trial and error method. It is not possible as yet to predict the template required for a given structure and a given composition. One should be able to estimate the free enthalpy difference between the hypothetical template in the microporous crystalline solid and in the synthesis solution . Some criteria, neverthelesss, can be used to select possible templates :
- the solubility in aqueous solutions
- the stability under synthesis conditions (pH, temperature...)
- the steric compatibility with the desired microporosity (size, form...)
- the possibility of stabilizing interactions (framework charges...)
- the possibility of removing the template without destroying the framework
- (economic aspects)

Crown-ethers as Templates for Silica-rich Faujasite Synthesis

Synthetic faujasite is an aluminosilicate zeolite referred to as acronyme X (1< Si/Al<1.5) and Y (1.5<Si/Al<3)[1]. Hydrated sodium cations play the role of templates in the large ($\approx$ 12 Å diam.) and small ($\approx$ 6.5 Å diam.) cages of the cubic structure. Fig. 10 illustrates, in a schematic way the thermodynamic problem which must be solved when synthesizing a silica-rich (Si/Al>3) faujasite. If Si/Al increases (the number of Al/ u. c. decreases) water replaces partially the hydrated cations balancing the framework charges. This involves a higher free enthalpy G_t level in the solid. In order to make the crystallization possible, one must raise the corresponding G_t level in the solution by increasing the activity of the dissolved species and, especially, that of the useful framework-forming species. These is, however, a practical limitation, corresponding to a Si/Al ratio of 3 in the framework in the case where only water and hydrated sodium cations are available as templates. ZSM20 is a hexagonal polytype of faujasite with probably some cubic non ordered intergrowths[24,25]. This material is obtained in the presence of tetraethylammonium and sodium hydroxide. A higher Si/Al ratio ($\leq$5) is achieved with this organic template, but this requires a high Et_4N^+/Si ratio and the specificity is low. Zeolite ß crystallizes under similar conditions [26].

Organic molecules of the crown-ether family were selected according to the above-mentioned criteria, among charge-free species, in order to favour low-loaded frameworks. Such species are known as phase-transfer catalysts because they complex the alkaline cations[27,28]. Among the tested molecules two of them led to very attractive results :
- 15-crown-5 : 1, 4, 7, 10, 13 pentaoxacyclopentadecane with a free diameter $\approx$ 7 Å (Fig. 11a)
- 18-crown-6 : 1, 4, 7, 10, 13, 16 hexaoxacyclooctadecane with a free diameter of $\approx$ 8 Å (fig. 11b)

A gel of molar composition
    10 SiO_2 : 1 Al_2O_3 ; 2.4 Na_2O ; 1 Crown-ether ; 140 H_2O
was prepared by mixing a solution containing the sodium aluminate, the crown-ether and the sodium hydroxyde, with a silica source (Aerosil or Ludox AS 40). The gel was aged one day at room temperature and, then, heated for 6 days at 115°C in a PTFE-lined autoclave. The yield of crystals was over 90 % with reference to the reactants introduced (silica, alumina, crown-ether). According to the type of crown-ether, the crystals were identified as cubic or hexagonal silica-rich faujasite. Without crown-ether, the crystallization would require 3.0 Na_2O for the abovementioned gel composition. Seeding and agitation drastically reduce the crystallization time.

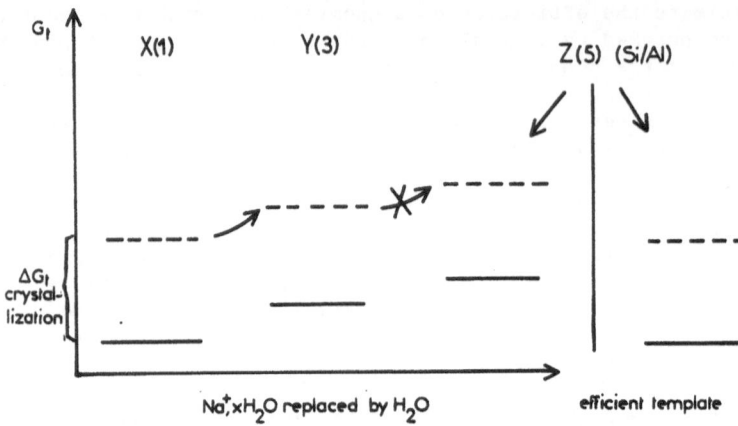

Fig. 10. Overall free enthalpy levels G_t (in arbitrary units) of the elements (framework-forming and micropore-filling elements), referred to a tetrahedron unit, in the solid (————) and in the solution (-----) for different Si/Al ratios in the solid and for different templates.

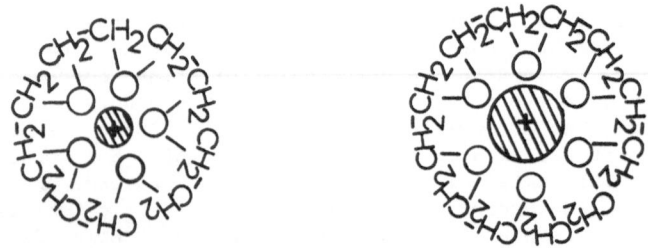

Fig. 11. (a) 15-crown-5, Ø ≈ Å and (b) 18-crown-6, Ø ≈ 8 Å with an alka-line cation

In order to get information about the mechanisms of synthesis, the changes of the pH and the Si/Al ratio were followed in the liquid and the solid phase of the gel, during the aging and the crystallization (Fig. 12). 18-crown-6 was used as an organic template and the conditions of the synthesis were the same as in the foregoing except for the value of Na_2O, which was 2.2.

Nucleation and crystallization begin in an "Al-rich" solution. ^{13}C CP MAS NMR spectroscopy results show that the molecules of organic templa-te are in this case in the liquid phase of the gel. The increases in pH value during the aging can be explained by the hydrolysis and the conden-sation reactions between the silicate and the aluminate anions

$$\equiv Si-O^- + HO-Al(OH)_3^- \; -> \; \equiv Si-O-Al\equiv^- + OH^-$$

This pH modification is normaly observed throughout the crystallization of the faujasite in the absence of crown ether[14]. The pH decrease during

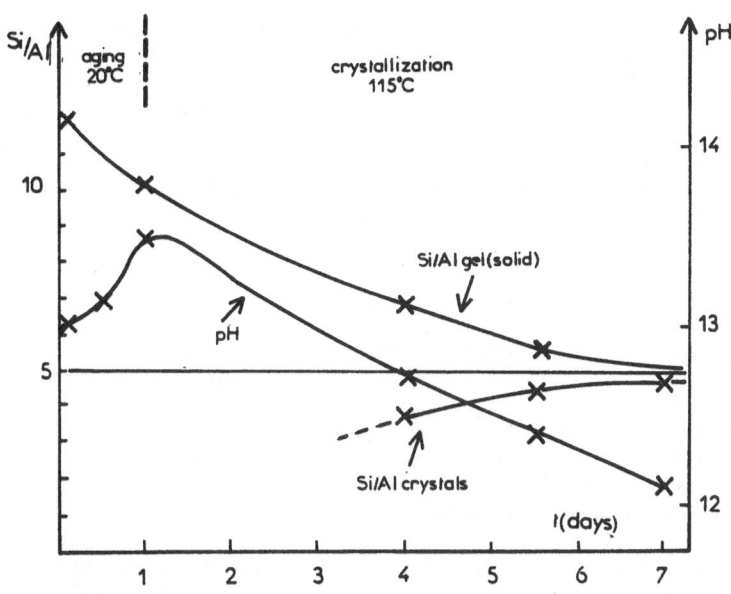

Fig. 12. Changes of the pH and the Si/Al ratio in the gel and in the crystals.

the crystallization at 115°C in the presence of the crown-ether is, therefore, explained by the formation of ion-pair complexes ([18-crown-6, Na]$^+$ OH$^-$) which are incorporated into the micropores of the zeolites. This hypothesis was confirmed by different results. The presence of the crown-ether in the structure of the zeolite was proven by means of TG and DTA curves, which show the loss of water between 25 and 150°C, and the thermal decomposition of the organic species around 300°C (when the sample is heated in air). The NMR chemical shift of ^{13}C (-69.5 ppm and - 71.5 ppm/TMS for the 15-crown-5 in the cubic zeolite and in its pure solid form ; -70.3 ppm and -73.2 ppm/TMS for the 18-crown-6 in the hexagonal zeolite and in its pure solid form) is in agreement with this conclusion. The formation of [crown-ether-sodium] complexes in the micropores is demonstrated by the ^{23}Na NMR chemical shifts observed for a series of compounds shown in Table 7. The chemical shift of ^{23}Na in the as-synthesized hexagonal faujasite (with 18-crown-6) is the same as in a non aqueous NaClO$_4$ (0.5 mol.1^{-1}) and 18-crown-6 (0.5 mol.1^{-1}) solution, where [18-crown-6, Na$^+$] complexes are present[29].

Table 7. ^{23}Na NMR (solution) and MAS NMR (solid) chemical shifts observed in Na$^+$-containing solutions and zeolites, in the presence and the absence of crown-ether.

Sample (*solid, **aqueous solution)	Si/Al	Na/Al	δppm
*NaCl	/	/	0
**(0.5M NaClO$_4$, 0.125M 18-Crown-6)	/	/	-10.3
**(0.5M NaClO$_4$, 0.5M 18-Crown-6)	/	/	-14.6
* As-synth. hex. Faujasite (18-Crown-6)	4.2	1.09	-14.0
* Calc. hex. Faujasite	4.2	1.09	-9.5
* As-synth. cub. Faujasite(15-Crown-5)	4.2	1.08	-12.9
* Calc. cub. Faujasite	4.2	1.08	-9.8
* As-synth. Na-Y	2.5	1.02	-9.8
* Calc. Na-Y	2.5	1.02	-9

81

Finaly, chemical analysis always yields Na/Al ratios above one. This is consistent with extra Na$^+$ cations, probably in the crown-ether ring, associated with OH$^-$ anions.

The Si/Al ratios were measured by electron microprobe analysis on sections of single crystals. The observed value distribution is a consequence of the pH decrease during the crystallization. The outer-shell of the crystals is richer in silica than the core.

The chemical composition and the structural characteristics of the two types of faujasite are shown in Table 8.

Scanning electron microscopy observations (Fig. 13) and X-ray powder diffraction data (Fig. 14) of a hexagonal faujasite polytype prepared according to recipe p. 11 reveal that the sample is a pure, gel-free phase with a high crystallinity (sharp and intense X-ray diffraction peaks were observed). Electron microdiffraction studies (Fig. 15) demonstrate the absence of stacking faults and that the measured c parameter is in agreement with the proposed AB sequence.

Table 8. Chemical composition and structural characteristics of the cubic and the hexagonal crown-ether containing faujasites.

	Cubic Faujasite	Hexagonal polytype
Morphology	Octahedra (2-10µm)	hex. plates (2-10µm)
Ideal unit-cell formula*	$Na_{32}(Si_{160}Al_{32}O_{384})$ $8(15$-crown-$5) \cdot 4NaOH$	$Na_{16}(Si_{80}Al_{16}O_{192})$ $4(18$-crown-$6) . 2NaOH$
Unit-cell parameters	a=24.55Å	a=17.37Å c=28.3Å
Stacking sequence	ABC	AB

* According to the original gel composition Si/Al varies between 3 and 5.2 and Na/Al between 1.07 and 1.12

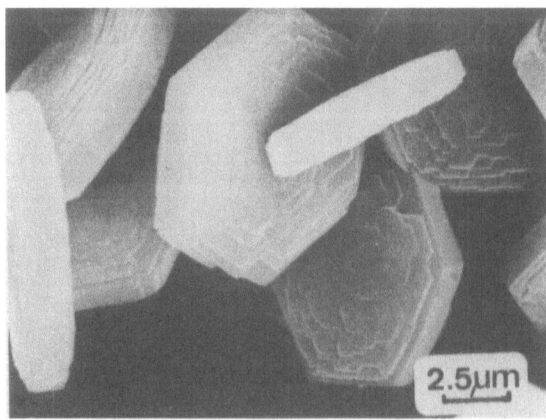

Fig. 13. Scanning electron microphotographs of the hexagonal faujasite polytype

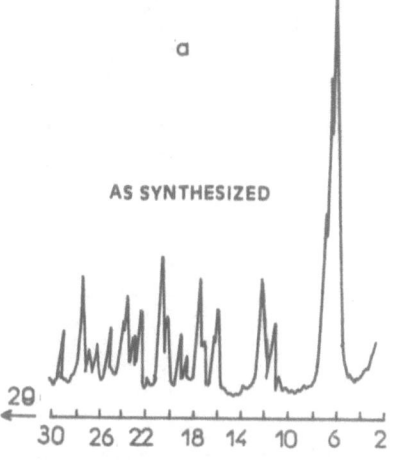
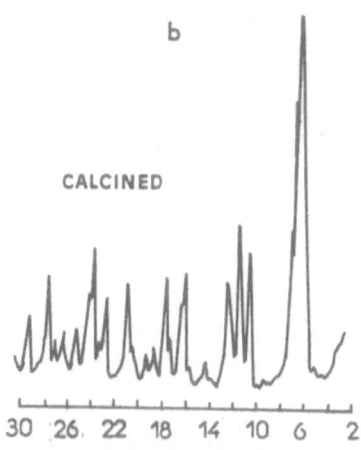

Fig. 14. X-ray diffraction pattern of the hexagonal faujasite
polytype, a) as-synthesized, b) calcined

Fig. 16. shows the hypothetical structure of the (18-crown-6)-sodium
hydroxide complexes in the 12-membered ring channel. This is in agreement
with the chemical composition, the value of the c parameter and the (^{13}C,
^{23}Na) NMR spectroscopy observations. In order to verify the hypothesis a
study of the structure based on X-ray powder diffraction data is in progress.

ACKNOWLEDGMENT

The authors gratefully acknowledge financial support from the Institut Français du Pétrole and the Société CECA, Groupe Elf-Aquitaine for
the study of the synthesis in fluoride media and preparation of new silica-rich faujasites respectively.

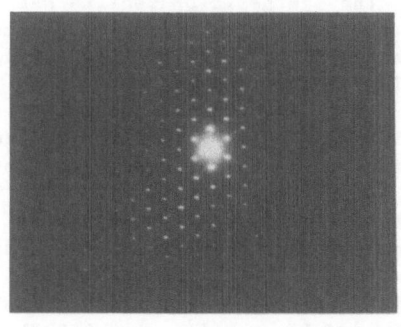

a   b

Fig. 15. Electron Microdiffraction patterns of a single crystal of
the calcined hexagonal polytype, a) (001)* and b) (100)*

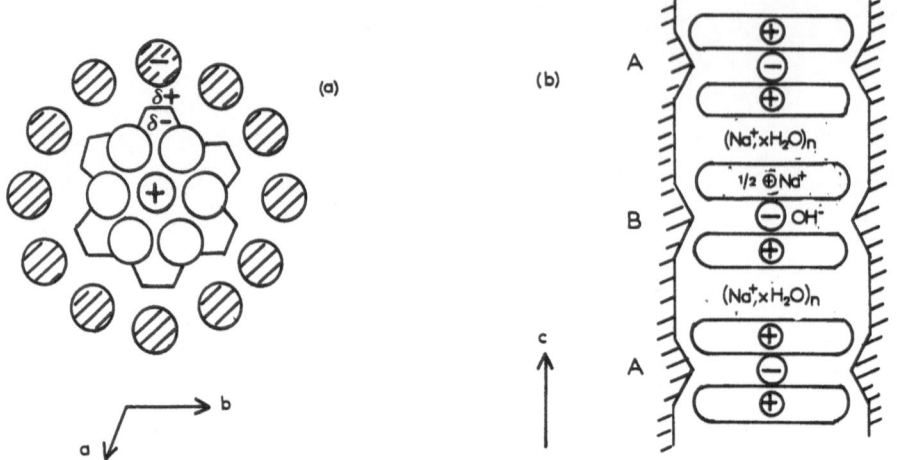

Fig. 16. Hypothetical location of the (18-crown-6)-sodium hydroxide
complexes in the 12-membered ring channel
(a) view along c   (b) view perpendicular to c

References

1. D. W. Breck, "Zeolite Molecular Sieves", John Wiley and Sons, New-York
    (1974)
2. A. Dyer, "An introduction to Zeolite Molecular Sieves", John Wiley
    and Sons, Chichester (1988)
3. P. Caullet, J. L. Guth and R. Wey, Solubilité et grandeurs thermodyna-
    miques de dissolution des zéolites 4A et 13X dans des solutions
    aqueuses basiques, Bull. Minéral., 103: 330 (1980)
4. P. Caullet, J. L. Guth and R. Wey, Solubilité de la zéolite Y et de
    l'analcite dans des solutions aqueuses basiques. Grandeurs thermody-
    namiques de dissolution, C. R. Acad. Sc. Paris, Série D, 291: 117 (1980)
5. G. T. Kerr, Chemistry of crystalline aluminosilicates. I. Factors
    affecting the formation of zeolite A, J. Phys. Chem., 70: 1047 (1966)
6. S. P. Zhdanov, Some problems of zeolite crystallization, in "Molecular
    Sieve Zeolites . I. Advances in chemistry series N°101", E. M. Flani-
    gen and L. B. Sand, ed., American Chemical Society, Washington (1971)
7. P. Wenqin, S. Ueda and M. Koizumi, The synthesis of zeolite NaA from
    homogeneous solutions and studies of its properties, in "New develop-
    ments in zeolite science and technology", Y. Murakami, A. Iijima, J.
    W. Ward, ed., Kodansha-Elsevier, Tokyo (1986)
8. P. Caullet, J. L. Guth, G. Hurtrez and R. Wey, Contribution à l'étude
    du mécanisme de formation des zéolites. V. Cristallisation de zéoli-
    tes à partir de solutions d'aluminosilicates caractérisées par un
    rapport Si/Al égal à 1, Bull. Soc. Chim. Fr., 7-8: 253 (1981)
9. P. Caullet and J. L. Guth, Contribution à l'étude du mécanisme de for-
    mation des zéolites: les équilibres solutions-zéolites ; les espèces
    et structures des solutions, Thesis, Mulhouse, France (1983)
10. G. Engelhardt and D. Michel, "High-resolution solid-state NMR of
    silicates and zeolites", John Wiley and Sons, Chichester (1987)
11. P. Caullet and J. L. Guth, 1989, Observed and calculated silicate and
    aluminosilicate concentrations in alkaline aqueous solutions, in
    "Zeolite synthesis", ACS Symposium Series n°398, M. L. Occelli and H. E.
    Robson, eds., American Chemical Society, Washington (1989)

12. R.M. Barrer, "Hydrothermal chemistry of zeolites", Academic Press, London (1982)

13. J.L. Casci and B.M. Lowe, Use of pH-measurements to monitor zeolite crystallization, Zeolites, 3:186 (1983)

14. J.L. Guth, P. Caullet and R. Wey, Variation du paramètre cristallin d'une zéolite Y au cours de sa cristallisation à partir d'un gel. Mise en évidence d'une hétérogénéité de composition, Bull. Soc. Fr. Minér. Cristall., 99:21 (1976)

15. P. Bodart, J.B Nagy, E.G. Derouane and Z. Gabelica, Study of mordenite crystallization III : Factors governing mordenite synthesis, in "Structure and reactivity of modified zeolites" (Studies in surface science and catalysis 18), P.A. Jacobs, N.I. Jaeger, P. Jiru, V.B. Kazansky and G. Schulz-Ekloff, ed., Elsevier, Amsterdam (1984)

16. R.W. Grose and E.M. Flanigen, Crystalline silica, U.S. Patent n°4061724 (1977)

17. G.L. Woolery, L.B. Alemany, R.M. Dessan and A.W. Chester, Spectroscopic evidence for the presence of internal silanols in highly siliceous ZSM5, Zeolites, 6:14 (1986)

18. R.H. Busey, E. Schwartz and R.E. Mesmer, Fluorosilicate equilibria in sodium chloride solutions from 0 to 60°C, Inorg. Chem., 19:758 (1980)

19. N.A. Matwiyoff and W.E. Wageman, Nuclear magnetic resonance studies of aluminium(III) fluoride ion complexes in aqueous solutions, Inorg. Chem., 9:1031 (1970)

20. D. Hass, S.P. Tetrosyants, Yu. A. Buslaev and I. Hartlab, Characteristics of the formation of aluminum and gallium fluoride complexes in solution, Doklady Akademii Nauk SSSR, 269:380 (1983)

21. A.M. Bond and G.T. Hefter, Critical survey of stability constants and related thermodynamic data of fluoride complexes in aqueous solution, Pergamon Press, Oxford (1980)

22. J.L. Guth, H. Kessler, J.M. Higel, J.M. Lamblin, J. Patarin, A. Seive, J.M. Chezeau and R. Wey, 1989, Zeolite synthesis in presence of fluoride ions. Comparison with conventional synthesis methods, in "Zeolite synthesis, ACS Symposium Series n° 398", M.L. Occelli and H.E. Robson, eds., American Chemical Society, Washington.

23. Z. Gabelica and J.L. Guth, A silicogermanate with Si:Ge ratio ≥ 2 — an MFI zeolite of novel composition, Angew. Chem. Int. Ed. Engl., 28 : 81 (1989)

24. E.W. Valyocsik, Improved method of preparing crystalline zeolite ZSM20, Eur. Pat. Appl. 0012522 (1980)

25. G.T. Kokotailo and J. Ciric, Synthesis and structural features of zeolite ZSM3, in "Molecular Sieve Zeolites. II. Advances in chemistry series n°102", E.M. Flanigen and L.B. Sand, ed., American Chemical Society, Washington (1971)

26. Z. Gabelica, N. Dewaele, L. Maistriau, J.B Nagy and E.C. Derouane, 1989, Directing parameters in the synthesis of zeolites ZSM-20 and Beta, in "Zeolite synthesis, ACS Symposium Series n° 398", M.L. Occelli and H.E. Robson, eds., American Chemical Society, Washington.

27. H.D. Durst, M. Milano, E.J. Kikta, Jr, S.A. Connelly and Eli Grushka, Phenacyl-esters of fatty acids via crown-ether catalysts for enhanced ultraviolet detection in liquid chromatography, Anal. Chem. 47:1797 (1975)

28. C.J. Pedersen, Cyclic polyethers and their complexes with metal salts, J. Am. Chem. Soc., 89:7017 (1967)

29. A. Delville, H.D.H. Stover and C. Detellier, Crown ether-cation decomplexation mechanics. ^{23}Na NMR studies of sodium cation complexes with dibenzo-24-crown-8 and Dibenzo-18-Crown-6 in nitromethane and acetonitrile, J. Am. Chem. Soc. 109:7293 (1987)

A ^{13}C - AND [129]Xe-NMR STUDY OF THE ROLE
OF TETRAALKYLAMMONIUM CATIONS
IN THE SYNTHESIS OF HIGH-SILICA ZEOLITES

Q. Chen[1], J. B.Nagy[2], J. Fraissard[1], J. El Hage-Al Asswad[2], Z. Gabelica[2], E.G.
Derouane[2], R. Aiello[3], F. Crea[3], G. Giordano[3] and A. Nastro[3]

1) Laboratoire de Chimie des Surfaces, URA CNRS 870
Université Pierre et Marie Curie
F-75252 Paris Cédex 05. France

2) Laboratoire de Catalyse, Facultés Universitaires Notre-Dame de la Paix,
Rue de Bruxelles 61
B-5000 Namur. Belgium

3) Dipartimento di Chimica, Università della Calabria,
Arcavacata di Rende
I-87030 Rende. Italy

ABSTRACT

The high silica gel is well organized around the tetraalkylammonium (TAA) cations
during prenucleation stage. By combined ^{13}C- and [129]Xe- NMR, two TAA+ states,
either hydrated or dehydrated, and three different TAA-gel associations can clearly be
distinguished:
1. dispersed monomeric or dimeric (either hydrated or dehydrated) TAA+ species
   (1-2 cations per cavity, cavity diameter ~11-16 Å)
2. less dispersed TAA+ ions (4-8 ions per cavity, cavity diameter ~17-23 Å)
3. large aggregates $(TAAX)_n$ (X=OH, Br) (n >15 per cavity, cavity diameter
   ~30-40 Å)
The addition of alkali cations can influence both the dispersion of TAA+ species and
silica gels, hence the crystallization of zeolites. A model is proposed to explain the role of
different species in gels for the crystallization of high silica zeolites.

INTRODUCTION

The role of organic directing agents in the synthesis of zeolites is well recognized
(1). While a given base or cation sometimes helps the formation of a specific zeolite,
e.g., ZSM-20 (2), a given zeolite (e.g.ZSM-5) can be crystallized in the presence of a
large variety of organic additives (1).

To gain a better understanding of the role of organic directing agents, precursor
species were recently identified by combined chemical trapping (silylation) and ^{29}Si-NMR
techniques (3). In contrast, the organization of the gel around the organic bases or cations
has received much less attention (4).

Early studies on zeolite precursor hydrogels dealt with the competition between
alkali and tetrapropylammonium (TPA) cations (5,6) for the neutralization of
aluminosilicate species. With the help of combined thermal analysis, ^{13}C-NMR and XRD

Guidelines for Mastering the Properties of Molecular Sieves
Edited by D. Barthomeuf *et al.*
Plenum Press, New York, 1990

87

data, the existence of the following organic species was proposed (4,7) : monomeric TPAX (X = OH, Br) entities in direct association with the gel, $TPA(H_2O)_x^+$ water clathrates that act as counter-cations to the gel negative charges, and crystalline $(TPAX)_n$ clusters.

Recently, we showed that ^{129}Xe-NMR of adsorbed xenon is capable to distinguish different types of cavities in calcined gel samples (8). In this paper, ^{13}C- and ^{129}Xe-NMR spectroscopies are used to characterize both the tetraalkylammonium (TAA) cations and the cavities they form, which are present during the nucleation, when the organic cations interact with the gel. The zeolitic phases formed after hydrothermal synthesis are identified by XRD analysis and by ^{129}Xe-NMR of adsorbed xenon.

EXPERIMENTAL

Three different types of gels were prepared from precipitated SiO_2 (BDH, containing 0.6 wt.% Na), TAABr, TAAOH, 30 % NaOH solution (reagent grade, Carlo Erba), and distilled water, with molar compositions: (I) 8 TAABr - 100 SiO_2 - 1000 H_2O, (II) 8 TAABr - 4 Na_2O-100 SiO_2 - 1000 H_2O, and (III) 4 $(TAA)_2O$ - 100 SiO_2 - 1000 H_2O (9). TAA compounds were tetramethylammonium TMABr, tetraethylammonium TEABr, tetrapropylammonium TPABr, and tetrabutylammonium TBABr (all Fluka purum), TMAOH (25 % aq. sol.), TEAOH (20 % aq. sol.), TPAOH (20 % aq. sol.), and TBAOH (40 % aq. sol.) (all Fluka). The hydrogels were dried immediately after their preparation at ca.373 K during 16 h. Before ^{129}Xe-NMR measurements the dried gel samples were further heated under dry N_2 to 673 K at 4 K per min. and maintained at this final temperature for 12 h. Dry air was then introduced for 6 h and the materials cooled down to 300 K.

The syntheses of various zeosilites (zeolites containing only traces of Al) were carried out in static conditions at 443 K under autogeneous pressure, using sealed Teflon-lined containers. An activation treatment similar to that used for the gels was carried out for the final crystalline samples, the highest temperature being in this case 823 K. The chemical composition of the materials was determined by atomic absorption (Si/Al $\geq$ 2000) (9).

The nature and the crystallinity of the solid phases were determined from the XRD patterns obtained with a Philips PW 1349/30 diffractometer monitored by an Olivetti M24 computer using the CuK_α radiation and a scan rate of 0.5 degree (2θ) per minute; $Pb(NO_3)_2$ (10 wt.%) was used as internal standard.

The CP-MAS or decoupled ^{13}C-NMR spectra were recorded on a BRUKER CXP-200 spectrometer. The CP-MAS conditions were : ν = 50.3 MHz, repetition time = 4 s, π/2 pulse = 5 μs, and contact time = 4 ms.

For the adsorption of xenon and the ^{129}Xe-NMR measurements, the samples were further treated under vacuum (10^{-5} Torr) at 673 K for 8 h. After cooling,the temperature was maintained at 300K. NMR spectra were recorded on a BRUKER CXP-100 spectrometer (ν = 24.9 MHz). The resonance frequency of gaseous xenon extrapolated to zero pressure was used as reference.

RESULTS AND DISCUSSION

Typical ^{129}Xe-NMR spectra of adsorbed xenon are reported in Figs. 1 and 2 for the three different gels (p_{Xe}=765 Torr). The chemical shift values of the different NMR lines extrapolated to zero xenon pressure (δ_s) are reported in Table 1. The observation of different NMR lines shows that xenon probes cavities of different sizes in conditions where the exchange is slow between the Xe sites. From the known dependence of the chemical shift δ_s on cavity size (10, 11), the dimension of the cavities left over after the elimination of organic cations and/or water is easily determined. These dimensions (d_L, M, S) are also reported in Table 1 (cavity L > cavity M > cavity S).

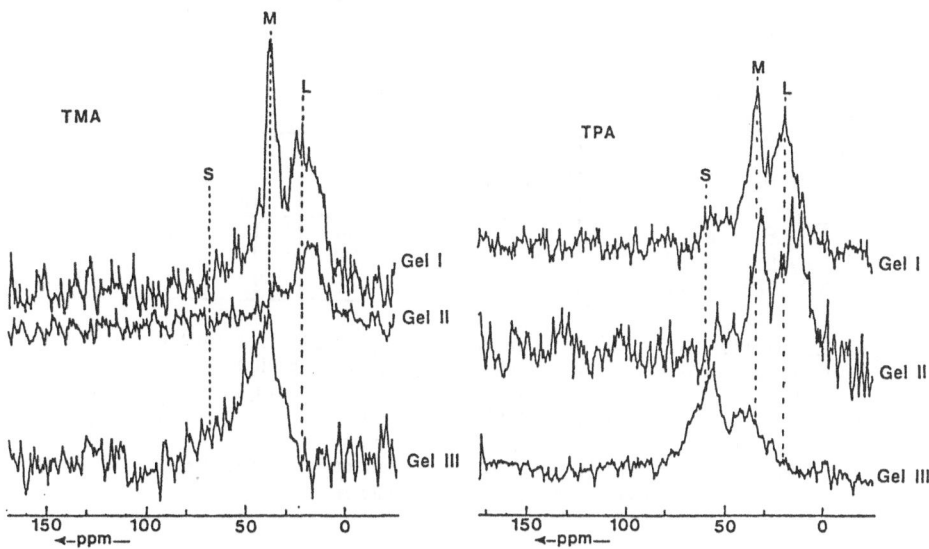

FIG. 1 ^{129}Xe-NMR spectra of xenon adsorbed in FIG. 2 ^{129}Xe-NMR spectra of xenon adsorbed in
cavities formed in gels   cavities formed in gels

Figure 3 illustrates the ^{13}C-NMR spectra of TPA$^+$ ions occluded in the hydrogel during the nucleation period. Spectrum $\underline{a}$ is characteristic of pure TPABr crystallized from the solution during evaporation (δ_{CH3} =12.5 ppm) (12). The same spectrum is obtained for gels TPABr-SiO$_2$ (I) and TPABr-SiO$_2$/Na$_2$O (II). The presence of crystalline TPABr was also detected by DTA methods (9). Spectrum $\underline{b}$ recorded for gels 4(TPA)$_2$O (III) shows a different chemical shift for the methyl group : δ_{CH3} = 11.4 ppm. This spectrum has been attributed to a water clathrate form of TPA$^+$ (4,7). However, except the methyl groups rotation, the mobility of the entire TPA$^+$ ions remains restricted (poorly resolved CH$_2$ lines) suggesting that the hydration of the ions is not complete. Indeed, in the gel having the composition 0.5 (TPA)$_2$O (spectrum $\underline{c}$), a higher amount of hydration water allows an intermediate mobility of the ion, all the lines being observed but still broadened with respect to spectrum $\underline{a}$. Finally, when the amount of available water in the cavities becomes sufficient, a well-resolved solution-like TPA$^+$ spectrum $\underline{d}$ is detected, corresponding probably to fully hydrated TPA$^+$ cations. Indeed, this sample was aged at 293 K for about three months before drying at 373 K.

Combining the ^{129}Xe-NMR results of adsorbed xenon in the calcined gel and the ^{13}C-NMR spectra of the organic cations in the non calcined hydrogel yields a more precise picture of the gel organization . Note first that the presence of the cavities detected by Xe probing is certainly not an artifact. These cavities are well created upon calcination, because of the reorganization of the gel. Indeed, when organic cations are not included in the gel (TEAOH in the ZSM-20 gel), no cavities are observed following calcination (8).

Table 1 .^{129}Xe-NMR δ_S values of adsorbed xenon in various silica gels.

Gels		L[a]		M[a]		S[a]	
		δ_S (ppm)	d_L (±5Å)	δ_S (ppm)	d_M (±3Å)	δ_S (ppm)	d_S (±2Å)
I	TMABr - SiO$_2$	30	29	45	18	-	-
II	TMABr - SiO$_2$-Na$_2$O	21	40	48	16	-	-
III	TMOH - SiO$_2$[b]	-	-	38	22	-	-
I	TEABr - SiO$_2$	-	-	36	23	50	16
II	TEABr - SiO$_2$-Na$_2$O	-	-	-	-	66	11
III	TEAOH - SiO$_2$[c]	-	-	-	-	58	13
I	TPABr - SiO$_2$	22	40	36	23	55	14
II	TPABr - SiO$_2$-Na$_2$O	28	31	37	22	-	-
III	TPAOH - SiO$_2$	-	-	45	18	58	13
I	TBABr - SiO$_2$	27	32	44	18	66	11
II	TBABr - SiO$_2$-Na$_2$O	27	32	39	21	-	-
III	TBAOH - SiO$_2$	-	-	38	22	55	14

a) Size for large cavity L : 30 - 40 Å   (more than 15 organic cations per cavity),
Size for medium cavity M : 17- 23 Å   (4 - 8 organic cations per cavity),
Size for small cavity S : 11 - 16 Å   (1 - 2 organic cations per cavity).
b) The small NMR-line at ca. 102 ppm (pXe=765 Torr) could correspond to very tiny ZSM-48 zeolitic crystals obtained after ca. 1 year at room temperature (pore dimension 5.5 Å).
c) The small NMR line at ca. 110 ppm (pXe=765 Torr) could correspond to tiny ZSM-12 crystals obtained after ca. 1 year at room temperature (pore dimension 5 Å).

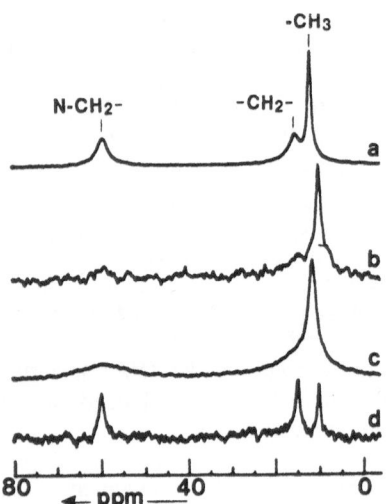

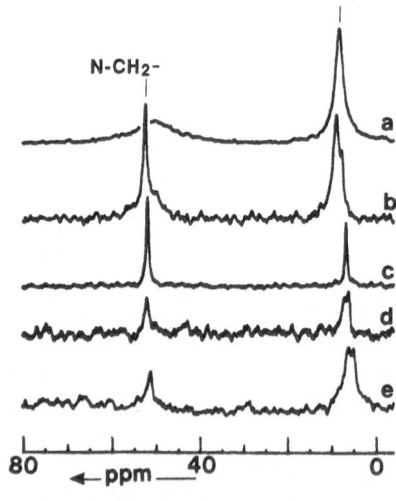

Fig. 3. ^{13}C-NMR spectra of adsorbed TPA ions in:
a) Gels I and II: crystalline TPABr
b) Gel III partially hydrated TPA
c) 0.5 (TPA)$_2$O-100 SiO$_2$-1000 H$_2$O: motionally broadened lines
d) Gel III (remained in contact with solution): hydrated mobile species

Fig. 4. ^{13}C-NMR spectra of adsorbed TEA ions in:
a) Crystalline TEABr
b) Gel I: crystalline TEABr and hydrated mobile TEA species
c) Gel II: hydrated mobile TEA species
d) Gel III: hydrated mobile TEA species and TEA in crystalline ZSM-12
e) TEA in crystalline ZSM-12

90

A large cavity L is formed in gels I and II. It corresponds to a void volume previously occupied by more than 15 organic cations, and having a diameter which varies between 30 and 40 Å (Table 1) (only TEABr does not form these larger aggregates). Gels I of composition 8 TAABr - 100 SiO$_2$ - 1000 H$_2$O however do not lead to the formation of any zeolite, because of low alkalinity. On the other hand, gels II having in addition 4 Na$_2$O in their composition do yield crystalline materials in hydrothermal conditions. The TAA$^+$ aggregates responsible for cavity L are characterized by ^{13}C-NMR spectra which are similar to those of TAA$^+$ species in the solid state (see for example Fig. 3a). We therefore believe that these aggregates are not true nucleation centers to zeolites.

A medium cavity M (d = 17-23 Å) may have contained 4-8 organic cations. These cavities are observed in all three types of gels, except again for the TEA-gels II and III. As seen from the ^{13}C-NMR spectra of the TPA species, these smaller aggregates also behave similarly to the crystalline TPA phase (Fig. 3a). Similar conclusions can be drawn for the TMA and TBA-gels. The TAA ions are not hydrated in these systems. The TEA-gel (I) shows in addition two different species (Fig. 4b) : one has a ^{13}C-NMR spectrum characteristic of the solid state spectrum (Fig. 4a) with δ_{CH2} = 52.4 ppm and δ_{CH3} = 9.9 ppm ; the other can be assigned to a more mobile species with δ_{CH2} = 53.5 ppm and δ_{CH3} = 8.7 ppm that occupies a smaller cavity S in the gel.

Indeed, cavities S (d = 11-16 Å) which appear in rather large amounts in gel III, can only contain 1 or 2 organic cations. The ^{13}C-NMR spectra suggest the presence of partially hydrated TPA species (Fig.3b) (vide supra). The TBAOH system behaves similarly. In other experimental conditions, at rather high water content, a completely mobile TPA$^+$ ion is detected in cavity S (Fig. 3d). The ^{13}C-NMR spectrum of TEA$^+$ in gel II also shows this mobile and fully hydrated species (Fig. 4c). In addition a new NMR line appears, for TEA-gel III, at δ_{CH3} = 7.6 ppm which could indicate the formation of very small ZSM-12 crystallites (Fig 4d and 4e). Indeed, smaller cavities of d~5 Å are observed by ^{129}Xe NMR in this gel aged for about one year at room temperature (Table 1, footnote c). These results emphasize the advantage to use both ^{13}C-NMR of the organic base and ^{129}Xe-NMR (16) of adsorbed xenon to unravel the presence of submicrocrystallites which remain undetected by powder XRD techniques under routine conditions (13-15).

In order to explain these observations for the three different gels, the following model is proposed :

$$
\begin{array}{ccc}
& \xrightarrow{\text{TAAOH}} & SiO_2 \text{ ......TAA}^+ \\
SiO_2 & & \text{depolymerized} \\
\text{agglomerated} & \xrightarrow{\text{TAABr + NaOH}} & SiO_2 \text{ ....Na}^+ \\
& & \text{depolymerized}
\end{array}
\qquad (1)
$$

$$
\begin{array}{ccccc}
\text{TAABr} & \rightleftharpoons & \text{TAABr} & \rightleftharpoons & \text{TAA}^+ + \text{Br}^- \\
\text{aggregated} & & \text{less aggregated} & & \text{partially or fully hydrated}
\end{array}
\qquad (2)
$$

The relative amount of TAA species depends on the basicity, the nature of the TAA$^+$ ions, and the presence of sodium ions in the medium.

In gel I (of low basicity), mainly cavities L and M are formed, which contain large aggregates of TAABr salts. The relative amount of cavities S is small. Interestingly, the

TEABr system is an exception, the TEA⁺ ions being less aggregated. The addition of sodium ions into gel I increases the relative amount of cavities L.

In gel III, the alkalinity being quite high, the TAA⁺ ions are the main counterions to the SiO⁻ groups in the gel and they are better dispersed. Entities corresponding to partially or fully hydrated monomeric or dimeric species are detected by both ^{13}C- and ^{129}Xe-NMR techniques (cavities M and S). However, the presence of sodium impedes partly the dispersion of TAA⁺ ions. Indeed, in gel II cavities L are also formed, while they are absent in gel III. Na⁺ is thus a better counterion to SiO⁻ groups than the bulkier organic cation and the competition is in favour of the former hard acid-hard base interaction with respect to a soft acid (organic cation)-hard base interaction. It is our proposal that the monomeric TAA⁺ species play a determinant role in the nucleation of zeolites.

Finally, the ^{129}Xe-NMR spectra also reveal their potential in characterizing the crystalline products. Figure 5 shows the variation of the ^{129}Xe-chemical shift as a function of adsorbed xenon (in atom.g⁻¹) for the ZSM-5 and ZSM-11 zeolites obtained from gels II and III containing TPA⁺ and TBA⁺ ions, respectively. The final crystalline materials show the same variation of δ values as a function of the amount adsorbed Xe. The less crystalline materials obtained from a gel II for TPA, containing eight times less Na₂O than the other gels or from a gel III for TBA but containing also NaCl, show a much larger variation of the chemical shift. This suggests that the available pore volume is much smaller in the latter case, reflecting the lower crystal perfection of the samples. More quantitative conclusions are arrived at by considering the initial slopes of the two curves :

$$\frac{\text{Slope of 100 \% crystalline material}}{\text{Slope of the less crystalline material}} = \frac{x}{100}$$

where x stands for the crystallinity of the less crystalline sample. For the cases which are illustrated, x = 20-25 %.

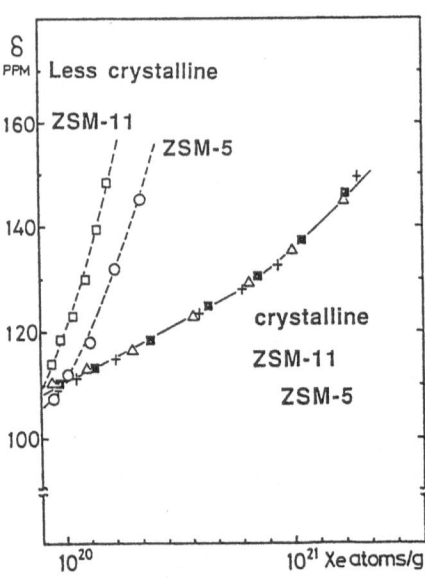

FIG. 5. Variation of ^{129}Xe-NMR chemical shifts as a function of adsorbed Xe (at g-1) for well crystalline ZSM-5 and ZSM-11 samples and for partially crystalline ZSM-5 and ZSM-11 samples

At low xenon pressure the NMR spectra of adsorbed xenon show a symmetrical line ($\delta_S = 94$ ppm) for the crystalline phases obtained from the TMA-gel II and III systems characterizing a ZSM-48 phase. However, at high pressure ($p_{Xe} > 600$ Torr), two adjacent lines were detected, which could correspond to possible intergrowths in the ZSM-48 structure (17).

Finally, the crystalline phases obtained from the TEA gels show, in addition to the ZSM-12 phase, an amorphous phase containing quite large cavities. The size of the crystallites must be very small, as the chemical shift decreases with increasing xenon pressure due to rapid exchange of Xe with the gaseous phase. The XRD spectra of ultrasound-treated samples show the probable presence of ZSM-12 zeolite and not of ZSM-48 as it was erronously interpreted, based on poor quality spectra (9).

CONCLUSIONS

The combination of ^{13}C-NMR of the occluded organic cations in hydrogels and [129]Xe-NMR of the adsorbed xenon in calcined gels and final products is a unique approach to characterize the initial gels as well as the zeolitic materials that are produced.

A model has been proposed to explain the state of the dispersion of the TAA[+] ions within the gel and the competition that exists between these ions and the Na[+] cations for the SiO[-] negative charges in the hydrogel. TAA[+] aggregates of different sizes are formed and it is felt that only the monomeric species are determinant for nucleation.

ACKNOWLEDGMENTS

J. El Hage-Al Asswad acknowledges financial support from Shell B.V., Amsterdam. This work was partly financially supported by the Italian National Research Council, Progetto Finalizzato Energetica 2. The authors are indebted to Mr. G. Daelen for technical assistance.

REFERENCES

1. See, e.g. B.M. Lok, T.R. Cannan and C.A. Messina , Zeolites, 3 : 282 (1983).
2. N. Dewaele, L. Maistriau, J. B.Nagy, Z. Gabelica and E.G. Derouane, Appl.Catal., 37 : 273 (1988).
3. J.P. Van den Berg, P.C. de Jong-Versloot, J. Keijsper and M.F.M. Post, Stud.Surf.Sci. Catal., 37 : 85 (1986).
4. J. B.Nagy, P. Bodart, E.G. Derouane, Z. Gabelica and A. Nastro, Stud.Surf.Sci.Catal., 28 : 231 (1986).
5. L.S. Dent Glasser and G. Harvey, J. Chem. Soc., Chem. Commun., 1250 (1984).
6. E.G. Derouane, J. B.Nagy, Z. Gabelica and N. Blom, Zeolites, 2 : 299 (1982).
7. Z. Gabelica, J. B.Nagy, P. Bodart, N. Dewaele and A. Nastro, Zeolites, 7 : 67 (1987).
8. T. Ito, J. Fraissard, J. B.Nagy, N. Dewaele, Z. Gabelica, A. Nastro and E.G. Derouane, in Proc. 8th Intern. Zeolite Conference, Amsterdam, 1989, in press.
9. F. Crea, J. B.Nagy, A. Nastro, G. Giordano and R. Aiello, Thermochim. Acta., 135 : 353 (1988).
10. J. Demarquay and J. Fraissard, Chem. Phys. Lett., 136 : 314 (1987).
11. E.G. Derouane and J. B.Nagy, Chem. Phys. Lett., 137 : 341 (1987).
12. J. B.Nagy, Z. Gabelica and E.G. Derouane, Zeolites, 3 : 43 (1983).
13. P.A. Jacobs, E.G. Derouane and J. Weitkamp, J. Chem. Soc., Chem. Commun. 591 (1981).
14. Z. Gabelica, J. B.Nagy and G. Debras, J. Catal., 84 : 256 (1983).
15. E.G. Derouane, S. Detremmerie, Z. Gabelica and N. Blom, Appl. Catal., 1 : 201 (1981)
16. T. Ito, M.A. Springuel-Huet and J. Fraissard, Zeolites, 9: 68 (1989)
17. J.L. Schlenker, W.J. Rohrbaugh, P. Chu, E.W. Valyocsik and G.T. Kokotailo, Zeolites, 5: 355 (1985)

PROMOTER EFFECTS ON PHASE GROWTH IN ALUMINOPHOSPHATE

AND SILICALUMINOPHOSPHATE GEL SYSTEMS

Dr. R. Szostak

Zeolite Research Program
Georgia Tech Research Institute
Georgia Institute of Technology
Atlanta, GA 30332

INTRODUCTION

In the synthesis of many aluminosilicate and silicate molecular sieves, organic quaternary amines and diamines are known to be extremely effective in controlling the type of structure crystallized. Little effect on structure direction by the neutral amines has been found based on the limited number of zeolites reported in the patent literature using such additives. ZSM-5 (1), ZSM-48 (2), ZSM-39 (3) and Theta-1 (4) are structures reported to crystallize in the presence of neutral amines, mainly the alkyl amines, dipropylamine and diethanolamine. On the other hand, the use of neutral amines has been found to be extremely effective in enhancing crystallization of a variety of microporous $AlPO_4$ based materials. The use of dipropylamine (dPA), for example, aids in the crystallization of $(M)AlPO_4$-11,31,39,41,43,46,50 of Union Carbide (5). VPI-5 (6), MCM-7 (7) and MCM-9 (8) are other aluminophosphate based materials claimed to form in the presence of dipropylamine.

The objective of this study was to examine the contribution of the neutral organic additive, dipropylamine, in the crystallization of two aluminophosphate systems, VPI-5 and MCM-9. The similarity between these two materials has been previously reported (9).

EXPERIMENTAL PROCEDURE

The synthesis of MCM-9 via a two-phase route was followed from the patent

Guidelines for Mastering the Properties of Molecular Sieves
Edited by D. Barthomeuf *et al.*
Plenum Press, New York, 1990

examples (8) except a static autoclave was used; the synthesis of VPI-5 followed the reported preparation of Davis, et.al. (6). Variations in the preparations are shown in Tables 1, 2 and 3. Little significant difference was noted in the resulting structures crystallized with or without the presence of the hexylalcohol as VPI-5 and MCM-9 appear to be structurally related (9). The exact role of hexylalcohol was not explored in detail in this work.

DISCUSSION

The synthesis of specific crystalline $AlPO_4$ structures in the dPA system is sensitive to the concentration of the amine as well as the temperature and precrystallyzation aging. $AlPO_4$-11 is identified at a range of dPA/Al_2O_3 ratios between 0.2 and 2, and at higher temperatures (170°C to 200°C). On the other hand, the VPI-5 phase is more sensitive to the combination of temperature and dPA concentration. Temperatures around 140-145°C appear necessary. The addition of amine greater or less than $[dPA]/Al_2O_3=1$ inhibits crystallization as shown in Table 2.

In the synthesis following the patent examples provided for MCM-9, the persistent presence of SAPO-11 is observed at the elevated synthesis temperatures employed. A high temperature aging (130°C) is also utilized. If no aging at 130°C is undertaken only pure SAPO-11 will form. Temperatures similar to those employed in the synthesis of VPI-5 (145°C) result in supressing SAPO-11 crystallization. However, in this range MCM-9 co-crystallizes with MCM-1 (7), the silicon containing aluminophosphate with x-ray pattern related to H3 of d'Vyoire (10). D'Vyoire has also observed a co-crystallization of H3 with a phase he calls H1 which has an x-ray diffraction pattern similar to VPI-5 (11). A short room temperature aging of the gel at room temperature or higher appears to improve the purity of the MCM-9 phase only if the crystallization temperature is lowered to 140°C - 145°C..

CONCLUSION

A pretreatment, or aging of the aluminophosphate gel appears to be necessary to improve the purity of the MCM-9 phase under the conditions employed in this study. MCM-9 co-crystallizes with either MCM-1 at low temperature or SAPO-11 at elevated temperatures in the presence of dPA. The presence of silica in the initial gel limits the range of formation of MCM-9 relative to the range observed for VPI-5. VPI-5 more readily forms under a wider range of conditions than that of MCM-9. Though dPA may influence the structures crystallized, it is not the only critical factor in determining which structure will ultimately result, indicating that it is not a strong structure director in this system.

TABLE 1. PARAMETER STUDY IN THE ALUMINOPHOSPHATE SYSTEM WHICH CRYSTALLIZES VPI-5.

$\dfrac{P_2O_5}{Al_2O_3}$	$\dfrac{H^{+*}}{Al_2O_3}$	$\dfrac{H_2O}{Al_2O_3}$	$\dfrac{DPA}{Al_2O_3}$	temp. ($^{\circ}$C)	phase produced
1	3.0	20.7	0.74	170	$AlPO_4$-11
1	3.0	50	0.5	200	$AlPO_4$-11 + quartz
1	3.0	50	1.0	170	$AlPO_4$-11
1	3.0	50	2.0	200	$AlPO_4$-11 + H3
1	0.95	20.7	0.74	170	UN**
1	0.03	20.7	0.74	170	UN**
1	-3.53	20.7	0.74	170	amorphous
1	-11.41	20.7	0.74	170	amorphous
1	-19.39	20.7	0.74	170	amorphous
1	3.0	50	0.5	142	H3 + VPI-5
1	3.0	50	1.0	142	VPI-5 + (H3)
1	3.0	50	2.0	142	amorphous + UN*

* [H^+] adjusted with H_2SO_4
** UN = unidentified phase, non-zeolitic

TABLE 2. SILICOALUMINOPHOSPHATE SYNTHESIS: MCM-9
COMPOSITION: $2DPA*SiO_2*2P_2O_5*2.7Al_2O_3*6H_2O*12C_6OH$

Aging	cryst.temp.	cryst.time	phase produced
130°C,24hr	200	24	MCM-9 + SAPO-11
No	142	0	amorphous
No	142	3	MCM-1
No	142	18	MCM-9 + MCM-1
No	142	24	MCM-9 + MCM-1
No	142	48	MCM-9 + MCM-1
No	142	60	MCM-9 + MCM-1
RT,2hr	142	24	MCM-9
RT,2hr	142	48	MCM-9
No	170	24	SAPO-11
No	170	48	SAPO-11
No	200	3	SAPO-11
No	200	18	SAPO-11
No	200	24	SAPO-11
No	200	48	cr*
No	200	60	cr*
RT,2hr	200	24	MCM-9 + SAPO-11
RT,2hr	200	48	MCM-9 + SAPO-11
No	130	24	MCM-1
No	130	48	MCM-1

*cr = crystobalite

TABLE 3. VARIATION IN GEL SILICA CONTENT ON PHASE PRODUCED

Aging	dPA/P_2O_5	SiO_2/Al_2O_3	cryst.temp.	phase produced
No	1.0	0	130	H3 + VPI-5
No	1.0	0	142	VPI-5 + (H$_3$)
RT,2hr	1.0	0	142	VPI-5
RT,6hr	1.0	0	142	VPI-5
No	1.0	0	170	AlPO$_4$-11
No	0.74	0.36	130	MCM-1
No	0.74	0.36	200	SAPO-11
RT,2hr	0.74	0.36	200	MCM-9 + SAPO-11
130°C,24hr	0.74	0.36	200	MCM-9 + SAPO-11
No	0.74	0.36	142	MCM-9 + MCM-1
RT,2hr	0.74	0.36	142	MCM-9

REFERENCES

1. Eur. Pat. 2900 (1979)
2. U.S. Pat. 4448675 (1984)
3. U.S. Pat. 4287166 (1981)
4. Eur. Pat. 57049 (1982)
5. U.S. Pat. 4310440; U.S. Pat. 4440871; U.S. Pat. 4544143; U.S. Pat. 4567029
6. M.E. Davis, C. Montes, J. Carces, in Symposium on Advances in Zeolite Synthesis, American Chemical Society Meeting, Los Angeles, Sept. 1988.
7. Eur. Pat. 146384 (1985)
8. Eur. Pat. 147991 (1985)
9. R. Szostak, T.L. Thomas, D.C. Shieh, Catalysis Letters, 2,63, (1989).
10. F. d'Yvoire, Bull. Soc. Chem. (France), 1762, (1961).
11. The similarity between VPI-5 and H1 was pointed out by Prof. P. Jacobs during the presentation of this paper.

RECENT ADVANCES IN TECHNIQUES FOR CHARACTERIZING

ZEOLITE STRUCTURES

D. E. W. Vaughan, M. M. J. Treacy and J. M. Newsam

Exxon Research and Engineering Company
Route 22 East, Annandale, NJ 08801

INTRODUCTION

In developing new zeolites for catalysis and separations,
the characterization of structure is an important step in the
identification of possible applications. By analogy with the
properties of zeolites for which structures are known (of which
there is a large and growing data base), possible applications
for a new zeolite can be identified once its essential
structural features are known. Structural details at several
degrees of precision and over a range of length scales are
usually sought. These range from local active site geometries
or framework cation orderings, through descriptions of the
framework structure and pore architecture, to questions of
crystal perfection, homogeneity and mesoporosity. Frequently
the most time-consuming step is the initial definition of the
framework topology. The unavailability of large single crystals
of practically all synthetic zeolites, the relative
complexities of zeolite structures, and a series of materials-
related difficulties combine to limit the usefulness of
conventional structure determination methods.

Some approaches to overcoming these limitations have
recently been overviewed [1]. Recent examples of the use of a
variety of different data in developing composite structural
models include the characterizations of ECR-1 [2,3] and beta
[4-6]. The objective in the present paper is to provide a
perspective on some recent advances in a number of techniques
that can contribute to the solving of new zeolite structures. A
crucial test of having correctly identified a new zeolite
structure is that the theoretical X-ray diffraction pattern for
the model optimized by distance least-squares (DLS) methods
[7,8] matches the experimental pattern. Advances in X-ray,
neutron and electron diffraction techniques play a major role
in an improving ability to recognize, define and identify new
zeolites. Improved high resolution electron microscopes (now
giving point-to-point resolutions of less than 2Å in high
vacuum environments) may provide direct data on zeolite sub-
structure. In situations where other structural information is
available, even "low" resolution structure images may provide

Guidelines for Mastering the Properties of Molecular Sieves
Edited by D. Barthomeuf *et al.*
Plenum Press, New York, 1990

sufficient detail to help structure definition, as has been demonstrated recently for CSZ-1 [9], ECR-1 [2,3], beta [4,5] and ZSM-20 [10]. The application of MAS-NMR to zeolites has been reviewed recently [11,12]. Noted here are the use of ^{13}C NMR of trapped templates [13,14] and of [129]Xe NMR of sorbed Xe [15] as indicators of channel and/or cavity sizes. Advances in methods of pore size analysis at very low pressures [16] facilitate definition of zeolite aperture dimensions without the extensive experimentation required by traditional molecular probing with a variety of large molecules.

METHODS OF SUB-STRUCTURE DEFINITION

^{13}C-MASNMR of Trapped Templates

The contributions of magic angle spinning (MAS) NMR to zeolite science are well known and the technique is now a "standard method" in the repertoire of zeolite characterization techniques [11,12]. The use of nuclei other than ^{29}Si (^{11}B, ^{27}Al, ^{31}P, etc.) is now providing insight into the framework chemistries of other zeolite-related microporous. materials such as the phosphates [17] – ALPO$_4$-n, SAPO-n, MeAPO-n and MeASPO-n phases – and of new compositions derived from T-atom replacement in conventional zeolites. These methods have been used most effectively to provide refined structural details for materials whose framework structures are known. For nearly pure SiO$_2$ compositions (including those generated by dealumination) they may clarify space group assignments [1,11,12].

Recent ^{13}C studies of trapped templates and sorbed ^{129}Xe indicate that information on pore architectures in unknown structures can be obtained. Our first interest [14] was in "calibrating" the intergrowth behavior in zeolite T and similar materials (which comprise offretite-erionite intergrowths). Synthetic products in this system, made using a tetramethylammonium (TMA) cation template, often yield "offretites" having similar X-ray diffraction patterns but variable sorption and catalytic properties. ^{13}C-MASNMR analyses of these materials showed limiting ^{13}C chemical shifts assigned to TMA cations in the gmelinite cages and extended channels of the offretite structure, with, in addition, intermediate structure in the ^{13}C spectrum implying that other TMA sites were also present. The ^{13}C chemical shifts for TMA trapped in the cages of known zeolites (FAU, LTA, MAZ, OFF-framework materials) provide a calibration curve that relates the ^{13}C chemical shifts to the reciprocal of the average cage dimension (Fig.1). Different ^{13}C chemical shifts indicate differing cage dimensions. For intergrowths in zeolite T, GME-> LOS-> CHA-> ERI cages represent increasing cage lengths in the series of structures derived from the stacking of parallel interlinked 6-rings. The number of zeolites that are synthesized containing significant amounts of occluded TMA is limited, but calibration curves of this type might possibly be extended to other templates (TMA being the simplest with only a single carbon atom in each alkyl chain), enabling template ^{13}C NMR generally to provide helpful data on the pore architectures of new materials. (The torsional vibration frequencies of the TMA$^+$ cation, which can be measured by inelastic neutron scattering, also vary depending on the size of the entrapping cage [18].)

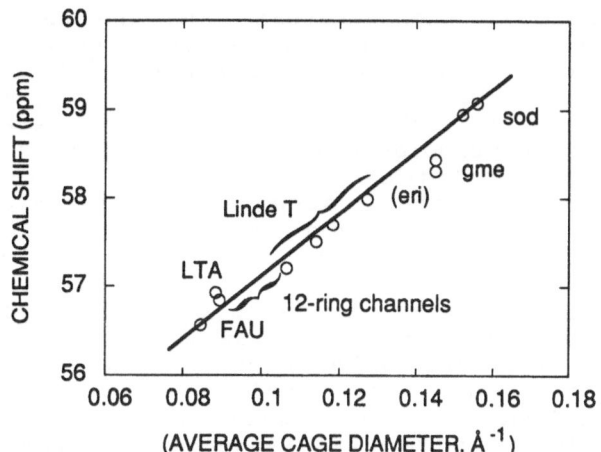

Fig. 1. A calibration showing the change in TMA ^{13}C NMR chemical shift as a function of the (reciprocal) average cage or channel diameters for TMA trapped in several zeolites. Observed "blocked offretite" intergrowths are indicated (Linde T).

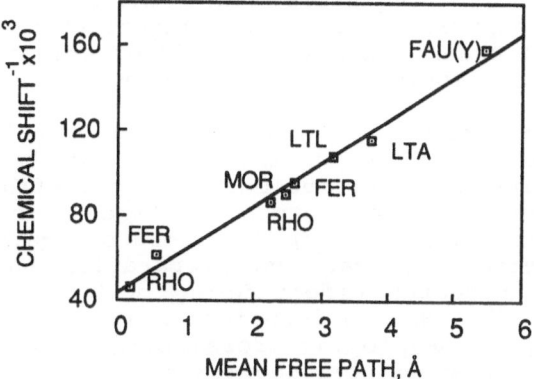

Fig. 2. Relationship of [129]Xe NMR chemical shift at 'zero' coverage to Xe 'mean free path' (mfp) in several zeolite structures. More than one point indicates multiple Xe sorption sites in the same zeolite [19].

[129]Xe MASNMR of Sorbed Xe

Zeolite applications of [129]Xe NMR have recently been reviewed by Fraissard and Ito [15]. The [129]Xe chemical shift depends on the xenon loading level, the character of the pore space available to the Xe atoms, and the nature of other species present within the accessible pore region. The contribution of Xe - Xe interactions make [129]Xe NMR data more difficult to interpret for unknown structures, particularly when trapped template molecules are present. However, the [129]Xe chemical shift extrapolated to zero coverage (when Xe - Xe interactions are negligible), like that of ^{13}C in TMA species does vary inversely with average pore dimension or mean free path (mfp) in an apparently simple fashion [19] (Fig. 2), with a chemical shift of over 200ppm for a near zero mfp and about 50ppm for a 7Å mfp. Extension of the method to encapsulated xenon [20] will allow the analysis of smaller pore and cage zeolites containing isolated Xe atoms. However, the single reported case of isolated Xe atoms in a cage, in Nu-3 (LEV) [21], indicates that truly isolated Xe atoms may have a different calibration curve compared to sorbed Xe. [129]Xe NMR will be applicable to a wide range of zeolites and related materials, and should prove helpful in defining pore and cage dimensions in new materials.

The approaches applied so far use estimated idealized spherical or cylindrical cavity dimensions based on crystallographic data. The chemical shift dependence might be further refined by investigating relationships with calculated vibration frequencies of the trapped or sorbed molecules obtained from potential energy calculations or measured equilibrium constants using the method of Barrer and Vaughan [22]. This approach is reasonable as long as the molecules are inert and near spherically symmetric (as are TMA, methane and Xe), and are in isolated sites. NMR methods are still evolving, but with more experience with their interpretation, they may be able to reveal channel details in new materials, such as lobe and pocket sizes, locations and connectivities.

Product distributions in a number of catalytic reactions are also sensitive to the details of the zeolite pore geometry. Analysis of product distributions from paraffin hydrocracking experiments on platinum-loaded zeolites can provide some structural data [23], but such experiments require a measure of skill and experience in catalyst preparation and testing [24]. A coordinated research effort using both hydrocracking test reactions and [129]Xe NMR on identical samples (with and without Pt) would help in developing cross correlations between these complementary probes of pore geometries.

Low Pressure Pore Size Analysis

The recent development of a new commercial instrument for micropore analysis [16,25] adds a powerful new tool to aid in zeolite structure characterization. Published data on the method are sparse at this time and much needs to be done to fully understand the method's capability. Fig. 3, and other published low pressure isotherm data, demonstrates clear differentiation of pore sizes not accessible with alternative instruments. An indication of diffusion kinetic effects is suggested in the Na,Ca-A "isotherm" (these are not necessarily equilibrium data in small pore materials); pore size dispersion is indicated in the tailing of the PILC isotherm. Apparent anomalies show that pore shape and "surface" affinity are important variables, exemplified by ZSM-5, a 10-ring channel

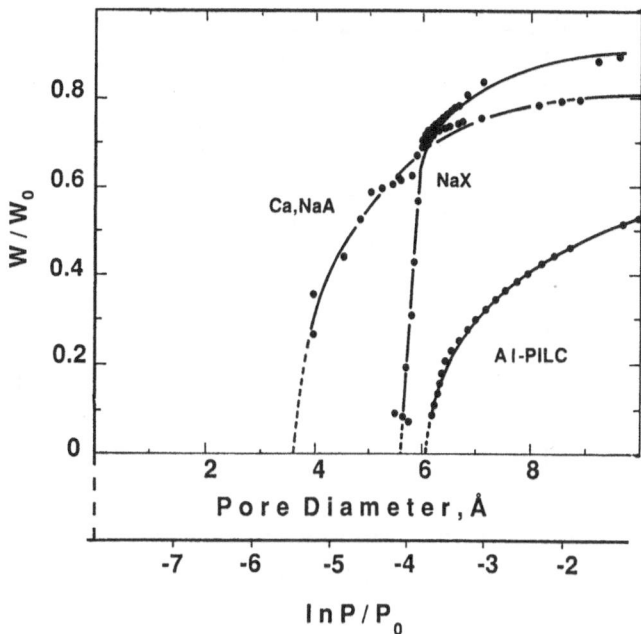

Fig. 3. Comparison of low pressure N_2 isotherms at 77K for
Ca,Na-zeolite A (5A), Na X and an Al-PILC, showing
distinctive differentiation of pore sizes, kinetic
limitations with partly blocked 8-ring windows in
Na,Ca-A, and the broader pore size spectrum in the
Al-PILC.

zeolite, which always gives a slightly smaller "pore size" than
Ca-LTA, an 8-ring cage zeolite (window diameter ~4.5Å; cage
diameter ~11Å) [26]. This may reflect a narrowing of the ZSM-5
channel at the low temperatures of the sorption experiments
(77°K or 88°K), possibly related to the orientation of residual
'OH' groups at the pore mouths. Analyses of several forms of
FAU-framework materials, show differences in pore size well
beyond the reproducibility of the technique.Argon sorption data
are given in Fig. 4. A trend of increasing pore size is
observed from LZ-210 (an Si/Al~ 5 material obtained by ammonium
silico-hexafluoride treatment of Y-FAU [27]), to Na-Y (Si/Al ~
2.5), to steam dealuminated US-Y (LZY-82) having an enlarged
average pore size due to a secondary pore system [28]. However,
conventional wisdom would predict the peak width for the more
perfectly crystalline LZ-210 to be narrower than that for the
partly amorphous US-Y - contrary to the experimental evidence.
Similarly the mesopore component in LZY-82 does not show, as
one might expect, as asymmetry, or tailing, in this peak.

Our experience with this technique is that the data are
specific and highly reproducible on zeolitic materials.
Although needing more thorough theoretical and experimental
exploration [eg. 29], it is already giving data of use in
differentiating and identifying zeolite pore sizes.

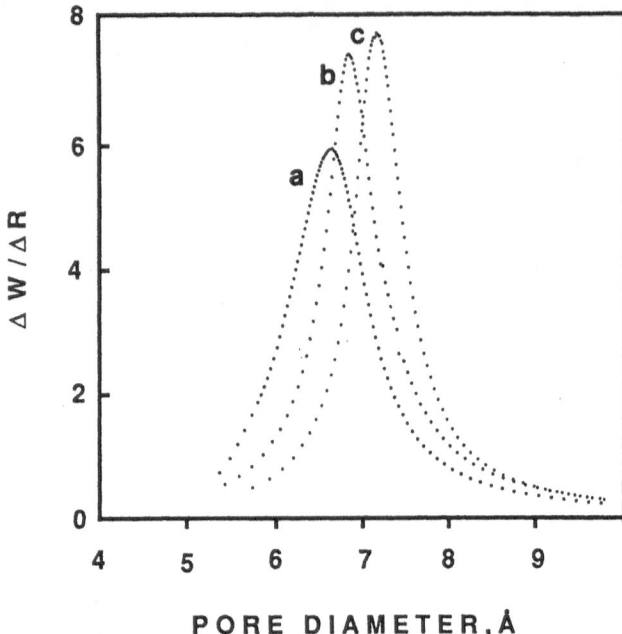

Fig. 4. Low pressure Ar pore size analyses (at 88K) for
LZ-210 (a), Na-Y (b) and LZY-82 (c) FAU-framework
zeolites.

The pore diameter analyses given in Figs. 3 and 4
represent a 4-10 potential model characteristic of parallel
plates, and is most appropriate for microporous carbons, clays
and PILCs. A 6-12 potential model may be more appropriate for
zeolites, as would the use of Slater-Kirkwood rather than
Kirkwood-Muller dispersion constants (16,20,30). A possible
drawback of the method is that argon and nitrogen, the
preferred probe molecules at this time, will not penetrate
windows less than about 4Å at 77°K - the diameter of a plane
regular 8-ring. Extension of studies to smaller molecules, such
as neon and helium, might therefore increase the utility of the
method in zeolite structure analyses, where distorted and
partly blocked windows are common.

DIFFRACTION METHODS

In House Powder X-ray Diffraction

The in-house powder X-ray diffractometer is the workhorse
in zeolite phase identification, and, in terms of hardware,
present instrumentation represents the fruits of a concerted
optimization effort that has taken place over more than four
decades. More reliable position-sensitive detectors, and energy
dispersive detectors have improved the quality and rate of
acquisition of data.

Recent progress in data analysis methodology has had major
impact in the utilization of this method. For phase
identification, the development of quantitative methods, rapid
peak searching algorithms and the ready availability of

reference patterns of standard materials (the full JCPDS file now being available on CD-ROM) greatly improves the speed and accuracy of analysis. For zeolite chemistry, the publication of the theoretical patterns volume [31] and ongoing developments to develop a compendium of reference patterns for hypothetical structures [32,33] also promises to be valuable in structure elucidation. Much still needs to be done in making reliable data files for all known zeolites widely available and up-to-date. There is a developing trend away from using decomposed peak positions and/or intensities toward treating the entire diffraction profile in a Rietveld type analysis [34]. Full pattern treatments (such as that illustrated in Fig. 5) readily indicate the quality of a trial unit cell, as well as yielding optimized lattice constants, and decomposed reflection intensities [35] for possible further treatment in structure refinement or solution. Full profile treatments also gives a better representation of intergrown phases that are common in zeolite systems, such as G (chabazite like), T (offretite-erionite), beta and ZSM-5/11. The number of structures solved using in-house data [e.g. 36] will likely increase. A primary limitation on such analyses is the limited instrumental resolution, which prevents the extraction of well defined peak intensities for overlapped reflections. As discussed below, developments at synchrotron X-ray facilities promise to enable such structure solutions and refinements that are intractable using in-house instrumentation.

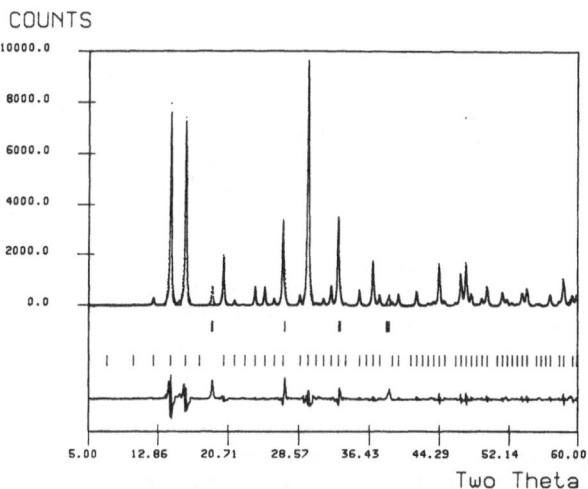

Fig. 5. Results of a full pattern fitting [35] of the PXD pattern of a sample (labelled phase B) of a hydrated sodium stannosilicate [47], based on a cubic unit cell with a = 12.437(1)Å. The upper reflection bars indicate the calculated positions of the most intense lines in the PXD pattern of lithium metasilicate present here as a second phase.

The rotating anode (RA) source provides a usable X-ray flux gain of ~5x, and current RA generators now offer stable and reliable operation. Evolution of hardware and software has significantly improved the stability and reliability of the newer instruments. Routine use for powder or single diffraction experiments is therefore possible, allowing, for example, studies of somewhat smaller crystals than has been possible with conventional laboratory X-ray equipment [e.g. 37].

Synchrotron X-ray Diffraction

Diffraction techniques at the synchrotron source are experimentally demanding, requiring stability in the source and experimental configuration over a time scale of several hours or even days, and progress in utilizing synchrotron sources has been steady over the last five years [38,39]. A schematic of the recently completed X10B Exxon beamline at the National Synchrotron Light Source at Brookhaven (NSLS) is shown in Fig. 6. The brightness and intrinsically low divergence of the synchrotron X-ray source permit high resolution, $\Delta d/d \sim 5.10^{-4}$, to be obtained in powder diffraction experiments. This increase in resolution implies a concomitant increase in the amount of information that can be extracted from the powder diffraction profile, enabling more complicated structures to be refined by Rietveld techniques [34], and enhancing the possibilities for *ab initio* structure solutions. Refinements [40,41] of the structure of ZSM-11 based on data from the X13A beamline at the NSLS provides the first refinement of the framework atomic coordinates, following the initial description of the structure based on only distance least squares modelling [42]. The structure refinement involved 8500 data points and 678 contributing reflections and yielded quite acceptable definition of the framework geometry, involving the adjustment of 76 atomic parameters. The structure of the clathrasil Sigma-2 [43] was solved based on decomposition of a synchrotron powder X-ray diffraction profile also accumulated on X13A at NSLS with the decomposed and corrected intensities used in a conventional Direct Methods process. Nine of the total of 11

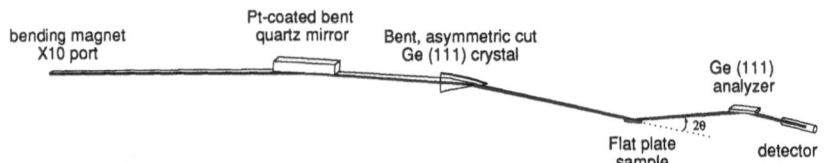

Fig. 6. Schematic representation of the Exxon beamline X10B at the NSLS. Radiation from a bending magnet port is reflected by a Pt-coated quartz mirror and monochromated by diffraction in the (near-) horizontal plane from a triangular, asymmetric cut Ge (111) crystal. The asymmetric cut provides collimation; the bending permits horizontal focussing of a narrow wavelength band at the sample position. For powder diffraction work, the diffracted beam resolution may be defined by slits or, as shown here, by an analyzer crystal.

unique framework atoms were located with the remainder of the structure then determined by Fourier synthesis and refinement. A small number of other structures have been successfully solved similarly using Direct Methods or Patterson techniques [e.g. 44-46]. Increasing ease of access to synchrotron powder X-ray diffraction facilities will also permit more routine use for phase identification [47]. An additional benefit, as yet little tapped, is the increased amount of information that can be extracted from the powder diffraction peak profiles themselves when the instrumental contribution to the broadening is dramatically reduced. Earlier studies suggested that strain can have a pervasive effect on bulk properties [9,48], although indications of strain in zeolite crystals have so far only been accessible by careful measurements in the electron microscope. Stacking disorder, typified by that observed in zeolite beta [4-6] (Fig. 7) and intergrown materials related to faujasite [9,10,48,49] (Fig. 8), can also be studied quantitatively by detailed quantitative analyses of the complete powder diffraction profile. A general purpose program, recently developed specifically for computing the effects of stacking disorder on diffraction patterns from complicated structures such as zeolites [50], greatly facilitates such analyses.

The availability of synchrotron X-ray sources also facilitates studies of poorly crystalline materials by radial distribution function work. Additionally, the wavelength tunability promises an increased ability to characterize metal catalyst particles dispersed on or throughout supports [51], and target metal cation species in framework or non-framework sites.

The study of zeolite ZSM-11 [40] illustrates that an optimal crystallite size is needed (with minimal contributions from inhomogeneity and strain) to utilize the potentially extremely high instrumental resolution offered by powder diffractometers at synchrotron X-ray sources. In this case [40] the effective particle size of ~0.5 μm produced a significant

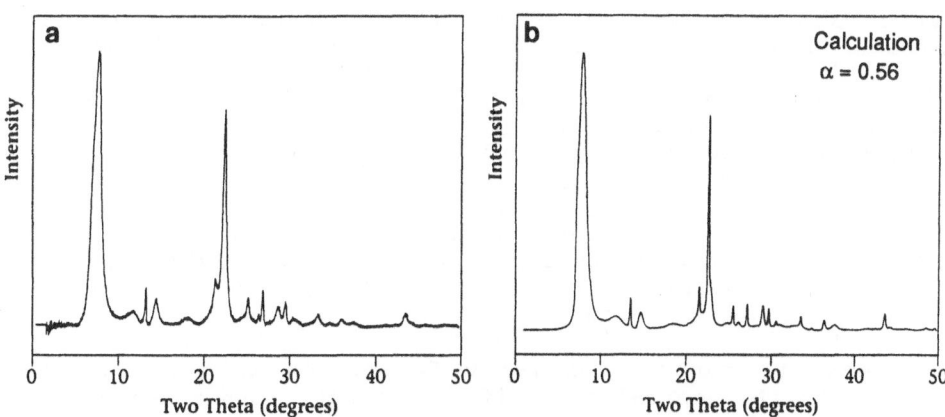

Fig. 7. Observed ('in house', Siemens D500 data) and simulated powder diffraction profiles for a typical sample of zeolite beta [50]. The best fit is for a probability 0.56 that layers will stack with alternating handedness.

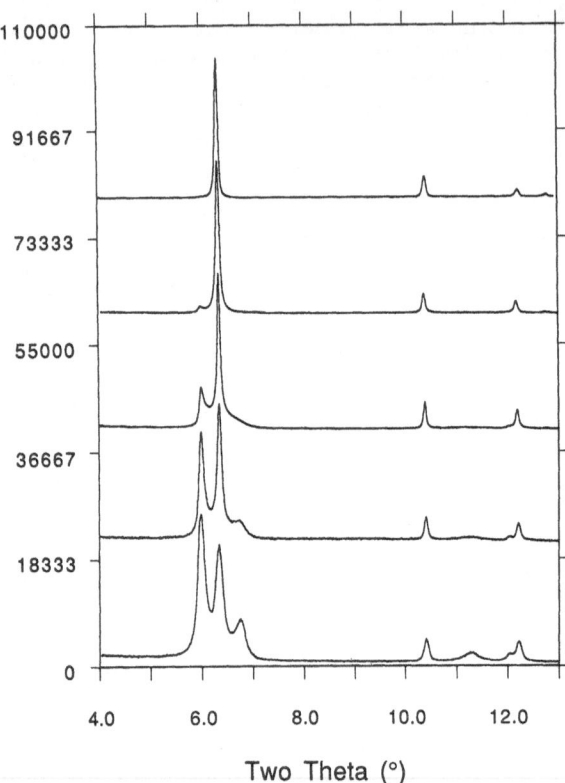

Two Theta (°)

Fig. 8. Low angle regions of the powder X-ray diffraction
profiles measured on X10B at NSLS for a cubic FAU-
framework material (top) and for materials showing
successively higher levels of the hexagonal mode
of stacking, representative of CSZ-3, ZSM-20, etc.

particle size broadening in the diffraction profile; individual
crystallites of ≥1 μm would be required for optimal resolution.
Particles in the micron size regime are accessible directly by
microcrystal techniques [39,47,52,53,54]. Microcrystal
diffraction experiments at the synchrotron suffer from the
temporal instabilities that hamper all diffraction experiments,
with the additional requirements of mounting and manipulating
tiny crystals, and defining their orientation on the
diffractometer. Many of these difficulties have been overcome
and it now proves routinely possible to manipulate, mount and
align crystals of ~5 μm [39,47]. (Fig. 9.) Although obtaining
fully internally consistent intensity data sets over several
hours or days currently remains a problem, ongoing developments
should shortly enable microcrystal diffraction to become a
widely applicable technique..

Access to synchrotron X-ray facilities is increasingly
available to the general scientific community. Other aspects of
zeolite research, not mentioned here, will also benefit from
synchrotron X-ray techniques, including studies of zeolite
crystallization and structural studies under non-ambient or *in
situ* catalytic conditions.

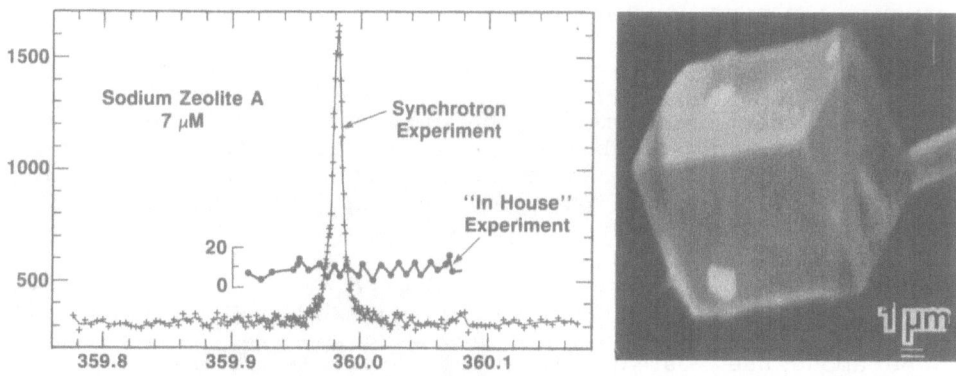

Fig. 9. SEM micrograph of a 7µm crystal of zeolite A (some 2500 unit cells on an edge) glued to a hollow glass fibre that tapers to a point of < 1µm (right), and a single reflection profile, (014)$_{subcell}$, for a similar crystal measured on the Exxon beamline X10A at the NSLS, (comparable 'in-house' data are also shown).

Neutron Diffraction

Although single crystal neutron diffraction is restricted to those materials that occur only as relatively large crystals, $\geq$ ~500 µm, acceptable specimens of several natural minerals are available, and studies on such materials have yielded precise information on framework geometries, as well as a detailed picture of the location and coordination of sorbed water molecules [55,56]. There are currently no practical prospects for neutron facilities with significantly enhanced fluxes such as would permit measurements on much smaller crystals. Although there has also been no outstanding progress in the production of large crystals of interesting synthetic aluminosilicate zeolites, the prospects for similar alumino-phosphate based materials is encouraging.

From relatively modest beginnings, powder neutron diffraction has continued to improve in its range of applicability, and in the degree of complexity in structural problems that can be tackled [57,58]. Initial studies were restricted to high symmetry, cubic materials, but recent applications include lower symmetry materials and more complicated structures. A series of gallosilicates have yielded data on the effects of framework cation substitution [59-61]. Several studies have yielded information about non-framework aluminate species generated by framework dealumination [62-64]. The complete hydrated structure of a polycrystalline sample of zeolite Li-A(BW) was determined [65,66]. A small number of complete studies of zeolite-hydrocarbon sorbate systems have also appeared [67-71], and although the precision of these results has been somewhat limited, they have yielded invaluable information on sorbate location and orientation.

TRANSMISSION ELECTRON MICROSCOPY

Structure imaging

The first experimental demonstrations of the crystal lattice resolving capabilities of the transmission electron microscope (TEM) were made more than 30 years ago using faujasite and platinum pthalocyanine as test specimens [72]. These materials were attractive because their large unit cells provided lattice spacings in excess of 1nm, suitable for the resolving power of TEM instruments of that time. The best modern TEMs have point-to-point resolutions approaching 0.15 nm. It is ironic that zeolites and organics are now considered to be among the most difficult of materials to image at high resolution due to the rapid structural disintegration which occurs under the irradiating electron beam. Radiation damage makes it difficult to resolve details finer than about 0.4 nm in zeolites without taking special precautions. Presumably, this electron beam sensitivity is partly responsible for why TEM tends to be a disfavored tool for zeolite characterization relative to the more central role that it enjoys in most other materials science fields.

Important progress was made when it was found that the radiation damage rate was influenced strongly by the water [73-77] and aluminum content of zeolites [78]. Significant improvements in stability under the electron beam can be achieved by thoroughly dehydrating the materials prior to analysis (by heating in vacuo) and by chemical dealumination [79,80]. The latter has the added benefit of expelling the charge-balancing cations, thus simplifying comparison between experimental and theoretical images. These important advances have facilitated the investigation of structural intergrowths and other defects which can affect zeolite properties. Millward et al. [81] demonstrated that the presence of coherent intergrowths of ZSM-5 and ZSM-11 could be identified from electron diffraction patterns (where streaking of certain spots occurs), and that individual intergrowth layers could be resolved in structure images taken down the [010] axis. The success of their image analysis revolved around the ability to determine the local projected symmetry about the 6-rings, which required sufficient structural stability of the material to distinguish two different types of 5-ring. The structural stability necessary for this study stemmed from the low aluminum content of as-synthesized ZSM-5/-11 materials. However, comparable results have been achieved in structural studies of chemically dealuminated zeolite L (LTL) and mazzite (MAZ) [79]. Dealumination has the disadvantage of potentially introducing defects into the zeolite, such as the 'estuarine' defects observed in LTL crystallites [79]. However, acceptable results can be obtained on non-dealuminated materials such as offretite, which can contain intergrowths of other members of the ABC-6 family of structures [82]. These studies demonstrate that, when used with care, TEM has an important role to play in the study of defect structures in zeolites. Notable recent studies include determinations of the framework structures of the zeolites ECR-1 [2], CSZ-1 [9,48], ZSM-23 [83], beta [4-6], EU-1 [84] and ZSM-20 [10,49], most of which contain defect structures. TEM played an important role in all of these studies.

Figure 10 shows a micrograph of ZSM-20 which is a disordered intergrowth of cubic and hexagonal stackings of faujasite sheets [49,85]. In a similar fashion to the ZSM-5/ZSM-11 materials, ZSM-20 can be considered as being built from sheets which interconnect either through an inversion or a mirror operation. A cubic FAU-framework (faujasite) results if the stacking of faujasite sheets is only through inversion operations; the hexagonal variant, 'Breck's structure 6' (*bss*) results if the stacking is only through mirror operations. ZSM-20 is an intergrowth of the two modes of stacking, as can be seen from the micrograph in figure 10, which is viewed down the cubic [110] axis. The [110] projection is included, showing the

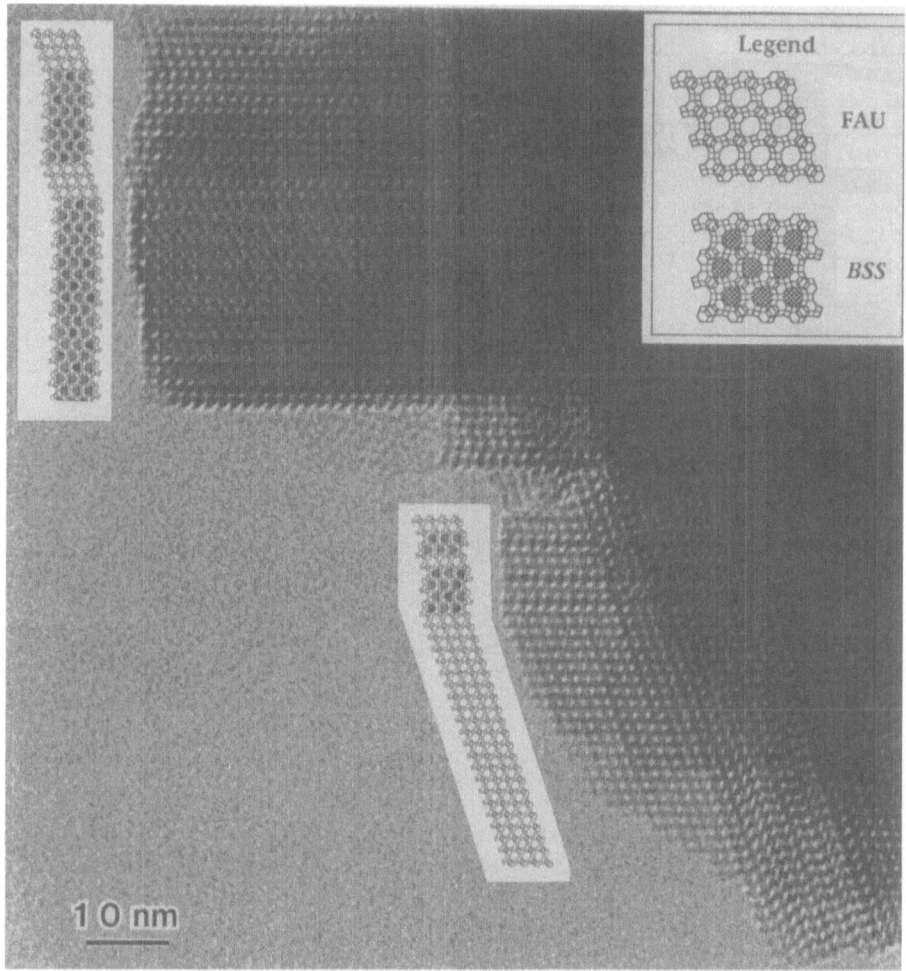

Fig. 10. [110] structure image of intergrown platelets in a sample of ZSM-20. The large bright spots are columns of supercages. Extensive faulting within each platelet is evident, revealing competition between the cubic (FAU), and hexagonal (*bss*) stackings. The projection of the corresponding framework model is superposed. The shaded pores represent the *bss* supercages.

effect of the stacking on the local cage symmetries. Comparison of the framework projection with the micrograph shows that, due to beam damage and associated specimen movement, the image resolution (about 0.4 nm) is not sufficient to resolve the secondary cage structure. However, the relative displacements of the supercages is sufficient to determine the character of the projected stacking vectors.

Structure determination is much more difficult if the structural properties of the material do not resemble any previously known structures, as was the case for zeolite beta [4-6]. Figure 11 shows a micrograph of zeolite beta (viewed down the tetragonal [100] axis). The micrograph indicates that zeolite beta is also highly disordered. In projection, the stacking disorder appears superficially similar to that in faujasite with inversions or mirror operations inducing relative lateral shifts in layers of ±1/3rd of a unit cell repeat. Electron diffraction supplies supplementary information indicating that the disorder arises from competition between 4_1 and 4_3 screw operations. The high silica to alumina ratio (>10) of zeolite beta materials means that they are relatively stable to the electron beam, and details of the secondary cage projected structure can be resolved, in particular the possible location of the 6-rings. The imaging and diffraction data, coupled with knowledge of the sorption and infra-red properties led to a computer reconstruction of the framework [5]. Zeolite

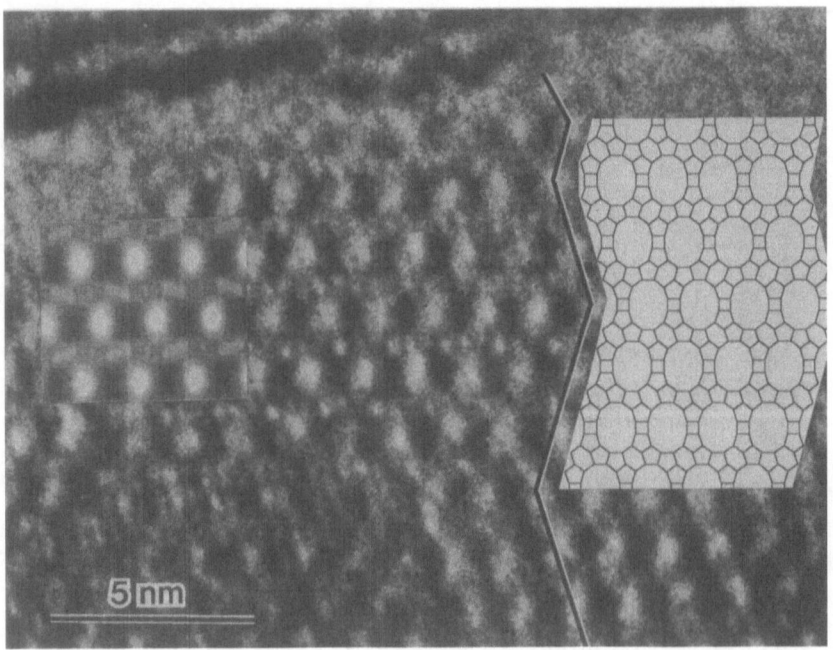

Fig. 11. [100] structure image of zeolite beta. The large bright spots reveal the location of the 12-ring channels aligned along [100]. There is extensive faulting along the vertical direction. The projected framework, and multi-slice image simulation are superposed for comparison.

beta is a three dimensional 12-ring system, that can be considered as an intergrowth between two end-member structures. One of the end-members, *bea*, is chiral, with either left-handed (laevo-*bea*, space group P4$_3$22) or right-handed (dextro-*bea*, space group P4$_1$22) channel connectivities along the c-axis. The other end-member, *beb*, (space group C2/c) is achiral, with layers stacking in alternating handedness. Typical zeolite beta materials have between 50 to 60% of the alternating mode of stacking (*beb*). This result has been confirmed by modelling of the PXD patterns (Fig 7.). To our knowledge, pure unfaulted forms of either end-member structure have not been synthesized.

It is important to note that neither of the micrographs shown in figures 10 and 11 display atomic resolution. The effective point-to-point resolution in each being little better than 0.4 nm. The critical information comes from determining the layer stacking vectors, and some insight into the secondary cage structure around the main pores (which are nearly always easily resolved). The stacking vectors are about 1.5nm in length for both beta and faujasite and are obliquely inclined to the stacking direction. Thus, faults are readily detected in these systems as kinks or reversals in the layer stackings. Furthermore, zeolite frameworks are open enough that the projection approximation holds well. The largest apertures show the largest contrast, and the smallest apertures the least. The smallest apertures may not be resolvable, but appear as dark heavily scattering areas in the micrograph. The 6-rings in zeolite beta are thus more contrasty than the 5-rings. The 4-rings, which are squeezed tightly between the 12-ring pores are revealed as the darkest patches in the micrograph (figure 11).

The projection approximation was used effectively in the TEM studies of intergrowths of ZSM-11 in ZSM-5 mentioned earlier. This is a difficult system in which to study faults, since the layer stacking vectors for the two materials are both equal to $a/2$, and are thus indistinguishable by TEM. Higher resolution images are required (~0.2 nm point-to-point) to resolve directly the two different types of 5-ring present. As shown in the studies of Millward and Thomas [81], the relative disposition of the two types of 5-ring reveal whether the layers are stacking through inversion or mirror operations.

Although in many respects it is fortunate that the projection approximation is valid for zeolites, permitting immediate interpretation of zeolite images, multi-slice simulation of high-resolution images is nowadays a routine (and, generally, a mandatory) exercise in any high resolution study. The projection approximation is applicable only for micrographs recorded at optimum focus in a well-aligned instrument. Multi-slice simulation is a safeguard against errors of interpretation. An example of a multi-slice simulation for zeolite beta is included in figure 11.

Scanning Electron Microscopy (SEM)

Clues as to the purity of a given material, or the identity of the various phases present can be obtained quickly by examining characteristic morphologies of the crystallites present in the SEM. The most significant recent advance in SEM technology is the increasing use of high brightness field emission sources. Although expensive, these electron sources enable resolutions of the order of 1nm in secondary imaging mode, whilst providing adequate beam current to ensure a useful signal-to-noise ratio. In many respects, such SEMs have comparable microanalytical performance to the dedicated scanning transmission electron microscope (STEM), which is capable of transmission imaging as well as the standard SEM modes.

Scanning Transmission Electron Microscopy (STEM)

In the STEM images are formed by scanning a finely-focused probe over a (usually) thin specimen. Probe sizes of ~0.2 nm have been achieved using 310 oriented tungsten cold field emission sources [86]. The reciprocity principle asserts that under identical imaging conditions, TEM and STEM images should be identical [87]. In practice, STEM images have slightly diminished resolution in bright field images, relative to those in TEM, because of the finite source size and diminished detection efficiency due to the collection geometry. However, the STEM has an important advantage. Since specimens are probed on a point-by-point basis, it offers a cornucopia of high resolution microanalytical capabilities which are unavailable to the conventional TEM. In addition, all the various microanalytical signals can be collected simultaneously. Thus, as demonstrated in the previous section, the TEM is principally a structure imaging tool, whereas the STEM is principally a microanalytical tool. Because of the advantages of both types of instrument, most modern microscopes can convert between the two modes. However, the state of the art instruments (offering the highest resolutions) are usually dedicated to a single mode. (For a clear review of STEM see ref [88]).

Microanalytical tools available to the STEM include X-ray spectroscopy (energy dispersive or wavelength dispersive), electron energy loss spectroscopy (EELS), secondary electron spectroscopy, Auger spectroscopy, microdiffraction, back-scattering, high angle scattering, bright field imaging and reflection electron microscopy. Each signal can (in principle) be displayed either as a spectrum per point or as a filtered image at each point. It is beyond the scope of this article to discuss each technique in detail. Instead we mention two techniques which have been used in studies of compositional heterogeneity in zeolites, Z contrast, and X-ray microanalysis.

Z Contrast Imaging

The Z contrast technique (or atomic number imaging) exploits the fact that high angle scattering of electrons from thin specimens follows the Z^2 dependence of Rutherford's law, where Z is atomic number [86,88-91]. Thus, one Pt atom scatters as strongly as about 100 oxygen atoms, or 32 silicon atoms. The technique is tailor-made for detecting clusters of catalytically active metals, such as Pt, on light catalyst supports such as alumina and zeolites [89,90].

Figure 12(a) shows a STEM high angle annular detector image of zeolite L which has Pt clusters dispersed throughout its channels. In this image the bright spots represent increased scattering and thus reveal the location of the Pt. On

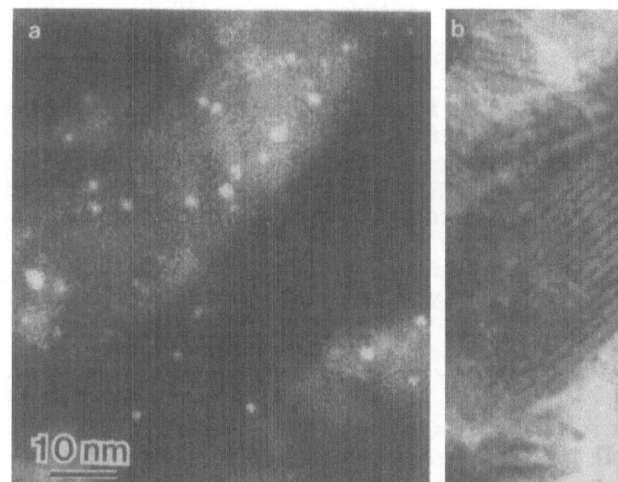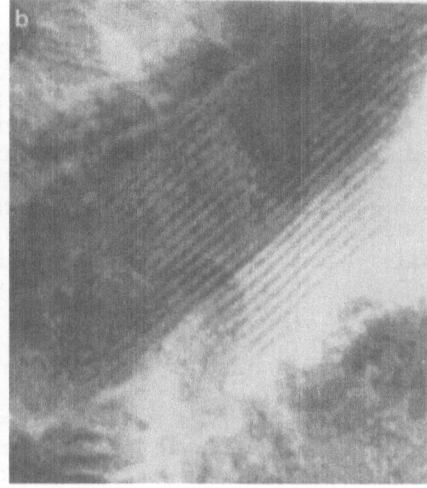

Fig. 12. (a) STEM Z contrast image of Pt in zeolite L. The bright dots represent regions of enhanced scattering and thus contain Pt. (b) STEM bright field image of the same area. Owing to the dominance of phase contrasts, the platinum is barely detectable.

a routine basis, this technique can detect clusters containing as few as three Pt atoms, but under optimum conditions (~3nm thick zeolite support) single Pt atoms can be detected. In bright field images (either STEM or conventional TEM), clusters smaller than about 0.7 nm (~10 - 15 atoms) cannot be discerned against the strong phase contrasts of the background. A STEM bright-field image of the same area is shown in figure 12(b). The zeolite L channels are easily resolved in this image; however, the smaller Pt clusters are not visible. Taken together, the two images provide valuable clues as to the disposition of Pt throughout the zeolite.

A recent refinement to the Z contrast technique, which takes advantage of advances in computer assisted digital image acquisition and processing, uses the total intensity scattered by each cluster to estimate the number of atoms present. In the high angle annular detector, intensity is proportional to the number of scattering atoms. This method is more reliable than directly measuring imaged diameters, since in the sub-1nm size regime, image diameters are exquisitely sensitive to the microscope operating parameters, such as focus [91]. However, at present, the technique relies heavily on expensive image processing capabilities and is more difficult to implement on a routine basis than conventional particle counting methods.

X-ray Microanalysis

The resolution of X-ray fluorescence signals in STEM is related to the total volume of material probed at any given point. This is governed by the probe size, by the probe spreading as it passes through the sample, and by beam-induced diffusion of atoms (such as the mobile cations in zeolites). In field emission STEM (and SEM), the probe size is a negligible factor relative to the others. In zeolites, beam damage is the principle resolution limiting factor. In order to obtain precise spectra, the doses required are sufficient to completely destroy the framework, induce pronounced cation migration, and reduce the sample to a glass [77]. For most practical zeolite studies, spatial resolution is thus limited to about 10 nm. Non-framework cation migration renders analysis of such species unreliable. However, studies show that the Si/Al ratio remains constant for a considerable irradiation period at 100kV [77]. This offers a valuable way to investigate compositional homogeneity across zeolites [92]. Several studies to date have exploited X-ray fluorescence in STEM to study Si/Al variations across as-synthesized, and dealuminated materials [93,94].

Figure 13 shows a STEM bright field image of a ZSM-20 platelet viewed down the [110] axis (to reveal the stacking faults). The lattice is not visible in the micrograph due to the rapid disintegration which occurs under the electron illumination conditions used for X-ray microanalysis. However, the platelet morphology shows that it has two extended, but distinct, regions; one is predominantly cubic FAU-type stacking, (here revealed by the tell-tale incline at the platelet edge), and the other with the perpendicular sides being predominantly *bss*. (This bi-phasic structural assignment was readily confirmed from the lattice fringes which were briefly, but clearly witnessed on the microscope screen prior to taking the micrograph.) Crystallites where the two phases are unambiguous and cleanly separated are uncommon in ZSM-20. Most crystallites exhibit extensive intergrowths as shown in the TEM image of figure 10. Analysis of 8 suitable specimen areas shows that the Si/Al ratio in the predominantly *bss* regions, is consistently 6±2% higher than in the predominantly FAU regions. This is consistent with the fact that the hexagonal *bss*-mode of stacking is observed progressively at higher Si:Al ratio gel compositions.

COMPUTER-BASED MODELLING

Molecular graphics packages have proven extremely helpful in studies of zeolite structure and structural chemistry [95-97]. The construction of models of zeolite structures, and the generation of new model structures by modification to existing ones is greatly facilitated [e.g. 5]. Total Energy Minimization techniques [98], the development of molecular mechanics parameters suitable for zeolite framework components [99], and, for example, the modification of the distance least squares optimization procedures to incorporate such energy terms, promise to further extend the utility of computer modelling methods in structural characterization efforts [100].

ACKNOWLEDGMENTS

The authors are grateful to Stephen B. Rice for permission to use the Z contrast and ZSM-20 microanalysis results, and to C. Z. Yang for help with the cited synchrotron PXD experiments. The authors are also grateful to Allan Jacobson for encouragement and many stimulating discussions.

116

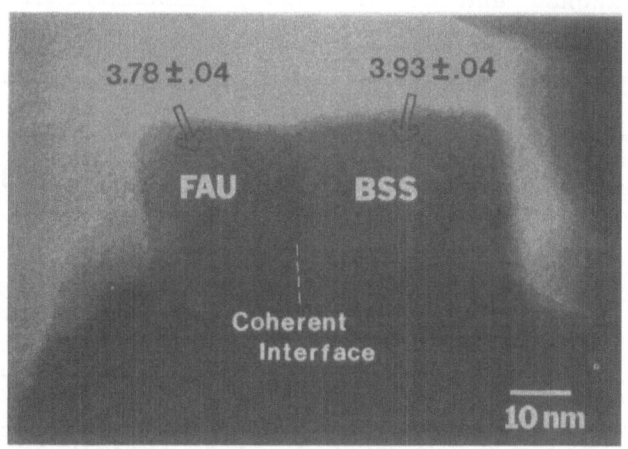

Fig. 13. STEM bright field image of a ZSM-20 platelet viewed down [110]. The lattice structure has been destroyed by the intense current density in the 0.4 nm probe. The platelet morphology (and lattice fringes prior to exposure) reveals regions which had predominantly the cubic FAU framework, and regions which had predominantly the hexagonal *bss* framework. X-ray fluorescence reveals that the *bss* regions consistently have a 6% higher Si/Al ratio relative to the FAU regions.

REFERENCES

1. J.M.Thomas and D.E.W.Vaughan, J.Phys.Chem.Solids, 50: 449 (1989).
2. M.E.Leonowicz and D.E.W.Vaughan, Nature, 329: 819 (1987).
3. D.E.W.Vaughan, K.G.Strohmaier and M.E.Leonowicz, Amer.Chem.Soc. Symp.Ser., Ed. S.A.Bradley, M.J.Gattuso and R.J.Bertolacini (1989), *in press*.
4. M.M.J.Treacy and J.M.Newsam, Nature, 332: 249 (1988).
5. J.M.Newsam, M.M.J.Treacy, C.B.DeGruyter and W.T.Koetsier, Proc.Roy.Soc., A420: 375 (1988).
6. J.B.Higgins, R.B.LaPierre, J.L.Schlenker, A.C.Rohrman, J.D.Wood, G.T.Kerr, and W.J.Rohrbaugh, Zeolites, 8: 446 (1988).
7. W.M.Meier and H.Villiger, Zeit. Kristallogr., 129: 411 (1969).
8. Ch.Baerlocher, A.Hepp and W.M.Meier, "DLS-76 - A Program for Simulation of Crystal Structures by Geometric Refinement", ETH Zurich Report (1977).
9. M.M.J.Treacy, J.M.Newsam, R.A.Beyerlein, M.E.Leonowicz and D.E.W.Vaughan, J.Chem.Soc. Chem.Commun., 1211 (1986).
10. J.M.Newsam, M.M.J.Treacy, D.E.W.Vaughan, K.G.Strohmaier and W.J.Mortier, J.Chem.Soc. Chem.Commun., 493 (1989).
11. J.M.Thomas and J.Klinowski, Adv.Catal., 33: 199 (1985).
12. G.Engelhardt and D.Michel, "High resolution solid state NMR of silicates and zeolites.", J.Wiley (1987).
13. R.H.Jarman and M.T.Melchior, J.Chem.Soc. Chem.Commun., 414 (1984).

14. D.E.W.Vaughan and M.T.Melchior, Mater.Res.Soc. Symp. Q, Boston, Dec.(1985), *unpublished*.
15. J.Fraissard and T.Ito, Zeolites, 8: 350 (1988).
16. G.Horvath and K.Kowazoe, J.Chem.Engrg.Japan, 16: 470 (1983).
17. C.S.Blackwell and R.L.Patton, J.Phys.Chem., 92: 3965 (1988).
18. T.O.Brun, L.A.Curtiss, L.E.Iton, R.Kleb, J.M.Newsam, R.A.Beyerlein and D.E.W. Vaughan, J.Amer.Chem.Soc., 109: 4118 (1987).
19. M.Springuel-Huet, J.Demarquay, T.Ito and J.Fraissard, "Innovations in zeolite materials science.", SSSC #37, Ed. P.J.Grobet, W.J.Mortier, E.F.Vansant and G.Schulz-Ekloff, Elsevier, 183 (1988).
20. R.M.Barrer and D.E.W.Vaughan, J.Phys.Chem.Solids, 32: 731 (1971).
21. D.W.Johnson and L.Griffiths, Zeolites, 7: 484 (1987).
22. R.M.Barrer and D.E.W.Vaughan, Trans.Faraday Soc., 63: 2275 (1967).
23. J.A.Martens, M.Tielen, P.A.Jacobs and J.Weitkamp, Zeolites, 4: 98 (1984).
24. J.Weitkamp, "Innovations in zeolite materials science.", SSSC #37, Elsevier, Eds. P.J.Grobet et al, 515 (1988).
25. W.J.M.Pieters and A.F.Venero, "Catalysis on the energy scene.", SSSC #19, Ed. S.Kaliaguine and A.Mahay, Elsevier, 155 (1984).
26. A.F.Venero and J.N.Chiou, Mater.Res.Soc. Symp.Proc., #111, Ed. M.M.J.Treacy, J.M.Thomas and J.M.White, 235 (1988).
27. D.W.Breck and G.Skeels, Proc.7th.Intl.Zeolite Conf., SSSC #34, Ed. Y.Murakami, A. Iijima and J. W. Ward, Kodansha/ Elsevier, 87 (1986).
28. U.Lohse, G.Engelhardt and V.Patzelova, Zeolites, 4: 163 (1984).
29. E.G.Derouane, J-M.Andre and A.A.Lucas, J.Catal., 110: 58 (1988).
30. R.M.Barrer and D.L.Peterson, Proc.Roy.Soc., A280: 466 (1964).
31. R.von Ballmoos, "Collection of Simulated XRD Powder Patterns for Zeolites.", Butterworths/ Intl.Zeolite Assoc. (1985).
32. G.Gottardi, private communication (1987).
33. J.V.Smith, Proc.8th. Intl Zeolite Conf., Ed. R.van Santen and P.A.Jacobs, Elsevier, *in press* (1989).
34. H.M.Rietveld, J.Appl.Cryst., 2: 65 (1969).
35. G.S.Pawley, J.Appl.Cryst., 14: 357 (1981).
36. P.R.Rudolf, C.Saldarriaga-Molina and A.Clearfield, J.Phys.Chem., 90: 6122 (1986).
37. J.J.Pluth, J.V.Smith and J.M.Bennett, J.Amer.Chem.Soc., 111: 1692 (1989).
38. J.M.Newsam, Science, 231: 1093 (1986).
39. J.M.Newsam, H.E.King,Jr. and K.S.Liang, Adv. X-ray Analysis (Plenum, New York), 32, *in press* (1989).
40. B.H.Toby, M.M.Eddy, C.A.Fyfe, G.T.Kokotailo, H.Strobl and D.E.Cox, J.Mater.Res., 3: 563 (1988).
41. C.A.Fyfe, H.Gies, G.T.Kokotailo, C.Pasztor, H.Strobl and D.E.Cox, J.Amer.Chem.Soc., 111: 2470 (1989).
42. G.T.Kokotailo, P.Chu, S.L.Lawton and W.M.Meier, Nature, 275: 119 (1978).
43. L.McKusker, J.Appl.Cryst., 21: 305 (1988).

44. J.P.Attfield, A.W.Sleight and A.K.Cheetham, _Nature_, 322: 620 (1986).
45. M.S.Lehmann, A.N.Christensen, H.Fjellvag, R.Feidenhans'l and M.Nielsen, _J.Appl.Cryst._, 20: 123 (1987).
46. A.N.Christensen, D.E.Cox and M.S.Lehmann, _Acta Chem.Scand._, 43: 19 (1989).
47. E.W.Corcoran, J.M.Newsam, H.E.King and D.E.W.Vaughan, "Zeolite Synthesis", _Amer.Chem.Soc. Symp.Ser._, #398, Ed. M.Occelli and H.E.Robson, _in press_ (1989).
48. M.M.J.Treacy, J.M.Newsam, D.E.W. Vaughan, R.A.Beyerlein, S.B.Rice and C.B.DeGruyter, _Mater.Res.Soc. Symp.Proc._, 111, Ed. M.M.J.Treacy, J.M.White and J.M.Thomas, 177 (1988).
49. D.E.W.Vaughan, M.M.J.Treacy, J.M.Newsam, K.G.Strohmaier and W.J.Mortier, _Amer.Chem.Soc. Symp.Proc._, #398, Ed. M.Occelli and H.E.Robson, _in press_ (1989).
50. M.M.J.Treacy, J.M.Newsam and M.W.Deem, "Disorder in Crystalline Materials", _Mater.Res.Soc. Symp.Proc._, 150, _submitted_ (1988).
51. K.S.Liang, S.S.Laderman and J.H.Sinfelt, _J.Chem.Phys._, 86: 2352 (1987).
52. P.M.Eisenberger, J.M.Newsam, M.E.Leonowicz and D.E.W. Vaughan, _Nature_, 309: 45 (1984).
53. J.M.Newsam and D.E.W.Vaughan, "ZEOLITES: Synthesis, Structure,Technology and Application", SSSC #24, Eds. B.Drzaj, S.Hocevar and S.Pejovnik, Elsevier, 239 (1985).
54. R. Bachmann, H. Kohler, H. Schulz and H-P Weber, _Acta Cryst._, A41: 35-40 (1985).
55. A. Kvick, _Trans.Amer.Crystallogr.Assoc._, 22: 97-106 (1986).
56. G. Gottardi, _Proc.7th Intl. Zeolite Conf._, SSSC #34, Eds. Y.Murakami, A.Iijima and J.W.Ward, Kodansha/Elsevier, 41 (1986).
57. J.M.Newsam, _Physica_, 136B: 213 (1986).
58. J.M.Newsam, _Materials Science Forum_, 27/28: 385 (1987).
59. J.M.Newsam and D.E.W.Vaughan, _Proc.7th Intl.Zeolite Conf._, SSSC #34, Ed. Y.Murakami, A.Iijima and J.W.Ward, Kodansha/Elsevier, 457 (1986).
60. D. Xie, J. M. Newsam, J. Yang, and W. B. Yelon, _Mater.Res.Soc. Symp.Proc._, 111, Ed. M.M.J.Treacy, J.M.White and J.M.Thomas, 147 (1988)
61. W. B. Yelon, D. Xie, J. M. Newsam and J. Dunn, _Zeolites_, _in press_ (1989).
62. J.B.Parise, T.R.Corbin, L.Abrams and D.E.Cox, _Acta Cryst._, C40: 1493 (1984).
63. R.X.Fischer, W.H.Baur, R.D.Shannon, R.H.Staley, A.J.Vega, L.Abrams and E.Prince, _J.Phys.Chem._, 90: 4414 (1986).
64. J.M.Adams and D.A.Haselden, _J. Solid State Chem._, 68: 351 (1987).
65. J.M.Newsam, _J.Phys.Chem._, 92: 445 (1987).
66. P.Norby, A.N.Christensen and I.G.Krogh Andersen, _Acta Chem.Scand._, A40: 500 (1986).
67. A.N.Fitch, H.Jobic and A.Renouprez, _J.Phys.Chem._, 90: 1311 (1986).
68. P.A.Wright, J.M.Thomas, A.K.Cheetham and A.K.Nowak, _Nature_, 318: 611 (1985).
69. J.M.Newsam, B.G.Silbernagel, A.R.Garcia and R.Hulme, _J.Chem.Soc. Chem.Commun._, 664 (1987).
70. J.C.Taylor, _Zeolites_, 7: 311 (1987).
71. J.M.Newsam, _J.Phys.Chem._, _in press_ (1989).
72. J.W.Menter, _Adv.Phys._, 7: 299 (1958).

73. L.A.Bursill, E.A.Lodge and J.M.Thomas, _Nature_, 286: 111 (1981).
74. L.A.Bursill and J.M.Thomas, _J.Phys.Chem._, 85: 3007 (1981).
75. L.A.Bursill, J.M.Thomas and K.J.Rao, _Nature_, 289: 157 (1981).
76. E.H.Hirsch, _Nature_, 293: 759 (1981).
77. M.M.J.Treacy and J.M.Newsam, _Ultramicroscopy_, 23: 411 (1987).
78. J.M.Thomas, _J.Molecular Catal._, 27: 59 (1983).
79. O.Terasaki, J.M.Thomas and R.G.Millward, _Proc.Roy.Soc._, A395: 153 (1984).
80. G.R.Millward, S.Ramdas and J.M.Thomas, _J.Chem.Soc. Faraday Trans 2_, 79: 1075 (1983).
81. G.R.Millward and J.M.Thomas, _J.Chem.Soc. Chem.Commun._, 77 (1984).
82. G.R.Millward, S.Ramdas and J.M.Thomas, _Proc.Roy.Soc._, A399: 57 (1985).
83. J.M.Thomas, G.R.Millward, D.White and S.Ramdas, _J.Chem.Soc. Chem.Commun._, 434 (1988).
84. N.A.Briscoe, D.W.Johnson, G.T.Kokotailo, L.B.McCusker and M.D.Shannon, _Zeolites_, 8: 74-76 (1988).
85. D.W.Breck, 'Zeolite Molecular Sieves', J.Wiley, 56 (1974).
86. A.V.Crewe, J.P.Langmore and M.S.Isaacson, "Physical Aspects of Electron Microscopy and Microbeam Analysis." Eds. B.M.Siegel and D.R.Beaman, J. Wiley, 47 (1975).
87. A.P.Pogany and P.S.Turner, _Acta Cryst._, A24: 103 (1968).
88. L.M.Brown, _J.Phys. F: Metal Physics_, 11: 1 (1981).
89. M.M.J.Treacy, A.Howie and C.J.Wilson, _Phil.Mag._, A38: 569 (1978).
90. M.M.J.Treacy, "Materials Problem Solving With The Transmission Electron Microscope.", Eds. L.W.Hobbs, K.H.Westmacott and D.B.Williams, _Mat.Res.Soc. Symp.Ser._, 62: 367 (1986).
91. M.M.J.Treacy and S.B.Rice, _J.Microscopy._, _in press_ (1989).
92. R. von Ballmoos and W.M.Meier, _Nature_, 389: 782 (1981).
93. C.E.Lyman, _J.Molec.Catal._, 20: 357 (1983).
94. C.E.Lyman, "Materials Problem Solving With The Transmission Electron Microscope.", Eds. L.W.Hobbs, K.H.Westmacott and D.B.Williams, _Mat.Res.Soc. Symp.Ser._, 62: 403 (1986).
95. S.Ramdas, J.M.Thomas, P.W.Betteridge, A.K.Cheetham and E.K.Davies, _Angew. Chemie (Int.Edn.)_, 23: 671 (1984).
96. A.K.Nowak and A.K.Cheetham, _Proc.7th Intl. Zeolite Conf._, SSSC #34, Eds. Y. Murakami, A.Iijima and J.W.Ward, Kodansha/Elsevier, 475 (1986).
97. E.G.Derouane and D.J.Vanderveken, _Appl.Catal._, 45: L15 (1988).
98. R.A.Jackson and C.R.A.Catlow, _Molec.Simulation_, _in press_.
99. A.Lasaga and G.V.Gibbs, _Phys.Chem.Miner._, 14: 107 (1987).
100.G.V.Gibbs, _private communication_ (1988).

RECENT PROGRESSES IN THE UNDERSTANDING

OF CHEMICAL PROPERTIES OF ZEOLITES

Jacques C. Védrine

Institut de Recherches sur la Catalyse, CNRS
2 Avenue Albert Einstein
69626 Villeurbanne, France

INTRODUCTION

Catalytic properties of zeolites and molecular sieves materials are very much depending on both their chemical and physical features. The former aspect correspond to acidic, metallic, basic or redox-type reactions since it involves active sites whose nature, chemical strength, density and distribution in strength have to be determined. The physical aspect is known to greatly influence shape selectivity (1), confinement (2) and diffusion (1) properties.

In the recent years isomorphous substitution of Si by foreign elements (3) as B(4-6), Fe(5), Ga(5), Ti(7), etc has been largely studied and has been shown to result in different catalytic properties. New materials as $AlPO_4$-n family has also opened new fields of interests and prospectives with the possibility of isomorphous substitution by many elements (8).

The purpose of obtaining isomorphous substitution and new molecular sieve materials was obviously to modify the chemical and physical properties of zeolitic-type materials and therefore their catalytic properties. In the previous paper by D. Vaughan et al (9) structural and crystalline properties of such samples have been described on the basis of NMR, TEM and XRD data.

Modification of chemical properties of "well established" zeolites may be carried out chemically. Dealumination by hydrothermal treatment followed by acid leaching or dealumination by chemical reactants as EDTA, acetyl acetonate, etc are well known processes. Treatment by chlorides as BCl_3 to substitute Al (10) or $SiCl_4$ (11) or $TiCl_4$ (12) to substitute Al+Na or Si has also been carried out. Another type of modification consists in grafting a given compound on the outer surface or in the inner pore of zeolite crystallites by chemical reaction with external or internal OH groups. This results in a decrease in acidic sites density but in an increase in physical constraints.
At last modification may be performed by ionic exchange of alkaline cations by other cations which exhibit different chemical properties as such or after reduction to metallic state forming metallic particles.

Guidelines for Mastering the Properties of Molecular Sieves
Edited by D. Barthomeuf *et al.*
Plenum Press, New York, 1990

121

In the present paper interest will be mainly devoted to chemical properties such as acidic, basic or redox features induced by either direct synthesis, or further modification, or cationic exchange, etc. The use of chemical probes for characterizing surface sites and of various physical techniques will be emphasized and examplified.

Chemical properties of zeolites

They correspond to acidity, basicity, redox or metallic features with their subsequent effects on catalytic properties as monofunctionnal or bifunctional type reactions.

Acidic properties

They constitute the greatest property for industrial interest. Many physical and chemical techniques have been used to try to characterize them (13-15). For liquids one defines the pK_a constant by the following relation :

$$AH \rightleftarrows A^- + H^+ \qquad K_a = [H^+] \; [A^-] \, / \, [AH] \qquad \text{for acidity}$$

$$\overline{BH}^+ \rightleftarrows \overline{B} + H^+ \qquad K_a = [\overline{B}] \; [H^+] \, / \, [\overline{BH}^+] \qquad \text{for basicity}$$

For liquids Hammett has defined the H_0 acidity function :

$$H_0 = - \log(a_{H^+} \cdot \gamma_{\overline{B}} / \gamma_{\overline{BH}^+})$$

where a_{H^+} is the activity of the proton and $\gamma_{\overline{B}} / \gamma_{\overline{BH}^+}$ the ratio of the activity coefficients of a neutral base B and its conjugate acid BH$^+$. One may also write :

$$H_0 = pK_{\overline{BH}^+} - \log [\overline{BH}^+] \, / \, [\overline{B}]$$

The $[\overline{BH}^+]/[\overline{B}]$ concentration ratio is directly determined from the Hammett indicator in its two different colored forms.

Acid catalysts are characterized by the negative value of H_0, the most negative corresponding to the strongest acidity. H_0 represents the trend of a site to give a proton to an indicator.

Hammett's indicators have been largely used for characterizing acidic properties of solid materials. However because of their bulky size their use for zeolite is limited because of accessibility limitations, except when one is interested in acidic properties of the <u>external surface</u> of the zeolite particles by contrast to the internal surface properties within the pores or cavities.

The main techniques which may be used for characterizing acidic properties of zeolites, including those of inner sites are summarized below :
(a) Infra red spectroscopy of functionnal groups as hydroxyl groups and of the effect of adsorbed basic probes as ammonia, pyridine, benzene, CO, C_2H_4 or others. The temperature of desorption of such molecules is indicative of acidity strength.
(b) Temperature programmed desorption of basic probes, mainly ammonia. The amount and the temperature(s) of desorption are related to the number, strength and distribution in strength of acidic sites.
(c) Microcalorimetry of adsorption to determine the differential heat

of adsorption of basic molecules versus coverage, i.e. the
distribution in acidity strength.
(d) UV-visible reflectance spectroscopy of adsorbed hydrocarbons
to characterize the acidity strength via the formation
of carbocations with an appropriate choice of hydrocarbons.

Several examples will be presented below in order to illustrate
different applications of the techniques described above.

Basic properties (16)

As for acidic properties several techniques may be employed for
their characterization. Basicity strength correspond to the proton
accepting ability of the surface with the Hammett and Deyrup H_- function
defined as follows :

$$BH + \bar{B} \rightleftharpoons B^- + \bar{B}H^+$$
$$H_- = pK_{BH} + \log [B^-] / [BH]$$

where $pK_{BH} = pK_a$, $[B^-]$ and $[BH]$ are concentrations of the adsorbed
indicators. However, as for acidity characterization Hammett's indicators
are usually too bulky to be used for zeolites.

The main techniques which may be used for zeolitic materials are :

a) Infra red spectroscopy of absorbed probe molecules as pyrrole,
benzene, etc.

b) Microcalorimetry of adsorption of probe molecules as CO_2, SO_2,
etc.
Basic sites may correspond to O^{2-} or OH^- species.

Redox properties

There is no direct technique for their characterization. One may use
suitable probe molecules as CO, NO etc and study their infra red spectra
which depend on the oxidation state of the cations.
UV - vis and ESR spectroscopies, which are techniques sensitive both
to oxidation state of the cations and to their environmental symmetry,
may also be used with benefit.

Metallic properties

As usually for metals, techniques which allow one to determine
metallic particle size and electron state density, may be used. XRD line
broadening, high resolution electron microscopy, H_2 - O_2 titration are
used for determining the particle size. Threshold X^2 ray absorption edge,
XPS an UPS may be used for characterizing electronic state density near
Fermi level. "Ship in a bottle" clusters formed in the zeolite cavity by
reacting for instance CO with Rh^{n+} (17) or Pd^{n+} (18) may be characterized
by infra red spectroscopy and EXAFS for comparison with usual clusters.

CHARACTERIZATION OF CHEMICAL PROPERTIES
Acidity of zeolites (13-14)

It is usually assigned to hydroxyl groups (Brönsted sites) attached
to framework oxygens bridging Si and Al tetrahedrally coordinated and to

uncompletely coordinated Al atoms designated as Lewis acid sites. Concerning hydroxyl groups as evidenced by infra red spectroscopy in the 3500-3750 cm^{-1} region, modern spectroscopic techniques in combination with sorption and ion exchange capacity permit a distinction between acid and non acid groups in zeolites.

For acidic OH groups one usually uses basic probes to neutralize them with further desorption at increasing outgassing temperatures to restore them. The restoring temperature is depending on the acidity of the sites. Different basic probes may be chosen following several criteria :

(a) basic strength with respect to acid site strength,
(b) thermal stability,
(c) steric hindrance with respect to pores or cavities openings,
(d) vapor pressure at experimental temperatures.

Several compounds may be used, namely :
    nitrogenous : pyridine, ammonia, piperidine, amines
    oxygenated : water, ethers, alcohols ...
    sulfidized : H_2S, mercaptans ...
    unsaturated hydrocarbons : C_2H_4, benzene, butene ...
    saturated hydrocarbons   : hexane ...
characterized by their pK_b (11 for piperidine, 8.75 for pyridine, 4.75 for ammonia, 3.34 for methylamine, etc.)

Strong bases as pyridine or ammonia are frequently used, particularly pyridine which results in nice infra red spectra. Pyridine interacts with Brönsted sites to give pyridinum ion : Py H^+, with Lewis site to give coordinated species Py L and with weakly acidic H to give H bonded pyridine characterized by their infra red bands summarized below.

Because of overlapping of bands one usually takes 1450 cm^{-1} for Py L, 1545 cm^{-1} for Py H^+. For other bases table 2 summarizes the bands chosen for the characterization.

Table 1. Some typical ir band for pyridine species. Frequencies given in cm^{-1}

vibrational modes	Physisorbed	Lewis	Brönsted	H bonded
19 b $\nu_{CC}(N)$	1439	1450	1545	1438
19 a $\nu_{CC}(N)$	1489	1490	1490	1490
8 b $\nu_{CC}(N)$	1572	1577	1620	1593
8 a $\nu_{CC}(N)$	1580	1620	1638	1614

Table 2. ir bands chosen for characterizing Brönsted (B) and Lewis (L) acid sites of zeolites. Values in cm^{-1}

probes	B	L
NH$_3$	1445-1475	1623-1630
piperidine	1610-1450	1450
n-butylamine] methylamine	1500-1540	1610

When weaker basic probes are used, one observes a more or less important shift of the ν_{OH} frequency towards lower values. Some examples are given in table 3.

The chemical shift was shown to increase with stronger acidic sites, which allows one to classify zeolite samples in acid strength order as follows :

$$H\text{-}AlZSM5 > H\text{-}FeZSM5 > HY > SAPO\text{-}5 > AlPO_4\text{-}5$$

It is important to note at this stage that probe molecules have not to be transformed chemically under experimental conditions. For instance C_2H_4 may polymerize even at room temperature, pyrrole which exhibits amphoteric properties may be transformed, etc. Such reaction may preclude some probe molecules to be used.

Molecular hydrogen may also be used as a probe at low temperatures ($-196°C$) (19). Three bands are usually observed for instance at 3980, 4110, and 4125 cm^{-1} on H-FeZSM5 (20). The latter two are assigned to diatomic hydrogen interacting with acidic and silanol groups (19). The chemical shifts are in the range 60-80 cm^{-1} with respect to gaseous H_2 for adsorption on protons of OH groups or on metal cations and in the range 100-180 cm^{-1} for H_2 coordinated to Lewis acidic sites. When H_2 is dissociated at 27°C new bands in the OH group region and in the hydride region (1600-2000 cm^{-1}) do appear. Such a dissociation occurs if the H-H bond polarization on acid-base pairs is sufficient as it was observed for η or γ-Al_2O_3 (19) but not for H-Y, HZSM-5 (19) and FeZSM-5 (20). The line at 3980 cm^{-1} corresponds to 184 cm^{-1} shift with respect to molecular H_2 (4164 cm^{-1} and is assigned to H_2 complexes coordinated to Lewis sites (20)).

Table 3. Shifts of ν_{OH} frequency upon adsorption of some weakly basic probes, expressed in cm^{-1}

Samples	ν OH	$\Delta\nu$ OH				ref.
		H_2S	CO	C_2H_4	C_6H_6	
		23°C	-196°C	23°C	23°C	
HY	3640	-	275	350	300	20
HY	3640	-	-	300	280-320	21
HY	3640	600[a]	-	440[b]	-	22
H-FeZSM5	3630		284	370	330	21
H-AlZSM5	3605		316	390	350	20,21
H-SAPO-5	3625		-	330	325	21
AlPO$_4$-5	3680		-	-	240	21
	3800				180	21

[a] a peak was detected at 2480 cm^{-1} / 2 δ_{OH}
[b] measured at -63°C

125

NMR studies of ^{1}H (23) and adsorbed probe molecules may also be used for characterizing acidity. For instance ^{15}N peaks for adsorbed pyridine correspond to 26 ppm (liquid pyridine), 80 ppm (Py L) and - 115 ppm (Py H$^+$) and for NH$_3$ and trimethylamine at - 353 and - 375 ppm respectively for ammonium ion on the two protonic sites of H-Y zeolite (24).

Temperature programmed desorption (TPD) of adsorbed probe molecules is also a very useful technique to characterize the amount and strength of acidic or basic sites. The principle is based upon the idea that at thermodynamical equilibrium the desorption temperature depends mainly on the strength of the adsorbing sites. Assuming that desorption rate is first-order, that diffusionnal limitations are negligible without readsorption and that adsorbing sites exhibit homogeneous strength, one may write :

$$2LnT_M - Ln\beta = (E_d/RT_M) + Ln (E_d/AR)$$

T_M is the temperature of the maximum in the curve, β the heating rate in °C.min^{-1}, E_d the desorption activation energy in J.mol^{-1}, A the preexponential factor of the rate constant of desorption and R the universal constant. The Ln (T_M^2/β) variations versus $1/T_M$ is a straight line with E_d/R as a slope.

Therefore the maximum temperature in the thermodesorption curve depends both on acid strength of the sample and on the desorption activation energy as shown in fig. 1 and on the desorption rate β.

Using the same probe molecules and adsorption rate it is then possible to get an order in the acidic strength of different samples as shown, in fig. 2 assuming no diffusional limitations.

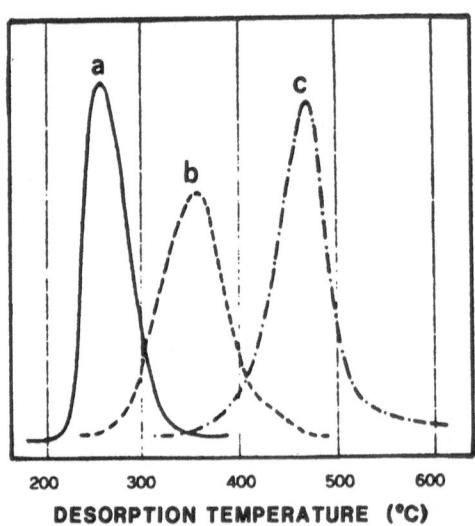

DESORPTION TEMPERATURE (°C)

Fig. 1. Thermoprogrammed desorption of basic probe molecules on HY zeolite sample from ref. 25.
(a) : CH$_3$CN (E$_d$ = 66.2 kJmol^{-1}) ; (b) : NH$_3$ (E$_d$ = 88.4 kJmol^{-1}) ; (c) : CH$_3$NC$_5$H$_8$ (E$_d$ = 210.6 kJmol^{-1}).

<u>Thermal analysis</u> (26). Several thermal analysis techniques may be used for characterizing zeolite matrices. It is possible for instance to use differential scanning calorimeter (DSC) or differential thermal analysis (DTA) to study zeolite samples after synthesis. By increasing the temperature under a gas flow at a given heating rate (e.g. 10 °C min^{-1}) one gets several thermal peaks corresponding to different peculiar features (27). For instance for a as-synthesized Fe-ZSM5 sample one gets two major exothermal peaks which were assigned to the decomposition of calcined template molecule (lower temperature) and of charge compensating template cation (higher temperature). The latter temperature was observed to increase with acidic strength, when comparing Fe-ZSM5 and Al-ZSM5 samples (28, 29) at 390 and 425°C against 395 and 510°C respectively.

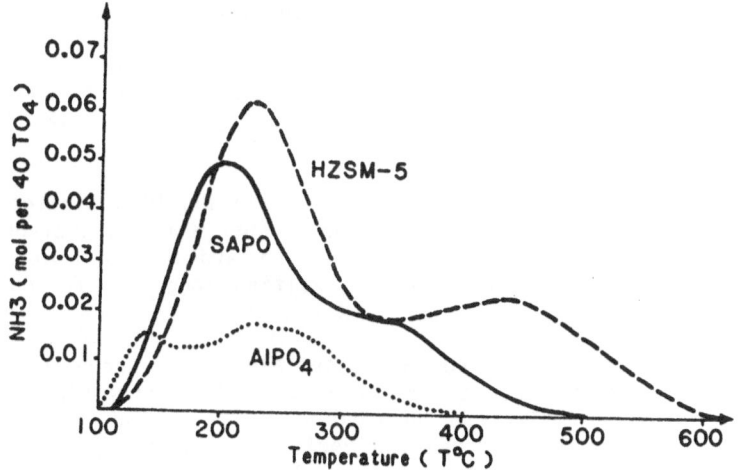

Fig. 2. Temperature programmed desorption of ammonia for different samples.

Another technique consists in the measure of the differential heat of adsorption of probes molecules versus coverage. Assuming that the heat of adsorption is directly depending on the acidic strength one then obtains information about the amount, strength and distribution in strength of such sites. In such a technique a vacuum line with gas manifolds is attached to a cell placed into a Tian-Calvet type calorimeter maintained at the desired temperature. Successive increments of the probe molecules are introduced over the sample placed in the calorimetric cell and the heat of adsorption is then measured. One may then plot the differential heats of adsorption versus coverage or even its derivative, i.e. the number of acid sites versus their strength. This is illustrated in fig. 3 when using NH_3 as a basic probe at 150°C. More detailed informations may be found in ref. 26, 30-33.

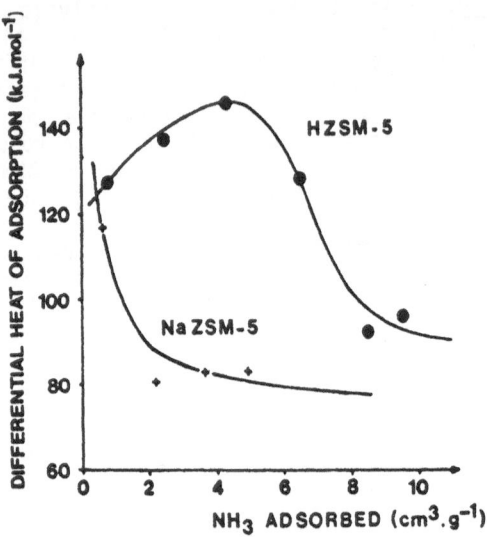

Fig. 3. Differential heat of NH_3 adsorption at 150°C versus
coverage for different ZSM5 samples outgassed at 400°C prior
adsorption : Na-ZSM5 (+) H-ZSM5 (•) from ref. 32.

The maximum in the curve of fig. 3 for H-ZSM5 indicates that NH_3 adsorption is diffusion limited and that inner sites are stronger that outer layers sites. Such a maximum may be avoided by contacting the samples with small NH_3 doses and/or by waiting long time for equilibrium between doses.

Combined uses of infra red and microcalorimetry techniques may also be performed by contacting a zeolite sample with increasing successive doses of NH_3 and analyzing the decrease in OH group band, the increase in NH_4^+ band (ν_{NH} = 1450 cm^{-1}) and the variations in the differential heat of NH_3 adsorption. This is shown in fig. 8 and 9 of ref. 33. The combined use of both microcalorimetry and infra red is complementary since it allows to determine the strength and amount and the nature (Bronsted or Lewis) of the acidic sites.

Recently Setaram has developped a new apparatus which allows to measure <u>simultaneously</u> the weight gain and the differential heat of adsorption versus coverage (TG-DSC 111). Comparison between static (usual) and temperature programmed adsorption and desorption on several zeolitic samples (H-FER, H-ZSM5, H-Y) has shown (34) that such techniques are in excellent agreement and allow to have informations about the strength, distribution in strength and number of acidic sites. Unfortunately the technique does not give informations concerning the nature of the sites.

<u>UV-visible spectroscopy</u> may also be useful to characterize the acidic properties of zeolites. Indeed protonation of aromatics may yield benzenium ion which may be easily detected by UV-vis. spectroscopy. It is then possible by using aromatics of different basic strength to obtain information about acidity strength distribution. It is known (35) that aromatics may form complexes with Lewis acid sites in presence of halides or with strong Brönsted sites as schematized below :

$$\bigcirc + HX + MX_3 \longrightarrow \left[\begin{array}{c} \text{H} \\ \text{H} \end{array} \right]^+ MX_4^-$$

Table 4. Relative basicity and UV-vis. reflectance absorption
bands of protonated methyl substituted benzene carbenium

compound	basicity relative to p-xylene	UV-vis. absorption band in nm	
		neutral	protonated
toluene	0.01	207-254	330
p-xylene	1	212-274	340
m-xylene	9	212-274	340
1,2,4 TMB	18	275	355
1,3,5 TMB	1400	268	355

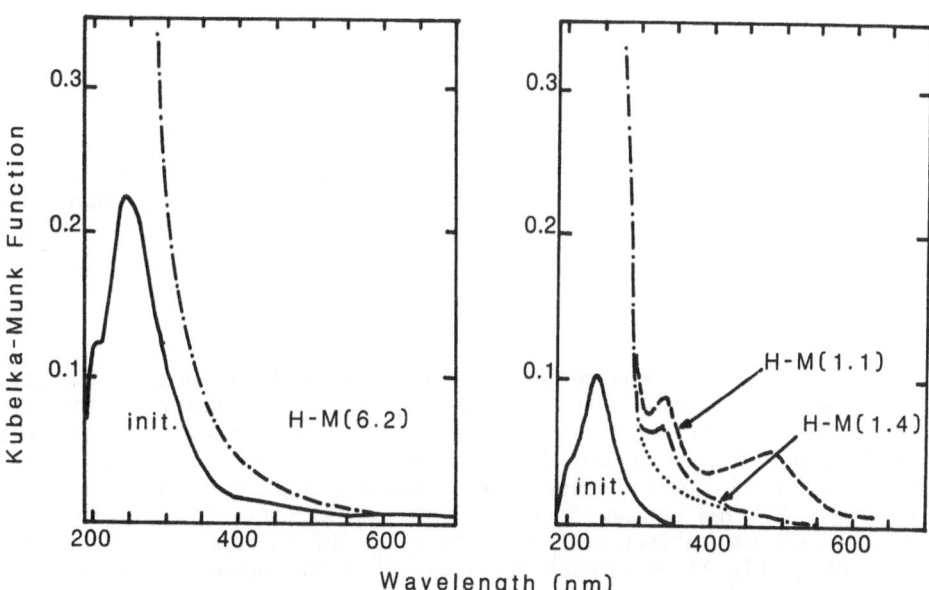

Fig. 4. UV-vis. diffuse reflectance spectra of m-xylene adsorbed on
H-M (6.2 Al/u.c.), and dealuminated mordenites (1.4 and
1.1 Al/u.c.) taken from ref. 36.

The basic strength of a series of substituted benzene compounds increases with the increase in methyl substitution of the aromatic ring. Basicity and UV-vis. reflectance bands of several methyl substituted protonated compounds are given in table 4.

Reflectance spectrum of H-mordenite sample exposed to m-xylene is shown in fig. 4 left. Only two bands near 210 and 262 nm are observed indicating that protonation of m-xylene did not occur. At variance for dealuminated mordenite samples an additionnal band was observed near 330 nm. It therefore appears that dealumination of mordenite (steaming + HCl leaching) resulted in the formation of strong acid sites able to protonate m-xylene hydrocarbons.

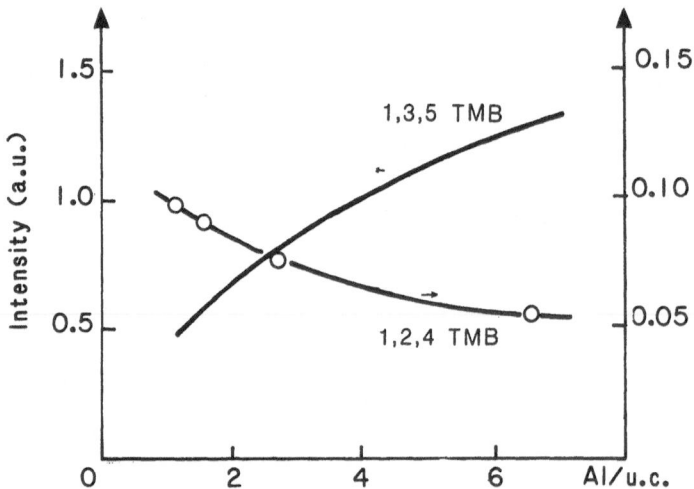

Fig. 5. Variations of the intensity of protonated 1,2,4 and 1,3,5 TMB UV-vis reflectance peaks versus Al content of H- and H- dealuminated mordenites samples taken from ref. 36.

If 1,2,4 and 1,3,5 trimethyl benzenes (TMB) are adsorbed on the same samples an additionnal UV-vis. band at 335 nm was observed for all samples. Such hydrocarbons are more basic than m-xylene i.e. are expected to be protonated by moderate or even weak acid sites. The variations of the UV-vis. band intensity versus Al content are plotted in fig. 5. The number of 1,2,4 TMB cations increased when the Al content decreased while that of 1,3,5 TMB cations decreased. As 1,3,5 TMB is much more basic than 1,2,4 TMB (table 4) it may be concluded that the number of weak sites increased with Al content increasing while that of moderate or strong acid sites decreased, in agreement with m-xylene adsorption data. The same behaviour was observed for H-Y and dealuminated H-Y zeolite samples (36). It was observed that the presence of residual Na cation or even RE cations greatly depressed the amount of strong and even moderate acid sites. The following ranking order in acid strength was observed :

NaHY < REY < USHY ∼ HM < deal. HM, H-ZSM5

130

Silanol Group Chemistry

Hydroxyl groups as described above correspond to Brönsted acidic sites. There also exist non acidic silanol groups which are necessary to terminate the external surface of the zeolite crystals. Their amount can be approximatively calculated from the particle size (37). However in dealuminated and highly siliceous zeolites this amount was found to be much higher (38,39). It was then suggested that internal silanol groups and hydroxyl nests are present at structural local defects or at missing T-atoms in the lattice, as isolated Si-OH or as pairs in broken Si-O-Si linkages (Si-OH HO-Si) or as clusters surrounding T-atom vacancies :

```
            Si
            |
            OH
  Si -OH         OH-Si
            OH
            |
            Si
```

Such silanol groups result in ir band near 3740 cm^{-1} and were evidenced by silylation reaction with e.g. trimethylchlorosilane at 300 to 400°C as schematized below with ^{29}Si MAS-NMR chemical shift (12).

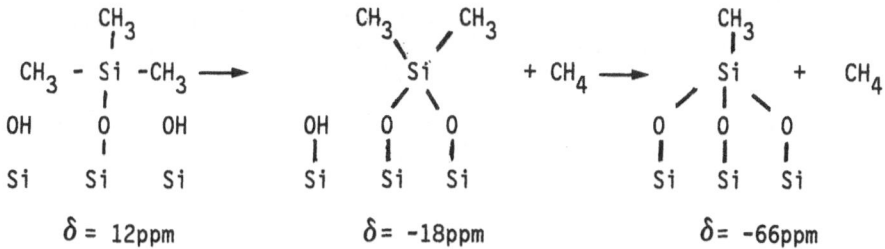

^{29}Si CP-MAS-NMR studies before and after silylation allow to evidence silylation products and therefore to characterize clusters of hydroxyl groups. The destruction of silylation products by cleavage of Si-C bonds may be achieved by reaction with strong bases such as t-butoxide ion (40). The bulky size of this ion allows also to identify external silanol groups for intermediate size pore zeolites as ZSM5.

Such silanol groups are shown by TPD of ammonia to be of weak acidity with temperature at the maximum of the curve near 200°C as for boron-silicalite (6). At variance for pure silicalite which exhibits only external silanol the acidity is still weaker (6).

The presence of such internal groups was also observed to affect catalytic properties. For instance intrinsic activity of H-ZSM5 in cracking of n-butane and n-hexane was shown to decrease with increasing the amount of internal silanol groups (12).

^{1}H MAS-NMR investigation was also carried out on H-ZSM5 zeolite samples (41) exhibiting internal structural defects and silanol groups. The increase in the number of silanol groups is accompanied both by a change in the ^{1}H chemical shift from 1.4 to 2.2 ppm and an increase in the residual linewidth from ca. 220 to 480 Hz. The chemical shift was interpreted following Lippmaa et al. findings (42) as due to a rearrangement of silanol groups into a gem Si(OH)$_2$ or Si-OH HO-Si arrangement.

The intracrystalline self diffusion and the molecular reorientation of sorbed molecules evidenced by ^{1}H pulsed field gradient NMR technique and ^{13}C line shape analysis respectively were shown to depend on the amount of internal silanol groups (structural defects) and to correspond to enhanced mobility of the guest molecules.

Note that the ^{1}H NMR chemical shift was observed to increase with increasing the acid strength (43-45).

Basicity of zeolites (13-16)

Much less work has been devoted to this subject although some industrial processes as the Aromax process (46) (aromatization on n-hexane) are obviously related to the strong basicity of Ba - K/L zeolite (16,47).

Infra red spectroscopy of adsorbed probe molecules has been used to characterize basicity. For instance CO_2 has been absorbed on Na-A, NaCa-A (48), NaCaMg-Y (49) zeolites. CO_2 is adsorbed on cations (lewis acid sites) and on oxide ions (basic sites) forming carbonate like species in the latter case absorbing in the $1300-1600 \ cm^{-1}$ region (48).

Pyrrole molecules which are amphoteric were also used as acid-type probe. In presence of basic sites the NH vibration frequency is shifted towards low wave numbers. The extent of the shift evaluates the basic strength. This is shown in fig. 6 for different zeolites versus the oxygen ion negative charge (50) evaluated using the Sanderson electronegativity equalization principle (51).

Other acidic probes as formic, benzoic or acetic acids were also used in infra red (S2) - Benzene which was shown to give a shift in OH

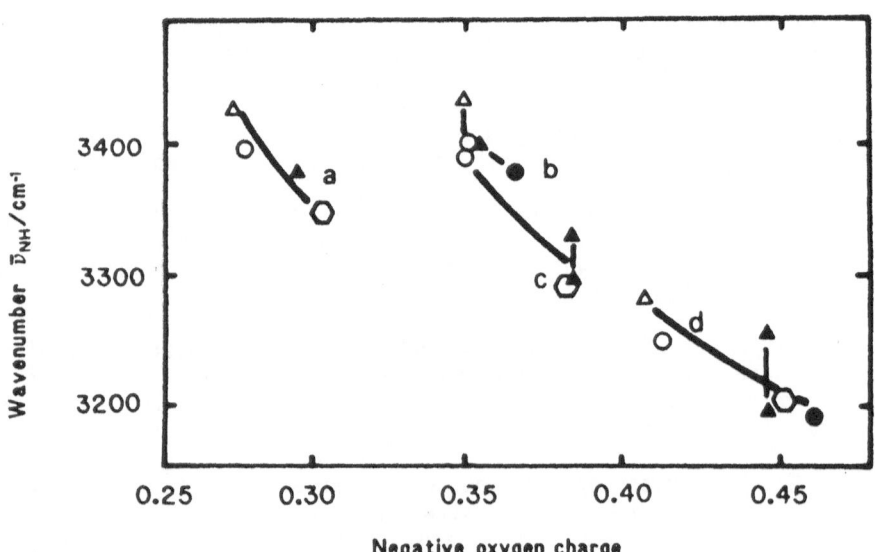

Fig. 6. Variation of NH frequency of pyrrole adsorbed on various zeolites versus negative oxygen ion charge from ref. 50 . Zeolites : Mordenite : a, L : b, Y : c and X : d , in forms Li (△), Na (o), K (▲), Rb o and Cs (●)

frequency may also be employed for basicity characterization. As a matter of fact the $_5^+$$_{17}$ and $_{10}^+$$_{17}$ combination bands corresponding to the CH out of plane stretching mode were observed to shift towards high wavenumbers upon interaction with cations and towards even higher wavenumbers upon interaction with oxygen ions, 12R windows in X-type zeolite samples (53) with Na cations being more or less exchanged by Cs cations. As the Cs content increased the window type adsorbed benzene increased in stability which was assigned to the increase of the framework oxygen charge and to the simultaneous decrease of the cation acidity.

Microcalorimetry of CO_2 adsorption was also studies on zeolitic samples as shown in fig. 7. The heat observed for H-Y sample at low CO_2 coverage presumably arises from carbonation of the crystallite surface.

Mechanism of reaction and coking features of zeolites

We have seen above how UV-vis. spectroscopy could bring some insight to the characterization of acidity. Such a technique may also be used via carbenium ion analysis to other types of information. The technique was for instance found to be very useful to characterize the mechanism of aromatics formation from methanol and light olefins conversion on H-ZSM5 zeolite via H transfer (54). Methanol or light olefins as C_2H_4 or C_3H_6 were contacted at room temperature with 400°C activated H-ZSM5 sample under ca. 100 Torr pressure. The samples were progressively heated at increasing temperatures (100, 200, 250, 300°C.). The UV-vis spectra were recorded and observed to be strongly modified. Olefins immediately reacted at room temperature giving rise to an intense absorption band near 280 nm. By heating, a band developped at 295 nm with shoulders ca 370 and 440 nm. For C_2H_4 broad bands appeared near 300 and 350 nm at 150°C and no further analysis (too large absorption) could be possible. For C_3H_6 bands appeared at 365 and 455 nm after heating at 200°C in addition to the 300 nm band. After heating at 300°C a broad and intense spectrum was obtained with peaks at 300 and 420 nm.

Methanol did not react at room temperature resulting in a slight shift of the charge transfer band. Strong bands appeared near 290 and 365 nm by heating at 200°C. Further heating at 250°C resulted in an increase of the 365 nm band with a shoulder at 260 nm. After heating at 300°C a broad spectrum was observed with maxima near 600, 500, 420, 330, 270 and 240 nm (54). We have seen above that such UV bands are characteristics of some carbocations. The following conclusions may be drawn.

(a) Olefins readily reacted at room temperature giving a cyclopentenyl carbocation (280 nm) for H-ZSM5 and a linear π-allyl carbocation (310 nm) for H-Y (55).
(b) The above cation underwent rearrangement into a cyclohexenyl carbocation at 100°C (295-300 nm).
(c) Alkyl aromatics (250-265 and 350-370 nm regions) were further formed, the cyclohexenyl carbocation being still present.
(d) Further heating above 200°C resulted in the formation of polyalkyl aromatics and polyaromatics (bands near 600, 500, 420, 330, 280, and 260 nm).

Such a study allowed one to propose a mechanism of methanol and olefins aromatization via cyclopentenyl⟶cyclohexenyl⟶cyclohexadiene ⟶protonated benzene species.

UV-vis spectroscopy was also further used for the study of benzene activation and cokefaction of H-ZSM5 and H-MOR zeolites samples (56). ESR technique was also used as a complementary technique since it is sensitive to radical cations (57,58). It was then possible to show that benzene (56,58) or olefins (57) or other aromatics as xylenes, TMB.. were ionized on H-MOR and H-ZSM5 zeolite samples giving a carbocation identified by ESR. Even more, benzene cation was observed as monomer in H-ZSM5 against a sandwich-like dimer in H-M sample (56) a_H = 4.6 oe against 2.3 oe in the latter cases with 7 (i.e. 6 equivalent protons) and 13 (i.e. 12 equivalent protons) hyperfine lines respectively. This result was interpreted (56) as due to differences in steric hindrance, i.e. in channel size between H-ZSM5 and H-MOR zeolites. The formation of benzene cation and of biphenyl could be followed kinetically by UV-vis. spectroscopy (absorption band near 255 nm). Upon heating only benzene cation and biphenyl were detected for H-ZSM5 sample while higher aromatic cations as pyrene$^+$ (430 nm) naphtalene$^+$ (380, 650-700 nm) were evidenced for H-MOR samples. However the role of oxygen treatment of the samples (activation) was observed to influence the formation of such carbocations (56).

Let us at last recall the use of ESR for characterizing charge transfer complexes formed by contacting hydrocarbons of low ionization potential or high electron affinity to probe the electron acceptor (acidic or oxidative) or electron donor (basic or reductive) properties of the surface (59,60). Probe molecules as anthracene, perylene in the former case or tetracyanoethylene, tri- or di-nitrobenzene in the latter case were often used. It clearly appears then that the use of small or in contrast bulky probe molecules with respect to the zeolite pore size may thus allow to characterize electron acceptor or electron donor properties within the pores or on the outer surface of the zeolite particles.

Isomorphous substitution of elements (3,5 ,61)

A large effort has been devoted to this subject rather recently. The main reason is that chemical and physical (pore size, mainly) properties may be modified by substituting Si or Al by another element in the framework. It is particularly important to determine if the "foreign" element has been actually substituted or if it is present as impregnated compound which either coats the inner pores neutralizing some active sites or is present on the external surface or at last if the element forms some tiny oxide particles entrapped within the pores or cavities. It is obvious that chemical and subsequently catalytic properties are different depending on the way the foreign element has been incorporated.

X-ray diffraction technique is certainly the most suitable technique for such a characterization but unfortunately the amount of foreign element is often too low for the data to be really conclusive particularly if hydrated extent is not well defined. The easiest way consists in determining the unit cell volume which is expected to vary depending on the size of the foreign element and of the T-O bonds. This was claimed for instance for dealuminated Y-type zeolite for which a formula was even proposed to relate the unit cell parameter a^o (in nm) and the Al content (62) as follows :

$$Si/Al^{IV} = \frac{192}{1124\,(a_0 - 2.42333)}$$

The unit cell volume was observed to decrease linearly with boron content (4, 6, 63) and to increase with titanium content. The B-O, Fe-O and Ti-O distances in tetracoordinated element are known to be smaller (0.14 nm) and larger (0.184 and 0.175 mm) respectively than Si-O (0.16 nm) or Al-O (0.17 nm).

However our own experience on Fe-ZSM5 zeolite sample with relatively low Fe content (Si/Fe>37) and on the effect of dehydration/hydration extent on the unit-cell volume values leads us to recommend the readers to be particularly cautions when interpreting unit-cell volume variations in terms of isomorphous substitution.

X-ray diffraction data may also be used in a more complex manner by developping the well-known Rietveld method applied to powder X-ray diffraction pattern. By substracting the total integrated Bragg surface intensity to the background measured for a standard as silicon powder one may calculate the crystallinity degree (66). Moreover using a refinement method (67) with the modified DBW 3.2 or DLS 76 programmes (68) and conventionnal difference-Fourier synthesis, in order to get a good fitting between calculated and experimental X-ray diffraction patterns (Rb should be less than 15 %), one may determine the topology of the material ($P2_1/n$ space group for MFI, Ima2 space group for AEL, etc), the spatial location of all T atoms and even more the presence of some hosts within the pores or cavities. It was then possible to locate p-xylene molecules within ZSM5 pores (69), di-n propylamine molecules in as-synthesized SAPO-11 (70), tetrapropylamine template molecules in silicalite (71) and even some non crystalline phase within the pores (70). The technique allows to postulate the presence of occupied sites but unfortunately does not give the exact nature of the occupant host.

Infrared spectroscopy appears also to be a very reliable technique for characterizing substitution in zeolites by analyzing the vibrational mode bands and the formation of additionnal OH group bands. For instance, for boron substitution, additional bands were observed at 1380, 920, 700 and 670 cm^{-1}, due to B-O asymmetric and symmetric stretching and to B-O-Si asymmetric and symmetric bending modes respectively. The 1380 cm^{-1} band was observed to be very sensitive to adsorbates as NH_3, H_2O, CH_3OH and to yield a new band at 1470 cm^{-1} to replace it. This phenomenon was reversible upon outgassing and was interpreted as indicative of tricoordination of boron atom even at lattice position i.e. in tetrahedral type environment. This feature was observed also by ^{11}B MAS-NMR study (72,73) and induced very weak acidity close to that of silanol groups (6,73).

UV-vis. Mössbauer, photoacoustic and ESR techniques appeared to be reliable techniques when applicable (28, 29, 74-79) for characterizing the environment of the substituted element.For instance for Fe-ZSM5 zeolite samples UV-vis. and photoacoustic techniques (29,74) showed that all Fe atoms are in tetrahedral environment even after calcination at 500°C since 10Dq values are in the range 8500-7500 cm^{-1} against 13,000-15,000 cm^{-1} for octahedral environments ($10Dq_{Td} = -4/9 \; 10Dq_{Oh}$). Mössbauer technique identified two types of Fe^{+3} ion environments with $= 0.34 \pm 0.04$ mms^{-1}, QS = 0.7 $\pm$ 0.4 ms^{-1} and $= 0.18 \pm 0.02$ mms^{-1}, QS = 0 mms^{-1} at room temperature, in a ratio 1:2. At last, ESR spectroscopy identified three types of Fe^{3+} (d^5) ions : I with $g_{eff} = 4.28$, II with $g_{eff} = 2.3$ and III with $g_{eff} = 2.0$. Signal I was identified as tetrahedrally coordinated lattice Fe atoms related to the formation of acidic OH groups. It corresponds to the so called rhombic symmetry with E/D = 1/3 due to the charge disymmetry surrounding Fe^{3+} ions (negative charge induced by Fe tetracoordination and positive compensating charge on Brönsted acid site). Signal II at $g_{eff} = 2.3$ was often assigned to amorphous tiny Fe_2O_3 particles entrapped in the pores. However this signal was observed to be very sensitive to adsorbates with g_{eff} values varying from 2.2 to 4. We are of the opinion (29) that this signal corresponds to tricoordinated lattice Fe atoms as described for B atoms and for Lewis acid sites. Signal III at $g_{eff} = 2.0$ was often assigned to Fe^{3+} ions in exchangeable cationic sites. Back exchange and H_2 reduction experiments indicated that such ions were not at exchangeable position for some samples (29) but were for other samples (80). The $g_{eff} = 2.0$ signal characterizes a rather highly symmetrical environment for Fe^{3+} whatever be tetrahedral or octahedral or other. For instance, it is observed for $FePO_4$ material where Fe is in tetrahedral coordination but without compensating charges (81).

Finally, one may suggest three different environments for Fe atoms as substituted element in zeolite matrices as described below :

Species I and III of higher local geometrical symmetry may correspond to the $\delta = 0.18$, QS = 0 mms^{-1} and species II to $\delta = 0.34$, QS = 0.7 mms^{-1} Mössbauer spectra respectively. Only a few Fe ions were detected as Fe_2O_3 particles (<5%).

For titanosilicate Ti-O distance in tetrahedrally coordinated Ti^{4+} compounds (0.175 nm) is larger than Si-O (0.16 nm) distance in ZSM5 resulting in an increase in unit cell volume as claimed by some authors (64,82,83). An infrared band near 960 cm^{-1} was assigned by the same authors to a Ti-O stretching mode and was observed to depend on dehydration extent (shifting to higher frequency upon H_2O, D_2O, NH_3 or CH_3OH adsorption (84). ^{29}Si MAS-NMR spectra exhibited an additional shoulder near -115 ppm relative to TMS. Moreover charge transfer band near 48,000 cm^{-1} as detected by UV spectroscopy was observed to be broadened and shifted to 42,000 cm^{-1} upon H_2O adsorption, value corresponding to octahedrally coordinated Ti^{4+} ions. It was then suggested (83) that substituted Ti are tricoordinated in a kind of titanyl species. Upon H_2O_2 adsorption a strong yellow colour developped while the material was claimed to exhibit very interesting oxidation reaction properties, particularly for phenol oxidation by H_2O_2 into hydroquinone and catechol. The mixture of silicalite and TiO_2 was shown not to exhibit such properties. The authors (83) suggested the following scheme with Si-O-Ti-O-Si entities which activate H_2O_2 while Ti-O-Ti-O entities were suggested to rapidly decompose H_2O_2.

$$\text{Si}(O)(OH) \quad \text{Ti}(O)(=O) \quad \text{Si}(HO)(O) \xrightarrow{H_2O_2} \text{Si}(O)(OH) \quad \text{Ti}(HO)(O)(OH) \quad \text{Si}(OH)(O)$$

$$\xrightarrow{-\ H_2O_2} \text{Si}(O)(OH) \quad \text{Ti}(O-O) \quad \text{Si}(HO)(O)$$

Tricoordination of Ti ions titanyl groups is rather unexpected. One may also imagine tetracoordination of Ti ions as for Si ions with the ability for hexacoordination of ligand molecules in a reversible way. In other words for Al, Fe and Ti one may imagine that, because of local disorder, tetrahedra highly distorted due to zeolitic type structure some additional coordination of tetracoordinated trivalent cations may occur resulting in a reversible change in neighbouring symmetry as evidenced by spectroscopic techniques and schematized below. In such a model the chemical feature of the cation is obviously modified.

$$\text{L} \cdots \text{Si} \cdots \text{Ti} \cdots \text{Si} \cdots \text{L}$$

Recall at last that infrared spectroscopy was used to determine Al content of dealuminated Y type zeolite via the shift of the 1100 and 790 cm^{-1} vibrational bands (85) towards higher frequencies. However, such variations, as those of unit cell volume as discussed above, are very sensitive to both the dealumination procedure (dealumination or replacement of Al atoms by Si) and the presence of adsorbates as template or water molecules.

XPS technique which is sensitive to the first top layers of the zeolitic particles may be useful to determine if surface layers exhibit the same concentration in foreign element than the bulk material. It may then result very different chemical and thus catalytic properties. For instance, when shape selectivity features are important the presence or not of Al and therefore of acid sites may preclude such properties because of the occurence of isomerization or other acid catalyzed reactions at the surface. If a material was impregnated for instance by H_3BO_3 (86) or phosphorus compound (87) it is important to determine if the compound has entered the pore and then neutralized some acid sites or has remained deposited on the particles surface. XPS and infrared of OH groups studies have shown (86,87) that for low amount of additionnal compound added, internal acidic groups are neutralized resulting in lower activity per Al atom, but the new compound attached within the pores created additionnal steric hindrance and therefore enhanced shape elective properties for instance in toluene methylation to p-xylene (86,88,89).

EDX-STEM technique on thin sections of zeolitic particles may also be useful to determine if the foreign element distribution within the grain is homogeneous or not when using high resolution electron microscope e.g. few nm in beam size. This was observed for instance for Al (90) or Fe (29) ZSM5 samples and was shown to greatly depend on the synthesis and/or activation conditions.

Ions in zeolite matrices

It is well admitted that because of tetracoordination of trivalent cations in the lattice, exchangeable cation sites exist in many zeolite-type materials. It is then possible to exchange cations stemming from the synthesis medium by other cations exhibiting peculiar chemical properties. For instance, Na^+ cations could be exchanged by cations as alkali ones (Cs^+, Rb^+, K^+, etc), alkaline earth or many others as Rh^{2+}, Pt^{2+}, Ni^{2+}, Co^{2+}, Cu^{2+}, ... The cationic sites have been localized via X-ray diffraction analysis long time ago (91). It follows obviously specific chemical properties as far as these cations are accessible to reactant molecules i.e. are located in the large cavities or moving from inaccessible sites upon effect of adsorbates as water, ammonia (92), etc. Moreover complexes could be formed by reacting these cations with reactant molecules with the eventual formation of unusual oxidation states (93-95).

Infrared, UV-vis. and ESR spectroscopies were used in addition to X-ray diffraction technique to characterize the oxidation states and complexes formed. For instance, ir bands at 2156, 2142 and 2127 cm^{-1} were assigned to Cu^+ CO complexes in S_I or $S_{I'}$, $S_{II'}$ and S_{II} sites respectively, while an additional band at 2178 cm^{-1} was assigned to CO interacting with Lewis acid sites.

Co^{2+} ion-exchanged Y zeolites treated with NaOH or NH_4OH solutions at various pH values were investigated by temperature programmed reduction technique. A significant hydrolysis of Co^{2+} ions occured at pH above 10.5 accompanying the migration of Co^{2+} ions in the sodalite cages and hexagonal prisms. Aggregated Co_3O_4 species were formed outside the cages while small cobalt oxide clusters were created inside the cages (96).

Highly dispersed metal oxides were formed by soaking CrY, NiY, FeY, MnY, CuY and CoY with NaOH solution at pH = 11.0 at room temperature followed by calcination at 300°C in air. Such dispersed oxides exhibit high catalytic activity particularly for total oxidation reactions (97) which was ascribed to less stable oxygen species than for bulky metal oxides. Such a property of well dispersed metal oxide may explain peculiar properties of zeolite for bifunctionnal type reactions as propane or butane aromatization in the cyclar process using Ga-ZSM5 zeolite (98). The formation of small metal oxide clusters within the cavities was suggested long time ago by ESR technique upon calcining Ca or Mg exchanged Y type zeolites at 500°C (99).

Exchange of Na^+ ions by other alkaline cations resulted in an increase in basic character as described in ref. 16. Such basic samples exhibit peculiar catalytic properties for instance for side chain alkylation of toluene by methanol (100) on Cs-X zeolite or for n-hexane aromatization in bifunctionnal reaction on Pt-Ba L zeolite (47) in the aromax process (46).

CONCLUSION

Chemical properties of zeolite materials could change in a very wide domain by many means, including :

(a) Synthesis of new zeolite-type material exhibiting different pore sizes and spatial pore arrangement networks.

(b) Isomorphous substitution of T atoms by other elements as trivalent Cr, Fe, B, Ga, etc or tetravalent as Ti, Ge, Zr, etc elements in zeolites or as multivalent, di or tetravalent in $AlPO_4$-n family.

(c) Modification of the zeolite by impregnating a suitable compound or by changing the Al content (dealumination by different processes).

(d) Exchange of Na or K cations by other cations exhibiting specific chemical properties as such or after reduction to lower oxidation state.

Several of these modifications may occur simultaneously and it is very difficult to differentiate the effect of each modification upon the catalytic properties. The use of several physical techniques allows to, at least partly, solve this problem and thus to have a better understanding of the chemical properties. One technique usually does not permit alone to solve the problem. It is therefore absolutely necessary to use complementary techniques. Moreover, it appears that a detailed analysis of the informations obtained by each technique is absolutely necessary for the understanding, particularly when several modifications do occur. Many examples and many techniques have been described but many others do exist and are not to be neglected. We insist on the necessity to broaden the analysis and the techniques used for the characterization to try to approach the reality.

REFERENCES

1. P. B. Weisz, Pure Appl. Chem 52 : 2091 (1980), J. M. Csicsery, in "Zeolite Chemistry and Catalysis", J. M. Rabo, ed., Monograph ACS 171 : 680 (1976).

2. E. G. Derouane, J. M. André and A. A. Lucas, J. Catal., 110 : 58 (1988).

3. R. M. Barrer, in "Hydrothermal Chemistry of Zeolite", Academic Press, London, UK, 1982 p 419.

4. M. Taramasso, G. Perego and B. Notari, in "Proceed. of 5 th Intern. Zeol. Confer., Napoli", L.V.C. Rees, ed., Heyden, London 1980 p 40.

5. C. T. W. Chu and C. D. Chang, J. Phys. Chem., 89 : 1569 (1985).

6. G. Coudurier and J. C. Vedrine, Pure Appl. Chem. 58 : 1389 (1986).

7. M. Taramasso, G. Perego and B. Notari, US Patent 4, 410, 501, (1983) assigned to ENI.

8. E. M. Flanigen, B. M. Lok, R. L. Patton and S. T. Wilson, Pure Appl.Chem. 58 : 1351 (1986).

9. D. E. W. Vaughan, M. M. J. Treacy, J. M. Newsam, this book, Page 99.

10. E. G. Derouane, L. Baltusis, R. M. Dessau and K. D. Schmitt in "Catalysis by acids and bases", Stud. in Surf. Sci. and Catal. Elsevier, Amsterdam 10 : 135 (1985).

11. H. K. Beyer and I. Belenkaja in "Catalysis by Zeolites", Stud. in Surf. Sci. and Catal., Elsevier, Amsterdam, 5:203 (1980).

12. B. Kraushaar-Czarnetzki, Thesis, Eindhoven, NL, april 1989
13. J. C. Vedrine, A. Auroux and G. Coudurier, in "Catalytic Materials, Relationship between structure and reactivity", T.E. Whyte et al, ed., ACS Sympos. Ser., Washington, 248 : 253 (1984).
14. D. Barthomeuf, Mater. Chem. and Phys. 17 : 49 (1987),
15. J. Kijenski and A. Baiker, Catal. today 5 : 1 (1989).
16. D. Barthomeuf, G. Coudurier and J. C. Vedrine, Mater. Chem. and Phys. 18 : 553 (1988).
17. G. Bergeret, P. Gallezot, P. Gelin, Y. Ben Taarit, F. Lefebvre and R.D. Shannon, Zeolites 6:392 (1986).
18. L. L. Sheu, H Knözinger and W.N Sachtler, Catal. Letters 2:129 (1989);
19. V. B. Kazansky, V. YU Borovkof and L.M. Kustov in "Proceed. of the 8 th Intern. Conf. on Catalysis", Berlin Dechema, Verlag Chemie, Basel 3 : 3 (1984).
20. L. M. Kustov, V. B. Kazansky and P. Ratnasamy, Zeolites 7 : 79 (1987).
21. S. G. Hegde, P. Ratnasamy, L. M. Kustov and V. B. Kazansky, Zeolites 8 : 137 (1988).
22. A. Macedo, PhD thesis, Paris VI, IFP 36875, 1988.
23. H. Pfeifer, D. Freude and M. Hunger, Zeolites 5 : 274 (1985)
24. W. L. Earl, P. O Fritz, A. V. Gibson and J. M. Lunsford, J. Phys. Chem. 91 : 2091 (1987)
25. B. Hunger and J. Hoffmann, React. Kinet. Catal. Letters 27:305 (1985).
26. A. Auroux in "Les techniques physiques d'Etudes des Catalyseurs "B. Imelik and J. C. Vedrine (ed.), Technip, Paris, chap. 24 p 823 (1988).
27. M. Soulard, S. Bilger, H. Kessler and J. L. Guth, Zeolites, 7 : 463 (1987).
28. J. Patarin, Thesis, Mulhouse, France (1988).
29. D. H. Lin, G. Coudurier and J. C. Vedrine, in "Proceed. of the 8 th Int. Zeol. Confer. Amsterdam july 1989", Stud. in Surf. Sci. Catal. Elsevier Amsterdam 1989, 49 : 1431 (1989).
30. A Auroux, J. C. Vedrine and P. C. Gravelle in "Adsorption at the gas-solid. and liquid-solid interface", Stud. in Surf. Sci. and Catal. J. Rouquerol and K. S. W. Sing (ed), Elsevier, Amsterdam 10 : 305 (1982).
31. A. Auroux in "Innovation in zeolite materials Science", Stud. in Surf. Sci. and Catal., P. J. Grobet et al.(ed.), Elsevier, Amsterdam, 35 : 385 (1988).
32. A. Auroux and J. C. Vedrine in "Catalysis by Acids and Bases", Stud. in Surf. Sci. and Catal., B. Imelik et al ((ed.), Elsevier, Amsterdam, 20 : 311 (1985).
33. Y. S. Jin, A. Auroux and J. C. Vedrine, Appl. Catal. 37 : 1 (1988).
34. A. Auroux, Y. S. Jin, J. C. Vedrine, and L. Benoist, Appl. Catal. 36 : 323 (1988).
35. G. A. Olah and C. U. Pittman Jr, in "Carbonium ions", G. A. Olah and P. Von Schleyer (ed), Interscience publish., New York, I : 153 (1968).
36. C. Naccache, F. R. Chen and G. Coudurier, in. "Proced. of the 8 th Intern. Zeolite Confer., Amsterdam". Stud. in Surf. Sci Series, Elsevier, Amsterdam 1989, 49 : 661 (1989).
37. J. B- Nagy, Z. Gabelica and t. G. Derouane, Chem. Letters 1105 (1982).
38. G. Boxhoorn, A. G. T. G. Kortbeek, G.R. Hays and N.C.M. Alma, Zeolites 4 : 15 (1984).
39. S. G. Fegan and B.M. Lowe, J. Chem. Soc., Chem. Commun. 437 (1984).
40. C. C. Price and J. R. Sowa, J. Amer. Chem. Soc. 32 : 4126 (1967).

41. M. Hunger, J. Kärger, H. Pfeifer, J. Caro, B. Zibrowius, M. Bülow and R. Mostowicz, J. Chem. Soc., Faraday Trans. I, 83 : 3459 (1987).

42. E. Lippmaa, A. Saloson, V. V. Brel and Y. I. Gorlov, Fiz. Chem. 259 : 403 (1981).

43. H. Pfeifer, D. Freude and M. Hunger, Zeolites 5 : 274 (1984).

44. J. Caro, M. Bülow, J. Richter-Mendau, J. Kärger, M. Hunger, D. Freude and L.V.C. Rees, J. Chem. Soc., Faraday Trans. I 83 : 1843 (1987)

45. D. Freude, M. Hunger and H. Pfeifer, Z. Phys. Chem. NF 152 : 171 (1987).

46. T. R. Hughes, R. L. Jacobson and P. W. Tamm, in "Catalysis", Stud in Surf. Sci. and Catal., J. W. Ward, ed., Elsevier, Amsterdam 38 : 317 (1988).

47. J. R. Bernard in "Proceed. of the 5 th Int. Conf. on Zeol". Napoli, L. V. C. Rees, ed., Heyden, London 681 (1980), C. Bezuhanova, J. Guidot, D. Barthomeuf, M. Breysse and J. R. Bernard, J. Chem. Soc., Faraday Trans. I, 77 : 1595 (1981).

48. P. A. Jacobs, F.H. van Cauwlaert, E.F. Vansant and J.B. Uytterhoeven, J. Chem. Soc. 69:1056 (1973).

49. Y. Delaval, R. Seloudoux and E. Cohen de Lara, J. Chem. Soc., Faraday Trans I, 82:365 (1986).

50. D. Barthomeuf, J. Phys. Chem. 88:44 (1984).

51. R. T. Sanderson, Chemical bonds and bond energy, Academic Press, New York, (1976).

52. A. Bielanski and J. Datka, J. Catal. 42 : 183 (1974).

53. A. de Mallmann and D. Barthomeuf, Zeolites 8 : 292 (1988).

54. J. C. Vedrine, P. Dejaifve, E. D. Garbowski and E. G. Derouane, in "Catalysis by Zeolites" Stud. in Surf. Sci. and Catal., B. Imelik et al (ed), Elsevier, Amsterdam, 5 : 29 (1980).

55. E. D. Garbowski and H. Praliaud, J. Chim. Phys. 76 : 687 (1979).

56. P. Wierzchowski, E. D. Garbowski and J. C. Vedrine J. Chim. Phys. 78 : 41 (1981).

57. A. A. Slinkin, A. V. Kucherov, D. A. Kondratyev, T. N. Bondarenko, A. M. Rubinstein and Kh. M. Minachev, J. Molec. Catal. 35 : 97 (1986).

58. A.V. Kucherov, A. A. Slinkin, D. A. Kondratyev, T. N. Bondarenko, A. M. Rubinstein and Kh. M. Minachev, J. Molec. Catal. 37 : 107 (1986).

59. Y. Kodratoff, C. Naccache and B. Imelik J. Chim. Phys. 65 : 562 (1968).

60. B. D. Flockhart in "Surface and defects properties of solids", The Chemical Society, London, 2 : 69 (1973).

61. M. Tielen, M. Geelen and P. A. Jacobs, Acta Phys. Chem. 31 : 1 (1985).

62. E. M. Flanigen, H. Khatami and M. A. Szymanski in "Molecular Sieve Zeolites", Adv. Chem. Ser. 101 : 201 (1971). M. Fichtner-Schmittler, V. Lohse, G. Engelhardt and U. Patzelova, Cryst. Res Techn. 9 : 11 (1984).

63. N. A. Kutz in "Proceed. of the 2nd Symp. of Industry-University cooperative programme", Texas A. et M. University Press, College Station 121 (1984).

64. M. Taramasso, G. Perego and B. Notari, U. S. Patent, 4, 410, 501 (1983).

65. R. B. Borade, Zeolites, 7:398 (1987).

66. R. Khouzami, G. Coudurier, B. F. Mentzen and J. C. Vedrine, in "Innovation in Zeolite Materials Science", Stud. in Surf. Sci. and Catal., P. J. Grobet et al (ed), Elsevier, Amsterdam, 37 : 355 (1987).

67. B. F. Mentzen C. R. Acad. Sci. Paris, Ser. II, 303 : 1289 (1986) Mater. Res. Bull. 22 : 309 (1987).

68. D. B. Wiles and R. A. Young, J. Appl. Cryst. 14 : 149 (1981).
69. B. F. Mentzen and J. C. Vedrine, C. R. Acad. Sci. Paris, Ser. II, 301 : 1017 (1985).
    J. C. Vedrine, G. Coudurier and B. F. Mentzen, in "Prospective in Molecular Sieve Science", W. H. Flank and T.E. Whyte (ed), ACS Symp. Ser., Washington, 368 : 66 (1988).
70. B. F. Mentzen, J. C. Vedrine and R. Khouzami, C. R. Acad. Sci. Paris, Ser. II, 304 : 11 (1987).
71. C. Baerlocher in "Proceed. of the 6 th Intern. Zeol. Conf." D. M. Olson and A. Bisio (ed), Butterworths, London, p 823 (1984).
72. K. F. M. G. J. Scholle, A. P. M. Kentgens, W. S. Veeman, P. Frenken and G. P. M. Van der Velden, J. Phys. Chem. 88 : 5 (1984).
73. G. Coudurier, A. Auroux, J. C. Vedrine, R. D. Shannon, L. Abrams and R. D. Farlee, J. Catal. 108 : 1 (1987).
74. L. E. Iton, R. B. Beal and D. T. Hodul, J. Molec. Catal. 21 : 151 (1983).
75. P. Ratnasamy, R. B. Borade, I. Sivasanker, V. P. Shiralkar and S. G. Hedge, Acta Phys. Chem. 31 : 137 (1985).
76. R. Szostak and T. L. Thomas, J. Catal. 100 : 555 (1986), R. Szostak, V. Nair and T. L. Thomas, J. Chem. Soc., Faraday Trans. I, 83 : 487 (1987).
77. G. C. Alis, P. Frenken, E. de Boer, A. Swolfs and M. A. Hefni, Zeolites 7 : 319 (1987).
78. B. Wichterlova, S. Beran, S. Bednarova, K. Nedomova, L. Dudinova and P. Jiru, in "Innovation in Zeolite Materials Science", P. J. Grobet et al, ed., Stud. in Surf. Sci. and Catal., Elsevier, Amsterdam 37 : 199 (1987).
79. G. Doppler, R. Lehnert, L. Marosi and A. X. Trautwein, Ibid p 215.
80. E. G. Derouane, M. Mestdagh and L. Vielvoye, J. Catal. 33 : 90 (1974).
81. Nat. Bur. Stand. (USA) Monogr. 25 Sec 15 (1978).
82. G. Perego, G. Belussi, C. Corno, M. Taramasso, F. Buonomo and A. Esposito, in "Proceed. of the 7 th Intern. Zeol. Conf., Tokyo 1986, Y. Murakami et al, ed, Elsevier, Amsterdam 129 (1986).
83. B. Notari in "Innovation in Zeolite Materials Science" Stud. in Surf. Sci. and Catal., P. J. Grobet et al, ed., Elsevier, Amsterdam, 37 : 413 (1988).
84. M. R. Boccuti, K. M. Rao, A. Zecchina, L. Leofanti and G. Petrini, in "Proceed. of the Sympos. on Reactivity of Solids", Trieste, 1988, in press 1989.
85. P. Pichat, R. Beaumont and D. Barthomeuf, J. Chem. Soc., Faraday Trans I, 70 : 1402 (1974).
86. M. B. Sayed, A. Auroux and J. C. Vedrine J. Catal. 101 : 43 (1986). M. B. Sayed, A. Auroux and J. C. Vedrine J. Catal. 116 : 1 (1989).
87. J. C. Vedrine, A. Auroux, P. Dejaifve, H. Hoser and S. B. Zhou, J. Catal. 73 : 147 (1982).
88. W. W. Kaeding, C. Chu, L. B. Young, B. Weinstein and S. A. Butter, J. Catal. 67 : 159 (1981).
89. J. A. Lercher and G. Rumplmayr, Appl. Catal. 25 : 215 (1986).
90. A. Auroux, H. Dexpert, C. Leclercq and J. C. Vedrine, Appl. Catal. 6 : 95 (1983).
91. P. Gallezot, Y. Ben Taarit and B. Imelik, C. R. Acad. Sci. Paris, Ser C, 272 : 261 (1971), J. Catal. 26 : 295 (1972).
92. C. Naccache and Y. Ben Taarit, Chem. Phys. Letters 11 :11 (1971).
93. M. Primet, J. C. Vedrine and C. Naccache J. Molec. Catal. 4 : 411 (1978).
94. J. C. Vedrine, E. G. Derouane and Y. Ben Taarit J. Phys. Chem. 78 : 531 (1974).
95. J. Howard and J. M. Nicol Zeolites 8 : 142 (1988).

96. M. Suzuki, K. Tsutsumi, H. Takamasmi and Y. Saito, <u>Zeolites</u> 8 : 381 (1981).
97. M. Suzuki, K. Tsutsumi, H. Takamasmi and Y. Saito <u>Zeolites</u> 8 : 387 (1988).
98. B. R. Gane and A. H. P. Hall, <u>Europ. Patent Appl</u>. 147, 111 (1984),
S. A. I. Barri and T. K. Mc Niff, <u>Europ. Patent appl</u>. 171, 981 (1985),
I. C. Bennett and A. H. P. Hall, <u>Europ. Patent Appl</u>. 202, 000 (1986) assigned to B. P. Corp.
99. A. Abou-Kaïs, C. Mirodatos, J. Massardier, D. Barthomeuf and J. C. Vedrine, <u>J. Phys. Chem</u>. 81 : 397 (1977).
100. M. Itoh, A. Miyamato and Y. Murakami, <u>J. Catal</u>. 64 : 284 (1980).

THE DETERMINATION OF (Si,Al) DISTRIBUTION IN ZEOLITES

Alberto Alberti, Glauco Gottardi * and Tiziana Lai

Istituto Geologico Mineralogico, Università di Sassari, Italy

* Istituto di Mineralogia, Università di Modena, Italy

ABSTRACT

The effects of the:

1. variation of the Si-O and Al-O distances depending on bonding forces on T and O atoms

2. variation of the measured T-O distances depending on disorder in the (Si,Al) distribution

3. variation of the measured T-O distances depending on the temperature of the diffraction data collection

4. shortening of the measured T-O distances as a result of averaging when apparent symmetry is higher than true symmetry

on the determination of the Al-content in the tetrahedral sites of zeolites has been shown. All these effects have a remarkable influence on the calculated Al-fraction. The differences in the (Si,Al) distribution found by comparing the results of a method proposed by Alberti and Gottardi[1], which takes into account effects no. 1 and no. 2, and of Jones'[2] determinative curve are stressed.

INTRODUCTION

The determination of (Si,Al) distribution in the tetrahedra is one of the most challenging and debated problems in silicate crystal chemistry.

The Si,Al ratio in a tetrahedral site can be obtained basically in two ways:

1. by the least square refinement of the two scattering curves ratio of the two atoms

2. from the dimension of the tetrahedron

The first way is not useful when X-ray diffraction is used, because the difference between the two scattering curves is very small, at least for $\sin\theta/\lambda$ values inside Ewald's sphere of Cu-radiation. This way is certainly more promising if neutron diffraction is used because Al and Si differ as much as 25 % in their neutron scattering amplitudes. Unfortunately, large crystals (nearly 1mm^3 in volume) are needed for neutron diffraction, therefore this method can only be applied to a restricted number of cases, and up

Guidelines for Mastering the Properties of Molecular Sieves
Edited by D. Barthomeuf *et al.*
Plenum Press, New York, 1990

Table 1 – Percent of Al in tetrahedral sites calculated according to AGm, AGl, and Jv models, average percent of Al, overall average percent of Al, minimum, maximum and mean values of the differences among AGm, AGl, and Jv for ordered zeolites.

Tetrah.	Model	Ami[15]	Bik[16]	Edi[17]	Gis[18]	Goo[19]	Mes[20]	Sco[21]	Tho[22]	Wai[23]	Wil[24]	Zeo.A[25]
	AGm	8.7	13.1	8.4	7.7	0.0	9.9	90.5	8.8	4.1	91.6	0.0
T1	AGl	9.2	12.0	7.8	9.0	0.0	9.2	90.2	8.1	1.7	91.2	0.0
	Jv	7.8	20.0	8.6	8.1	1.2	9.6	93.7	8.3	3.1	90.3	0.0
	AGm	5.9	2.9	7.6	6.2	0.6	3.8	91.7	5.6	6.0	4.9	87.8
T2	AGl	5.5	0.2	7.2	7.4	0.0	2.9	91.4	5.0	4.1	4.4	85.2
	Jv	6.7	2.5	11.6	5.9	0.0	9.9	89.7	12.9	9.4	7.6	81.4
	AGm	91.9	77.8	90.0	96.8	10.0	5.3	7.2	3.3	3.2	91.6	
T3	AGl	91.6	76.6	89.7	96.8	9.5	4.5	6.5	2.4	1.2	90.8	
	Jv	85.0	77.8	89.0	89.2	12.0	9.4	8.1	10.5	5.5	90.8	
	AGm	93.8	75.8		96.1	2.7	6.4	3.9	94.7	3.0	6.7	
T4	AGl	93.8	74.6		96.1	2.1	5.7	3.4	94.6	1.6	6.5	
	Jv	89.2	78.0		91.4	5.7	11.8	9.7	89.0	5.5	9.7	
	AGm		4.0			0.0	9.1	4.8	86.8	83.9	92.7	
T5	AGl		1.3			0.0	7.9	4.3	86.2	81.7	92.4	
	Jv		3.3			0.9	9.6	10.2	89.4	78.6	90.6	
	AGm		16.7			1.3	85.8		87.2	83.2	2.8	
T6	AGl		15.8			1.2	85.1		86.7	81.5	1.7	
	Jv		20.6			4.3	87.0		90.5	82.2	7.8	
	AGm					84.9	91.0					
T7	AGl					83.0	90.8					
	Jv					79.3	92.9					
	AGm					92.3	85.9					
T8	AGl					91.3	85.1					
	Jv					87.4	84.6					
Mean	AGm	50.1	31.7	40.7	51.7	24.0	39.0	39.6	47.7	30.6	48.4	43.9
Mean	AGl	50.1	30.1	40.3	51.7	23.4	38.2	39.2	47.2	28.6	47.8	42.6
Mean	Jv	47.2	33.7	42.0	48.7	23.8	41.3	42.3	50.1	30.7	49.5	38.6
Mean	Ch.An.	48.8	33.5	40.6	48.6	25.5	40.0	40.0	45.-50.	32.1	48.8	49.2

$AGm = 39.6$ $AGl = 38.8$ $Jv = 39.9$ $Chem.An. = 39.8$

$\overline{\Delta}(AGm - AGl) = 0.8$ $\Delta(AGm - AGl) = (0.0 - +2.7)$

$\overline{\Delta}|AGm - Jv| = 2.9$ $\Delta(AGm - Jv) = (-7.3 - +7.6)$

$\overline{\Delta}|AGl - Jv| = 3.1$ $\Delta(AGl - Jv) = (-8.1 - +9.0)$

Chem. An. of Zeo.A probably overestimated. Tho and Zeo.A are not considered in the Mean values.
References of the Tables can be requested to the authors.

to now it has been applied in only one case (brewsterite[3]) with an intermediate Si/Al ratio in the tetrahedra.

At present the Al-fraction in the tetrahedra of zeolites is normally deduced from the dimensions of the tetrahedra, or better said, from the average T-O distance $< T - O >$, for instance using Jones'[2] or Ribbe and Gibbs'[4] determinative curve.

These curves relate linearly $< T - O >$ with the Al-content. In particular the widely used relationship proposed by Jones is

$$Al - fraction = 6.4116 \cdot < T - O > -10.282 \qquad (1)$$

Hill and Gibbs[5] showed that the Si-O bond length depends on the Si-O-T angle; Baur[6] showed that the mean Si-O bond length depends on the tetrahedral sites according to the coordination number of oxygen atoms and on the mean value per SiO_4 tetrahedron of the negative secant of the angle Si-O-T; Geisinger et al. [7] pointed out that the Si-O and Al-O distances also depend on the tetrahedral O-T-O angles, and obtained, via regression analysis, two relationships which linearly relate the Si-O and Al-O distances to the sum of the Pauling bond strengths on the bridging oxygen, to the bridging T-O-T angle and to the tetrahedral O-T-O angles. These results indicate clearly that Jones'[2] or Ribbe and Gibbs'[4] linear relationships cannot accurately estimate the Al-fraction in individual tetrahedra because the Si-O and Al-O distances are assumed to be a constant, whereas the effects of local environment can remarkably change these distances. Alberti and Gottardi[1] showed that the variation of the T-O distances as a function of the bridging T-O-T angles can dramatically change the Al-content given by the linear relationship.

It is known that the average Al-content in a zeolite deduced via Jones' curve agrees satisfactorily with the Al-content given by the chemical analysis, if the (Si,Al) distribution is ordered, but it is systematically lower (~ 0.05 of the Al/(Si+Al) ratio) in the case of a disordered distribution. Ribbe[8] and Harlow and Brown[9] pointed out that the T-O distances averaged over all the tetrahedra of ordered felspars - especially alkali felspars and sodic plagioclases - are slightly larger (0.002-0.003Å) than those of their disordered equivalents, so their Al-content is apparently 1.5-2.0 % higher than that of the disordered ones. Alberti and Gottardi[10] showed that this discrepancy occurs because the positions of the atoms given by the X-ray structure refinement do not correspond, when disorder is present, to the true position of the atoms but to the centre of gravity of the sites occupied in the different unit cells, each site having a weight depending on its occupancy and on its scattering power.

Liebau[11] showed that high temperature factors, whether reflecting static disorder [e.g. caused by (Si,Al) distribution] or dynamic disorder (thermal motion), cause a shortening of the observed interatomic distances with respect to the actual bond lengths. He formulated a correlation between the isotropic temperature factor of oxygen atoms and the observed interatomic distances. As a result the observed interatomic distances, and consequently the calculated Al-content in the tetrahedral sites, depend on the temperature of the X-ray or neutron diffraction data collection.

Meier[12] emphasized that the observed T-O distances are shorter than the true values as a result of averaging. Such an averaging may occur when the refinement of a structure is carried out in a space group which has a symmetry higher than the true symmetry of the crystal. Alberti[13] showed that in zeolites of the mordenite group the true symmetry is a subgroup of the apparent symmetry. Nevertheless for these zeolites the structure refinements are normally carried out in the space group of the apparent symmetry, and in this case the underestimation of the Al-content is particularly high, as shown by Alberti and Gottardi[10].

Table 2 – Percent of Al in the tetrahdral sites calculated according to the AGm, AGl, and Jv models, average percent of Al, overall average percent of Al, minimum, maximum and mean values of the difference among AGm, AGl, and Jv for natrolite and zeolite X.

Tetrah.	Model	NatII[27]	NatI[28]	Nat[29]	Nat[30]	Nat[31]	NatIII[28]	NatII[28]	Gon[32]	Z.X[33]	Z.X[34]	Z.X[35]	Z.X[36]	Z.X[37]
T1	AGm	10.8	10.0	13.1	17.0	17.7	23.0	23.5	44.7	7.8	8.8	8.6	40.3	44.3
	AGl	10.0	8.9	12.0	15.7	16.0	21.2	21.6	43.4	5.5	7.6	6.9	39.3	41.7
	Jv	9.6	8.6	11.2	14.7	15.3	20.1	20.5	40.6	8.9	9.9	10.5	38.6	39.4
T2	AGm	5.8	7.8	9.6	13.5	13.7	18.9	20.7	46.6	77.3	83.5	79.7		
	AGl	5.1	7.1	9.0	12.5	12.4	17.5	19.	45.6	75.6	82.7	78.4		
	Jv	10.7	12.6	14.0	17.4	17.6	22.2	23.8	47.7	74.6	80.5	77.8		
T3	AGm	87.4	86.2	85.4	79.9	75.8	70.9	69.2						
	AGl	87.0	85.7	85.1	79.3	75.0	70.1	68.3						
	Jv	87.9	86.6	85.8	80.4	76.5	71.71	70.0						
Mean	AGm	39.4	39.6	40.6	40.7	39.2	40.5	40.6	46.2	42.6	46.2	44.2	40.3	44.3
Mean	AGl	38.8	38.9	40.0	39.9	38.2	39.3	39.3	45.2	40.6	45.2	42.6	39.3	41.7
Mean	Jv	41.4	41.4	42.2	42.1	40.7	41.6	41.6	46.3	41.8	45.2	44.2	38.6	39.4
Mean	Ch. An.	39.4	40.0	41.6	39.5	39.0	40.0	40.0	46.2	47.9	45.8	45.8	42.2	45.8

Chem.An. = 40.7
Chem.An. = 45.3

| Natrolite | AGm = 40.8 | AGl = 39.9 | Jv = 42.2 |
| ZeoliteX | AGm = 43.5 | AGl = 41.9 | Jv = 41.8 |

$\overline{\Delta}(AGm - AGl) = 1.5$ $\Delta(AGm - AGl) = (0.3 - +2.6)$

$\overline{\Delta}|AGm - Jv| = 1.9$ $\Delta(AGm - AGl) = (-4.9 - +4.9)$

$\overline{\Delta}|AGl - Jv| = 2.2$ $\Delta(AGl - Jv) = (-5.6 - +2.8)$

To summarize, among the sources of error in the determination of the Al-content in the tetrahedral sites of zeolites four seem to be particularly remarkable:

1. the variation of the Si-O and Al-O distances depending on bonding forces on T and O atoms

2. the variation of the measured T-O distances depending on disorder in the (Si,Al) distribution

3. the variation of the measured T-O distances depending on the temperature of the diffraction data collection

4. the shortening of the measured T-O distances as a result of averaging when apparent symmetry is higher than true symmetry.

The aim of this work is to evaluate quantitatively these sources of error in the determination of the (Si,Al) distribution in several zeolites chosen as examples and to show the results obtained with a computing method proposed by Alberti and Gottardi[1].

DISCUSSION

Starting from the statements summarized in points no 1. and 2., Alberti and Gottardi[1] proposed a method which instead of calculating the Al-content from the dimensions of the tetrahedra obtained by the structure refinement, as is normally done by the other methods, calculates the Al-content which gives the measured dimensions of the tetrahedra.

Basic assumptions of the method are:

a) the distances Si-O and Al-O are not constant but are given by the relationships:

$$Si-O = 1.3816 + 0.0530 \cdot (-sec[TOT]) + 0.0681 \cdot BSTT + 0.0271 \cdot BSCAT + 0.0048 \cdot NLSIA +$$

$$0.0088 \cdot (-sec[OSiO]) \qquad (2)$$

$$Al-O = 1.6609 + 0.0447 \cdot (-sec[TOT]) + 0.0215 \cdot BSCAT + 0.0105 \cdot (-sec[OAlO]) \quad (3)$$

where:
    TOT = T-O-T angle of the T-O bond
    OTO = average of the three O-T-O angles formed with the bridging T-O bond
    BSTT = bond strength of the tetrahedral cations on framework oxygen
    BSCAT = bond strength of the extraframework cations on framework oxygen
    NLSIA = no. of Si-O-Al bridges of the tetrahedron.

These relationships have been found by multiple stepwise regression using Si-O and Al-O distances experimentally determined in 13 zeolites with an ordered, or almost ordered, (Si,Al) distribution.

The Al-content in a tetrahedron obtained from a linear relationship but where the distances Si-O and Al-O are calculated according to equations (2) and (3) can be interpreted, when compared with the Al-content given by the Jones method, as the effect of the dependency of the T-O distances from the bonding forces on atoms T and O, i.e. the source of error no. 1. From now on the Al-fraction calculated in such a way will be indicated with the symbol AGl.

b) distance vectors are used instead of their absolute values. This means that the true T-O distance, on the left of the equation

Table 3 – Percent of Al in tetrahedral sites calculated according to AGm, AGl, and Jv models, average percent of Al, overall average percent of Al, minimum, maximum and mean values of the differences among AGm, AGl, and Jv for heulandites and clinoptilolites.

Tetrah.	Model	Heu[38]	Heu[39]	Cli[40]	Cli[40]	Heu[41]	Heu[41]	Cli[42]	Cli[42]	Heu[43]	heu[44]	Heu[45]
	AGm	15.8	19.3	11.5	3.7	12.6	9.7	13.5	11.7	19.2	12.1	15.3
T1	AGl	13.6	17.1	10.0	2.2	10.3	8.0	11.9	9.9	16.8	9.7	12.9
	Jv	16.9	20.0	13.7	5.7	13.4	11.0	15.0	13.1	19.2	11.6	17.4
	AGm	38.2	38.1	29.5	44.5	39.0	36.5	39.2	29.2	40.6	42.5	36.9
T2	AGl	34.3	34.1	25.0	41.7	35.0	32.2	35.4	25.9	36.9	38.8	33.0
	Jv	34.6	34.7	26.5	41.0	34.9	32.5	35.5	26.5	36.8	37.3	34.2
	AGm	23.6	17.5	7.6	11.6	22.0	9.3	5.7	11.5	21.8	18.6	19.5
T3	AGl	19.8	12.5	5.5	9.0	17.1	6.2	3.2	8.8	16.6	15.3	16.2
	Jv	20.1	12.8	5.7	8.9	16.9	6.7	4.1	9.1	16.3	14.0	15.5
	AGm	12.2	21.5	3.3	10.6	16.1	17.6	6.7	10.2	21.8	14.3	14.7
ᵗ T4	AGl	7.8	16.4	1.8	7.8	11.2	14.4	4.5	7.8	16.5	10.8	11.0
	Jv	7.3	16.0	2.5	7.3	10.7	13.7	1.5	7.5	15.6	9.6	10.0
	AGm	24.1	22.0	3.5	5.0	14.1	8.9	7.5	5.1	20.7	13.9	24.3
T5	AGl	21.7	19.4	2.7	3.8	11.9	7.1	6.1	3.8	17.9	11.7	21.6
	Jv	23.3	21.1	7.3	7.3	14.7	9.6	8.9	6.7	19.5	13.4	23.3
Mean	AGm	22.6	23.9	11.9	16.2	21.5	17.2	15.3	14.5	25.3	21.0	21.9
Mean	AGl	19.2	20.0	9.7	13.9	17.7	14.3	12.9	12.1	21.3	17.9	18.6
Mean	Jv	20.1	20.9	11.6	13.2	18.5	15.3	13.5	13.2	21.6	17.6	19.7
Mean	Ch.An.	26.0	26.3	18.0	21.3	21.1	21.7	18.7	17.5	21.7	20.0	25.8

AGm = 19.2   AGl = 16.1   Jv = 16.8   Chem.An. = 21.6

$\overline{\Delta}(AGm - AGl) = 3.0$ $\Delta(AGm - AGl) = (0.8 - +5.3)$

$\overline{\Delta}|AGm - Jv| = 2.8$ $\Delta(AGm - Jv) = (-3.8 - +6.2)$

$\overline{\Delta}|AGl - Jv| = 1.5$ $\Delta(AGl - Jv) = (-4.6 - +3.0)$

Table 4 - Percent of Al in tetrahdral sites calculated according to AGm model, and average Al-contents for some zeolites at different temperatures.

Tetrah.	Bikitaite		Gismondina		Scolecite		Natrolite		
	$13°K^{16}$	$295°K^{16}$	$15°K^{46}$	$293°K^{18}$	$20°K^{47}$	$293°K^{21}$	$20°K^{48}$	$293°K^{26}$	$361°K^{26}$
T1	13.8	13.1	10.5	7.7	91.2	90.5	11.4	8.7	7.9
T2	6.2	2.9	9.0	6.2	91.8	91.7	6.6	6.3	5.3
T3	78.9	77.8	95.4	96.8	8.4	7.2	91.3	88.1	87.5
T4	77.8	75.8	97.1	96.1	5.5	3.9			
T5	7.1	4.0			6.1	4.8			
T6	18.1	16.7							
Mean	33.6	31.7	53.0	51.7	40.6	39.6	41.5	39.5	38.7

Tetrah.	Natrolite		Heulandite				
	$419°K^{26}$	$471°K^{26}$	$293°K^{45}$	$373°K^{45}$	$593°K^{45}$	$293°K^{43}$	$343°K^{43}$
T1	7.1	6.6	15.3	12.7	20.2	19.2	20.0
T2	4.2	3.9	36.9	26.7	21.9	40.6	37.3
T3	87.3	88.2	19.5	17.0	17.3	21.8	16.2
T4			14.7	22.4	0.0	21.8	16.1
T5			24.3	18.3	0.0	20.7	14.4
Mean	38.0	38.2	21.9	19.5	13.2	25.3	21.5

$$\sum_{j=1}^{n} |d_j(T - O)|/n \geq |\sum_{j=1}^{n} d_j(T - O)|/n \qquad (4)$$

is used instead of the shorter observed T-O distance on the right of the equation.

The Al-content in a tetrahedron is the average of the four Al-contents deduced by the four T-O bonds of the tetrahedron.

The Al-content, calculated according to assumptions a) and b), will be from now on indicated with the symbol AGm. Its difference from the AGl value of the Al-content can be interpreted as the effect of choice of the term on the left of equation (4) instead of the term on the right of the same equation, i.e. the source of error no. 2. Obviously the difference between the AGm value and that calculated via Jones' method (from now on Jv) can be interpreted as the sum of the sources of errors no. 1 and no. 2.

Table 1 shows the AGm, AGl and Jv values for many ordered, or almost ordered, zeolites. In Table 2 the same values for some natrolites and zeolites X, from an ordered to a complete disordered (Si,Al) distribution, are shown. Table 3 reports the AGm, AGl and Jv values for many heulandites and clinoptilolites, zeolites with a strongly disordered (Si,Al) distribution. Note that the Jv model is very similar to the other linear models like that of Ribbe and Gibbs[4].

From this data some interesting conclusions can be drawn:

1. the differences between AGl and Jv vary in a wide range. In fact the Al/(Si+Al) ratio calculated with the AGl model can be up to 8 % less or up to 9 % more than the same ratio calculated with the Jv model. Considering that in zeolites this ratio is always ≤ 0.5, the disagreement in the calculated Al-content between the two models can be as high as 20 % of the Al-fraction.

The absolute value of the mean difference between AGl and Jv ($\overline{\Delta}|AGl - Jv|$) is 3.1, 2.2 and 1.5 % in Tables 1, 2 and 3 respectively so that we can expect a mean error in the Al-content calculated with the Jv model around 4-5 %.

2. the mean difference between AGm and AGl ($\overline{\Delta}(AGm - AGl)$ increases rapidly with the increase of disorder in the (Si,Al) distribution, as expected by Alberti and Gottardi[10]. In fact this difference, which is of about 1.0 % for the almost ordered zeolites, is 3.0 % in the disordered heulandites.

3. the difference (AGm-Jv) varies in the range ±7% whereas the absolute value of the mean difference $\overline{\Delta}|AGm - Jv|$ is around 2.0-3.0 %. The mean difference $\overline{\Delta}(AGm - Jv)$ is about zero or slightly negative for ordered zeolites but becomes positive in the case of a disordered (Si,Al) distribution. As a result, the Al-content given by AGm matches the Al-content given by chemical analysis much better than the Al-content given by Jv, above all for disordered zeolites.

Table 4 shows the variations of the Al-content calculated with the AGm method as a function of the temperature of the diffraction data collection. The thermal motion causes a dynamic disorder and consequently an apparent variation in the calculated T-O distances. Equations (2) and (3) have been calculated from the data of structures at room temperature, i.e. they are "tared" at 293° K, therefore we may expect from eq. (4) that, in the case of lower temperatures, and consequently lower thermal motion, the calculated T-O distances tend to increase, whereas for temperatures higher than 293° K, and higher thermal motion, the T-O distances tend to decrease.

Consequently, the Al-content in the tetrahedra found at very low temperature should be too high, and at high temperature too low. As expected, at low temperature

Table 5 - Percent of Al calculated according to the AGm and Jv models, and average Al-content for some mordenites, ferrierites and dachiardites.

	Mordenite										Ferrierite	
Tetrah.	AGm^{49}	Jv^{49}	AGm^{50}	Jv^{50}	AGm^{51}	Jv^{51}	AGm^{52}	Jv^{52}	AGm^{53}	Jv^{53}	AGm^{54}	Jv^{54}
T1	11	10	13	11	8	7	12	10	7	7	9	4
T2	4	0	6	2	5	0	9	5	0	0	30	21
T3	25	20	28	23	21	17	27	19	39	33	11	10
T4	16	11	18	14	15	11	22	16	7	5	2	0
Mean	11.6	8.8	13.9	10.5	10.2	7.1	15.0	10.9	10.0	8.8	12.1	7.4
Ch.An.	18.7		16.7		16.5		16.7		15.0		19.2	

	Ferrierite						Dachiardite					
Tetrah.	AGm^{55}	Jv^{55}	AGm^{56}	Jv^{56}	AGm^{57}	Jv^{57}	AGm^{58}	Jv^{58}	AGm^{58}	Jv^{58}	AGm^{59}	Jv^{59}
T1	7	3	19	14	14	11	8	7	3	0	4	0
T2	22	15	25	17	1	1	7	7	4	0	0	0
T3	0	0	7	7	8	7	18	12	22	15	27	18
T4	0	0	0	0	4	3	33	28	38	31	30	22
Mean	5.6	3.6	9.2	6.0	5.4	4.3	13.5	11.2	12.5	7.6	10.6	7.2
Ch.An.	16.1		19.3		14.3		20.4		20.4		18.9	

AGm = 10.5   Jv = 7.5   Chem.An. = 17.4

the Al-content is apparently higher (see Table 4) and the differences in the Al/(Si+Al) ratio are in the order of 1-2 %. At high temperature, even though for single tetrahedra the results seem to conflict with the expectation, the apparent lowering of the Al-content is extremely evident from the average values.

Table 5 reports the Al-contents in some mordenites, ferrierites and dachiardites, calculated with the AGm and the Jv methods. Alberti[13] showed that in these phases the real symmetry is lower than the topological symmetry but the structure refinement is normally carried out in the higher topological symmetry. Only the structure refinement of Elba dachiardite[14] has been carried out in the real symmetry Cm instead of the topological symmetry $C2/m$. In all these refinements the calculated Al-content is very low when compared with the Al-content given by the chemical analysis. In fact against an average value of the Al/(Si+Al) ratio of around 17.5 %, this value calculated with the AGm method is only 10.5 %, and the Jv value is as low as 7.5 %.

The $\overline{\Delta}(AGm - Jv) = 3.0\%$ can be attributed to the source of error no. 2, but the residual $\overline{\Delta}(ChemicalAn. - AGm) = 7.0\%$ must be largely arributed to the averaging of the true ion site coordinates, i.e. to the source of error no. 4. If we compare the Al-fraction of Elba dachiardite calculated in the $C2/m$ space group with the Al-fraction calculated in the true space group, the better matching of this last value with the chemical analysis is evident, even if the difference between these two values remains very high.

CONCLUSIONS

The determination of the Al-content in the tetrahedra of zeolites is strongly influenced by the four sources of error previously discussed. The Alberti and Gottardi[1] method strongly reduces the negative effects on results of the non-constant T-O distances and of the disorder in the (Si,Al) distribution. Temperature effects on the calculated Al-fraction could be corrected to a certain extent by a relationship which relates the observed T-O distances to the temperature factors. What is more difficult, probably, is to annul the serious discrepancy due to a wrong choice of symmetry. Also in this case, however, an opportune utilization of the thermal ellipsoids, could be, despite the difficulties, useful.

ACKNOWLEDGEMENTS

The authors thank the C.I.C.A.I.A. of the University of Modena for computing facilities. The Consiglio Nazionale delle Ricerche and the Ministero della Pubblica Istruzione are also acknowledged for finantial support.

REFERENCES

1. A.Alberti and G.Gottardi, The determination of the Al-content in the tetrahedra of framework silicates, Z.Kristallogr., 184:49 (1988)
2. J.B.Jones, Al-O and Si-O tetrahedral distances in aluminosilicate framework structure, Acta Cryst., B24:355 (1968)
3. G.Artioli,G.V.Smith and Ake Kvick, Multiple Hydrogen position in the zeolite Brewsterite, Acta Cryst., C41:492 (1985)
4. P.H.Ribbe and G.V.Gibbs, Statistical analysis and discussion of mean Al/Si-O bond distances and the aluminum content of tetrahedra in felspars, Am.Mineral., 54:85 (1969)

5. R.J.Hill and G.V.Gibbs, Variation in d(T-O), d(T...T) and TOT in silica and silicate minerals, phosphates and aluminates, Acta Cryst., B35:25 (1979)

6. W.H.Baur, Variation of mean Si-O bond lenghts in Silicon-Oxygen tetrahedra, Acta Cryst., B34:1751 (1978)

7. K.L.Geisinger,G.V.Gibbs and A.Navrotsky, A Molecular Orbital study of Bond Length and Angle Variations in Framework Structures, Phys. Chem. Minerals, 11:266 (1985)

8. P.H.Ribbe, Average structures of alkali and plagioclase felspars: systematics and applications, in:"Felspars and Feldspathoids" W.L.Brown,ed., Reidel,Dordrechts (1984)

9. G.E.Harlow and G.E.Brown, Low albite: an X-ray and neutron diffraction study, Am. Mineral., 65:986 (1980)

10. A.Alberti and G.Gottardi, On the influence of (Si,Al) disorder on the T-O distances measured in zeolites, in:"Zeolites - Synthesis, Structure, Technology and Application" B.Drzaj, S.Hocevar, S.Pejovnik, eds., Elsevier, Amsterdam (1985)

11. F.Liebau, "Structural Chemistry of Silicates" Springer, Heidelberg (1985)

12. W.M.Meier, Symmetry aspects of zeolite frameworks, in: "Molecular Sieves" W.M.Meier, J.B.Uytterhoeven, eds., Adv. in Chem. Series 121. Am. Chem. Society, Washington D.C. (1973)

13. A.Alberti, The absence of T-O-T angles of 180° in zeolites, in: "New developments in zeolite science and technology" Y.Muramaki, A.Jijima, J.W.Wards, eds., Kodansha, Tokio (1986)

14. G.Vezzalini, A refinement of Elba dachiardite: opposite acentric domains simulating a centric structure, Z.Kristallogr., 166:63 (1984)

SOLID-STATE ION EXCHANGE - PHENOMENON AND MECHANISM

H.G. Karge, V. Mavrodinova*, Z. Zheng** and H.K. Beyer***

Fritz-Haber-Institut der Max-Planck-Gesellschaft
Faradayweg 4-6, 1000 Berlin 33 (West)

ABSTRACT

Highly exchanged Me^I-Y (Me^I = Li, Na, K, Rb, Cs) as well as La-Y zeolites were prepared through solid-state ion exchange between NH_4-Y (98%) and the respective chlorides. It was shown that, except for the case of Li-Y, both low-temperature and high-temperature exchange processes occurred. With the system $Me^I Cl/NH_4$-Y the high-temperature reaction seemed to proceed more easily the lower the lattice energy of the alkaline chloride. In La-Y, the bare La^{3+} ions introduced via solid-state ion exchange were catalytically inactive. However, La-Y obtained via solid-state ion exchange was rendered an active catalyst for both ethylbenzene disproportionation and n-decane cracking after a brief contact with traces of water vapor. For the solid-state reaction itself, the presence of water is not a prerequisite.

INTRODUCTION

In the past there have been only a few cursory reports on solid-state ion exchange[1-3]. More recently, systematic studies on this topic were advanced by Karge et al.[4-7] and Kucherov et al.[8-9] It was the aim of the present study to elucidate whether (i) in principle, higher degrees of exchange can be obtained by solid-state reaction than by conventional ion exchange in salt solution; (ii) solid-state ion exchange might result in catalysts comparable or even superior to those prepared via conventional ion exchange, and (iii) the presence of traces of water is required for solid-state ion exchange to occur.

---

*   On leave from Bulgarian Academy of Sciences, Institute of Organic
      Chemistry, Sofia 1040, Bulgaria
**   On leave from Changchun Institute of Applied Chemistry, Academia Sinica,
      China
***   On leave from Hungarian Academy of Sciences, Central Research Institute
      for Chemistry, 1525 Budapest, Hungary

Guidelines for Mastering the Properties of Molecular Sieves
Edited by D. Barthomeuf *et al.*
Plenum Press, New York, 1990

EXPERIMENT

The zeolites used are listed in Table 1.

Table 1

	Zeolites used	Si/Al
NH_4-Y	$(NH_4)_{44.3}\,Na_{7.7}\,Al_{52.2}\,Si_{140.7}\,O_{384}$	2.7
NH_4-Y	$(NH_4)_{53.5}\,Na_{1.3}\,Al_{54.9}\,Si_{137.1}\,O_{384}$	2.5
H-MOR	$H_{2.42}\,Na_{0.12}\,Al_{2.5}\,Si_{45.5}\,O_{96}$	18.2
H-ZSM-5 (1)	$H_{2.68}\,Na_{0.12}\,Al_{2.8}\,Si_{93.2}\,O_{192}$	33.3
H-ZSM-5 (2)	$H_{3.93}\,Na_{0.07}\,Al_{4.0}\,Si_{92.0}\,O_{192}$	23.0

NH_4-Y with a 85% and 98% degree of exchange was obtained by repeated exchange (5 and 15 times, respectively) of NaY (Union Carbide) in 1 M NH_4Cl solution. ZEOLON 100 (Norton Comp.) was dealuminated by leaching with 0.1 M HNO_3 to result in H-MOR with an overall Si/Al = 18.2. Both H-ZSM-5 samples were provided by DEGUSSA and used without further treatment as obtained.

All the chlorides used for intimate grinding with the zeolite samples were analytical grade and purchased from MERCK, Darmstadt. The techniques employed (apparatus and procedures) have been described in earlier publications [4-7]

RESULTS AND DISCUSSION

Solid-state ion exchange between Me^ICl and NH_4-Y (98%) was easily monitored by the changes in the IR spectra. For $LiCl/NH_4$-Y this is demonstrated by Figure 1. The lower part shows the OH stretching region of pure NH_4-Y after treatment at 625 K in high vacuum and (right) the bands obtained after pyridine adsorption. Although, the prominent bands at 3640 and 3550 cm^{-1} appeared after heat-treatment of NH_4-Y under 10^{-5} Pa at 625 K, they were completely missing after similar treatment of a solid $LiCl/NH_4$-Y mixture. NH_4^+ was displaced by Li^+ cations. The degree of exchange was 98% (confirmed also by chemical analysis), whereas after five-times-repeated exchange in solution only 53% was achieved. This is most likely due to the fact that penetration of Li into the small cavities is, in water solution, impeded by the hydration shell. Pyridine adsorption subsequent to dehydration and deammoniation of pure NH_4-Y provided evidence that even at a pretreatment temperature as low as 625 K some dehydroxylation and removal of Al from the framework occurred. This is indicated by the band at 1453 due to pyridine coordinatively bound to so-called true Lewis sites[10]. In contrast, pyridine adsorption onto Li-Y produced via solid-state ion exchange (upper part of Figure 1, right side) gave rise to the band at 1446, indicative of pyridine coordinated to Li^+. Thus, the presence of Li during the solid-state reaction obviously suppressed the dehydroxylation step. However, under similar conditions, treatment of Me^ICl/NH_4-Y with Me^I = Na, K, Rb, Cs yielded materials which also showed some dehydroxylation. Moreover, the degree of exchange obtained via solid-state ion exchange was less than 98%, although still higher than after five-times-repeated exchange in 1M solutions of the respective chlorides.

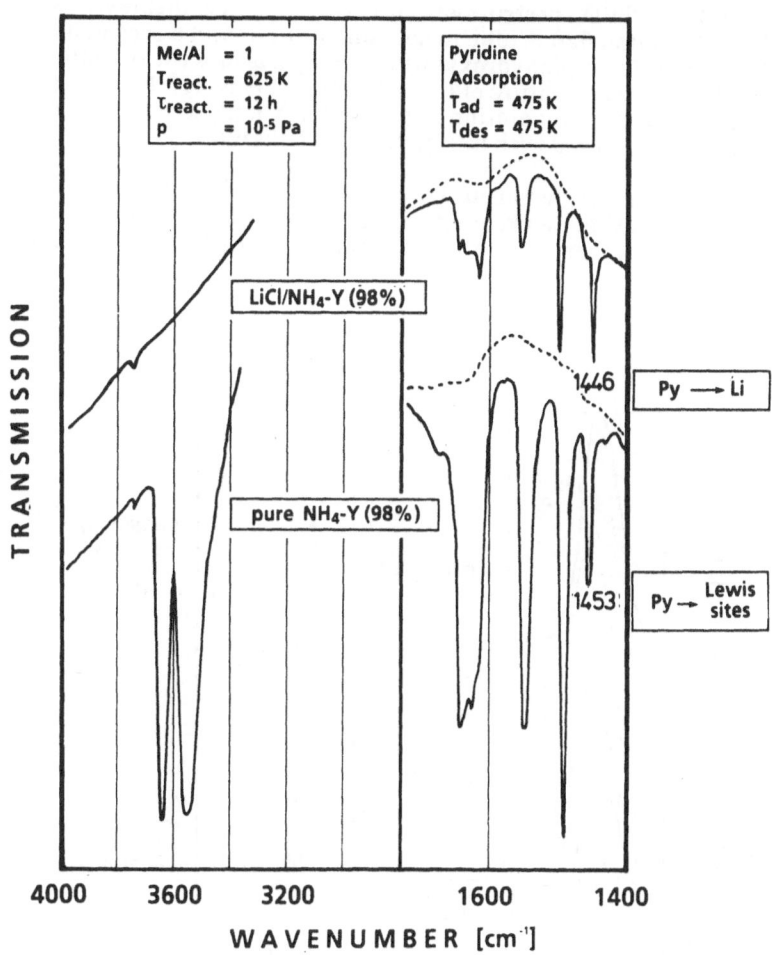

Fig. 1. IR spectra of H-Y and heat-treated Me^ICl/NH_4-Y mixtures before (left) and after (right) adsorption of pyridine

Temperature-programmed heating of solid Me^ICl/NH_4-Y mixtures was accompanied by parallel mass spectrometric measurement of evolved gases (NH_3, H_2O, HCl). As demonstrated in Figure 2, an intense low-temperature exchange (peak around 500 K) was usually followed by a high-temperature process (peak at 827 to 915 K) replacing residual protons by alkaline metal cations. The exchange proceeded most easily with LiCl during which the high-temperature process did not appear (no high-temperature peak of evolved HCl). In the other cases, where a high-temperature exchange of Me^+ for H^+ was observed, the peak temperature decreased in the same sequence ($Na^+ > K^+ > Rb^+ > Cs^+$) in that the lattice energies of the corresponding chlorides Me^ICl change[4].

In Figure 2 the TPD spectrum of pure NH_4-Y is also displayed with a first broad NH_3 peak around 520 K (due to deammoniation) and a second one (H_2O) at 870 K being indicative of dehydroxylation. Comparison with the other patterns shows, first, that the temperature of the replacement of NH_4^+ by Me^+ (first peak) is lower than that of deammoniation of pure NH_4-Y and, secondly, practically no dehydroxylation is observed with mixtures of $MeCl/NH_4$-Y.

Solid-state reaction between $LaCl_3$ and NH_4-Y (98%) provides, rather elegantly, almost 100% of La exchange, i.e. nearly one La^{3+} per three framework aluminum atoms, whereas conventional exchange in La^{3+} salt solution must be

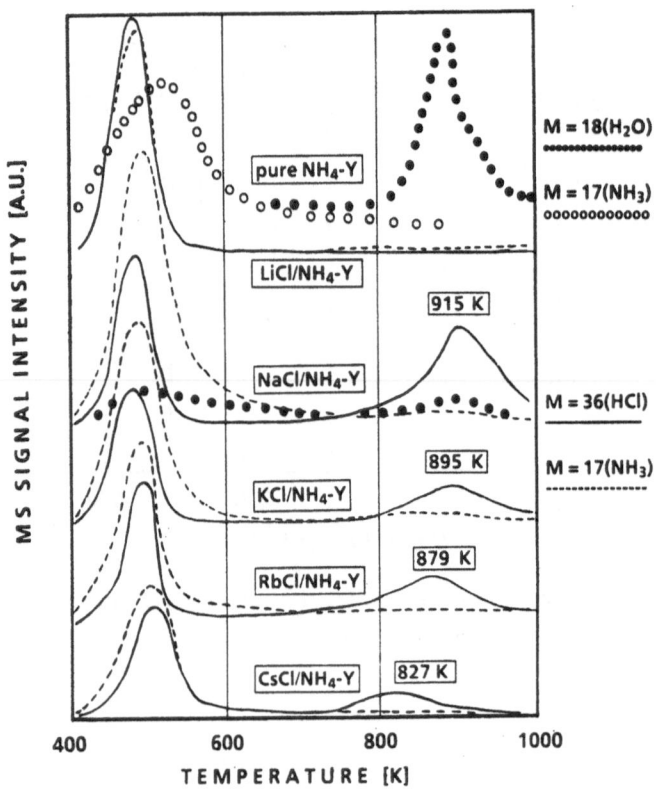

Fig. 2. Temperature-programmed evolution of HCl ($M = 36$, ———) from $MeCl/NH_4$-Y mixtures; additionally, evolution of H_2O ($M = 18$, ●●●) and NH_3 ($M = 17$, ○○○ or - - -) is shown.

repeated many times with intermittent calcination at high temperatures. After contact with water vapor, the catalyst obtained via solid-state ion exchange exhibited even superior performance with respect to the test reaction (ethylbenzene disproportionation, compare Ref. 11), shown in Figure 3. Comparison with La-Y catalysts formed via the procedure described by Rees et al.[12] or via exchange at temperatures higher than 375 K under autogeneous pressure[13] has not yet been carried out.

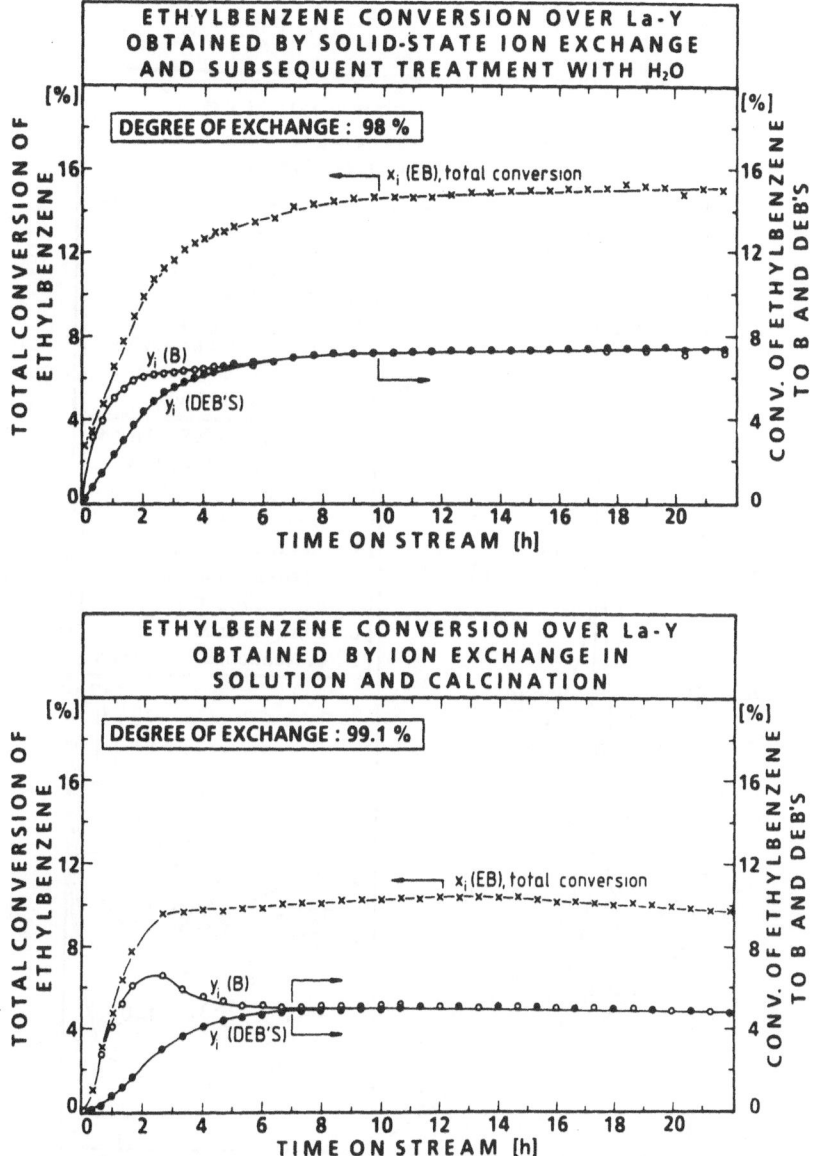

Fig. 3. Comparison of the catalytic performance of La-Y catalysts with a high degree of exchange obtained by solid-state ion exchange (upper part, La/NH$_4$ = 0.33) or by conventional exchange in solution (lower part).

$T_{act} = 725$ K; $m_{cat} = 0.25$g; $T_{react} = 425$ K.

A La-Y (85%) catalyst, obtained by solid-state ion exchange, was also active in cracking of n-decane but again only after brief contact with small amounts of water vapor. This is shown by Figures 4a and 4b which describe the results from an in situ experiment in an IR reactor-cell connected with a gas chromatograph .

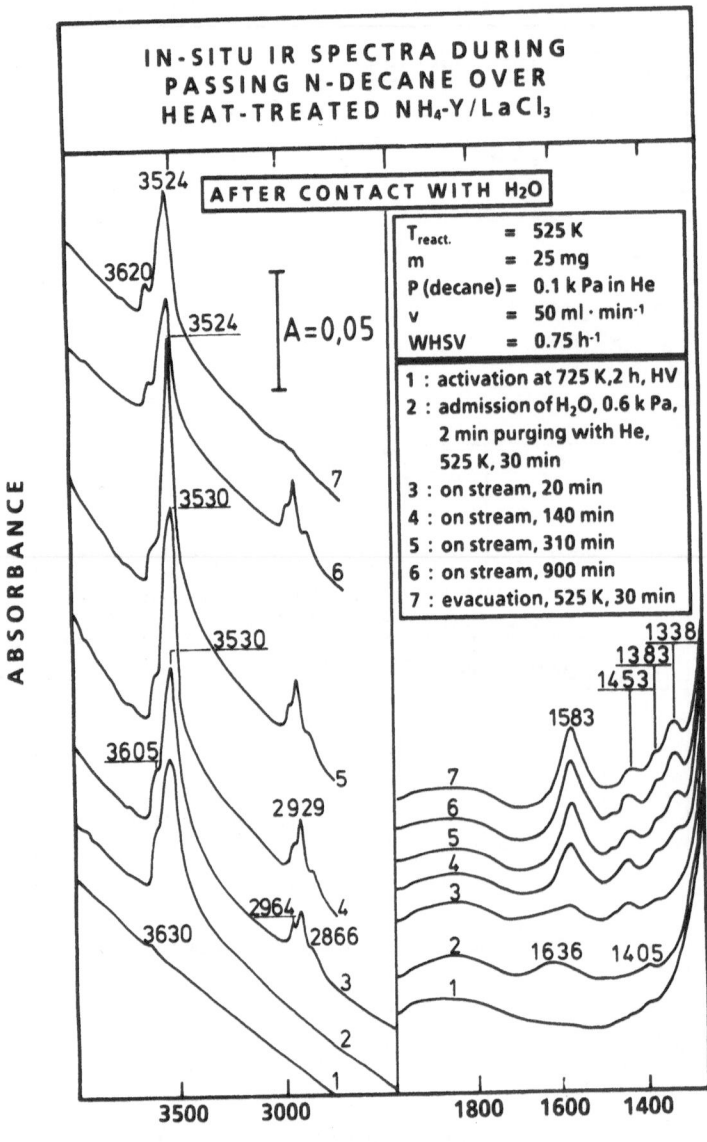

Fig. 4a. In-situ IR spectroscopic measurements in an IR flow reactor cell, employing a La-Y catalyst with a high degree of exchange obtained by solid-state ion exchange

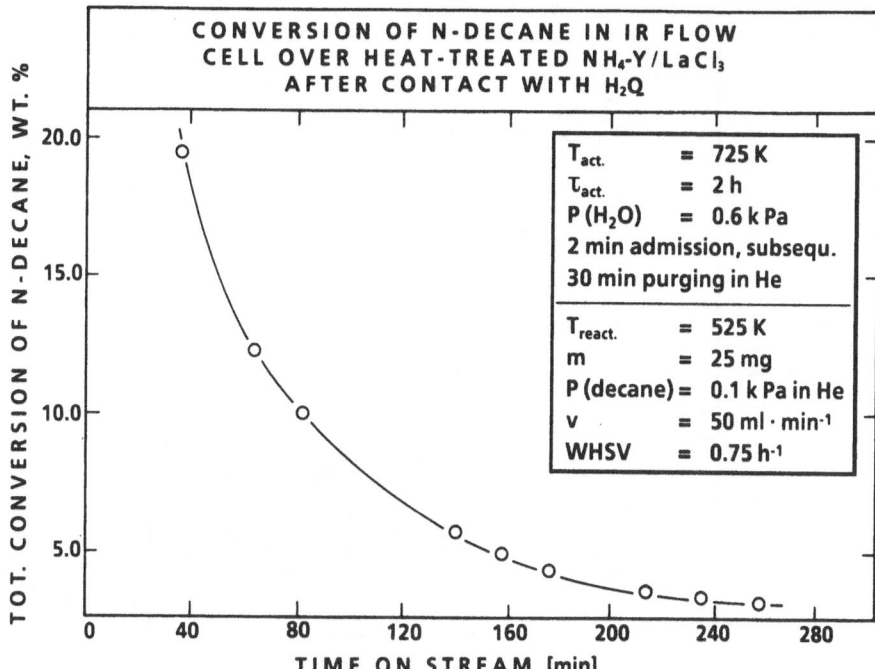

Fig. 4b. Conversion measurements of n-decane cracking in an IR flow reactor cell over a La-Y catalyst with a high degree of exchange obtained by solid-state ion exchange.

After heating the wafer of a stoichiometric $LaCl_3/NH_4$-Y mixture in high vacuum (10^{-5} Pa) at 725 K the OH bands typical of H-Y were completely removed (spectrum 1 in Figure 4a). Upon subsequent short (2 min) contact with H_2O vapor at low pressure (600 Pa) and 525 K, followed by removal of most of the excess H_2O at the same temperature, the OH bands typical of La-Y appeared (3605 → 3620 cm^{-1}, 3524 cm^{-1}). Admission of the feed stream (n-decane in helium) resulted immediately in a high initial conversion (Figure 4b). In fact, the activity rapidly decreased (Figure 4b). This was obviously due to severe coke formation, indicated by the growth of the so-called coke IR band at 1583 cm^{-1} (Figure 4a). Samples without contact to water vapor prior to feeding the reactant proved to be completely inactive.

While contact with water is necessary to transform the Y zeolite containing trivalent (or divalent) cations into an active catalyst, the presence of water does not seem to be a prerequisite for the solid-station ion exchange itself. This was demonstrated by an experiment in which preparation of the sample for solid-state ion exchange in the system NaCl / dealuminated H-MOR was carried out with extreme exclusion of water, viz. in a glove box under a very carefully dried atmosphere of helium (see explanation to Figure 5).

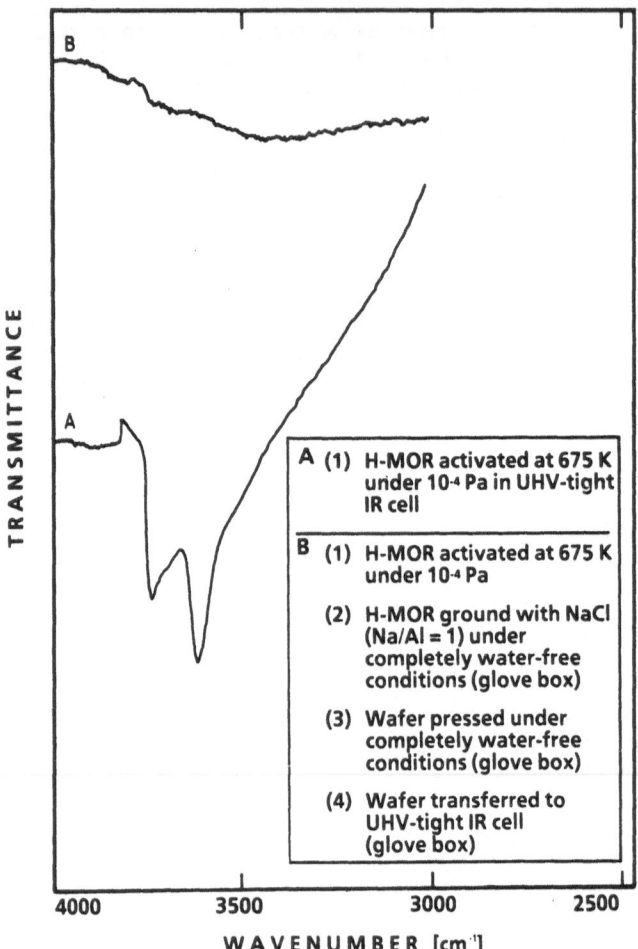

Fig. 5. Solid-state ion exchange in a mixture of dealuminated H-MOR with careful exclusion of water (details see text).

After transfer to an ultrahigh-vacuum-tight IR cell and heating of the cell to 675 K, the OH bands of a pure H-MOR sample, activated at 675 K, did not appear (compare spectra A and B in Figure 5). It appears that NaCl molecules were separated from the tiny halide crystallites (intimately mixed with zeolite crystallites), i.e. they evaporated from kinks ("Halbkristallagen", see Ref.14) and moved to the opening of the zeolite structure. From there they may penetrate into the pores as molecules or, more likely, dissociate into ions due to the electrolytic solvent effect of the water-free, porous zeolite structure. In any case, when the sodium reaches a hydroxyl group it reacts, replaces the proton and a HCl molecule is evolved. Since HCl is easily removed, the equilibrium is rapidly shifted to the alkaline form of the zeolite.

This result, i.e. that the presence of H_2O is not required for the solid-state reaction to proceed, is supported by the following experiments with H-ZSM-5/AgCl and H-ZSM-5/Hg_2Cl_2 mixtures.

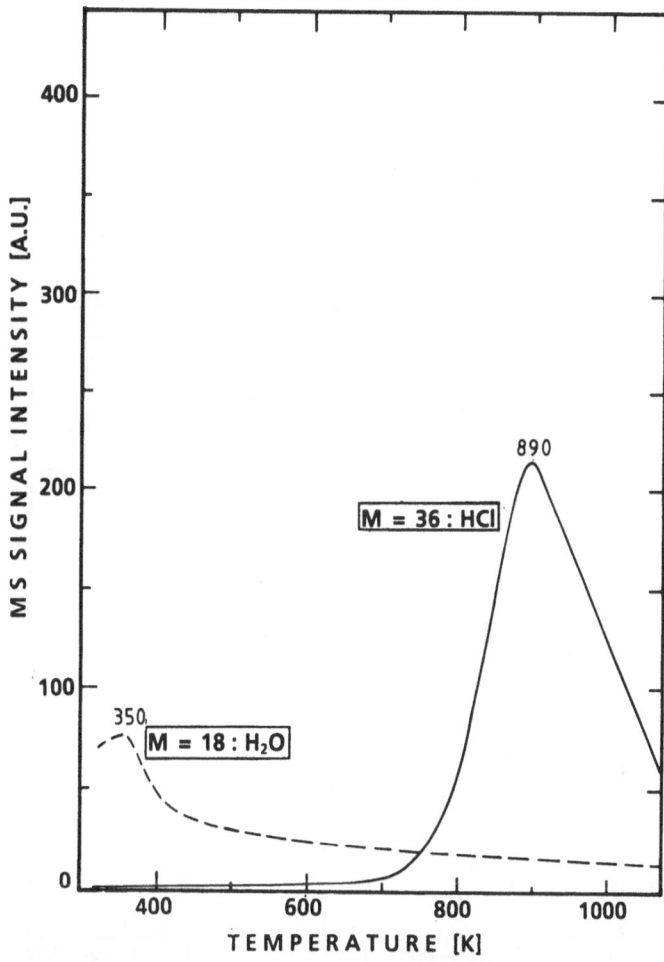

Fig. 6. Solid-state ion exchange in the system H-ZSM-5/AgCl monitored by mass spectrometric analysis of the evolved gases as a function of temperature. Zeolite mixture H-ZSM-5 (1) + AgCl, Ag/Al = 1; heating rate: 10 K · min^{-1} (details see text).

Figure 6 shows a temperature-programmed heat-treatment of a H-ZSM-5/AgCl mixture. The evolving gases (H_2O, M = 18) and HCl (M = 36) were monitored by a mass spectrometer. A very pronounced HCl peak occurred at 890 K but no dehydroxylation peak was observed. This demonstrates that the solid-state ion exchange (Ag^+ for H^+) proceeds with salts which are insoluble in water indicating that this solvent is unnecessary for the exchange mechanism.

Similarly, the IR spectra of Figure 7 confirm that solid-state ion exchange occurs with water-insoluble salts (AgCl, Hg_2Cl_2) even though under the conditions of these experiments the exchange reached only about 70% (compare the absorbances of the bands at 3610 cm^{-1} of pure H-ZSM-5, shown in the upper part of Figure 7 with the corresponding signals of the heat-treated H-ZSM-5/salt mixtures displayed in the lower part of the same figure).

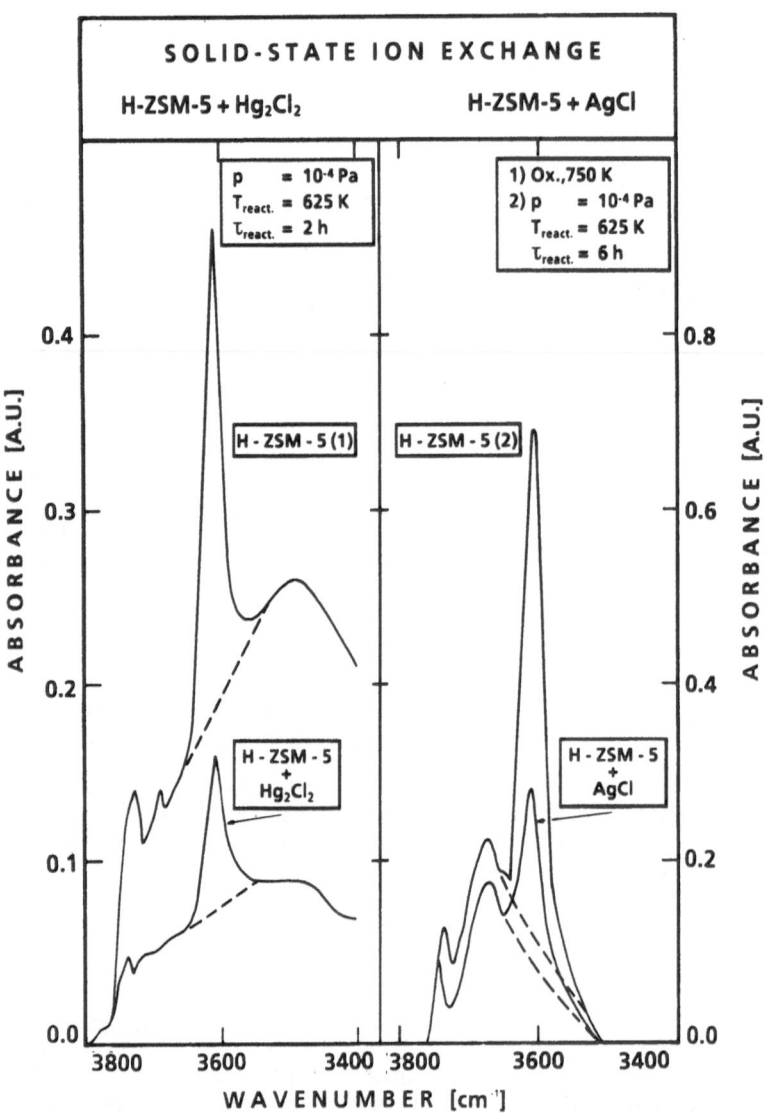

Fig. 7. Solid-state ion exchange in the systems H-ZSM-5/Hg_2Cl_2 and H-ZSM-5/ AgCl shown by the decrease in the intensity of the OH bands of the parent zeolite (details see text).

CONCLUSIONS

Using hydrogen zeolites and metal halides, a higher degree of exchange is, in general, obtained more easily via solid-state ion exchange than via conventional ion exchange. This is most likely due to the fact that in solid-state ion exchange the migration of cations is not hindered by a hydration shell of the cations.

Solid-state ion exchange of NH_4-Y with $LaCl_3$ easily provides an almost 100% exchanged La-Y zeolite. However, the bare La^{3+} cations are not active sites in hydrocarbon conversions.

After contact with small amounts of water, La-Y obtained via solid-state ion exchange is converted to a catalyst which is very active in ethylbenzene conversion and n-decane cracking. The activity is equivalent or even superior to that of a conventionally prepared highly exchanged La-Y sample.

Solid-state ion exchange proceeds under conditions where even traces of water were carefully excluded or where chlorides not soluble in water (such as Hg_2Cl_2 or AgCl) were used as reactants. Therefore, the presence of water is not a prerequisite for solid-state reaction between hydrogen forms of zeolites and metal chloride.

ACKNOWLEDGEMENTS

The authors thank Mrs. Erika Popović, Mr. Walter Wachsmann und Dipl.-Phys. Wilhelm Nießen for considerable experimental help. They are also indebted to Prof. Dr. Jens Weitkamp (Stuttgart) for providing the conventionally exchanged La-Y (99.1%) sample. Financial Support by the Bundesminister für Forschung und Technologie (BMFT, Project No. 03C 25 2A7) is gratefully acknowledged.

REFERENCES

1. J.A. Rabo, M. Poutsma, and G.W. Skeels, New zeolite-salt adducts and their catalytic properties, in: "Proc. 5th Int. Congr. Catalysis", J.W. Hightower, ed., North-Holland Publ., New York (1973).
2. . A. Clearfield, C.H. Saldariagga, and R.C. Buckley, Ion exchange between solids. II. Zeolites, in: "Proc. 3rd Int. Conf. Molecular Sieves", J.B. Uytterhoeven, ed., Leuven University Press, Leuven (1973).
3. J.A. Rabo, "Salt occlusion in zeolite crystals", ACS Monograph, The American Chemical Society, Washington (1976).
4. H.K. Beyer, G. Borbély, and H.G. Karge, Solid-state ion exchange in zeolites. I. Alkaline chlorides / ZSM-5, Zeolites 8:79 (1988).
5. H.G. Karge, H.K. Beyer, and G. Borbély, Solid-state ion exchange in zeolites. II. Alkaline earth chlorides / mordenite, Catalysis Today 3:41 (1988).
6. H.G. Karge, H.K. Beyer, G. Borbély, and G. Onyestyak, Solid-state ion exchange in zeolites. III. Preparation and test of lanthanum zeolite catalysts, in: "Proc. 9th Int. Congr. Catalysis", M.J. Phillips and M. Ternan, eds., The Chemical Institute of Canada, Ottawa (1988).
7. G. Borbély, H.K. Beyer, H.G. Karge, L. Radics, and P. Sándor, Solid-state ion exchange in zeolites. IV. Evidence for contact-induced ion exchange between hydrated NaY zeolite and metal chlorides, Zeolites accepted.
8. A.V. Kucherov and A.A. Slinkin, Introduction of transition metal ions in cationic positions of high-silica zeolites by a solid-state reaction. Interaction of copper compounds with H-mordenite or H-ZSM-5, Zeolites 6:175 (1986).

9. A.V. Kucherov and A.A. Slinkin, Introduction of Cr(V), Mo(V) and V(IV) ions in cationic positions of high-silica zeolites by a solid-state reaction, Zeolites 7:38 (1987).

10. G.H. Kühl, A study of aluminium coordination in zeolites using the K_β line, in: " Proc. 3rd Int. Conf. Molecular Sieves", J.B. Uytterhoeven, ed., Leuven University Press, Leuven (1973).

11. H.G. Karge, K. Hatada, Y. Zhang, and R. Fiedorow, Conversion of alkylbenzenes over zeolite catalysts. II. Disproportionation of ethylbenzene over faujasite-type zeolites, Zeolites 3:13 (1983).

12. D. Keir, E.F.T. Lee, and L.V.C. Rees, Catalytic activity of differently prepared, fully exchanged lanthanum Y zeolites, Zeolites 8:228 (1988).

13. St. Hočevar and B. Držaj, Ion exchange in zeolites at high temperatures, in: "Proc. 5th Int. Conf. Zeolites", L.V. Rees, ed., Heyden, London (1980).

14. I.N. Stranski, Zur Theorie des Kristallwachstums, Z. phys. Chem. 136:259 (1928).

TEMPERATURE DEPENDENCE OF THE MOBILITY OF MOLECULES SORBED IN

TYPE A ZEOLITES

E. Cohen De Lara, R. Kahn*

Laboratoire de Recherches Physiques associé au CNRS
Université Pierre et Marie Curie
4, Place Jussieu, Tour 22, 75252 Paris Cedex 05

*Laboratoire Léon Brillouin, CEN, Saclay, 91191 Gif-sur-Yvette

Introduction

"When a certain number of molecules strike continually upon a surface and stay there for a certain length of time before re-evaporating,
we shall find a higher concentration of the gas. This is the phenomenon
we call adsorption" [1]. The "certain length of time" is one of the principal notion emphasized by J.H. de Boer. It governs the adsorption and
depends on the strength of the interaction between the molecule and the
surface. The Frenkel relation gives the order of magnitude of this "time
of adsorption", τ, according to the value of the heat of adsorption, Q.
At room temperature, τ is in the picosecond scale when Q is around
10 kJ/mole, and longer than a week for 100 kJ/mole,[1].

The concept of mobility has been introduced by Einstein about the
diffusion process in liquid. The forces acting on the particules come
from the interaction with the surrounding and can be seen as a friction
force, F. The mobility, μ , is defined by the ratio of the diffusion
velocity of a particule to the force [2].

When a molecule is adsorbed, we can assume in a first approximation
that the interactions with the solid are independent of the temperature.
Therefore the mobility depends on temperature such as the kinetic energy.

It is then interesting to observe the effects of the temperature
on the adsorbed molecules in order to follow the temperature dependence
of their mobility and then evaluate the order of magnitude of the
interaction forces. The molecular motions must be observed in the
picosecond scale. By infrared absorption and neutron scattering, we have
studied several molecules - diatomic (H_2, O_2, N_2) and polyatomic
(N_2O, CO_2, C_2H_2, CH_4) - adsorbed in type A zeolites in a large range of
temperature.

Guidelines for Mastering the Properties of Molecular Sieves
Edited by D. Barthomeuf *et al.*
Plenum Press, New York, 1990

Theoretical

A - INFRARED

The molecular vibrations occur in the frequency range 4000-200 cm^{-1}. Infrared absorption spectra are characterized not only by the peak frequencies but also by the band shapes and intensities. The band width is related to the relaxation of the vibrational motion, i.e. the perturbations affecting this motion either in direction or in amplitude. The Heisenberg picture of spectroscopy leads to consider a spectrum as the Fourier transform of an appropriate time correlation function. In infrared this function is the correlation function of the transition dipole moment μ : $\Phi(t) = \langle \overrightarrow{\mu(0)}\ \overrightarrow{\mu(t)} \rangle$ [3]. $G(\omega)$ being the infrared band profile, $\Phi(t)$ is given by :

$$\Phi(t) = \int_{band} G(\omega)\ e^{i(\omega-\omega_0)t}\ d\omega$$

where ω_0 is the peak frequency.

$\Phi(t)$ contains the orientational correlation function Φ_R of the transition dipole moment, and the vibrational correlation function Φ_V expressing the variation of the modulus of the transition moment [4]. If the bandshapes are Lorentzian, the functions are exponential, then the widths and the relaxation times are related by the equation :

$$\Delta \nu_{1/2} = \frac{1}{\tau \Pi c}$$

In condensed phases the band widths are generally larger than 0.5 cm^{-1} corresponding to times lower than 20 picosecondes [5].

On the other hand the interaction with the surroundings affects the vibrational frequency as the perturbation changes the internal energy of the molecule [6]. Infrared allows to evaluate the strength of the interaction forces by measuring the frequencies. It is well known that I.R. is useful to distinguish chemical and physical adsorption [7]. When the frequency shifts are small compared to the frequencies, that means that the molecule is not strongly perturbed. This is the case in the liquid phase and in the physisorbed state. The shifts are related to the derivatives of the perturbing potential [6], therefore the shift-interaction relation is not direct but in some way for a given vibrational mode the shift increases with the interaction. For instance the frequency shifts for a given molecule are larger when the molecule is adsorbed in zeolite than in the liquid phase [8].

Finally the interaction acts on the intensity of the bands. This is noticeable especially on forbidden bands. A vibrational mode absorbes infrared radiation when the dipole moment varies. This is not the case for instance for diatomic homopolar molecules and for some symmetric vibrations on polyatomic molecules. When the molecule is submitted to an electric field, an induced moment appears ($\overrightarrow{\mu} = \overline{\overline{\alpha}}\ \overrightarrow{E}$) which varies with the vibration; an induced band is then observed. From a theoretical development performed by Buckingham we have shown how the intensity of a forbidden band is related to the inducing field [9]. The infrared spectra of molecules adsorbed in NaA zeolites such as diatomic homonuclear molecules, methane and acetylene show the occurence of bands forbidden by the selection rules. Their intensity allows an experimental determination of the field strength. The latter varies from point to point in the cavity. The molecules at different positions experience different field strength and thus contribute differently to the intensity of the forbidden bands.

The informations on the mobility come from the analysis of the band shape : frequency distribution and width. The field in a zeolitic cavity varies rapidly with the distance to the surface. During the translation of the molecule, the potential acting on it changes; if the lifetime of this motion is large with respect to the period of the vibrational motion, then, for each position of the molecule, there is a vibrational frequency corresponding to the potential in this position. The band profile can thus be related to the potential distribution inside the cavity. In the center of the cage, the field is very weak and the frequency is near to the gas frequency; inversely, when the molecule is close to the surface, the frequency shift is significant. On the other hand, the intensity of the induced bands is also a quantity related to the mobility of the molecule : it reflects the average of the squared field along the trajectory [10].

B - INCOHERENT NEUTRON SCATTERING

When a neutron monokinetic beam impiges on the target nuclei, one measures the angular distribution of the scattered neutrons and their speed distribution in a given direction. Both analyses (angular direction, and speed hence energy) allow a determination of the static and dynamic properties of the target nuclei. The measured signal is the differential cross section $\dfrac{d^2\sigma}{d\Omega\,dE}$ related to the scattering function $S(q,\omega)$. This latter is the probability for a neutron to change its direction as well as its energy during the scattering event [11].

The incoherent scattering factor of the hydrogen atom is about 10^2 time larger than that of other atoms. Hence when the sample contains hydrogen, $S(q,\omega)$ is proportional to the Fourier transform on space and time variables of the autocorrelation function of the nuclei $G(\vec{R},t)$.

$$G(\vec{R},t) = \langle\, \exp -i\ \vec{Q}.\vec{R}_0(t_0).\ \exp i\ \vec{Q}.\vec{R}(t)\,\rangle$$

is the probability for a nucleus located in $\vec{R}_0$ at time t_0 to be found in $\vec{R}$ at time t. A model describing the motion of the nuclei is necessary for calculating the theoretical spectra.

With usual wavelengths $(1 < \lambda < 10\ \text{Å})$, one can follow the motion on distances comprised between 2 to 60 Å and during times shorter than

10^{-9} sec. This limit in time comes from the resolution of the spectrometers.

Experimental

A - INFRARED ANALYSIS

The study of IR bands of adsorbed molecules with respect to the temperature provides informations on the mobility of the molecules at short times, that means essentially inside the zeolite cavities.

We will see now some examples on different molecules adsorbed in A zeolites : N_2O, CO_2, CH_4, N_2, O_2... The experimental cells have been presented in previous papers. They allow to dehydrate the zeolite at 600 K under 10^{-5} to 10^{-6} torr, and to record the spectra at different temperatures from 450 K to 50 K, on a large scale of gas pressures. In most of the experiments presented here, at each temperature the pressure is adjusted in order to adsorb less than one molecule per cavity (n<1). The zeolite powder is compressed in pellets of thickness comprised between 60 and 200 μ. A thermocouple is sealed on the pellet.

The first analysis of the evolution of the band shape with temperature has been carried out on N_2O in NaA and CaNaA [8] and further investigations have been taken up in detail for CO_2, CH_4, O_2, N_2, H_2. As the absolute intensity of the bands of N_2O and CO_2 is large and the adsorbed amount is sufficient at high temperature, these molecules can be studied from 450 K. For the others, the band intensity is too weak for the amount adsorbed at room temperature and the spectra are recorded at temperatures lower than 200 K.

N_2O - CO_2

The parallel band ν_3 and ν_1 of N_2O were recorded at 313 and 180 K [8]. The broad asymmetric ν_3 band becomes narrower and is shifted to high frequency at 180 K (fig.1). The profile points out an absence of rotation, as supported by the height of the potential barrier to rotation. The ν_1 vibration gives two bands on both side of the gas frequency, shifted away at low temperature (table I). Then considering that the frequency shift $\Delta \nu$ increases in absolute value with the interaction, we infer that N_2O remains closer to the adsorption site at 180 K. The wide frequency distribution of ν_3 at 300 K shows that the molecule is able to move at this temperature.

For CO_2, the ν_3 band profile was investigated from 450 to 180 K for a small and nearly constant number of molecules per cavity (n ~ 0.3) [12] (fig.2). Two main components appear at high temperature, both shifted to high frequency. Their intensities vary in opposite way with T. At 180 K the most shifted component is 85 % of the total intensity. The components have been attributed to molecules in different

Table I Frequencies of vibrational bands for molecules in 3 states : gas, solid, adsorbed in NaA

		gas	solid	Adsorbed in NaA		
N_2O	ν_3	2224	2238(a)	313 K	2260	
				183 K	2272	
	ν_1	1285	1293(a)	313 K	1303	1264
				183 K	1305	1255
CO_2	ν_3	2349	(b)	423 K	2357	2349
				178 K	2365	2353
N_2		2330	2326(b)	223 K	2339	
O_2		1556	1552(d)	93 K	1561.6	
					1553.8	
CH_4	ν_1	2914	2902(e)	200 K	2889	
				50 K	2882	
	ν_3	3020	3009	200 K	2999	
				50 K	3002	2980

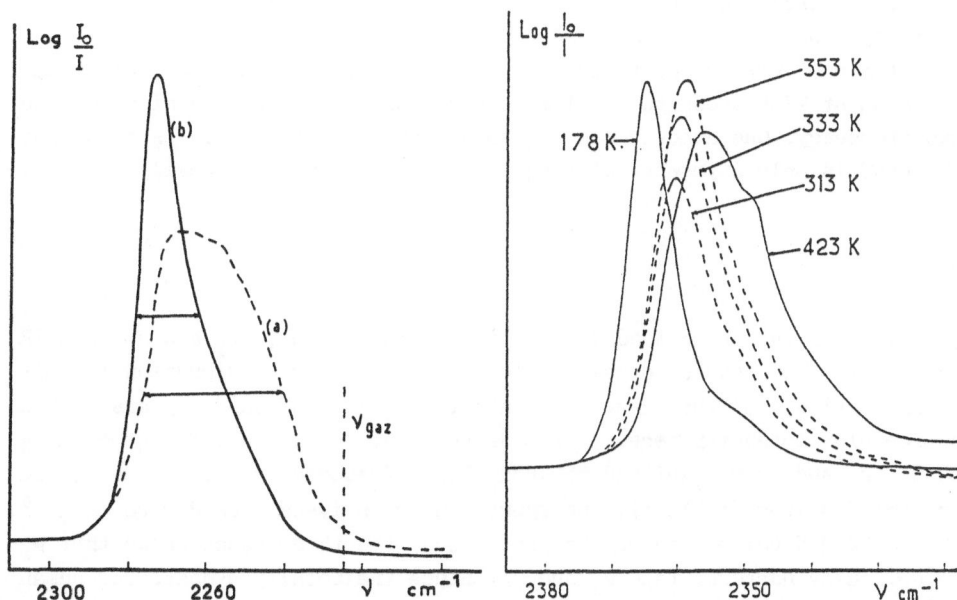

Fig.1 The ν_3 band of N_2O in NaA at 313 (a) and 180 K (b)

Fig.2 IR ν_3 band of CO_2 in NaA at different temperatures (n~0.3 mol/cavity)

energetic sites. Therefore even with less than one molecule per cavity, CO_2 occupies simultaneously several sites. The occupation probability of the less energetic sites increases with T while at low temperature the molecules remain a longer time in the most energetic one.

These two molecules behave in the same way in a similar range of temperature : a noticeable slowing down of their mobility inside the cavity occurs around 170 K.

$N_2 - O_2$

The observation of the vibrational modes of diatomic homonuclear molecules are forbidden by the IR selection rules. The band appears when the molecule is adsorbed in zeolite [13,14] (Table I).

First of all for a same n value the intensity increases inversely with temperature, and secondly at low temperature, it is not proportional to n above n = 1. This first remark agrees with the observations concerning N_2O and CO_2. Lower is the temperature, closer to the adsorption site is the mean position of the molecule. On the contrary the mobility of the molecule increases with T and the intensity is lowered, reflecting an average of the field along the trajectory of the molecule inside the cavity. The second remark can be easily explained, owing to the presence in the NaA cavity of one "near-zero coordinate" Na^+ cation [15] This cation is probably the most energetic site and the field near this cation has the highest value. When n > 1, one molecule occupies this site, the others consequently are located in parts of the cavity where the field is weaker.

The strong enhancement of the intensity and the effect of the mobility appears on the spectrum of O_2. This molecule may have two orientations with respect to the surface (potential calculation) [14] and the band at 93 K shows two well separated components on both sides of the gas frequency. One hundred degrees above, the mobility of O_2 is such that the band is only asymmetrical, the molecule rotating more easily (fig.3).

CH_4

Two of the four fundamental vibrations of CH_4, ν_3 and ν_4 are IR active in the gaseous phase while the ν_1 symmetric vibration is only Raman active. In the adsorbed state the ν_1 induced band appears with a noticeable intensity because the Raman band is strong. The stretching bands ν_1 and ν_3 are shifted towards lower frequencies, as in the liquid and solid states (Table I). The spectra were analysed from 200 to 50 K [10].

At 200 K comparison of the profiles of the three bands shows that ν_1 is obviously narrower than ν_3 and ν_4. Since the totally symmetrical Raman bands of a spherical rotor are not accompanied by rotational wings, the ν_1 profile is strictly due to the distribution of the vibrational frequency. On the contrary, the presence of wings on ν_3 leads one to infer the CH_4 retains some rotational freedom inside the cavity.

At 50 K, the effect of the temperature is remarkable : concurrently with the increase of the induced band ν_1, appears the splitting of the allowed band ν_3 : this splitting shows that the molecule is oriented on the surface in a tripod configuration (fig.4). The evolution of the band shapes (frequency, width, intensity) with respect to T can be completely correlated to the variation of mobility in the cavity.

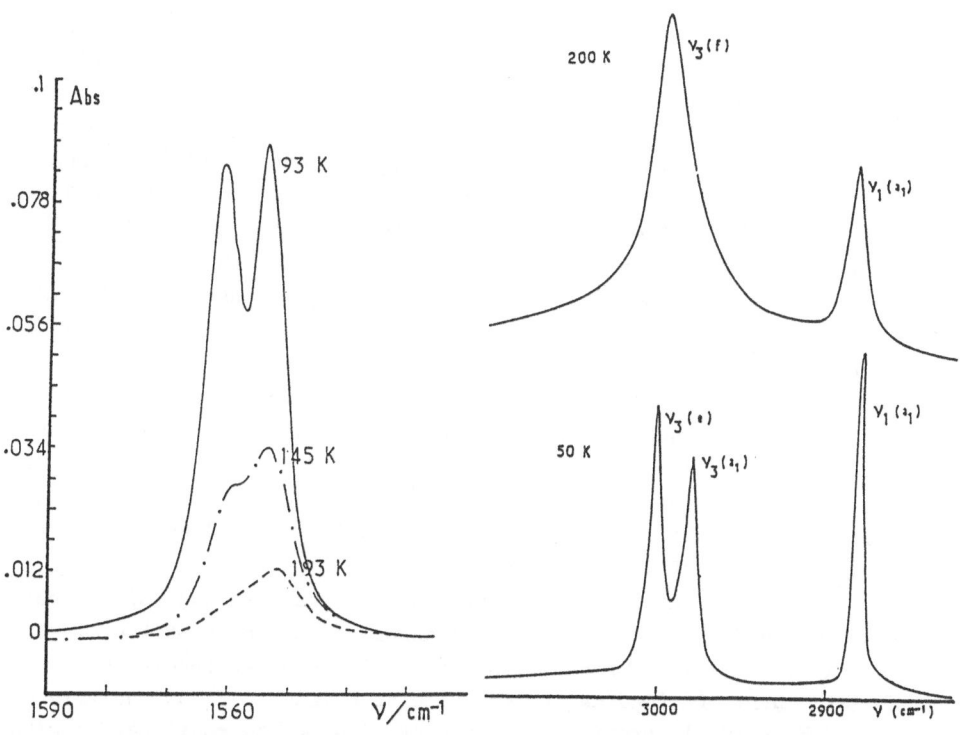

Fig.3 Induced IR band of O_2 in NaA at 3 temperatures with the same number of adsorbed molecules (n ~ 1 mol/cavity)

Fig.4 IR spectrum of CH_4 in NaA at 200 and 50 K

C_2H_2

On this molecule, the results on IR spectra with respect to temperature are not published. We observe very few changes on the frequency and on the width of the induced band ν_4, appearing at 1950 cm^{-1}. The other vibrational modes are difficult to analyse because of strongle Fermi resonance.

A neutron diffraction analysis of C_2H_2 in NaA has shown that, inversely to CH_4, the molecule is localized in its adsorption site at room temperature [16].

Neutron scattering results on CH_4 and H_2 adsorbed in NaA are presented. The samples are prepared in sealed quartz containers, in order to have around 1 molecule/cavity at different temperatures.

1° Methane

Experiments showed that intra and inter cavities motions occur on very different time scales. Therefore their studies make use of different techniques, and when one of these components is measured, everything is going as if the other motions could be ignored.

The intracrystalline (inter cavities) motion has been studied by neutron spin echo (N.S.E) technique on CH_4 in NaA and CaNaA [17] and by NMR on CH_4 in NaA [18]. The NSE experiment performed at 300 and 350 K showed that this motion can only be measured for CH_4 in CaNaA; the residence time of CH_4 in NaA cavity is too long for this technique. In CaNaA the model of random-walk on a cubic lattice of parameter $\ell_0 = 12$ Å gives the residence time τ_0 between two instantaneous jumps $\tau_0 = 4 \ 10^{-10}$ sec. N.M.R measurements performed on a large range of temperature (300-200 K) showed that the residence time τ_0 of CH_4 in a NaA cavity obeys an Arrhenius law with an activation energy of 21 kJ/mole; this leads to : $\tau_0 = 7 \ 10^{-5}$ sec at 300 K

Therefore a quasielastic neutron scattering (QENS) experiment, allowing measurements on a time scale 10^{-12} to 10^{-10} sec , gives the dynamical behaviour of CH_4 inside a NaA cavity [19].

At each temperature the scattered spectra are made of a purely elastic part and inelastic parts. The angular dependence of the elastic scattering (E.I.S.F) represent the spatial Fourier transform of the probability $p(\vec{r})$ of finding the molecule at point $\vec{r}$ [11]. The analysis of the E.I.S.F. (fig.5) provides a representation of the modification with temperature of the trajectory. At 300 K, $p(\vec{r})=c^{te}$ inside a sphere of 3.5 Å radius, the molecule moves in the whole cavity. The narrowing of the E.I.S.F. when the temperature decreases gives clear indication of a higher probability to find the molecule in the vicinity of the cavity walls; for T < 200 K it remains close to the walls. In the same time a non zero minimum becomes visible below 240 K; this points out some trapping effect corresponding to sites, where CH_4 rests for longer and longer times as the temperature decreases. This can be seen on fig 5 where α is a "trapping time coefficient": $\alpha = \dfrac{\tau_1}{\tau_1 + \tau_2}$ where τ_1 is the time the molecule is trapped and τ_2 the time it moves along the walls.

The quasi elastic scattering, which comes from the slow motion of the molecule ($\sim 10^{-11}$ sec) , is well described by a single Lorentzian function of width Λ : the corresponding motion is of diffusive type with a correlation time $\tau_c = \dfrac{2\hbar}{\Lambda}$; τ_c obeys an Arrhenius law with an activation energy of 5.8 kJ/mole. Lastly, the inelastic part of the spectrum gives information on the rapid motion ($\sim 10^{-13}$ sec); it concern primarily

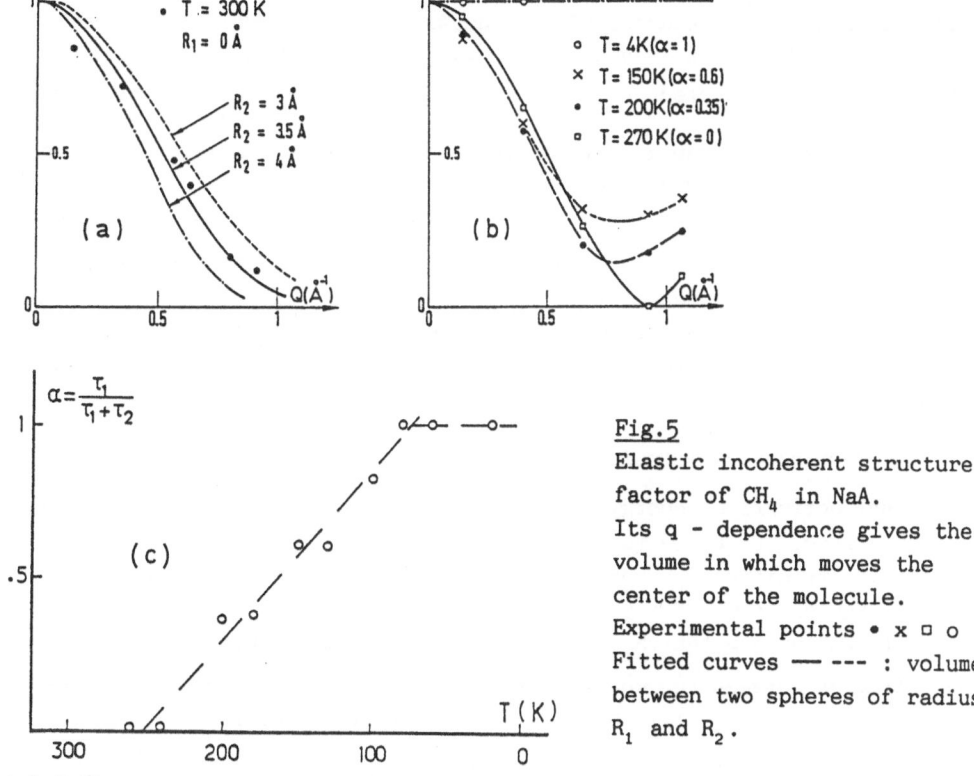

Fig.5
Elastic incoherent structure factor of CH_4 in NaA.
Its q - dependence gives the volume in which moves the center of the molecule.
Experimental points • x □ o
Fitted curves —— --- : volume between two spheres of radius R_1 and R_2.

(a) full sphere $R_1 = 0$ (b) Empty sphere $R_1 \neq 0$. From 200 K the minimum is not zero. The fit contains the trapping time coefficient α
(c) Variation of α versus T

a component which can be describe as a distribution $g(\omega_0)$ of vibration like motions of the molecule with respect to the cavity wall, (external vibrations). At high temperature $g(\omega_0)$ is almost flat between 0 and 5 THz in agreement with the picture of a molecule diffusing in the entire cavity. Indeed, on one hand the interaction potential varies from one point to another inside the cavity, leading to a continuous distribution of ω_0; on the other hand each frequency is broadened by its short life-time. When temperature decreases the distribution $g(\omega_0)$ gradually transforms into a bell-shaped curve and around 150 K a relatively well defined peak becomes visible around 1.5 THz; with respect to the trapping parameter, this temperature corresponds roughly to one half of the molecules beeing trapped.

2° Hydrogen

A rather different image comes out from experiments on sorbed hydrogen [20]. From the analysis of Q.E.N.S. in the temperature range 70-150 K we come to the following conclusions :

i) The H_2 molecule has a translational motion in a non restricted volume.
ii) The angular dependence of the quasi elastic width shows that the molecule undergoes a random walk with isotropic jumps of length ℓ statiscally distributed according to $a(\ell) = \dfrac{\ell}{\ell_0^2} \exp\left(-\dfrac{\ell}{\ell_0}\right)$. Between two jumps it has very rapid motions (external vibration, rotations) around sites where it stays during a time τ_0.
iii) The mean jump length $\bar{\ell}$ between two sites is temperature independent. The best fit is obtained with $\bar{\ell} = \dfrac{5\,\ell_0}{3} = 3.9$ Å. On the other hand the trapping time τ_0 increases from 4 psec at 150 K to 18 psec at 77 K. This shows that the translational motion gradually slows down. Taking in account the spectrometer resolution (~ 20 psec), below 70 K this time becomes to large to be measured.

iv) The macroscopic diffusion coefficient $D = \dfrac{\ell_0^2}{\tau_0}$ follows an Arrhenius law with an activation energy of 2 kJ/mole and a value of D_0 equal to $6\ 10^{-4}$ cm^2/sec. Compared to other molecules H_2 is rather free in the zeolite. Its diffusion coefficient is of the same order of magnitude as in the liquid phase. ($D = 5\ 10^{-5}$ cm^2/sec at 15-18 K [21]).

If we suppose that H_2 in NaA behaves as in the liquid, we can apply the relation between the mobility and the diffusion coefficient $\mu = \dfrac{D}{K\,T}$ [21]. From our measurements (fig.6), it seems that the mobility in NaA tends to zero around 20 K.

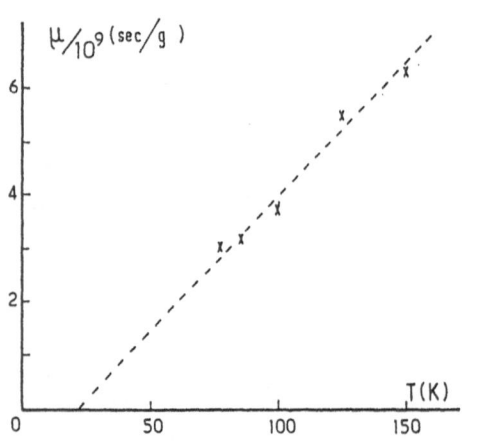

Fig.6
Mobility of H_2 in NaA versus T

Potential energy calculation

A classical potential model has been applied to the different molecules in order to compare their interaction with zeolite. The potential is expressed in term of molecular quantities : size, permanent electric moments and polarisability. The zeolite is a distribution of point charges located on the oxygens and the Na cations of the framework (9,19).

178

We remind in table II the molecular quantities of the molecules and we give the value of the minimum of the potential in NaA cavities. This minimum is located in the vicinity of the Na_{III} cation, and the orientation of the molecule with respect to it depends on the values and the signs of the molecular quantities.

Table II   Molecular quantities and potential energy of the adsorbed molecules

	$\sigma_m/10^{-8}$	$\mu/10^{-18}$	$Q/10^{-26}$	$\alpha_{\parallel}/10^{-24}$	$\alpha_{\perp}/10^{-24}$	Φ_{min} kJ/mole
H_2	2.9	0	+0.66	0.97	0.69	27
O_2	3.5	0	-0.39	2.35	1.21	33
N_2	3.8	0	-1.52	2.38	1.45	32
CO_2	4.2	0	-4.3	4.05	1.95	43
N_2O	4.2	0.17	-3.0	4.86	2.07	42
C_2H_2	4.2	0	+3.0	5.12	2.43	85
CH_4	3.8	0	0	2.6	2.6	26

Taking in mind that these values are only orders of magnitude (point charge model - simplest representation of the zeolite charge distribution) we can see that the calculated interactions energy is quite similar respectively for H_2 and CH_4, for N_2 and O_2, for N_2O and CO_2 and increase progressively from the H_2 to CO_2, becoming very high for C_2H_2.

Numerical simulation of the dynamics

Several experimental results on methane in NaA zeolite have been compared to the results of molecular dynamics (M.D) simulation [22]. The method consists in integrating the equation of motion of the molecule, and the trajectory is found step by step, by using algorithms which give the position and the velocity at each time step ($\Delta t = 5 \cdot 10^{-15}$ sec). The forces acting on the molecule are deduced from the potential previously mentionned. The simulation is carried out in the microcanonical ensemble: the total energy E_T is conserved. At each step we know the electric field in which is the molecule, its velocity, its potential and kinetic energies. After a large number of steps ($N \sim 50\ 000$), we compute the average values of these quantities, and the correlation functions of the position and of the velocity.

In the case of CH_4 in NaA, the comparison with experiments is quite satisfactory as the shape of the position correlation function is comparable to the shape of the elastic incoherent structure factor. The simulation shows the progressive localisation of the molecule as the energy decreases (fig.7) and the increase of the mean electric field as noticed experimentally on the intensity of the induced band.

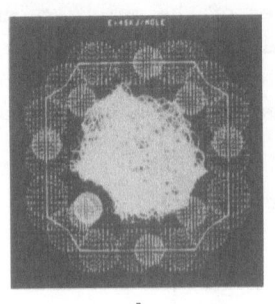  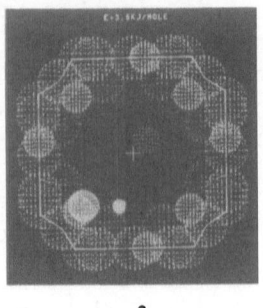

| 1 | 2 | 3 |

Fig.7 Trajectory of CH$_4$ in NaA cavity by molecular graphics.
1) E$_T$ = 40 kJ/mole   2) E$_T$ = 28 kJ/mole   3) E$_T$ = 4 kJ/mole

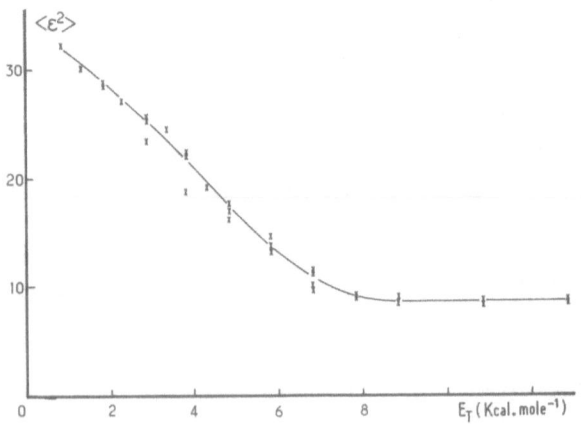

Fig.8
Mean squared field versus
the energy

Conclusion

Experiments on different molecules absorbed in the same zeolite show that the dynamical behaviour, on time scale comprised between 10^{-13} sec to 10^{-10} sec , depends on the temperature range. Some molecules are strongly localised at room temperature (C_2H_2), or are still mobile in the cavity down to 150 or 100 K (CO_2 - N_2O - N_2 - O_2); for others, trapping occurs at lower temperatures (CH_4 - H_2).

The calculation of the interaction of these molecules with the zeolite, as well as the values of the heat of adsorption show that higher is the interaction, higher is the temperature for which the molecule is seen at rest. As the diffusivity depends on the "length of time" the molecule stays on the surface : longer this time, smaller is the mobility.

Nevetherless this effect is not the only one in zeolites : the porous structure play an important role illustrated here by the comparison between hydrogen and methane. As a matter of fact their interaction with the zeolite ions is similar; the activation energy of the diffusion of H_2 in the bulk crystal (2kJ/mole) and the one of the translational motion of CH_4 inside a cavity (5.8 kJ/mole) are roughly comparable. However the activation energy of the diffusion of CH_4 in the crystal is ten time higher (21 kJ/mole) than the one of H_2 [18]. This is due to the steric hindrance of CH_4 to cross the windows of NaA as shown by this rough evaluation : J. de Boer gives a formula expressed by Clausius for the calculation of "the average time $\bar{t}$ which a molecule needs in order to proceed through a capillary of a length ℓ and a diameter d" [1] :

$$\bar{t} = \frac{\ell^2}{2 \, d \, \bar{u}} + \frac{\ell^2 \tau}{2 \, d^2} \qquad [1]$$

where $\bar{u}$ is the mean velocity. The first term represents the capillary effect, the second accounts for the residence time τ on the surface.

Taking for d the window aperture (4 Å) and the same τ value for H_2 and CH_4 (10^{-11} sec) , the terms 1/2d and $\tau/2 \, d^2$ are respectively 10^2 and $3 \, 10^3$ .On the other hand the diffusion theory leads to $\bar{t} = \ell^2/D$. For H_2 at 77 K, 1/D is equal to $3 \, 10^4$ while is equal to $3 \, 10^{10}$ for CH_4 at 300 K. Therefore the diffusion of H_2 is mainly governs by τ. On the contrary for CH_4 even the capillary effect is not sufficient to explain the very small value of the diffusion coefficient. The expression [1] does not take in account the steric hindrance.

References

1. de Boer, J. H., 1968, "The dynamical character of adsorption", Clarendon Press, Oxford.
2. Egelstaff, P. A., 1967, "An introduction to the liquid state", 1967, Academic Press, London and New York.
3. Gordon, R. G., "Advance in magnetic resonance" vol.3, 1968, Academic press, New York.
4. Bratos, S., Rios, J., and Guissani, Y.,1970, J. Chem. Phys., 52:439
5. Vincent-Geisse, J., in "Vibrational spectroscopy of molecular liquids and solids", 1970, S. Bratos and R.M. Pick, Plenum Publishing Corporation
6. Buckingham, A. D., 1960, Trans. of the Faraday Soc. 56:753.
7. Little, L.H., 1966, "Infrared spectra of adsorbed species", Academic Press., London.
8. Cohen De Lara, E., 1972, Molecular Physics, 23:555.
9. Cohen De Lara, E., and Delaval, Y., 1978, J. Chem. Soc. Faraday Trans II, 74:790
10. Cohen De Lara, E., Kahn, R., and Seloudoux, R., 1985, J. Chem. Phys 83:2646.

Kahn, R., Cohen De Lara, E. and Moeller, K. D., 1985, J. Chem. Phys., 83:2653.

11. Marshall, W., and Lovesey, S., "Theory of thermal neutron scattering", 1971, Clarendon Press, Oxford.

12. Delaval, Y., and Cohen De Lara, E., 1981, J. Chem. Soc. Faraday Trans I, 77:869.

13. Barrachin, B., and Cohen De Lara, E., 1986, J. Chem. Soc. Faraday Trans II, 83:1953

14. Soussen-Jacob, J., Cohen De Lara, E., and Tsakiris, J., 1989, J. Chem. Phys.

15. Yamagida, R.Y., Amaro, A. A. and Seff, K., 1973; J. Phys. Chem. 77:805.

16. Kahn, R., Cohen De Lara, E., Thorel, P., and Ginoux, J. L., 1982 Zeolites, 2:260.

17. Cohen De Lara, E., Kahn, R., and Mezei, F., J. Chem. Soc. Faraday Trans I, 79:1911.

18. Alloneau, J. M. and Volino, F., 1986, Zeolites, 6:431.

19. Cohen De Lara, E., and Kahn, R., 1981, J. Physique (Paris) 42:1029

20. Kahn, R., Cohen De Lara, E., and Viennet, E., 1989, J. Chem. Phys.

21. Egelstaff, P. A., "An introduction to the liquid state", 1967, Academic Press.

22. Cohen De Lara, E., Kahn, R., and Goulay, A. M., 1989 J. Chem. Phys.

NMR-STUDIES OF MOLECULAR MOTION OF HYDROCARBONS IN DIFFERENT FAUJASITES

Hans T. Lechert, Wolf D. Basler and Mendong Jia

Institute of Physical Chemistry, University of Hamburg
Bundesstraße 45, D-2000 Hamburg 13, Fed. Rep. of Germany

INTRODUCTION

The peculiarities of molecular motion in zeolite cavities in different frameworks determine the activity and especially the selectivity of
catalytic reactions. The molecular motion is determined by the interaction
of the sorbed molecules with the zeolite walls and spatial restrictions in
the cavity system.

Detailed informations about the dynamics of the motion of sorbed
molecules can be obtained, within certain limits of time scale, by the
different techniques of NMR-spectroscopy.

Most important for studies of motional phenomena are the relaxation
times T_1 and T_2 and their temperature functions /1-4/.

The transverse relaxation time T_2 proportional to the inverse of the
line width and is determined by the static part of the local field or more
exactly by the spectral density $J(0)$ of the correlation function of the
interaction at zero frequency.

The longitudinal or spin-lattice relaxation time T_1 describes the
energy exchange between the spin system and the other degrees of freedom
of energy. It is caused by the spectral density $J(\omega_o)$ at the Larmor frequency ω_o. If the correlation function is one single exponential which is
more or less exactly true for isotropic reorientations, its time constant
τ_c is called correlation time. τ_c may be regarded as the inverse jump frequency of the molecular motion.

The relaxation times T_1 and T_2 are given in the theory of BLOEMBER-
GEN, PURCELL and POUND (BPP) /5/ by

$$1/T_{1D}(I-I) = (2/3) M_2(I-I) [J(\omega_I) + 4J(2\omega_I)] \qquad (1)$$

$$1/T_{2D}(I-I) = (1/3) M_2(I-I) [3J(0) + 5J(\omega_I) + 2J(2\omega_I)] \qquad (2)$$

with the spectral density function

$$J(\omega) = \tau_c/(1 + \omega^2\tau_c^2) \qquad (3)$$

The strength of the magnetic dipole-dipole-interaction is expressed
by the second moment

$$M_2(I-I) = (3/5)(\mu_o/4\pi)^2(h/2\pi)^2 \gamma_I^4 I(I+1) \Sigma_j r_j^{-6} \qquad (4)$$

Guidelines for Mastering the Properties of Molecular Sieves
Edited by D. Barthomeuf *et al.*
Plenum Press, New York, 1990

183

ω_I and γ_I are the angular frequency and the magnetogyric ratio of spin i, r_i are the distances between the observed and the interacting spins, μ_0 is the induction constant, and h is Planck's constant. For the interaction of a spin I with unlike spins S, slightly different equations are obtained containing terms of the sums and the differences of the resonance frequencies of both nuclei and a second moment describing the interaction of the unlike nuclei. The relations

$$\tau_c \omega_0 \approx 1 \qquad (5)$$

at the minimum of T_1 and

$$\tau_c \approx T_2 \qquad (6)$$

at the onset of the motional narrowing of the line-width permit the evaluation of the correlation time from relaxation data. The temperature function of τ_c is usually an Arrhenius function with good accuracy, so that the activation energy can be obtained from the slopes of log T_1, T_2 versus the reciprocal temperature $1/T$.

A further possibility to estimate the activation energy of the motion has been pointed out by WAUGH and FEDIN /6/. From Eq. 6 and a reasonable preexponential factor $\tau_{co} = 10^{-13}$ s and a static $T_2 \approx 10^{-5}$ s the semiempirical relation

$$E_A = 0.156 \, T_C \; kJ/mole \cdot K \qquad (7)$$

can be derived, where T_C is the temperature of the onset of the motional narrowing.

For sorbed molecules the observed temperature functions of T_1 and T_2 deviate more or less from the ideal BPP theory with one correlation time demonstrated in Fig. 1. Generally the minimum of T_1 is broadened and the

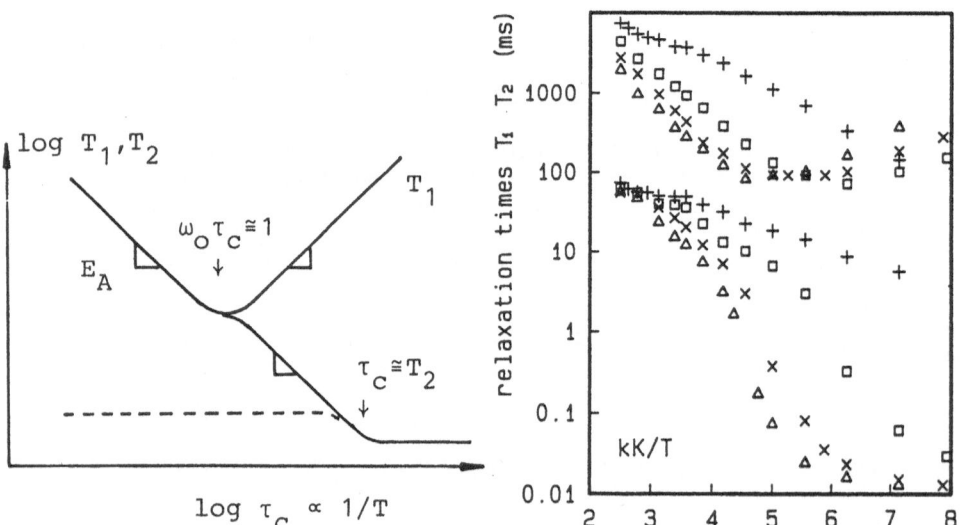

Fig. 1 (left). Temperature dependence of the relaxation time T_1 and T_2 for an isotropic (solid line) and an anisotropic (broken line) motion of correlation time $\tau_c = \tau_{co} \exp(E_a/RT)$.

Fig. 2 (right). Temperature dependence of H-NMR-relaxation times T_1 and T_2 of n-butane (+), n-hexane (□), n-octane (x) and n-decane (Δ) in the a weakly dealuminated sample of Y-zeolite with Si/Al=3.1.

value of T_1 at the minimum is increased. But the position of the minimum of T_1 remains at the same temperature. The curves of T_2 are shifted almost parallel to lower values.

It has been shown by RESING et al./7-10/ that a wide range of experimental data can be explained by the assumption of a logarithmic normal distribution of τ_c. This corresponds to a normal distribution of the activation energies. The ratio of T_1/T_2 at the T_1-minimum which is 1.6 for a single correlation time can be used to obtain the width of the distribution. As the slope of the temperature function of T_2 in the range of motional narrowing is not very much affected by the presence of a distribution of correlation times, this slope may be used for the determination of the activation energy.

If the reorientation is anisotropic around a fixed axis, a finite and temperature independent line width is expected even for rapid rotation. This is indicated by the broken line in Fig.1. The static line width is reduced by a factor

$$R = (3\cos^2\theta - 1)/2 \qquad (8)$$

where θ is the angle between the direction of the interaction and the axis of rotation. T_1 of isotropic and anisotropic reorientations are only slightly different.

For distincly different modes of reorientation, e.g. for plane molecules like benzene, different minima of T_1 and steps in the variation of T_2 with temperature are observed.

In most cases various interactions contribute to the observed T_1 and T_2 and a separation is needed which is called "relaxation analysis" /11/. In H-NMR, mainly intra- and intermolecular H-H interactions and relaxation by the electronic moments of paramagnetic iron impurities are important. The H-Fe-relaxation is, however, only effective near the Fe-ions and the molecules have to meet the Fe-centers within the corresponding relaxation time to perform a space averaging. Hence the translational diffusion coefficient D may be estimated from the relation

$$D \geq l^2/6\,t \qquad (9)$$

with l = mean distance between the Fe ions and $t = T_1$, T_2 respectively.

The intra- and intermolecular dipolar as well as the paramagnetic contributions can be separated by using fully and partially deuterated molecules.

Often various sites with distinctly different τ_c are present in the sample and the molecules exchange more or less rapid beween them. As has been pointed out by MURDAY et al. /12/, in sorption systems often regions with short τ_c are in fast exchange with one region with a long τ_c. For slow exchange two signals with a long and a short T_2 are observed.

If the exchange becomes faster the nuclei with the long T_2 are progressively affected by the region with the short correlation time. Here T_2 decreases with increasing temperature until the normal increase dominates. In such systems a characteristic T_2-minimum can be observed.

EXPERIMENTAL

The faujasite samples have been prepared from a batch with the composition 6.45 SiO_2 $NaAlO_2$ 4.77 NaOH 478 H_2O using a nucleation gel as seeding material. The crystallization temperature was 360 K. After 6 days pure Y-zeolite with Si/Al = 2.5 was obtained.

For the removal of the Na-ions 30 g NaY were treated two times with 300 ml 2n NH_4Cl solution at 370 K. Thereafter they were heated for 2 hours to 900 K and refluxed for 2 hours in a NH_4Cl solution which was adjusted to pH = 2. Gradual dealumination was achieved by subsequent calcination at

980 K and acid leaching. The degree of dealumination was determined via the lattice constant and by AAS-analysis of the samples.

For the NMR-measurements the samples were dehydrated by heating to 673 K and pumping to 10^{-5} torr for 12 hours. The degree of dehydration was checked by NMR. Oxygen was removed from the sorptives by the "freeze-pump-thaw"-technique. The sorbed quantity was controlled by weighting the samples.

H-NMR relaxation times were measured with a Bruker BKS 322 spectrometer at 60 MHz using conventional pulse sequences.

RESULTS AND DISCUSSION

General Considerations

Molecules of intermediate size are especially suited to study peculiarities of the motion under the influence of an interaction in a restricted space of the different zeolite cavities. Furthermore, hydrocarbons of this size are involved in most reactions catalyzed by zeolites.

Hydrocarbon may interact with the cations at the walls of the zeolites, with the Broensted- and the Lewis-acid centers and with the oxygens of the zeolite framework.

Relaxation and self-diffusion measurements on zeolites containing different cations have been carried out already in earlier investigations. The relaxation times of paraffins, olefins and of benzene have been studied in dealuminated faujasites of different degrees of dealumination to obtain the relative importance of the different interactions. The influence of spacial restriction has been studied using coked samples.

Relaxation and Self-Diffusion of Aliphatic Hydrocarbons in Faujasites

The motion of paraffins in zeolite pores is determined primarily by their rather weak interaction with the cavity walls. T_1, T_2 and the self-diffusion coefficients of the n-paraffins from methane to octadecane in NaX have been carried out in a wide range of temperatures and coverages by KÄRGER and PFEIFER /13-17/.

The temperatures of the T_1-minima for n-butane, n-hexane, n-heptane and cyclohexane are almost independent of coverage. Only a broadening with increasing coverage is observed. The minima vary between about 120 K for butane and 170 K for heptane. The activation energy is 7.1 kJ/mole for n-butane and 11 - 13 kJ/mole for the other hydrocarbons. At high coverages the activation energies decrease to 7 - 9 kJ/mole. These results are interpreted by a jump process with jump lengths decreasing with increasing coverage.

The self-diffusion coefficients of the n-paraffins decrease monotonously by two orders of magnitude with increasing coverage.

From NMR-relaxation measurements /11,18/ as well as from thermodynamic considerations it is suggested that the n-paraffins behave as an intracrystalline liquid at higher coverages. Here the interaction with the cavity walls is comparable with the van-der-Waals interactions between the molecules. Assuming the COHEN-TURNBULL free volume theory for pure liquids /19/ self-diffusion coefficients have been calculated which are in fairly good agreement with the measured values.

In the dealuminated samples of the present study the cations are removed and the hydrophobic interaction is increased. Further, interactions with the acid centers should play a more important role which can be studied by varying the degree of dealumination.

Fig.2 shows the relaxation data of n-butane, n-hexane, n-octane and n-decane in a DHY-zeolite with Si/Al = 4.3. The coverages corresponded to about 12 C-atoms/cavity.

No T_1-minimum can be observed for n-butane. The T_1-minima of the other paraffins are at the same temperatures as the minima in NaX observed by PFEIFER et al /17/. The T_1-minima are rather broad and asymmetric so that a determination of the activation energy from the temperature of the T_1-minimum is not feasible. Low temperature T_2 of $n-C_6$, $n-C_8$ and $n-C_{10}$ equal the values which are calculated for a rotational reorientation of the CH_2-groups.

The following activation energies are calculated from the WAUGH-FEDIN-relation (Eq. 7) and from the slope of log T_2 versus 1/T:

Table 1.   Activation Energies (kJ/mole) of n-Paraffins in DHY

	from WAUGH-FEDIN	from slope of T_2
n-butane	-	5
n-hexane	20	20
n-octane	27	30
n-decane	30	38

These values are higher than those obtained by PFEIFER et al. /17/ for NaX, which have been taken from the slope of log T_1 versus 1/T which is probably affected by a distribution of correlation times.

The relatively small ratios of T_1/T_2 at the T_1-minima show that the mobility of the hydrocarbons is fairly isotropic even for n-decane. The activation energies increase with increasing chain length. Therefore, the activation energy may be correlated with the hydrophobic interaction of the paraffins with the zeolite walls.

The activation energies of the translational diffusion obtained by KÄRGER /13/ first increase with the chain length, too, but remain constant near 15 kJ/mole above C_5.

To study the influence of these centers the relaxation times of n-butane and n-octane in dealuminated samples with different Si/Al-ratios have been compared. The results are demonstrated in the Figs. 3 and 4. It

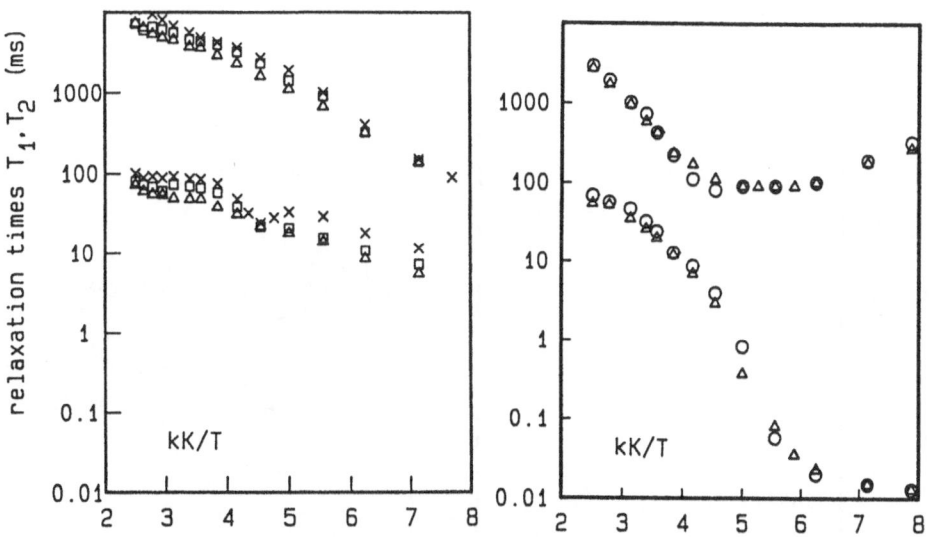

Fig.3 (left).   Temperature dependence of H-NMR-relaxation times T_1 and T_2 of about 3 molecules of n-butane in a supercage in Y-zeolites with different degree of dealumination. (x) Si/Al=3.1, (□) Si/Al=6.4, (Δ) Si/Al=14

Fig.4 (right). Temperature dependence of H-NMR-relaxation times T_1 and T_2 of 3 molecules n-octane in a large cavity in Y-zeolites with different degree of dealumination. (o) Si/Al=3.1, (Δ) Si/Al=14.

can be seen that the Si/Al-ratio and the concentration of acid centers has no influence on the relaxation times and hence on the mobility of paraffins in dealuminated faujasites.

Furtheron, our data agree with those found by PFEIFER et al. for NaX /17/. Therefore, the interaction with the walls which is expressed by the activation energy, seems to be given mainly by a hydrophobic interaction with the walls of the cavities. The models for the structure of the sorbate discussed for NaX may be true with slight changes.

The mobility of olefins in NaX is governed by an interaction of the double bonds with the cations. As has been shown by NAGEL, PFEIFER and WINKLER /20/ the mobility of cyclohexane, cyclohexene, cyclohexadiene, and benzene decreases monotously. The largest change is observed for the introduction of the first double bond.

To study the influence of the acid centers on a olefin molecule which should preferably interact with these centers the relaxation times of 3.3 molecules 1-butene/cavity in the sample with Si/Al = 4.3 were studied. The results are shown in Fig. 5. It can be seen that the motion of 1-butene is strongly restricted compared with butane where no T_1- minimum is observed.

The activation energy is 25 kJ/mole from the WAUGH-FEDIN-relation and 35 kJ/mole from the slope of log T_2. This is much larger than the value of 5 kJ/mole obtained for butane. The difference may be attributed to the interaction with the acid centers.

Relaxation of Aromatic Hydrocarbons in Faujasites

Beside the olefins strong interactions with the acid centers can be expected for aromatic molecules. The gradual transition from cyclohexene to benzene has been described in the paper of NAGEL, PFEIFER and WINKLER /20/ mentioned above. Thorough investigations of aromatics in faujasites with different cations have been carried out in our group /21-24/.

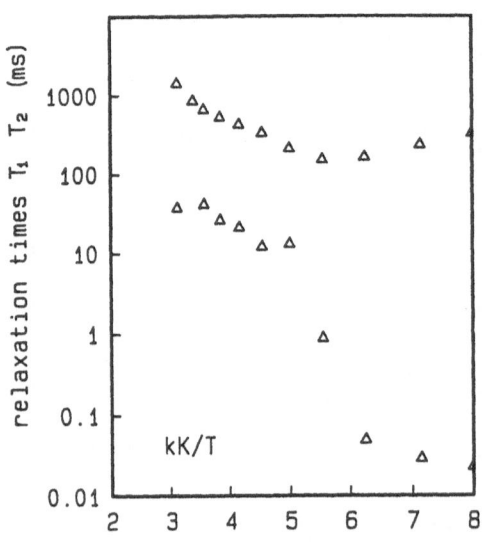

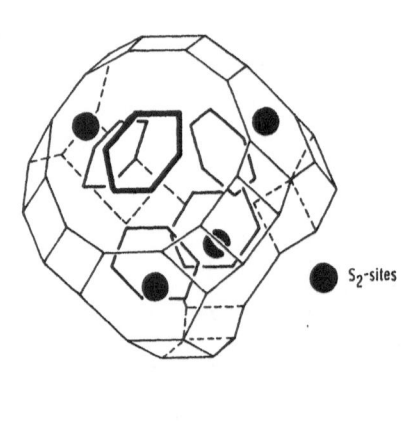

Fig. 5 (left). Temperature dependence of H-NMR-relaxation times T_1 and T_2 of 3 molecules/cavity 1-butene on a dealuminated Y-zeolite with Si/Al=4.3.

Fig. 6 (right). Model of the arrangement of benzene molecules in the supercages of the faujasite structure as obtained for NaX from proton resonance spectra /21/.

From the intra- and the intermolecular dipolar contributions it
followed that benzene occupies sites above the S_2-ions rotating around its
hexagonal axis even at low temperatures. At higher coverages additional
sites in the windows between the cavities are occupied which is shown in
Fig. 6. Later on this model has been proved by de MALLMANN and BARTHOMEUF
/25/ by IR-experiments and by RENOUPREZ et al. /26/ by neutron diffraction investigations.

The available sites are not completely occupied even at maximum
coverage. With increasing temperatures jumps into the remaining free sites
take place accompanied by reorientations around the twofold axes of the
benzene molecules, which can be observed as a motional narrowing in T_2.
These jumps lead to a self-diffusion of the benzene molecules which can be
estimated using Eq. 13 and samples of low paramagnetic impurity to be
larger than $6\ 10^{-13} m^2 s^{-1}$ /27/.

In Y-zeolites the rotation at low temperatures is indicated by a
second T_1-minimum at about 100 K.

In Fig. 7 the relaxation data of 4 benzene/cavity in NaY, Si/Al=2.5
are compared with those in a dealuminated sampled with Si/Al=3.1. It can
be seen that the activation energy of the reorientation around the twofold
axes which is connected with the translational jumps is appreciably decreased in the dealuminated sample where the cations are replaced by OH-
groups. This shows that the interaction of benzene with the Na-ions is
stronger than the interaction with the acid centers. The activation energies are 45 and 20 kJ/mole for NaY resp. DHY-3.1.

The minimum of T_1 in the dealuminated sample is rather flat, resulting from a distribution of the correlation time as well as from deviations
from isotropic motion. According to /25/ the interaction of the benzene
molecules with the cavity walls is strongly reduced when the cations are
missing. This and the more randomly distributed acid centers lead to a
distribution of correlation time. The interaction with the acid centers
causes a deviation from the isotropic motion.

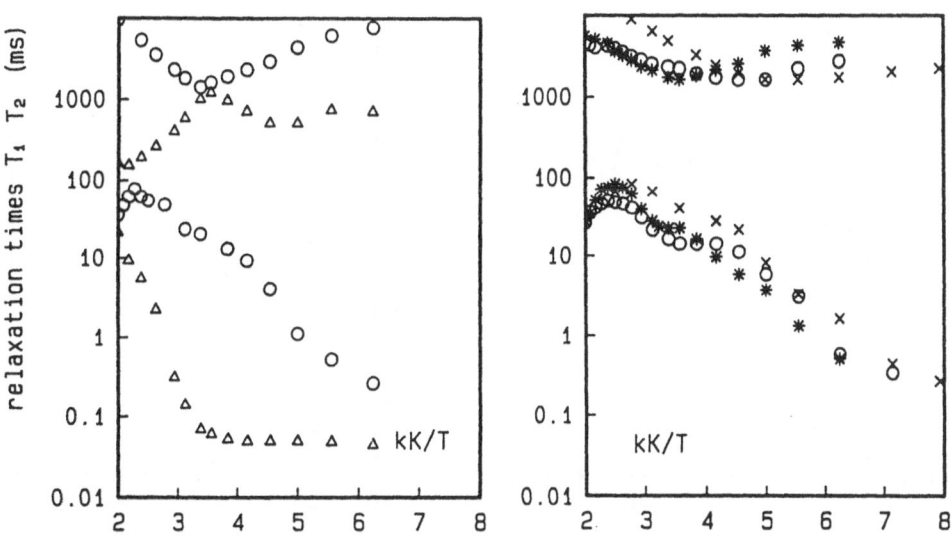

Fig. 7 (left).   Temperature dependence of H-NMR-relaxation times T_1 and T_2
                 of 4 benzene molecules/large cavity in NaY Si/Al=2.5 (Δ)
                 and in dealuminated Y Si/Al=3.1 (o).
Fig. 8 (right). Temperature dependence of H-NMR-relaxation times T_1 and T_2
                 of 2.6 benzene molecules/large cavity in Y-zeolites of
                 different degree of dealumination.
                 (*) Si/Al=4.0, (o) Si/Al=14, (x) Si/Al=34.

To obtain the influence of these acid centers samples with different degrees of dealumination have been studied. The results for three of these samples at intermediate coverages are shown in Fig. 8. All results are summarized in the following table:

Table 2. Temperatures of T_1-minima of Benzene in dealuminated Y-zeolites with different Si/Al-ratios at intermediate and maximum coverage

Sample	Si/Al	Coverage mol/cage	T_1-minimum at T/K	Maximum Coverage mol/cage	T_1-minimum at T/K
DHY-3.1	3.1	2.5	295	4.5	330
DHY-4.3	4.3	2.6	280	--	--
DHY-4.8	4.8	2.0	280	4.5	320
DHY-6.4	6.4	--	--	4.8	310
DHY-14	14.0	2.6	210	4.5	260
DHY-36	36.0	2.4	180	--	--

It can be seen that the T_1-minima are found at lower temperatures with increasing degree of dealumination, being connected with a decrease of the number of OH-groups available for the interaction with benzene. The activation energy is always near 20 kJ/mole from the WAUGH-FEDIN relation.

Generally it is accepted that the hydrophilic character of zeolites decreases with increasing Si/Al.

The decrease of T_2 at higher temperatures shows that obviously at higher temperatures an exchange with sites of stronger interaction and low correlation times takes place. These sites may be correlated with the presence of acidic centers. The curve with Si/Al = 34 cannot be followed up to that temperature region because benzene shows already an appreciable desorption.

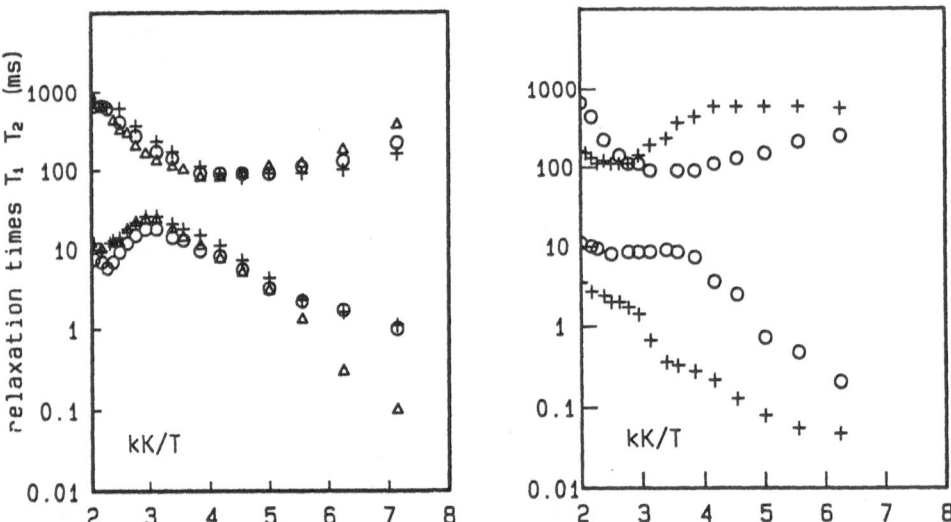

Fig. 9 (left). Temperature dependence of H-NMR-relaxation times T_1 and T_2 of benzene in dealuminated LaY-zeolites with Si/Al=6.4 and 6.4 % La-content at different coverages.
(o) 1.1, (+) 2.7, (Δ) 4.7 molecules benzene/large cavity.

Fig. 10(right). Temperature dependence of H-NMR-relaxation times T_1 and T_2 of 3.5 benzene molecules/large cavity in dealuminated Y-zeolite with Si/Al=6.4 and 660 ppm Fe^{3+}-impurity before (o) and after (+) coking of butadiene to 8.5% coke content

For further studies of the influence of acidic centers from the sample with Si/Al = 6.4 a partially La-exchanged form has been prepared to increase the number and the strength of the acidic centers. The relaxation times of the benzene molecules at different coverages is shown in Fig. 9.

T_1 is decreased by about one order of magnitude because of the paramagnetic impurity of the La-ions. The minimum in the T_2 is shifted to nearly ambient temperature. In the range of the decreasing T_2 the average life time of a molecules in one the two regions is approximately equal to T_2 of 10-20 ms. This relatively long time suggests a strong binding of the molecules to the sorption centers.

In another series of experiments the influence of coke on the motion of sorbed molecules has been studied. For the coking reaction the dealuminated samples have been treated with butadiene at 523 K for 1 hour .This treatment gave a quantity of 9 weight% coke. The coke had a ratio of C/H of about 2 and showed the typical IR "coke band" at 1585 cm^{-1}.

In Table 3 the relative sorption capacities and the temperatures of the T_1-minima for benzene at saturation coverage are given:

Table 3.   Temperatures of the T_1-minima of Benzene in Dealuminated Faujasites uncoked and coked with Butadiene at 523 K

Sample	Si/Al	uncoked		coke %	coked	
		coverage mol/cage	T_1-minimum at T/K		coverage mol/cage	T_1-minimum at T/K
DHY-3.1	3.1	4.5	330	9.2	3.0	420-440
DHY-6.4	6.4	4.8	310	8.5	3.5	380-400
DHY-14	14.0	4.5	260	8.8	3.5	400

The samples showed also an ESR-signal. To compare the relaxation behaviour of the coked and the uncoked samples the relaxation times have been studied in a sample with 660 ppm Fe^{3+}-impurity, in which the relaxation is no more determined by the relaxation with the coke. The results are shown in Fig. 10. It can be seen that the T_1-minima as well as the temperatures of the beginning motional narrowing are shifted about 100 K to higher temperatures. Comparing these results with Fig. 7 it can be seen that the mobility of the benzene molecules in the coked samples is comparable with that in untreated NaY showing that the coke deposits have a comparatively small influence on the mobility of benzene.

SUMMARY

The proton relaxation times of different hydrocarbons in faujasites with a different degree of dealumination show that the relxation times T_1 and T_2 for paraffins increase with the chain length but not depend on the Si/Al-ratio which leads to the conclusion that the hydrophobic interaction is most important for these substances.

Introducing a double bond the relaxation times are determined by the interaction with the cationic centers. This is shown for 1-butene.

Thorough experiments have been carried out with benzene. At first it can be seen that the mobility in the dealuminated forms is increased compared with the NaY showing that the interaction with the cation is most important for these molecules. The mobility decreases with decreasing degree of dealumination because of the increasing number of acid centers.

At higher temperatures a minimum of T_2 is indicated in all samples which can be shifted in a La-containing samples to about 300 K. This minimum shows that a part of the benzene molecules is sorbed in a phase with rather tight binding.

Coked samples show a reduced motion of the benzene molecules. The remaining mobility shows, however, a rather low influence of the coke.

REFERENCES

1. A. Abragam, "The Principles of Nuclear Magnetism," Oxford University Press, Oxford (1961).
2. C. P. Slichter, "Principles of Nuclear Magnetism," Harper and Row, New York (1964).
3. A. Lösche, "Kerninduktion," VEB Deutscher Verlag der Wissenschaften, Berlin (1957).
4. D. Michel, "Grundlagen und Methoden der kernmagnetischen Resonanz," Akademie Verlag, Berlin (1981).
5. N. Bloembergen, E. M. Purcell, and R. V. Pound, Phys.Rev. 78:679 (1948).
6. J. S. Waugh and E. I. Fedin, Solid State Phys. 4:2233 (1962).
7. H. A. Resing, J.Chem.Phys. 43:669 (1965).
8. H. A. Resing, Adv. Molec. Relaxation Processes 1:109 (1968).
9. H. A. Resing, Adv. Molec. Relaxation Processes 1:282 (1968).
10. H. A. Resing, Adv. Molec. Relaxation Processes 3:199 (1972).
11. H. Pfeifer, Nuclear Magnetic Resonance and Relaxation of Molecules Adsorbed on Solids, in: "NMR, Basic Principles and Progress," Vol.7, Diehl et al., eds., Springer Verlag, Berlin, Heidelberg, New York (1972).
12. J. S. Murday, R. L. Patterson, H. A. Resing, J. K. Thompson, and N. H. Turner, J.Phys.Chem. 79:2674 (1975).
13. H. Pfeifer and J. Kärger, Self-Diffusion of n-Paraffins in NaX Zeolites, in: Preprints of the Workshop "Adsorption of Hydro-carbons in Zeolites," Academy of Science of the GDR, Berlin (1979).
14. J. Kärger and A. Walter, Z.phys.Chem. Leipzig 255:142 (1974).
15. J. Kärger, M. Bülow, P. Struve, M. Kociric, and A. Zikanova, J.C.S. Faraday I 74:1210 (1978).
16. J. Kärger, M. Bülow, and R. Haberlandt, J.Colloid Interf. Sc. 60:386 (1977).
17. H. Pfeifer and J. Kärger, Nuclear Magnetic Resonance Study of Trans-lational Motion in Porous Systems, in: "Magnetic Resonance in Colloid and Interface Science," J.P.Fraissard and H.Resing, eds., Nato Advanced Study Institutes Series Dordrecht, Boston, London (1980).
18. H. Pfeifer, ACS Symp. Series 34:36 (1976).
19. M. H. Cohen and D. Turnbull, J.Chem.Phys. 31:1164 (1959).
20. M. Nagel, H. Pfeifer, and H. Winkler, Z.phys.Chem. Leipzig 255:283 (1974).
21. H. Lechert and K. P. Wittern, Ber.Bunsenges.physik.Chemie 82:1054 (1978).
22. W. D. Basler, H. Lechert, and H. Weyda, Catalysis Today 3:31 (1988).
23. H. Lechert, J. Wienecke, W. Schweitzer, and W. Maiwald, Z. Physik. Chem. NF 152:185 (1987).
24. H. Lechert, W. D. Basler, and K.-P. Wittern, in: "New Developments in Zeolite Science and Technology," Proceedings of the 7th Inter-national Zeolite Conference, Tokyo 1986, Y. Murakami,A. Iima and J. Ward, eds., Kodansha, Elsevier, Tokyo (1986).
25. A. de Mallmann and D. Barthomeuf, Four Different States of Benzene Adsorbed in Faujasites, in: "New Developments in Zeolite Science and Technology," Proceedings of the 7th International Zeolite Conference, Tokyo 1986, Y. Murakami,A. Iima and J. Ward, eds., Kodansha, Elsevier, Tokyo (1986).
26. A. J. Renouprez, H. Jobic, and R. C. Oberthür, Zeolites 5:222 (1985).
27. H. Lechert and K.-P. Wittern, Ber.Bunsenges.physik.Chemie 83:596 (1979).

ZEOLITIC HETEROPOLY OXOMETALATES

J.B. Moffat*, G.B. McGarvey, J.B. McMonagle, V. Nayak and H. Nishi

Department of Chemistry and
Guelph-Waterloo Centre for Graduate Work in Chemistry
University of Waterloo
Waterloo, Ontario N2L 3G1

INTRODUCTION

The definition of zeolite has evolved from that of J.V. Smith[1] in 1963 "an aluminosilicate framework structure enclosing cavities occupied by large ions and water molecules, both of which have considerable freedom of movement, permitting ion exchange and reversible dehydration" to encompass microporous aluminophosphates in 1980 and "nous ne savons quoi" by the year 2000. As Galen Stucky has prophesized in his Preface to Intrazeolite Chemistry[1] it is probable that zeolites with compositions from many of the elements of the Periodic Table will soon be reported.

The present work describes another class of microporous inorganic solids which the definition of zeolite may be expanded to encompass. These are heteropoly oxometalates[2], but of particular interest in the present work are those with anions of Keggin structure. Unlike the most familiar zeolites, those of aluminosilicate composition, with which the cation as prepared or found naturally is frequently sodium, with the heteropoly oxometalates of Keggin structure the primary form contains protons as cations and these may, if desired, be converted to the corresponding salts. The Keggin anion has the stoichiometric formula

$$XM_{12}O_{40}{}^{-n}$$

where X is the central atom (e.g. P, Si) in a tetrahedron with oxygen atoms at the vertices which is surrounded by and shares oxygen atoms with twelve octahedra with oxygen atoms at their vertices and peripheral metal atoms M (e.g. W, Mo) at their centres. Thus, in contrast with the network structure generally found in aluminosilicates the heteropoly oxometalates possess discrete cagelike anions. Further, the crystal structure of the heteropoly oxometalates is essentially the result of the interactions between the cations and anions.

While the anions of the heteropoly oxometalates are nearly spherical in shape the structures are packed and consequently unable to permit the entry of foreign molecules. The solid acids themselves have low surface areas, of the order of 10 m^2/g, as measured by applications of BET theory to N_2 adsorption isotherms[3]. However polar molecules such as ammonia[4], pyridine[5] and methanol[6] will diffuse into the bulk structure of the heteropoly oxometalates but not into the anions, while nonpolar molecules such as methane are unable to do so.

Guidelines for Mastering the Properties of Molecular Sieves
Edited by D. Barthomeuf *et al.*
Plenum Press, New York, 1990

CATION	ANION			
	$SiW_{12}O_{40}^{-4}$	$PMo_{12}O_{40}^{-3}$	$PW_{12}O_{40}^{-3}$	$AsW_{12}O_{40}^{-3}$
H	3.2	(8)	(8)	6.5
Na	1.3	3.5	3.7	-
Ag	-	-	3.0	-
K	3.3	39.9	90.0	46.0
NH_4	116.9	193.4	128.2	82.1
Rb	116.3	-	-	101.4
Cs	153.4	145.5	162.9	65.4
$MeNH_3$	3.0	1.3	3.0	2.1
Me_4N	-	-	4.5	-

Work in this laboratory has shown that certain of the monovalent cation salts of heteropoly oxometalates display evidences of microporous structures[7-10]. The present report describes the characterization of these catalysts through the use of N_2 physisorption isotherms, X-ray diffraction data, measurements of diffusivity and sorption capacities of organic molecules and the reaction of probe molecules.

EXPERIMENTAL

The parent acids, 12-tungstophosphoric $(H_3PW_{12}O_{40})$, 12-molybdophosphoric $(H_3PMo_{12}O_{40})$, 12-tungstosilicic $(H_4SiW_{12}O_{40})$ and 12-tungstoarsenic $(H_3AsW_{12}O_{40})$ acids, abbreviated to HPW, HPMo, HSiW and HAsW, respectively were recrystallized before use. The heteropoly salts were prepared by precipitation or crystallization from aqueous solutions[11-13].

Nitrogen adsorption isotherms were measured at 77.4 K using a standard glass volumetric system. Samples were pumped at 10^{-5} Torr and 473°K prior to use. The MP method[14] together with a technique for obtaining t as a function of P/Po[15] was employed to calculate the micropore distributions.

X-ray powder diffraction patterns were obtained with a Phillips diffractometer (Model PW-1011/60) with nickel filtered CuKα radiation.

Sorption rates and equilibria were measured by a gravimetric method using a Cahn electrobalance fitted with a high-vacuum system. Pressures were obtained with a Texas Instruments fused quartz precision pressure gauge. Reaction studies were carried out in a micropulse reactor equipped with an onstream gas chromatograph (HP 5880).

RESULTS

The BET N_2 surface areas for the various monovalent salts of HPW, HPMo, HSiW and HAsW are collected in Table 1 and are shown in Fig. 1 as a function of cation diameter. It is evident that the surface areas of the parent acids and the salts of sodium and silver are small but values for the potassium, ammonium, rubidium and cesium salts with the exception of KSiW are considerably higher.

Table 2. BET C parameter for heteropoly salts.

CATION	ANION			
	$SiW_{12}O_{40}^{-4}$	$PMo_{12}O_{40}^{-3}$	$PW_{12}O_{40}^{-3}$	$ASW_{12}O_{40}^{-3}$
H	100	-	-	58
Na	473	142	18	-
Ag	-	-	101	-
K	127	1070	2430	22
NH_4	1919	877	760	1292
Rb	489	-	-	149
Cs	-	568	1720	1359
$MeNH_3$	-	-	261	292
Me_4N	-	-	113	-

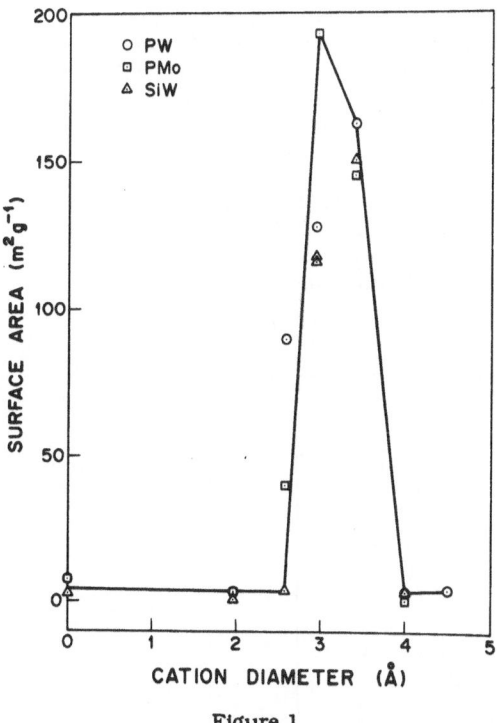

Figure 1

Surface area vs. cation diameter
for parent acids and monovalent
salts of HPW, HPMo and HSiW

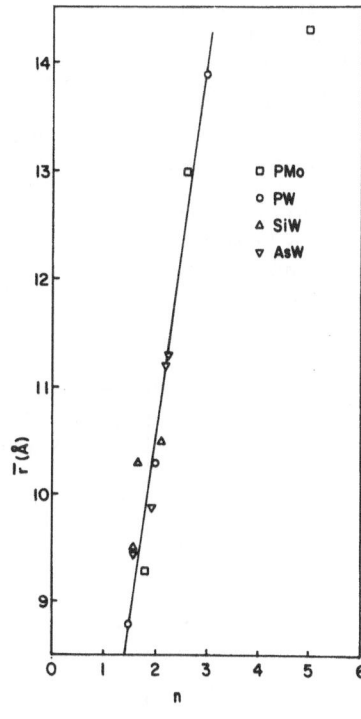

Figure 2

Average micropore radius vs.
optimized n value from n-layer
BET equation for microporous
heteropoly oxometalates

Those salts of the heteropoly acids having higher surface areas also yield relatively large values for the C parameter of the BET equation (Table 2). This, along with the high surface areas, is indicative of microporosity.

The average values of the micropore radii calculated by the MP method[14] are summarized in Table 3. The experimentally measured nitrogen adsorption isotherms were fitted to the n-layer BET equation and the values for the various $\bar{r}$ are plotted versus the optimized n values in Fig. 2. The mean micropore radius is seen to be an exceptably good linear function of the n values, providing support for the validity of the micropore analysis. Further the slope of the line is 3.5 Å, very similar to the diameter of the nitrogen molecule.

Unlike many of the aluminosilicate zeolites the microporous heteropoly oxometalates have relatively broad pore size distributions. As an example that for NH_4PW is shown in Fig. 3. While the average micropore radius is calculated as 10.3 Å, it is evident that the pores range in radius from approximately 8 to 13 Å. These broad pore size distributions inject an additional complexity not found with many of the aluminosilicate zeolites and increase the difficulty in characterizing the micropore structure of the heteropoly oxometalates.

Typical values for the sorption capacities and diffusivities of aromatic hydrocarbons, alkanes, alkenes and alcohols are provided in Tables 4 and 5. The sorption capacity of NHSiW is greater than that of either NHPW or NHPMo for all sorbates, while that for the latter two is quite similar, in spite of the observation from the nitrogen adsorption isotherm analyses, that the micropore volumes are 40, 50 and 40 (x 10^{-3}) cm^3 liquid per gram of solid. It is also interesting to note that the sorption capacities for all the organic molecules shown, except for mesitylene, differ relatively little for a given sorbent and at least for NHPW and NHPMo are in reasonably good agreement with the values found from the micropore analysis for N_2. In contrast the values for mesitylene are considerably smaller than those observed with the remaining organic compounds with all sorbents.

The diffusivities (Table 5) are largest with each sorbent for n-hexane followed by 1-hexene, benzene, isooctane, 2-methyl-2-butanol, 2,3-dimethyl-1-butene, methanol and mesitylene in order of decreasing diffusivity. For a given sorbate, the diffusivities are highest with NHSiW followed by NHPMo and NHPW in that order.

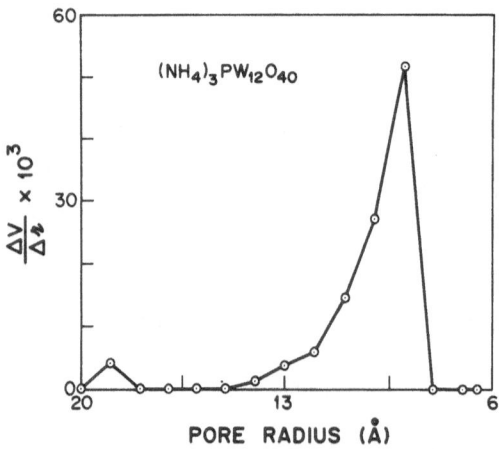

Figure 3

Micropore size distribution for ammonium 12-tungstophosphate.

Table 3. Micropore Average Radii (Å) of High Surface Area Heteropoly Oxometalates

CATION	ANION			
	$SiW_{12}O_{40}^{-4}$	$PMo_{12}O_{40}^{-3}$	$PW_{12}O_{40}^{-3}$	$AsW_{12}O_{40}^{-3}$
K	-	9.3	8.8	11.2
NH_4	9.5	13.0	10.3	11.3
Rb	10.3	-	-	9.4
Cs	10.5	14.3	13.9	9.9

As noted above, the presence of relatively broad pore size distributions add a further complexity to the study of the microporous heteropoly oxometalates. However, reactions in which the intrusion of shape selectivity is possible can supply additional information. The alkylation of toluene with methanol is a useful reaction for this purpose and has been studied by many workers for at least twenty years, largely in the presence of aluminosilicate zeolites. Space restrictions prevent a full discussion of the previous work on this reaction. It may suffice to note two recent publications which refer to a large portion of the earlier work. IR spectroscopic studies have been employed to determine the structures formed on CsX, RbX and KX zeolites during the alkylation of toluene with methanol[16]. The alkylation of toluene has also been compared on the zeolites theta-1 and ZSM-5[17].

Typical results from the alkylation of toluene with methanol on NHPW and NHSiW are summarized in Table 6. Perhaps the most interesting observations relate to the xylene composition. Ashton et al.[17] have noted that H-theta-1 is a highly selective catalyst for the production of p-xylene and it can be seen that the present catalysts are capable of attaining values for the percentage of p-xylene in the xylene isomers as high as those found by Ashton et al[17] with H-theta-1 and ZSM-5..

DISCUSSION

The most definitive information on the microporous structure of the heteropoly oxometalates is obtained from the analysis of nitrogen adsorption isotherms. The high values for surface area, large BET C parameter values, the values of $\bar{r}$ and of the number of adsorbed layers, n, all point to the presence of micropores. The application of the MP analysis to generate pore size distributions provides direct evidence of the existence of such microporous structures but provides little or no information on the cross-sectional or longitudinal uniformity of the pores.

The general agreement between the total micropore volumes calculated through use of the MP method with the N_2 adsorption data and the sorption capacities of organic molecules provides further confirmation not only of the microporous structures but also of the internal self-consistency between the two methods.

The diffusivities of the various organic molecules in the ammonium salts of the heteropoly oxometalates provide further evidence for the presence of micropores and in particular the relatively low values for mesitylene, one of the larger molecules examined, seem indicative of the presence of small pores. However the irreversible nature of the sorption of this molecule on NHSiW and the somewhat higher than expected sorption capacities of all the organic species on this solid is not understood at this time.

Table 4. Sorption Capacities[a] of Ammonium Heteropoly Oxometalates

Sorbate	Sorbent		
	NHPW	NHSiW	NHPMo
Benzene	32.0	57.5	40.9
Mesitylene	16.7	irrev.	7.0
n-hexane	39.2	69.2	43.1
isooctane	40.6	61.8	43.9
1-hexene	47.5	67.5	45.0
2,3-dimethyl-1-butene	38.4	64.3	42.1
methanol	36.2	68.4	40.3
2-methyl-2-butanol	43.6	69.7	47.9

[a]   293 K, cm^3/g ($x10^3$)

Table 5. Diffusivities on Ammonium Heteropoly Oxometalates

Sorbate	Sorbent		
	NHPW	NHSiW	NHPMo
Benzene	2.4	59	11
Mesitylene	0.22	irrev.	0.84
n-hexane	4.8	110	25
isooctane	2.2	45	8.8
1-hexene	2.9	54	11
2,3-dimethyl-1-butene	1.9	33	6.6
methanol	1.0	24	4.2
2-methyl-2-butanol	2.2	54	10

[a]   Square centimeters per second x 10^{11}, 293 K.

The product distribution from the alkylation of toluene with methanol also suggests a size and/or shape restriction. However, considering the relatively large sizes of the micropores in the heteropoly oxometalates and the size of toluene and the xylenes this observation is somewhat surprising. However in separate experiments in which methanol was passed over the catalyst before the toluene it was evident that similar products resulting from an alkylation process were also observed, although additional hydrocarbons were also present in the product. Earlier work in this laboratory has shown that methanol is converted to hydrocarbons on certain of the heteropoly

Table 6. Xylene compositions from the alkylation of toluene with methanol on NH_4PW, NH_4SiW, THETA-1* and ZSM-5* Catalyst

	NH_4PW	NH_4SiW	NH_4PW	NH_4PW†	THETA-1	ZSM-5
Conditions						
Temperature/°C	200	250	200	160	550	600
Tol/MeOH molar	1/1	1/1	2/1	1/1	2/1	2/1
W/F [mg. cat. min/ml.He]	1.43	1.43	1.43	1.43	1(h^{-1})	4.8(h^{-1})
Conversion/%						
Toluene	17.3	17.0	20.5	0.9	23.9	31.8
Xylene Composition/%						
p-xy	36.2	45.1	30.1	49.9	44.9	41.4
m-xy	33.7	36.0	42.6	31.3	28.6	41.4
o-xy	30.1	18.9	27.3	18.8	26.5	18.2

*   Ref (17)
†   200 µl NH_3/50 mg. cat. added to catalyst at 160°C prior to reaction.

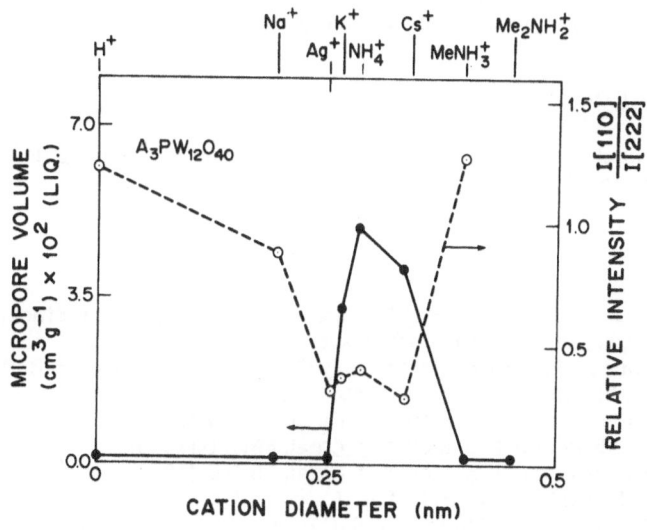

Figure 4

Total Micropore Volume and X-ray Diffraction Intensity vs. Cation Diameter

oxometalates and preferably the salts in particular the ammonium salt of HPW[13]. Parenthetically it is relevant to note that over this latter catalyst the quantities of saturated hydrocarbons are higher than those found in the absence of microporosity. This may also be taken as additional evidence for the existence of a microporous structure. Photoacoustic FTIR spectroscopy has shown that methanol first interacts with the proton in these catalysts, forming $CH_3OH_2^+$ which subsequently, at higher temperatures decomposes to water and the methyl ion which in turn becomes attached to an anionic oxygen atom as

$$H^{+\bullet} + CH_3OH \rightarrow CH_3OH_2^{+\bullet}$$
$$CH_3OH_2^{+\bullet} \rightarrow CH_3^{+\bullet} + H_2O$$

where the asterisk refers to the anion of the heteropoly oxometalate[18]. It may then be surmised that the alkylation process occurs via the methylated anions even in the cases where methanol and toluene are passed over the catalyst concurrently. Thus the methylation of the anions may be sufficient to reduce, at least from an operational viewpoint, the effective micropore sizes so that the existence of geometrical constraints becomes manifested in the observations.

Finally some comments on the apparent source of the micropore structure in the heteropoly oxometalates may be of value. A comparison of the manner in which the total micropore volumes and the ratio of the X-ray diffraction intensities in the [110] and [222] planes change with cation diameter shows an interesting approximately reciprocal relationship (Fig. 4). Although not shown, similar relationships are found for all four anions studied. Interstitial voids separated from each other by the terminal oxygen atoms are present in the parent acids. The terminal oxygen atoms occupy the [110] planes and produce the relatively high values for the [110]/[222] ratios. Replacement of the protons by larger cations results in a translation and rotation of the anions so that the terminal oxygen atoms no longer separate the interstitial voids from each other and molecules such as N_2 are able to pass down the channels. Concomitantly the [110]/[222] ratio will decrease since the oxygen atoms will be moved from the [110] plane.

It is evident that, while the results presented here provide positive evidence for the existence of micropores in certain of the salts of heteropoly oxometalates, much is yet to be learned about the nature of these micropores, their geometries and stabilities. This work is currently in progress in this laboratory.

ACKNOWLEDGEMENT

The financial support of the Natural Sciences and Engineering Research Council of Canada is gratefully acknowledged.

REFERENCES

1. G. D. Stucky in the Preface to "Intrazeolite Chemistry", G. D. Stucky and F. G. Dwyer, eds. American Chemical Society, Washington (1983).
2. M. T. Pope in "Heteropoly and Isopoly Oxometalates", Springer, Berlin (1973).
3. H. Hayashi and J. B. Moffat, J. Catal. 77:473 (1982).
4. J. G. Highfield and J. B. Moffat, J. Catal. 88:177 (1984).
5. J. G. Highfield and J. B. Moffat, J. Catal. 89:185 (1984).
6. J. G. Highfield and J. B. Moffat, J. Catal. 95:108 (1985).
7. J. B. McMonagle and J. B. Moffat, J. Colloid Interf. Sci. 101:479 (1984).
8. D. B. Taylor, J. B. McMonagle and J. B. Moffat, J. Colloid Interf. Sci. 108:278 (1985).
9. G. B. McGarvey and J. B. Moffat, J. Colloid Interf. Sci. 125:51 (1988).
10. J. B. Moffat, J. B. McMonagle and D. Taylor, Solid State Ionics 26:101 (1988).
11. G. A. Tsigdinos, Topics Current Chem. 76:1 (1978).
12. H. Hayashi and J. B. Moffat, J. Catal. 81:61 (1983).
13. H. Hayashi and J. B. Moffat, J. Catal. 83:198 (1983).
14. R. Sh. Mikhail, S. Brunauer and E.E. Bodor, J. Colloid Interface Sci. 26:45 (1968).
15. A. Lecloux and J. P. Pirard, J. Colloid Interface Sci. 70:265 (1979).
16. S. T. King and J. M. Garces, J. Catal. 104:59 (1987).
17. A. G. Ashton, S. A. I. Barri, S. Cartlidge and J. Dwyer in "Chemical Reactions in Organic and Inorganic Constrained Systems", R. Setton, ed., Reidel, Amsterdam (1986).
18. J. G. Highfield and J. B. Moffat, J. Catalysis 98:245 (1986).

ON LATTICE DYNAMICS, STABILITY AND ACIDITY OF ZEOLITES

R.A. van Santen[°+], B.W.H. van Beest[+], A.J.M. de Man[°]

[°] Eindhoven University of Technology
Eindhoven
The Netherlands
[+] Koninklijke/Shell Laboratory
Amsterdam
The Netherlands

INTRODUCTION

Theoretical approaches to chemical bonding and the stability of zeolites can be distinguished in quantum mechanical methods based on electronic structure calculations[1,2] and methods that start with known potentials[3].

Because of the large elementary unit cell of zeolite structures no quantum chemical calculations are available for zeolites, but a full quantum mechanical ab-initio calculation for quartz is available[4]. Otherwise the quantum mechanical approach limits itself to the study of silicate rings or clusters[1]. Quantum chemical ab-initio or semi-empirical techniques have been applied to study the relative stability of such clusters[1] as well as the acidity of protons as a function of composition[5].

In solid state chemistry extensive use is made of infinite lattice techniques, based on electrostatic potentials, modified by empirical short range potentials[3c]. The minimum lattice energy and corresponding lattice configuration are determined by relaxing the structure. Pure valence force field calculations[6] have been applied too and a combination of valence force fields and electrostatic forces has recently been used to simulate an isolated sodalite cage[7].

The empirical potentials used in infinite lattice techniques have usually been determined by fitting them such that computed structure and elasticity constants reproduce experimental values optimally. For chemical purposes potentials are required that give correct potential energy surfaces. Infrared and Raman spectra provide sensitive probes for the potentials used by comparison of computed and measured spectra.

Here we will present such an analysis for the results of infinite lattice calculations based on empirical potentials, derived for α-quartz using rigid ion as well as shell model calculations. Structures considered will be limited to zeolitic polymorphs of silica, containing only silicon and oxygen atoms. From the difference in average energy of the longitudinal and transverse optical frequencies a deduction can be made of the effective charges on the atoms forming a zeolite. However, experimental data on this difference are only available for quartz. From this the effective electrostatic field in

Guidelines for Mastering the Properties of Molecular Sieves
Edited by D. Barthomeuf *et al.*
Plenum Press, New York, 1990

201

the micropores of the zeolite can be estimated. The results appear to agree rather well with those based on quantum mechanical estimates[5].

We will conclude the paper with a short discussion of the impact of our results on ideas concerning the factors that determine the acidity of zeolites.

THE VIBRATIONAL DENSITY OF STATES (VDOS)

Notwithstanding the large number of SiO_2 tetrahedral units per elementary unit cell in zeolite structures, a useful approach to the basic understanding of the mode distribution can be derived from a model based on the Bethe lattice approximation. In its most simple version only a system with one SiO_2 tetrahedron per unit cell is considered. Extension to the prescribed number of SiO_2 tetrahedra per unit cell for a particular zeolite structure in combination with a factor group analysis enables a prediction of the frequency distribution of vibrational stretching modes. This has been discussed elsewhere[8]. Here we will summarize the results of the Bethe lattice considerations and use the result to analyse computed spectra from rigid ion and shell model calculations on α-quartz, silica-sodalite and silica-faujasite[38].

The VDOS is best understood realising that the coupling of the individual SiO_2 tetrahedral modes depends on the Si-O-Si-angle ϕ. If this angle is 90° and the Si-O-Si threebody potential is ignored, the modes of the tetrahedra will not couple. Coupling is maximum at $\phi=180°$. For quartz the average angle ϕ is 144°, so coupling of the modes will be significant.

Quantum mechanical[1a,1b,1p] as well as spectroscopic studies indicate that the frequency of the Si-O-Si bending modes is two orders of magnitude less than that of Si-O stretch modes. For this reason in the Bethe lattice calculations to be discussed the Si-O-Si threebody potential is ignored. The situation is different for the O-Si-O bending modes. This bending mode frequency is of the order of 300 cm^{-1}. The tetrahedral coordination of oxygen atoms around silicon requires the O-Si-O threebody to be rather stiff. In the Bethe lattice calculations to be discussed it is assumed that

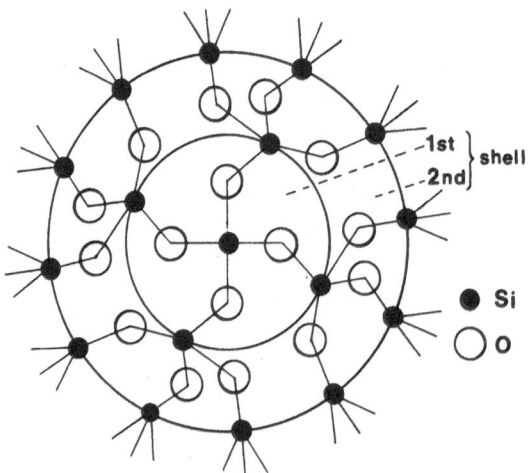

Figure 1. The Bethe lattice of Si-O-Si oscillators.

the O-Si-O angle is rigid. A discussion of the calculational details is given in ref. (8). Here we will give an outline of the results.

In the Bethe lattice method the fundamental bonding unit is the Si-O-Si bond. Figure 1 illustrates the Bethe lattice approximation, figures 2 and 3 the resulting VDOS's. Symmetric and asymmetric Si-O-Si stretching modes are formed with frequency differences depending on the Si-O-Si angle. The average frequency depends on the Si-O stretch frequency in the uncoupled system. The symmetric and asymmetric Si-O-Si modes mix according to tetrahedral symmetry because they are coupled by

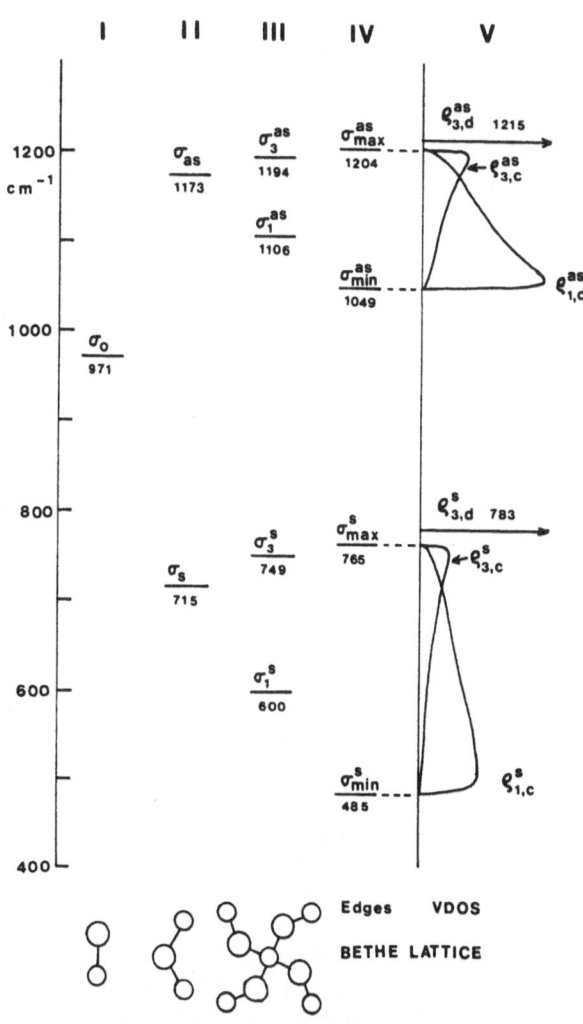

Figure 2. The VDOS of stretching modes for a Si-O-Si angle ϕ of 149°.

        Column I: Wave number σ_0 of the uncoupled Si-O oscillator.

        Column II: Wave numbers σ_{as} and σ_s of the "free" Si-O-Si oscillator.

        Column III: Wave numbers of four tetrahedrally arranged Si-O-Si oscillators.

        Column IV: VDOS of the Bethe lattice of Si-O-Si oscillators.

the same silicon atom. Because of the large energy difference between the symmetric and asymmetric Si-O-Si modes and the small coupling of the modes, due to the closeness of the O-Si-O angle to 90°, the symmetric and asymmetric Si-O-Si stretching modes can be considered to remain independent from each other in the tetrahedral coordination around the silicon atom. Eight modes are found, split according to T_d symmetry (figs. 2 and 3).

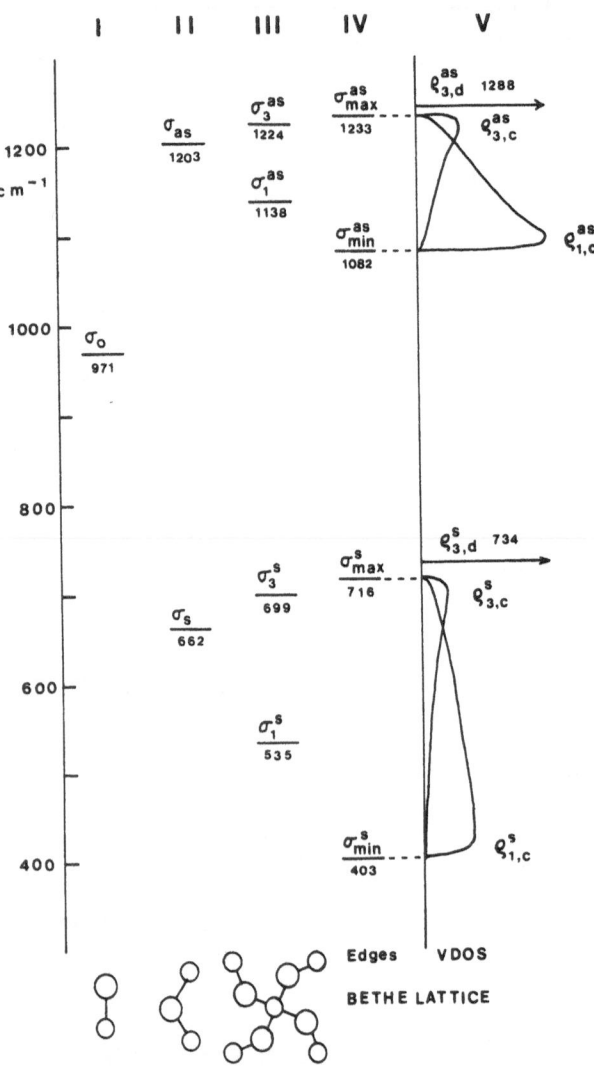

Figure 3. The VDOS of stretching modes for a Si-O-Si angle ϕ of 180°.
Column I: Wave number σ_0 of the uncoupled Si-O oscillator.
Column II: Wave numbers σ_{as} and σ_s of the "free" Si-O-Si oscillator.
Column III: Wave numbers of four tetrahedrally arranged Si-O-Si oscillators.
Column IV: VDOS of the Bethe lattice of Si-O-Si oscillators.

To form the Bethe lattice the Si-O-Si modes have to be connected to the other modes *via* silicon atoms in a second shell. This is continued *ad infinitum* with the essential approximation that no cross connections are made between the Si-O-Si modes. So ring formation between the tetrahedra is ignored. This seems a rather unphysical approximation. However it can be shown[9,10] that one can systematically improve on it to incorporate structural details. In the absence of cross connections and maintaining the approximation that the asymmetric and symmetric Si-O-Si modes remain uncoupled, the resulting set of dynamic equations can be analytically solved, with the resulting VDOS given in figures 2d and 3d. One finds that the four Si-O stretching vibrations of the isolated SiO_4 tetrahedron transform into two localized and two delocalized stretching vibrations. One set of localized and delocalized modes can be considered to originate from the antisymmetric Si-O-Si stretching vibrations located between 900 and 1200 cm^{-1}. The other set of localized and delocalized modes can be considered to be derived from the symmetric Si-O-Si stretching vibrations and is found between 450 and 850 cm^{-1}. In the case a structure contains n SiO_2 tetrahedral units per elementary unit cell $2n$ localized and $2n$ delocalized modes are found. The number of optically active modes is found by an adapted factor group analysis method[8,3g].

We will now analyse experimental and computed spectra on the basis of Bethe lattice model results. Figures 4 and 5 show a comparison of experimental and computed infrared spectra for α-quartz. Details of the computations are given in ref. (3g). In figure 4 results of rigid ion calculations are compared, in figure 5 results of shell model calculations.

Clearly computed and experimental spectra do not agree very well if results of the rigid ion calculations are compared. Especially the appearance of intensity in the

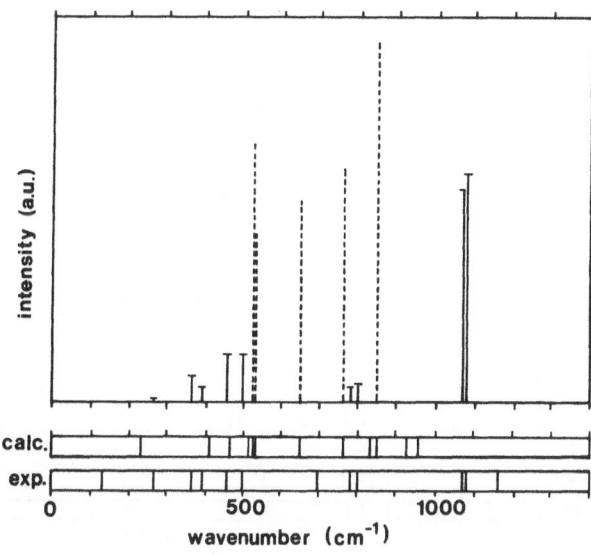

Figure 4. Infrared spectra of α-quartz. Solid lines: experimental[11] frequencies; Broken lines: calculated according to the rigid ion model.

"quartz frequency gap" between 500 and 700 cm^{-1} is of concern.

The shell model calculations (fig. 5) reproduce an intensity free "quartz frequency gap". Also the grouping of computed and measured bands corresponds rather well. Agreement of modes below 500 cm^{-1}, the vibrations derived from the O-Si-O bending modes, is satisfactory. The average position of the calculated stretching modes above 800 cm^{-1} appears to be too low. Also the difference in energy of the two high frequency stretching modes ($\sim$900 cm^{-1}) and the low frequency stretching modes ($\sim$800 cm^{-1}) is much smaller than the experimental value. The corresponding experimental high frequency stretching modes are at approximately 1080 cm^{-1} whereas the experimental low frequency stretching modes are found at approximately 800 cm^{-1}. The high frequency modes correspond to the asymmetrically coupled Si-O stretching modes. The low frequency modes correspond to the symmetrically coupled Si-O stretching modes. The difference as well as the average value of these modes depends on the value of the Si-O stretch frequency. One concludes that bending and stretch frequencies computed according to the shell model are in much better agreement with experiment than those derived from the rigid ion calculations. However, the Si-O stretch mode frequency computed according to the shell model is still too small.

In figures 6 and 7 a comparison between computed shell model and experimental lattice infrared spectra is given for silica-sodalite and silica-faujasite. The discrepancies between experimental and computed data are similar for α-quartz and again agreement is rather good in the bending frequency region.

In the next section a detailed analysis of the reasons for the differences in results of rigid ion and shell model calculations is given. Also of interest for this discussion is a comparison of computed differences in frequency of longitudinal (LO) optical and transversal optical (TO) modes. For α-quartz such a comparison is shown in fig. 8. One observes a much larger difference in frequencies of corresponding LO and TO

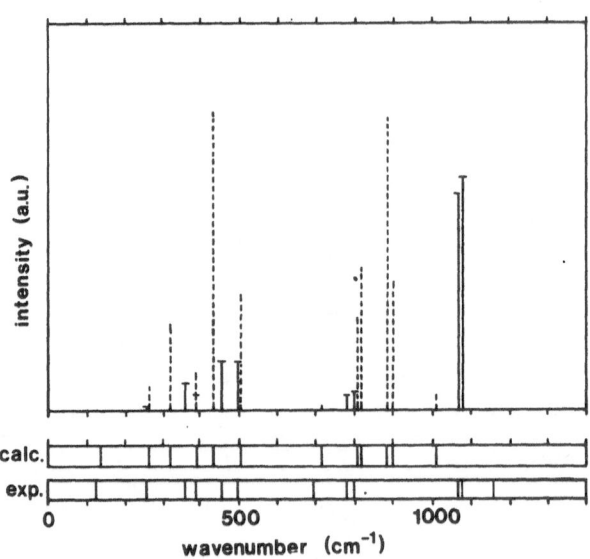

Figure 5. Infrared spectra of α-quartz. Solid lines: experimental[11] frequencies; Broken lines: calculated according to the shell model.

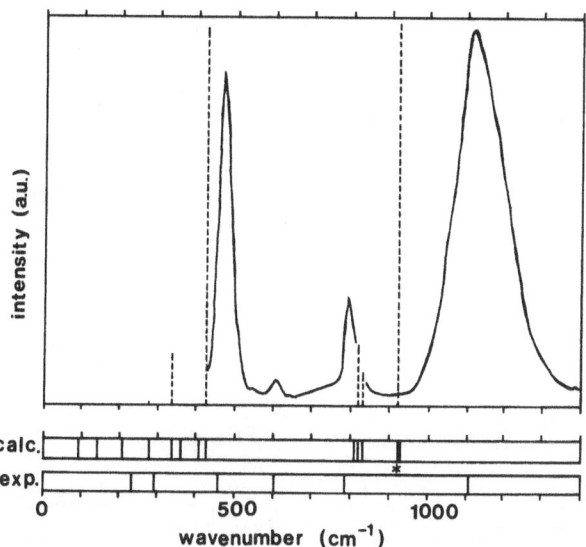

Figure 6. Infrared spectra of sodalite (spacegroup P$\overline{4}$3n or T$_d^4$).
Solid lines: experimental frequencies;
Broken lines: calculated according to the shell model.

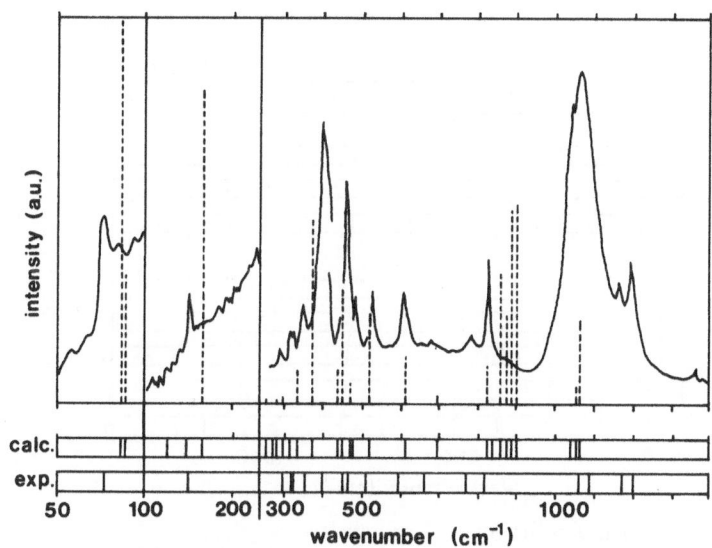

Figure 7. Infrared spectra of faujasite (spacegroup Fd3m or O$_h^7$).
Solid lines: experimental frequencies;
Broken lines: calculated according to the shell model.

modes for the rigid ion and shell model calculations than seen in the experiment. The largest differences are found in the stretching frequency regime.

RIGID ION versus SHELL MODEL CALCULATIONS

For a detailed derivation of the equations of motion to be used in the rigid ion and the shell model, we refer to the excellent monograph of P. Brüesch[12]. An extensive treatment can be found in the book of Maraduddin e.a.[13]. Here we will summarize some essential results and use them to interpret the difference between rigid ion and shell model results. This will be used to indicate how further improvements

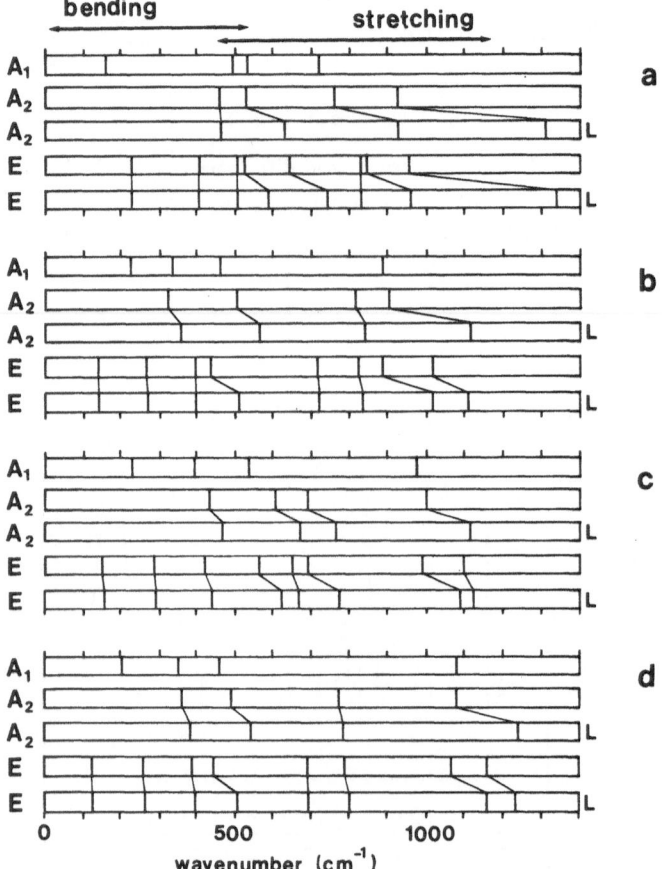

Figure 8. TO-LO splitting of α-quartz.
    a: Calculated according to the rigid ion model;
    b: Calculated according to the shell ion model;
    c: Calculated according to the potential derived[15] from ab-initio calculations of Tsuneyuki[14];
    d: Experimental values, as collected by Striefler and Barsch[16];

can be made. In the rigid ion as well as the shell model calculations used here, full formal charges on the cations are used to derive the electrostatic potential. In the rigid ion method to the Coulomb potential a Born repulsion and a van der Waals attraction term is added. The parameters used are given in ref. (3e), and were found by optimisation studies of calculated lattice constants and elastic constants for computed minimum energy structures and corresponding experimental values of α-quartz. The vibrational modes were calculated within the harmonic approximation.

It is useful to quote[12] the resulting rigid ion equations of motion for a diatomic cubic alkali-halide crystal having only a Coulomb- and a nearest neighbour short range interaction.

One finds (for a zero wavevector $\vec{k}$):

$$\mu\ddot{w}_{LO} + Sw_{LO} = Ze\beta_{LO}P = -\frac{8\pi}{3}\frac{Z^2e^2}{v_a}w_{LO} \tag{1a}$$

$$\mu\ddot{w}_{TO} + Sw_{TO} = Ze\beta_{TO}P = \frac{4\pi}{3}\frac{Z^2e^2}{v_a}w_{TO} \tag{1b}$$

where w is the relative displacement of the two ions, μ the reduced mass, S the force constant derived from the short range potential, Z the charge of the cations, P the polarization and v_a is the volume per oscillator.
β and P are given by:

$$\beta_{LO} = -\frac{8\pi}{3} \tag{2a}$$

$$\beta_{TO} = \frac{4\pi}{3} \tag{2b}$$

and:

$$P = \frac{Zew}{v_a} \tag{3}$$

The expressions for the LO and TO frequencies ω are:

$$\mu\omega_{LO}^2 = S + \frac{8\pi Z^2e^2}{3v_a} \tag{4a}$$

$$\mu\omega_{TO}^2 = S - \frac{4\pi Z^2e^2}{3v_a} \tag{4b}$$

According to the rigid ion method, the difference between longitudinal and transversal optical frequencies relates directly to the charge of the ions.

The corresponding expressions according to the shell model are:

$$\mu\omega_{LO}^2 = S' + \frac{8\pi(Z'_{\text{eff}})^2e^2}{3v_a\epsilon(\infty)} \tag{5a}$$

$$\mu\omega_{TO}^2 = S' - \frac{4\pi(Z'_{\text{eff}})^2e^2}{3v_a\epsilon(\infty)} \tag{5b}$$

Two new constants are introduced: the effective charge Z'_{eff} and the dielectric constant for infinite frequency $\epsilon(\infty)$ (see Appendix for a derivation).

The equations of motion become :

$$\mu \ddot{w}_{LO} + S' w_{LO} = Z'_{eff} E_{LO} = \frac{8\pi}{3} \frac{(Z'_{eff})^2 e^2}{\epsilon(\infty) v_a} \tag{6a}$$

$$\mu \ddot{w}_{TO} + S' w_{TO} = Z'_{eff} E_{TO} = -\frac{4\pi}{3} \frac{(Z'_{eff})^2 e^2}{\epsilon(\infty) v_a} \tag{6b}$$

and

$$E_{LO} = \beta_{LO} P_{LO} = -\frac{4\pi}{3} \frac{(Z'_{eff})^2 e^2}{\epsilon(\infty) v_a} w_{LO} \tag{7a}$$

$$E_{TO} = \beta_{TO} P_{TO} = \frac{8\pi}{3} \frac{(Z'_{eff})^2 e^2}{\epsilon(\infty) v_a} w_{TO} \tag{7b}$$

According to the shell model the expression for the dielectric constant becomes:

$$\epsilon(\omega) = \epsilon(\infty) + \frac{4\pi(Z'_{eff})^2 e^2}{\mu v_a (\omega_{TO}^2 - \omega^2)} \tag{8}$$

$\epsilon(\infty)$ follows from the Clausius-Mosotti relation:

$$\frac{\epsilon(\infty) - 1}{\epsilon(\infty) + 2} = \frac{4\pi \alpha^*_{eff}}{3 v_a} \tag{9}$$

α^*_{eff} being the effective polarizability per cation-anion pair.
Since

$$\omega_{LO}^2 - \omega_{TO}^2 = \frac{4\pi(Z'_{eff})^2}{\mu v_a \epsilon(\infty)} = \omega_{plasmon}^2 \tag{10}$$

and $\epsilon(\infty)$ follows from the calculations, Z'_{eff} can be derived. In table 1 $\omega_{plasmon}$, Z'_{eff} and $\epsilon(\infty)$ are listed as computed and used for α-quartz, silica-sodalite and silica-faujasite.

As outlined in the appendix the value of Z'_{eff} results from a comparison of rigid ion and shell model calculations. The main difference between the shell model approximation and the rigid ion model is the introduction of an effective polarizability (eq. 9) in the shell model. Since in the rigid ion model $\epsilon(\infty) = 1$, the electrostatic interactions decrease in the shell model because of the larger dielectric constant. The electrostatic interaction is also reduced in the shell model by the lower value of Z'_{eff}. In the shell model polarization is included by coupling the negatively charged electron clouds by force constants k with the positively charged cores. The electron clouds of different ions are coupled with force constants f.
Z'_{eff} relates to the rigid ion charges by:

$$Z'_{eff} = \frac{\epsilon(\infty) + 2}{3} \left(Z_{core} + \frac{Z_{electrons} \frac{k}{f}}{1 + \frac{k}{f}} \right) = \frac{\epsilon(\infty) + 2}{3} Z^* \tag{11}$$

Since in the rigid ion model $\epsilon(\infty) = 1$ and $k = \infty$, Z'_{eff} reduces to $Z_{core} + Z_{electrons}$. Clearly in the shell model Z'_{eff} is significantly reduced. Table 1 shows this reduction of Z'_{eff}.

The improvements of the shell model calculations compared to the rigid ion calculations derive from the significant reduction of the Coulomb term. As a result

Table 1. Calculation of Z'_{eff}/Z.

Species	Shell model			Rigid ion model			Z'_{eff}/Z
	v_a	$\epsilon(\infty)$	$\omega_{plasmon}$	v_a	$\epsilon(\infty)$	$\omega_{plasmon}$	
	Å^3		cm^{-1}	Å^3		cm^{-1}	
α-quartz*	36.107	2.1163	366.80	42.367	1.0000†	557.80	0.8831
α-quartz°	36.107	2.1397	254.24	42.367	1.0000†	394.43	0.8704
Sodalite	56.168	1.6992	166.73	59.110	1.0000†	248.50	0.8525
Faujasite	74.066	1.5030	107.53	62.51	1.0000†	149.98	0.8384
α-quartz‡*	40.35	1.0000†	342.94				
α-quartz‡°	40.35	1.0000†	242.49				
α-quartz×*	37.66	2.383	139.57				
α-quartz×°	37.66	2.356	98.55				

† : In the rigid ion model the high frequency dielectric constants are exactly one.
* : The zz-component of the dielectric tensor and the shift of the A_2 modes are taken;
° : The xx-component of the dielectric tensor and the splitting of the E modes are taken.
‡ : Calculated[15] with the potential of ref. 14.
× : Experimental[46] value.
The calculations are made with a length of the wavevectors $\vec{k}$ of about 0.001 reciprocal lattice units.

the potential fitting procedure becomes much more sensitive to a proper choice of S, with the improved spectral predictions as a result.

It is interesting that the form of the equations of motion to solve according to the rigid ion or shell model is very similar. The shell model provides an equation of motion if the values of Z'_{eff} and $\epsilon(\infty)$ are known. Comparison of Z'_{eff} used in the calculations and that derived from the values found in quantum mechanical calculations (see table 1 and figure 11a), show considerable discrepancies. Interestingly however the shell model effective charges have significantly decreased compared to the full formal charges. The effective charge decreases with decreasing density. This is expected on the basis of the lower Madelung potentials.

The infrared spectra of α-quartz computed according to the rigid ion model with an effective charge Z'_{eff} and force constants derived from an ab-initio calculation on a $SiO_4H_4^+$ cluster[14] gives improved results[15] compared to the rigid ion model calculations based on empirical potentials (see fig. 9). Comparing experimental and computed plasmon frequencies shows rather good agreement if the computed values are divided by $\epsilon(\infty)$.

The small experimental values for the plasmon frequencies of α-quartz[16] (see table 1) indicate that chemical bonding in the silica polymorphs is dominated by the short range interactions. One can estimate the ratio of the Coulomb force constant and short range (two-body) force constant from the average energy of the longitudinal and transverse optical frequencies. For α-quartz this ratio is about 0.008. Indeed the calculated shell model lattice energies for α-quartz, silica-sodalite and silica-faujasite indicate a maximum difference in lattice energy of $\sim$20 kJ/mol SiO_2 (table 2). In view

Table 2. Contributions to lattice energy (kJ/mol SiO_2).

Species	Model	Coulomb	Two-body	Three body	Total
α-quartz	ref. 14[†]	-5419.6	5419.56	0.0[†]	-5183.6
	Rigid ion	-15412.8	3447.0	9.8	-11956.1
	Shell	-15298.7	2998.2	116.8	-12417.3
	(diff.)	-114.1	448.8	-107.0	461.2
Faujasite	Rigid ion	-15310.9	3402.6	3.4	-11911.7
	Shell	-15235.6	3020.1	204.0	-12397.5
	(diff.)	-102.3	382.5	-200.6	485.5

(diff.): Difference between Rigid ion and Shell model.
[†]: The potential of Tsuneyuki does not contain a threebody term and uses partial charges (0.6 times the formal charge).

of the much smaller differences in energy found for optimized silicon hydroxide ring systems[1b] in ab-initio calculations, the differences in lattice energy of silica zeolites mainly derive from electrostatic interactions. This is confirmed by analysis of the differences in energy using rigid ion and shell model calculations.

In a recent paper Johnson e.a.[17] and Patarin e.a.[18], comparing the experimental heats of formation of silicalite and α-quartz, find a difference of only 6 kJ/mol SiO_2. This is in agreement with the theoretical results.

In the last section implications of our finding that the long range electrostatic field has only a very small contribution to binding in silica-zeolites to the understanding of acidity will be discussed.

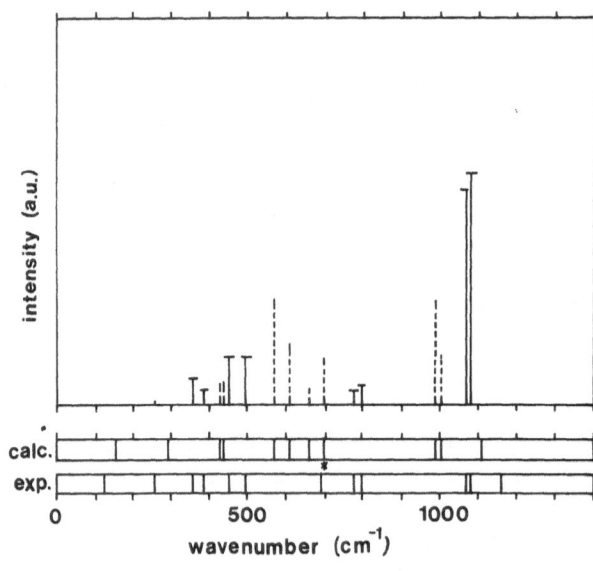

Figure 9. Infrared spectra of α-quartz. Solid lines: experimental[11] frequencies; Broken lines: calculated according to the potential derived[15] from ab-initio calculations of Tsuneyuki[14];

IMPLICATIONS FOR ACIDITY THEORY

The disagreement between the effective charges found on the basis of empirical shell model calculations and those based on potentials derived from ab-initio calculations indicates the need to use quantum chemical calculations to get reliable electrostatic field estimates.

An interesting approach along these lines is due to Goursot *e.a.*[5]. Using ab-initio quantum chemical cluster calculations they arrive at typical values of electrostatic stabilization of a proton next to an Al^{3+} containing tetrahedron of the order of 250 kcal/gat H^+.

In order to compute hydrogen bond forming properly, care has to be taken that local electric gradients are accurately computed. In addition the covalent nature of the O-H bond has to be accounted for. In order to do so for zeolites rather large clusters or embedding methods may have to be used.

To an analysis of acidity a breakdown of overall protonation into the following imaginary steps is helpful (see also ref. (2b)):

$$
\begin{array}{lclr}
BH & \longrightarrow & B & + H & \quad\quad I \\
B & + H & \longrightarrow & B^- & + H^+ & \quad\quad II \\
H^+ & + A & \longrightarrow & [HA]^+ & \quad\quad III \\
B^- & + H^+A & \longrightarrow & B^-\cdots[HA]^+ & \quad\quad IV \\
\end{array}
$$

$$BH + A \longrightarrow B^-\cdots[HA]^+$$

Step I is covalent dissociation, step II is ionization, step III protonation and step IV stabilization of the protonated complex. Step II, ionization, depends strongly on the difference in electrostatic potential of an electron on the proton position, compared to an electron on the oxygen in the zeolite lattice. So for protonation, in this step not only the proton potential, but also stabilization of the negative charge on the zeolite lattice is important. The difference in potential relates to the Madelung potential of the B^-H^+ pair. Earlier we estimated differences in acidity of protons in mordenite and montmorillonite on the basis of differences in Madelung potential[19]. Of course differences in the electrostatic field the $[HA]^+$ complex experiences are of importance, but also proper calculation of the Born repulsion energy[20] of the $B^-\cdots[HA]^+$ complex. That stabilization of the protonated complex by the negatively charged zeolite wall is important, has been proposed by Kazansky[1f,2d]. Stabilization of the protonated complex by the negatively charged zeolite wall replaces hydratation of the generated ions that will occur in water.

In order to discuss the importance of electrostatic effects it is useful to consider the effect of cluster choice and zeolite lattice composition on the ionization step. Then we are interested in the difference in potential energy of an electron on the proton position and the oxygen atom.

Whereas electrostatic interactions are long range, it appears that changes in Madelung potential by the generation of surfaces[21] or vacancies[3c] remain limited to finite disturbances. Changes in electrostatic potential disappear in the third or fourth coordination shell with respect to the disturbance. Evjen[22] demonstrated that reasonable estimates of the Madelung potential can be made by choosing the charges on the neighbouring atoms such, that the total clusters remain neutral. We have shown[21] that differences in Lewis acidity of TiO_2 surfaces can be understood in a purely electrostatic model from the differences in local environment of the oxygen atoms coordinated to the Ti ions reacting with the base. Also Madelung calculations of the lattice energy of the zeolite lattices with varying cation content indicated the dominance of short range effects[3a]. Only when cations enclosed in the zeolite channel directly contact each other large repulsive effects are observed.

Semi-empirical calculations demonstrated[23] that the effect of aluminum concentration on acidity is dominated by the number of aluminum tetrahedra neighbouring the $[_{Si}-\overset{H}{\underset{}{O}}-_{Al}]$ unit. The larger the number of aluminum tetrahedra next to the silicon tetrahedron, the lower the acidity. This was rationalized on basis of step II. The negative charge of the Si-O-Al bridging oxygen atom is better accomodated if its local environment becomes electron deficient. Substitution of Si^{4+} lattice ions by Al^{3+} cations has demonstrated more highly charged oxygen atoms next to neighbour tetrahedron Si ions. As a result the charge of Si decreases and it is more difficult to accomodate the negative lattice charge left upon protonation.

This is illustrated in figure 10.

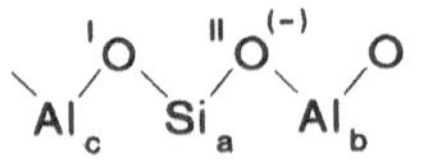

Figure 10. Oxygen sites in a alumino-silicate.

With an aluminum ion in position **c**, instead of a Si ion, the negative charge on the oxygen ion in position II will become larger. As a result the positive charge on the Si atom in position **a** will become reduced. The increased negative charge in position II and the decreased positive charge on position **a** will destabilize the negative charge on position I. This reduces the stability of B^- in step II. This will unfavourably affect acidity. Mortier e.a.[2i] have also emphasized the importance of the negative charges on the zeolite lattice oxygens.

We will analyse this using the results of fully angle and distance optimized STO-3G ab-initio calculations on three tetrahedra containing silica and alumino-silicate rings.

In figures 11 and 12 the charge distributions and equilibrium configurations are shown in the absence and presence of aluminum and compensating ions for the three-tetrahedron ring systems. The tetrahedra are connected with bridging oxygen atoms and are closed by two OH-groups. For details of the computations we refer to refs. 24 and 1b.

Relevant to theoretical computations are not only the different charges of the bridging oxygen atoms and the increase in charge with a neighbouring aluminum atom (compare e.g. fig. 11a, b), but also the very different charges of oxygen atoms in the hydroxyl groups. The electrostatic potential in the proton position as well as oxygen atom to which it is attached will significantly depend on the charges of the hydroxyl groups.

The OH terminated cluster has Si and Al charges nearly twice that of the hydrogen terminated clusters. The large charge differences computed for oxygen-hydroxyl charges and bridging oxygen atoms raise the question whether part of the embedding approaches using effective Madelung fields[2c,j,k] to enhance acidity, partiallly compen-

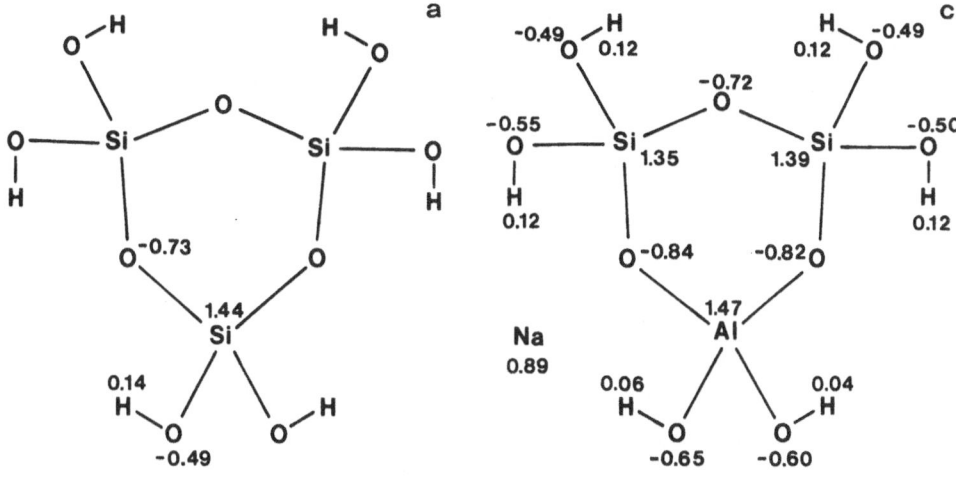

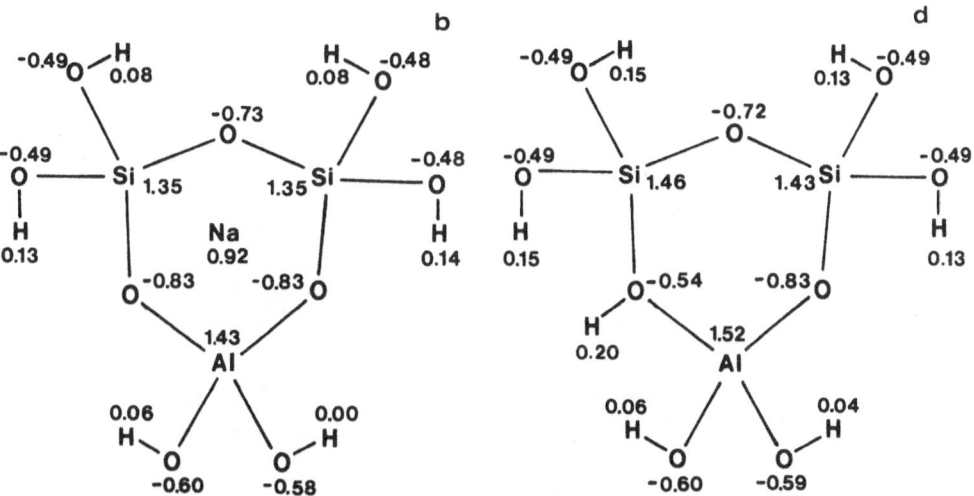

Figure 11: Charge distributions (schematic structure).
a: Aluminumfree ringsystem: $Si_3O_3(OH)_6$.
b: Na-aluminosilicate ring. Na^+ symmetrically coordinated:
$AlSi_2O_3(OH)_6^- + Na^+$.
c: Na-aluminosilicate ring. Na^+ asymmetrically coordinated:
$AlSi_2O_3(OH)_6^- + Na^+$.
d: The protonated aluminosilicate ring:
$AlSi_2O_3(OH)_6^- + H^+$.

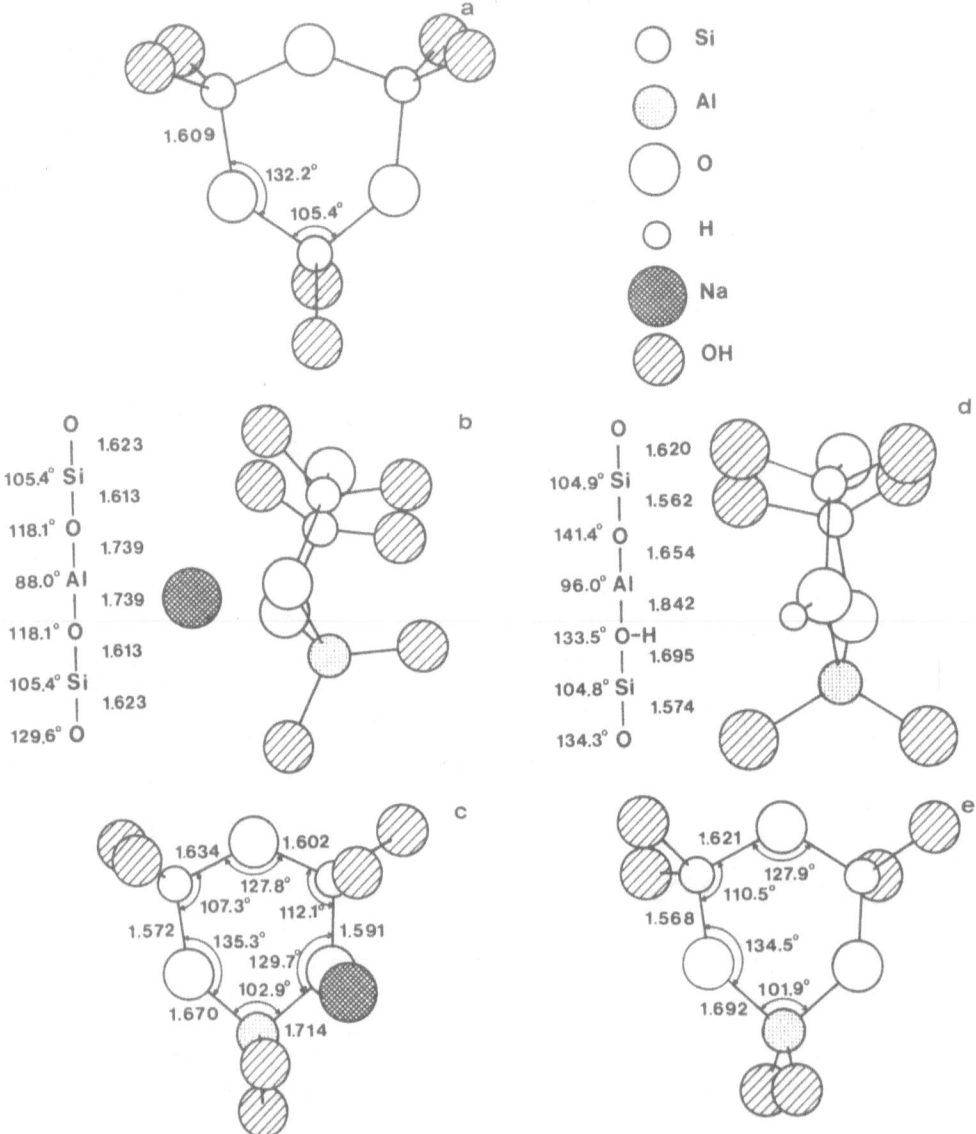

Figure 12. The optimized geometries of the three tetrahedron ring systems (angles in degrees and distances in Å).

a: Aluminumfree ringsystem: $Si_3O_3(OH)_6$ (C_{3v}). Top view.

b: Na-aluminosilicate ring. Na^+ symmetrically coordinated: $AlSi_2O_3(OH)_6^- + Na^+$ (C_s). Side view.

c: Na-aluminosilicate ring. Na^+ asymmetrically coordinated: $AlSi_2O_3(OH)_6^- + Na^+$. Top view.

d: The protonated aluminosilicate ring: $AlSi_2O_3(OH)_6^- + H^+$. Side view.

e: Aluminosilicate ring geometry in the absence of compensating positive charge: $AlSi_2O_3(OH)_6^-$ (C_s). Top view.

sate for the inadequacies of the electrostatic charge distributions in the OH, H or O^- terminated clusters sometimes used.

Noteworthy is the near invariance of the charge on the oxygen atoms if no change occurs in their nearest neighbour coordination. The charge on the Si bridging oxygen atom remains -0.73 ± 0.01 for all clusters. The charge on the Si-Al bridging oxygen atoms becomes -0.83 ± 0.01 in all clusters, except on the oxygen atom to which the proton is coordinated. Now the bridging oxygen charge becomes -0.54, because charge is transferred to the covalently bonded proton. Note that this proton charge is ~0.05 higher than that of the protons of the hydroxyl groups coordinated to Si.

The computed effective charge on the aluminum atoms becomes larger than that on the silicon atoms, because the aluminum atoms have a larger electron donation tendency and a lower formal charge. The charge on the silicon atoms in the aluminosilicate clusters is reduced. This remains also the case for the silicon atom in the tetrahedron that is not bounded with the proton, if the cluster becomes protonated. It illustrates the changes of the charge distribution in a silica tetrahedron due to aluminum substitution in the next tetrahedron. However the changes in the charge distribution are relatively small. As figures 12 show and will be discussed shortly changes in covalent bonding may also contribute to the weaker acidity at high aluminum concentration.

Inspecting figures 12 one observes also that apart from the changes in local electrostatic potential due to altered electron distributions on oxygen atoms sharing the same tetrahedra as the bridging oxygen atoms to which the proton is attached, proton transfer causes significant relaxation of atom distances. The relaxation of distances comparing smaller protonated and non-protonated clusters have also been calculated by Mortier e.a.[2i]. As figures 12 show, in comparison to the all silicon cluster, the negatively charged cluster with an aluminum atom has the expected larger Al-O distance (Al-O = 1.692 Å; Si-O = 1.609 Å). The atom distances are slightly altered by the presence of Na^+ ions. The Na^+ ions find minima in two positions. One symmetrically with respect to the cluster (fig. 12b), the other in between an OH group and a bridging oxygen atom (fig. 12c). In the symmetric position the cluster buckles. The bending away of the highly charged cations indicates the importance of electrostatic interactions between the Na ion and Si and Al ions.

The changes in bond lengths are much larger if the cluster becomes bonded to a hydrogen atom (fig. 12d). This is indicative for the strong O-H bond strength, with a resulting weakening of the neighbouring Si-O and Al-O bonds, expected on the basis of Bond Order Conservation[25]. The OH-group is located in the Si-O-Al plane and no buckling of the clusters is seen (fig. 12d) in contrast to the Na^+ case (fig. 12b). Clearly relaxation of the distances and angles upon protonation is important.

Figure 12d shows the long Al-O distance (1.842 Å) if the bridging oxygen atom has the proton attached to it. The other Al-Si tetrahedron bridging oxygen atom has a much shorter Al-O distance (1.65 Å). These changes have been explained above. The Si-O bond sharing the same oxygen atom becomes much shorter than in the aluminumfree ring system (1.56 Å versus 1.60 Å). Again because of the Bond Order Conservation principle the other tetrahedron connecting Si-O bond lengthens (1.64 Å versus 1.60 Å). If the bridging oxygen atom at the end of this bond would have been connected to another aluminum containing tetrahedron and would be protonated, (the imaginary case not studied here), the weakening of the Si-O bond due to the presence of the other neighbouring aluminum containing tetrahedron would strengthen the OH bond compared to the case that the latter tetrahedron would contain silicon. This implies a lower Brønsted acidity. So surrounding $[_{Si-\overset{H}{O}-Al}]$ sites with aluminum containing tetrahedra will strengthen the OH bond, also because of changes in covalency. The changes in T-O bond length indicate that this may be considerable. Clearly we are dealing with effects of rather short range nature.

Note that the effects of alternating bond lengthening and bond shortening are quantum mechanical. They relate to Friedel oscillations, familiar of impurity scattering in metals[26,27]. The use of classical two body terms would predict a different behaviour from that found in the cluster calculations. The minimal basis set used in our cluster calculations implies that our results have to be considered of qualitative rather than quantitative nature. especially the charges computed will differ if larger basis sets are used.

Whereas bond length and bond angle relaxation may occur also if silica-alumina rings are embedded in a zeolite lattice, restraints may decrease the bond distance relaxation effects found for the free clusters. This has been suggested to be the case in zeolites[1e]. Our calculations however indicate significant driving forces for deformation. The experimentally[28] observed changes in unit cell dimensions of dealuminated zeolite lattices indicate that the zeolite lattice may adjust to the differences in Si-O and Al-O distance. However this may vary for different zeolite lattices (figure 13 and 14).

In systems with little possibility for relaxation, the Si-O-Al angle ϕ may be constrained by its environment, e.g. the size of the rings it shares. In such a case hybridization of the oxygen atom electrons depends on the bond angle $\phi^{(1g,29)}$. As a result the OH bond strength as well as the oxygen charge are less if the Si-O-Al angle is 180° (sp-hybridization). Hence one expects the higher acidity for the largest bond angles. Indeed, ab-initio calculations by O'Malley and Dwyer[28] show such a lowering of the bridging OH frequencies in hydrogen terminated Al, Si ditetrahedral clusters with increasing angle ϕ. Beran[1h,1i,1j,1k,1l,1m,1n], using semiempirical methods, found smaller OH dissociation energies in silica-aluminate rings modelling ZSM-5 (five-rings) than faujasite, containing dominantly four-rings. As found from ab-initio calculations[1b] as well as experiment[30,31], the T-O-T angle for the five ring is larger, so indeed on the basis of hybridization arguments a weaker bond is expected. Dwyer[23a] as well as Sohn e.a.[28b], using infrared spectroscopy, have shown that dealumination of faujasite structures results in protons with an acidity slightly weaker than that of ZSM-5.

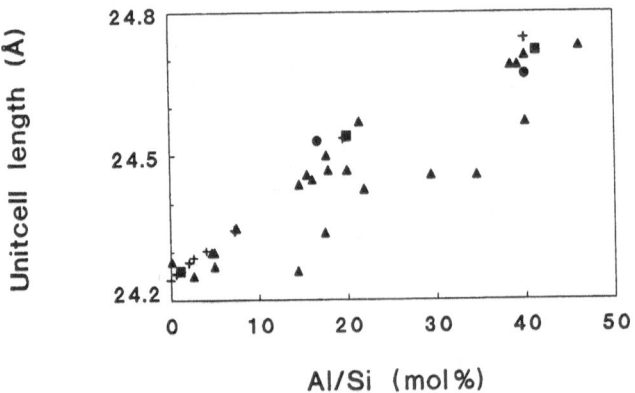

Figure 13. Unit cell dimensions *versus* framework Al/Si ratio for faujasite.
•: ref. 28a; +: ref. 28b; ■: ref. 28c; ▲: ref. 28d.

The arguments presented indicate that in addition to electrostatic contributions to the changes in energy of protonation reactions changes of covalent nature also may

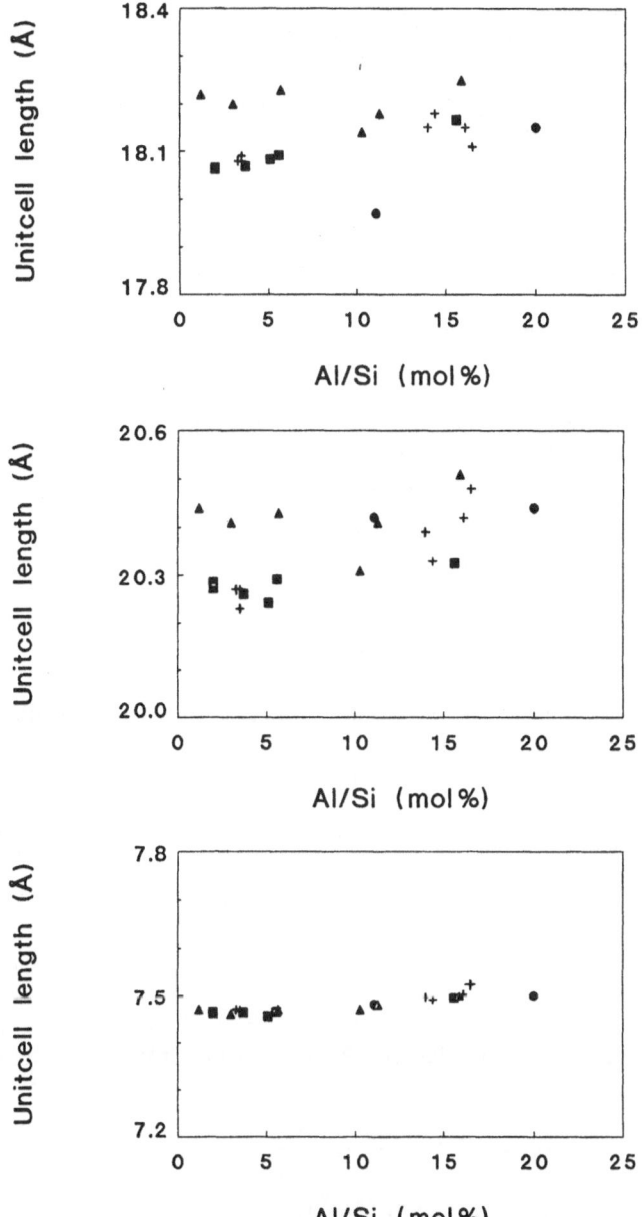

Figure 14. Unit cell dimensions *versus* framework Al/Si ratio for mordenite.
●: ref. 28e; +: ref. 28f; ■: ref. 28g; ▲: ref. 28h.
a: Length of a-axis;
b: Length of b-axis;
c: Length of c-axis;

be significant (see also ref. (20)). Both will be predominantly of a local nature limited to a range of a few lattice atom distances. The presence of cations in the cavity close to the protonation site may also effect protonation. If close they will destabilize the protonated molecule formed in step IV.

According to the view presented zeolite acidity may be affected by four factors:

1: Framework composition;
2: Channel cation composition;
3: Framework relaxability and ring size;
4: Channel dimension.

Our analysis has not considered acidity due to the presence of hydrolizable H_2O [32].

Experimental evidence for the importance of the first two factors is abundant[33]. In the first case changes in lattice oxygen charge occur but also changes in covalent bonding. In the second case destabilization of protonated molecules by the close presence of cations may become important.

Little is known about the presence of lattice strain in zeolites. However proof of lattice relaxation under the influence of adsorbed organic molecules does exist[34]. Stabilization of carbenium ions due to zeolite lattice walls has not yet been demonstrated experimentally. However theoretical (Derouane[35]) as well as experimental evidence (Stach[28c]) is available of lattice wall stabilization of adsorbed organic molecules.

APPENDIX

In this appendix the relation between the formulas of Brüesch[12] for alkali-halides and formulas (5a) and (5b) will be shown.

Also an outline is given of the procedure to be used to extract Z'_{eff} from the calculations.

Brüesch[12] finds for the shell model:

$$\mu\omega_{LO}^2 = S^* + \frac{8\pi(Z^*e)^2}{3v_a} \frac{1}{1 + (8\pi\alpha^*/3v_a)} \tag{A1a}$$

$$\mu\omega_{TO}^2 = S^* - \frac{4\pi(Z^*e)^2}{3v_a} \frac{1}{1 - (4\pi\alpha^*/3v_a)} \tag{A1b}$$

With the relation of Clausius-Mosotti (9), these two relations can be rewritten as:

$$\mu\omega_{LO}^2 = S^* + \frac{8\pi(Z^*e)^2}{3v_a} \frac{(\epsilon(\infty) + 2)}{3\epsilon(\infty)} \tag{A2a}$$

$$\mu\omega_{TO}^2 = S^* - \frac{4\pi(Z^*e)^2}{3v_a} \frac{(\epsilon(\infty) + 2)}{3} \tag{A2b}$$

We wish to rewrite formulas (A2a) and (A2b) in the form

$$\mu\omega_{LO}^2 = S' + \frac{8\pi(Z'_{\text{eff}}e)^2}{3v_a\epsilon(\infty)} \tag{A3a}$$

$$\mu\omega_{TO}^2 = S' - \frac{4\pi(Z'_{\text{eff}}e)^2}{3v_a\epsilon(\infty)} \tag{A3b}$$

These formulas are equal to (5a) and (5b).

Substracting (A2b) from (A2a) and (A3b) from (A3a) gives:

$$\mu\omega_{LO}^2 - \mu\omega_{TO}^2 = \frac{4\pi(Z^*e)^2}{9v_a}\frac{(\epsilon(\infty)+2)^2}{\epsilon(\infty)} = \frac{4\pi(Z'_{eff}e)^2}{v_a\epsilon(\infty)} \tag{A4}$$

so one obtains an expression for Z'_{eff}:

$$(Z'_{eff})^2 = \left(\frac{(\epsilon(\infty)+2)Z^*}{3}\right)^2 \tag{A5}$$

Z'_{eff} is also called the *transverse effective charge*, Z^* is known as the *Szigeti charge*. Adding formula (A2a) and two times (A2b), and (A3a) and two times (A3b) respectively, results in:

$$\mu(\omega_{LO}^2 + 2\omega_{TO}^2) = 3S^* + \frac{8\pi(Z^*e)^2}{3v_a}\frac{(\epsilon(\infty)+2)}{3}\frac{(1-\epsilon(\infty))}{\epsilon(\infty)} = 3S' \tag{A6}$$

finally giving

$$S' = S^* + \frac{8\pi(Z^*e)^2}{9v_a}\frac{(\epsilon(\infty)+2)}{3}\frac{(1-\epsilon(\infty))}{\epsilon(\infty)} \tag{A6}$$

So one derives that it is possible to put the shell model formulas (A1a) and (A1b) of Brüesch in a form (formulas (5a) and (5b)) that is very similar to the rigid ion model formulas (4a) and (4b).

The equations given above apply to diatomic cubic systems. In a more general case the plasmon frequency $\omega_{plasmon}$ is given by[36]:

$$\langle\omega_{LO}^2\rangle - \langle\omega_{TO}^2\rangle = \frac{4\pi}{v_a\epsilon(\infty)}\sum_i\frac{(Z'_{eff})_i^2 e^2}{m_i} = \omega_{plasmon}^2 \tag{A7}$$

where $(Z'_{eff})_i$ and m_i are the effective charge and the mass of ion i, respectively, and the summation runs over all the ions in the unit cell. Equation A7 can be rewritten as

$$\langle\omega_{LO}^2\rangle - \langle\omega_{TO}^2\rangle = \frac{4\pi(Z'_{eff})^2}{\mu'v_a\epsilon(\infty)} = \omega_{plasmon}^2 \tag{A8}$$

where μ', **not** the ordinary reduced mass, is fixed for a given stoichiometry. The ratio Z'_{eff}/Z can now be obtained from a combination of a rigid ion and a shell model phonon calculation.

$$\frac{(\langle\omega_{LO}^2\rangle - \langle\omega_{TO}^2\rangle)^{shell}}{(\langle\omega_{LO}^2\rangle - \langle\omega_{TO}^2\rangle)^{rigid}} = \frac{1}{\epsilon_L(\infty)^{shell}}\frac{(Z'_{eff})^2 v_a^{rigid}}{Z^2 v_a^{shell}} \tag{A9}$$

Because the relaxed lattice is different for the rigid ion and shell model calculations two different values of v_a are used. The results of these calculations are given in table 1.

References

1a   M.D. Newton, M. O'Keeffe, and G.V. Gibbs, <u>Phys. and Chem. of Miner.</u> 6:305 (1980).

b   B.W.H. van Beest, J. Verbeek, and R.A. van Santen, <u>Catal. Lett.</u> 1:147 (1988).

c   J.G. Fripiat, F. Berger-André, J. André, and E.G. Derouane, <u>Zeolites</u> 3:306 (1983).

d   J. Sauer, and R. Zahradnik, <u>Int. J. Quantum Chem.</u> 26:793 (1984).

e   A.G. Pelmenshchikov, E.A. Paukshtis, V.S. Stepanov, K.G. Ione, G.M. Zhidomirov, and K.I. Zamaraev, in: "Proc. 9th Int. Congr. on Catalysis," M.J. Philips and M. Ternane ed., The Chemical Institute of Canada, 1:404 (1988).

f   G.M. Zhidomirov, and V.B. Kazansky, <u>Adv. Catal.</u> 34:131 (1986).

g   G. Ooms, and R.A. van Santen,. <u>Recl. Trav. Chim. Pays Bas</u> 106:69 (1987).

h   S. Beran, P. Jírŭ, and B. Wichterlová, <u>J. Phys. Chem.</u> 85:1951 (1981).

i   S. Beran, <u>J. Phys. Chem.</u> 85:1956 (1981).

j   S. Beran, P. Jírŭ, B. Wichterlová, <u>React. Kinet. Catal. Lett.</u> 18:51 (1981).

k   S. Beran, <u>J. Phys. Chem.</u> 86:111 (1982).

l   S. Beran, <u>Chem. Phys. Lett.</u> 91:86 (1982).

m   S. Beran, <u>Z. Phys. Chem. Neue Folge</u> 130:81 (1982).

n   S. Beran, <u>Z. Phys. Chem. Neue Folge</u> 137:89 (1983).

p   G.V. Gibbs, <u>Amer. Miner.</u> 67:421 (1982).

2a   J. Sauer, <u>J. Phys. Chem.</u> 91:2315 (1987).

b   J. Sauer, and W. Shirmer, <u>Stud. Surf. Sci. Catal.</u> 37:323 (1988).

c   E. Kassab, K. Seiti, and M. Allavena, <u>J. Phys. Chem.</u> 59 (1989), in press.

d   V.B. Kazansky, <u>Bulg. Acad. of Sciences Comm. Rep. Chem.</u> 13:19 (1980).

e   P.J. O'Malley, and J. Dwyer, <u>Chem. Phys. Lett.</u> 143:97 (1988).

f   P.J. O'Malley, and J. Dwyer, <u>Zeolites</u> 8:317 (1988).

g   P.J. O'Malley, and J. Dwyer, <u>J. Phys. Chem.</u> 92:3005 (1988).

h   W.J. Mortier, P. Geerlings, C. van Alsenoy, and H.P. Figeys, <u>J. Phys. Chem.</u> 83:3257 (1979).

i   W.J. Mortier, J. Sauer, J.A. Lercher, and H. Noller, <u>J. Phys. Chem.</u> 88:905 (1984).

j   R. Vetrivel, C.R.A. Catlow, and E.A. Colbourn, <u>Stud. Surf. Sci. Catal.</u> 37:309 (1988).

k   R. Vetrivel, C.R.A. Catlow, and E.A. Colbourn, to be published.

l   I.D. Mikheikin, I.A. Abronin, G.M. Zhidomirov, and V.B. Kazansky, <u>Kinet. Katal.</u> 18:1580 (1977).

m   I.D. Mikheikin, I.A. Abronin, A.I. Lumpuv, and G.M. Zhidomirov, <u>Kinet. Katal.</u> 19:1050 (1978).

n   K. Seiti, Thesis, Univ. Paris VI, 1988.

p   K.A. van Genechten, Thesis Univ. Leuven (1987).

q   K.A. van Genechten, and W.J. Mortier, <u>Zeolites</u> 8:273 (1988).

3a   G. Ooms, R.A. van Santen, C.J.J. den Ouden, R.A. Jackson, and C.R.A. Catlow, <u>J. Phys. Chem.</u> 92:4462 (1988).

b   E. Dempsey, <u>J. Phys. Chem.</u> 75:3660 (1969).

c   C.R.A. Catlow, and W.C. Mackrodt, "Computer Simulation of Solids," Lecture Notes on Physics, Springer Verlag, Berlin (1982).

d   G. Ooms, R.A. van Santen, R.A. Jackson, and C.R.A. Catlow, <u>Stud. Surf. Sci. Catal.</u> 37:317 (1988).

e   M.J. Sanders, "Computer Simulation of Framework Structured Minerals," PhD Thesis Univ. of London (1984).

f   R.A. Jackson, and C.R.A. Catlow, <u>Molec. Simulation</u> 1:207 (1988).

g   A.J.M. de Man, B.W.H. van Beest, M. Leslie, and R.A. van Santen, submitted.

h   C.R.A. Catlow, M. Doherty, G.D. Price, M.J. Sanders, and S.C. Parker, <u>Mater. Sci. Forum</u>, 7:163 (1986).

4   C. Pisani, R. Dovesi, and C. Roetti, "Hartree-Fock ab initio treatment of crystalline systems," Lecture Notes in Chemistry 48, Springer Verlag, Berlin (1988).

5   A. Goursot, F. Fajula, C. Daul, and J. Weber, <u>J. Phys. Chem</u> 92:4456 (1988).

6a   C.S. Blackwell, <u>J. Phys. Chem.</u> 83:3251 (1979).

b   C.S. Blackwell, <u>J. Phys. Chem.</u> 83:3257 (1979).

c   P. Demontis, G.B. Suffritti, S. Quartieri, E.S. Fois, and A. Gamba, <u>Zeolites</u> 7:522 (1987).

d   P. Demontis, G.B. Suffritti, S. Quartieri, E.S. Fois, and A. Gamba, <u>J. Phys. Chem.</u> 92:867 (1988).

e   K.T. No, D.H. Bae, and M.S. Jhon, <u>J. Phys. Chem.</u> 90:1772 (1986).

7   M. Mabilia, R.A. Pearlstein, and A.J. Hopfinger, <u>J. Am. Chem. Soc.</u> 109:7960 (1987).

8   R.A. van Santen, and D.L. Vogel, <u>Adv. Solid State Chem.</u> I:151 (1988).

9   R. Haydock, V. Heine, and M.J. Kelley, <u>J. Phys. C: Solid State Physics</u> 5:2845 (1972).

10   R. Haydock, V. Heine, and M.J. Kelley <u>J. Phys. C: Solid State Physics</u> 8:2591 (1975).

11   J. Etchepare, and M. Marian, <u>J. Chem. Phys.</u> 60:1873 (1974).

12   P. Brüesh, "Phonons: Theory and experiments. I," Springer Series Solid State Sciences 34, Springer Verlag, Berlin (1982).

13   A.A. Maraduddin, E.W. Montroll, G.H. Weiss, and I.P. Ipatova "Theory of lattice dynamics in the harmonic approximation", Academic Press, New York (1971).

14   S. Tsuneyuki, M. Tsukada, H. Aoki, and Y. Matsui, <u>Phys. Rev. Letters</u> 61:869 (1988).

15   G.J. Kramer, and A.J.M. de Man, unpublished results.

16   M.E. Striefler, and G.R. Barsch, <u>Phys. Rev. B</u> 12:4553 (1975).

17   G.K. Johnson, I.R. Tasker, D.A. Howell, and J.V. Smith, <u>J. Chem. Thermodynamics</u> 19:617 (1987).

18   J. Patarin, M. Soulard, H. Kessler, J.L. Guth, and M. Diot, <u>Thermochimica Acta</u>, in press.

19   R.A. van Santen, <u>Recl. Trav. Chim. Pays-Bas</u> 101:157 (1982).

20   C.E. Dijkstra, <u>Acc. Chem. Res.</u> 21:355 (1988).

21   J. Woning, and R.A. van Santen, <u>Chem. Phys. Lett.</u> 101:541 (1983).

22   H.M. Evjen, <u>Phys. Rev.</u> 39:675 (1932).

23a   J. Dwyer, <u>Stud. Surf. Sci. Catal.</u> 37:333 (1988).

b   P.A. Jacobs, <u>Catal. Rev.-Sci. Eng.</u> 24:415 (1982).

24   B.W.H. van Beest, J. van Lenthe, and R.A. van Santen, to be published.

25   E. Shusterovich, <u>Surf. Sci. Rep.</u> 6:1 (1986).

26   R.A. van Santen, <u>J. Chem. Soc. Faraday Trans.</u> 83:1915 (1987).

27   J. Friedel, <u>Nuovo Cimento Suppl.</u> 7:286 (1958).

28a   P. Gallezot, R. Beaumont, and D. Barthomeuf, <u>J. Phys. Chem.</u> 78:1550 (1974).

b   J.R. Sohn, S.J. DeCanio, J.H. Lunsford, and D.J. O'Donnel, <u>Zeolites</u> 6:225 (1986).

c   H. Stach, U. Lohse, H. Thamm, and W. Schirmer, <u>Zeolites</u> 6:74 (1986).

d L. Kubelková, V. Seidl, G. Borbély, and H.K. Beyer, <u>J. Chem. Soc. Faraday Trans.</u> 84:1447 (1988).

e B.H. Ha, J. Guidot, and D. Barthomeuf, <u>J. Chem. Soc. Faraday Trans. I</u> 75:1245 (1979).

f R.W. Olsson, and L.D. Rollmann, <u>Inorganic Chemistry</u> 16:651 (1977).

g B.L. Meyers, T.H. Fleisch, G.J. Ray, J.T. Miller, and J.B. Hall, <u>J. Catalysis</u> 110:82 (1988).

h M. Musa, V. Tarină, A.D. Stoica, E. Ivanov, E. Ploştinaru, E. Pop, Gr. Pop, R. Ganea, B. Bîrjega, G. Muscă, and E.A. Paukshtis, <u>Zeolites</u> 7:427 (1987).

29 J.R. Durig, M.J. Flanigan, and V.F. Kalasinsky, <u>J. Chem. Phys.</u> 66:2775 (1977).

30 C. Baerlocher, <u>in</u>: "Proc. 6th Int. Zeol. Conf. Reno," 823 (1983).

31 W.H. Baur, <u>Amer. Miner.</u> 49:697 (1964).

32 P.A. Jacobs, "Carboniogenic Acidity of Zeolites," Elsevier, (1977).

33 P.A. Jacobs, and W.J. Mortier, <u>Zeolites</u> 2:226 (1982).

34a J. Klinowski, <u>Progr. NMR Spectr.</u> 16:237 (1984).

  b C.A. Fyfe, H. Strobl, G.T. Kokotailo, G.J. Kennedy, and G.E. Barlow, <u>J. Am. Chem. Soc.</u> 110:3373 (1988).

35a E.G. Derouane, <u>J. Catal.</u> 100:541 (1986).

35b E.G. Derouane, J.M. André, and A.A. Lucas, <u>J. Catal.</u> 110:58 (1988).

36 M.F. Thorpe, and S.W. de Leeuw, <u>Phys. Rev. B</u> 33:8490 (1986).

CONFINEMENT EFFECTS IN SORPTION AND CATALYSIS BY ZEOLITES

Eric G. Derouane

Facultés Universitaires N.-D. de la Paix, Laboratoire de Catalyse
Rue de Bruxelles, 61
B-5000 Namur. Belgium

ABSTRACT

Molecular sieves, including zeolites, are distinct from other sorbents and catalysts by the curvature at the atomic scale of the internal surface which they offer to the molecules they host. Surface curvature leads to confinement effects when the size of the host structure (the pores of the framework) and that of the guest molecule (reactant, reaction intermediate, product) become comparable. Following a quantitative description of confinement and its consequences, topics and examples selected from the literature on physisorption and catalysis are discussed. It is shown that intimate relationships may exist between the host molecular sieve framework and the conformation of the guest molecules in the intracrystalline free volume. It is proposed that zeolite should be considered as solid enzymes, i.e., systems in which non-bonding or supermolecular interactions with guest molecules play a major role in sorption and catalytic activation.

INTRODUCTION

The sorption and catalytic properties of zeolites result from an efficiency and a selectivity which are intimately dependent on the way by which molecular sieves operate[1-3]. These properties depend namely on the number, the quality, and the accessibility of the sorption and catalytic sites which are available. They are determined by the zeolite framework (topology and chemical composition) and by the sorbate-sorbent (guest-host) interactions.

We recently proposed that for sorbates in microporous solids such as zeolites, molecules and their environment tend to optimize their van der Waals interaction[4]. We showed in particular that van der Waals interactions are amplified by the curvature of the pore walls with which the molecules interact[3,5]. If this leads to recognition of or preorganization on specific sites[6] by the sorbates or other molecules (reactants, intermediates, products), it is expected that zeolites may be the locus of remarkable effects which we will refer to as *confinement sorption and/or catalysis*. These effects, as presented in the discussion below, resemble in some of their aspects those encountered in the action of enzymes[2,7]. They are distinct from the well-known molecular shape selectivity of zeolites[8-11].

Consequently, an unitary view emerges for the properties of zeolites and molecular sieves as selective sorbents and catalysts. *Molecules in zeolites* must be considered as *solvated* by the surrounding framework (the solvent), which affects their physical and chemical properties, and the molecular sieve itself shows some *resemblance to a solid enzyme* . Indeed, its site properties can be varied both in quantity and quality, and the site environment can also be optimized by choosing the most appropriate zeolite or by

Guidelines for Mastering the Properties of Molecular Sieves
Edited by D. Barthomeuf *et al.*
Plenum Press, New York, 1990

225

molecularly engineering a given molecular sieve[12]. We offer in this paper a status review of the above proposals which treat the molecule(s)-zeolite system in terms of guest-host relationship.

A SIMPLE MODEL FOR SURFACE CURVATURE AND CONFINEMENT EFFECTS

Molecules within the free intracrystalline volume of zeolites interact primarily (via van der Waals interaction) with the anionic oxygens of the pore walls, which are the most polarizable species of the framework[13]. The latter may be considered as a weak but soft base[14]. The molecules must be considered as solvated by the framework as their sizes compare to those of the pores. Such a picture is a simple extension of the reaction field concept of Onsager[15] in which the system is depicted as a solute molecule immersed in a solvent continuum. Spontaneous electric moments which arise in the solute induce a reaction field in the solvent. As a response, the reaction field affects the properties of the molecule.

Considering the interaction of a point molecule of polarizability $\alpha(\omega)$ with a spherical cavity in a dielectric zeolite continuum, the van der Waals physisorption energy $W(s)$ of the molecule in this curved environment is given within reasonable approximation by[2,3,5]:

$$W(s) = -(C/4d^3)(1-s/2)^{-3} \tag{1}$$

with $s = d/a$ the surface curvature factor, d the sorption distance, and a the effective (spherical) pore radius. $(C/4d^3)$ is a molecular constant which can be factorized in two terms, C_z and C_m, characteristic of the sorbent and of the sorbate, respectively. Thus:

$$W(s) = -C_z.C_m.W_r(s) = W(0).W_r(s) \tag{2}$$

in which $W(0)$ is the physisorption energy for the flat surface (s=0) case and:

$$W_r(s) = (1-s/2)^{-3} \tag{3}$$

Similarly, the radial sticking force $F(s)$ acting on the sorbed molecule, relative to the flat surface value $F(0)$, is given by[2,3,5]:

$$F(s) = F(0).F_r(s) \tag{4}$$

with:

$$F_r(s) = (1-s)(1-s/2)^{-4} \tag{5}$$

The following equations may be used to evaluate s or approximate its value for non-spherical pores[2,16]:

$$d°(Å) = r_m(Å) + 1.38 \tag{6}$$

$$a(Å) = r_p(Å) + 1.38 \tag{7}$$

with 1.38 Å the O^{2-} anion radius, r_m and r_p the sorbate critical dimension and the crystallographic pore radius, respectively. d in the curvature factor s may be related to d° approximately as $d = 2^{1/6}d° \approx 1.12246d°$ for a (6,12) Lennard-Jones interaction potential. r_p is the crystallographic radius for a spherical cavity or its corrected value for a cylindrical pore of crystallographic radius r_c, i.e., $r_p = 1.5^{1/3}r_c$. For elliptical pores, the average radius r_c is given by[17]:

$$r_c = (a^2/b)(1 - 3m/4 + 9m^2/64 + 5m^3/256 + \varepsilon) \tag{8}$$

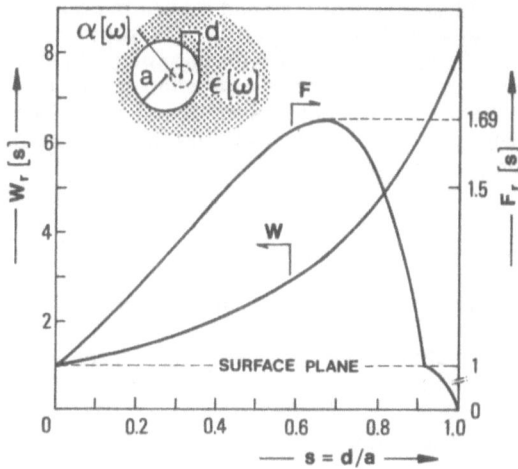

Fig.1. Dependence of $W_r(s) = W(s)/W(0)$ and $F_r(s) = F(s)/F(0)$ on the curvature parameters $= d/a$, i.e., as a function of pore size for a fixed sorption distance d. The degree of confinement increases with the value of s. (Adapted from ref. 5).

with $m = 1- (b/a)^2$ and ε smaller than 2.10^{-2}, where a and b are the ellipse long and short axes, respectively.

Figure 1 shows the dependence of $W_r(s)$ and $F_r(s)$ on the surface curvature s.

The major conclusions which emerge from this simple model and emphasize the molecule(s)-zeolite guest(s)-host relationships are that [2,3]:

i.   confinement effects resulting from the amplification of the physisorption energy become important when the value of s exceeds about 0.25; then, the molecule(s) and the zeolite will adapt their conformations in order to minimize the system energy; confinement is thus at the origin of the molecule-zeolite guest-host relationship;

ii.   the surface curvature effect has both energetic ($W_r(s)$) and dynamic ($F_r(s)$) implications;

iii. the center of a pore is never a stable equilibrium position unless both the pore and the molecule have similar size (s=1); in the latter (tight fit) case, $F_r(1) = 0$ and molecules may be considered as supermobile or "floating";

iv.   for s < 0.9, given the large increase in sorption energy relative to the flat surface situation, molecular displacement in the pores for a polysegmental molecule will be a "creeping" motion effected by local deformation of molecular segments while the molecule sticks to the pore wall; spherical molecules will hop over the atomic corrugation barriers of the pore wall; these considerations hold for $|W(s)|/k > T$;

v.   defining the "specific surface area" of a zeolite has no significance, the sorbed molecule interacting with its whole environment; what matters is the sorption volume. The molecule is "solvated" by its surroundings.

Recently, the model has been extended and refined to account quantitatively for the effect of repulsion, i.e., by considering a repulsive r^{-12} interaction between the molecule and the pore wall[18]. It suffices to mention for the purpose of the following discussion that all of the above general conclusions remain valid although the analytical expressions of $W_r(s)$ and $F_r(s)$ become less aesthetical, more complicated, and tedious to use and rationalize without the support of computing facilities. In the following discussion, we will use eqn.3 to explicit the variation of $W_r(s)$.

TESTING THE MODEL

Support for the aforementioned model is gained both from an analysis of the energetics of sorption of various sorbates and from the quantification of the NMR chemical shifts of atoms (Xe) or molecules (tetramethylammonium, TMA$^+$, and tetramethylphosphonium, TMP$^+$, cations) occluded in zeolite pores.

Energetics of Molecules in Zeolite Pores

If $W(s)$ is the interaction energy at the equilibrium sorption separation d in a pore of size a, then the isosteric heat of *initial* sorption $Q_{iso}(0)$ can be defined as[16]:

$$Q_{iso}(0) = RT + C_z.C_m.W_r(s) \tag{9}$$

in which C_z is a molecular constant characterizing the zeolite and $C_m = (\alpha/d^3)$ combines molecular variables for the sorbed molecule only.

Scaling of eqn. 9, i.e., the evaluation of C_z, has been achieved by a linear regression analysis of the interaction energies of rare gases sorbed by faujasite (FAU, X or Y) type zeolites, leading to[2,16]:

$$Q_{iso} \ (kJ.mol^{-1}) = 2.4_4 + 112.7_5 \ C_m W_r(s) \tag{10}$$

Equation 10 does not take into account the effect of chemical composition (Si/Al ratio, presence of counterions). This problem is dealt with below. Table 1 lists experimental (Q_{exp}) and computed (Q_{theo}) physisorption heat values for a variety of sorbates.

Good agreement is generally observed between experimental and theoretical results, even for rather large molecules such as hexane or benzene, or sorbates for which higher

Table 1. Experimental vs. calculated sorption heats ($kJ.mol^{-1}$) for simple sorbate molecules in various aluminosilicate zeolites[16].

System	Q_{exp}	Q_{theo}	System	Q_{exp}	Q_{theo}
He/FAU	4.3	4.7	Ne/MOR	5.2	6.6
Ar/FER	17.1	24.9	Ar/MOR	18.8	17.1
Ar/FAU	12.8	12.6	Kr/MOR	23.0	23.6
Kr/FAU	17.6	16.9	Xe/FER	31.5	54.8
Xe/MOR	33.9	34.4	Xe/FAU	23.1	23.6
O_2/MOR	16.7	18.2	N_2/MOR	23.4	20.0
NO/MOR	23.4	19.4	NO/MFI	18.4	17.8
CO/MOR	28.8	23.3	CO/MFI	21.3	21.4
CH_4/MOR	23.0	22.8	CH_4/LTL	19.0	19.8
CH_4/MFI	28.0	30.7	CH_4/FAU	15.2	15.8
C_2H_6/FAU	22.0	22.7	C_3H_8/FAU	29.0	29.9
C_6H_{14}/FAU	50.0	53.9	C_6H_{14}/LTL	60.8	61.7
C_6H_{14}/MFI	68.8	74.4	C_6H_6/MFI	73.0	73.7
NH_3/MFI	-	26.6	NH_3/FAU	-	13.9

order electrostatic interactions are expected, e.g., N_2 and benzene which have a molecular quadrupole moment. It supports the generalization of eqn. 10 to a variety of rather simple molecules possessing no permanent dipolar moment. As van der Waals interactions are pairwise and additive, computations using eqn. 10 can also be used to estimate the effect of surface curvature on the non-specific component of the sorbate interaction for polar molecules as illustrated by the case of NH_3[16].

The a priori and correct recognition and evaluation of surface curvature effects in physisorption bears on our quantitative interpretation of temperature programmed desorption phenomena, e.g., NH_3 desorption where differences in desorption activation energy should arise in part from non-negligible variations of physisorption energy as shown by the NH_3/MFI and NH_3/FAU examples in Table 1. Thus, it does affect the signification and use of NH_3 desorption activation energies for acid strength evaluation.

From eqn. 9, it also expected that physisorption heats will depend on the chemical composition of molecular sieves via the molecular constant C_z. Polarizability data for zeolites are generally not available at this moment. However, as the sorbed molecules interact principally with the anionic framework oxygens, it seems reasonable to assume that C_z will vary with the effective average absolute charge $<Q_O>$ of the formally dinegative oxygen framework atoms, the polarizability increasing for higher $<Q_O>$ values. If x_{Si}, x_{Al}, and x_P are the molar fractions of Si, Al, and P in molecular sieves, ab initio quantum mechanical calculations show that $<Q_O>$ is given by:

$$<Q_O> = 0.695 + 0.010\, x_{Al} \tag{11}$$

for (the hydrogen forms of) (alumino)silicates[14] and by:

$$<Q_O> = 0.613 + 0.174\, x_{Si} \tag{12}$$

for (silico)aluminophosphates in which Si preferentially replaces P[19]. Little dependence is expected in the aluminosilicate series (eqn. 11), which explains the good agreement of the values in Table 1 that considers zeolites with greatly varying Al content. Equations 9, 11, and 12 predict however that a composition effect should be observable when comparing the sorption heats of a given sorbate on isomorphous aluminosilicate and aluminophosphate (ALPO)-based molecular sieves. Table 2 shows such data[13]. It has also been determined experimentally that the ratio C_z(aluminosilicate)/C_z(ALPO) was equal to about 1.33 for various sorbates and that intermediate values were observed for silicoaluminophosphates[20].

Table 2.   n-Hexane sorption heats (Q_{iso}, $kJ.mol^{-1}$) on isostructural molecular sieves[a] with different chemical composition[13].

Sieve	Q_{iso}	$<Q_O>$	Sieve	Q_{iso}	$<Q_O>$
SAPO-37	48.5	0.622-0.648	H-Y	54.8	0.698
ALPO-5	54.6	0.613	H-MOR	68.0	0.697

a. SAPO-37 and H-Y are strictly isomorphous and possess the FAU topology. Both ALPO-5 and H-MOR have unidimensional 12-membered ring channels.

Docking of Molecules in Zeolites

When considering the sorption of a variety of molecules on microporous sieves, other striking observations also emerge. Firstly, the initial adsorption heat rises rapidly when the molecule size becomes comparable to the pore size[21]. Secondly, initial sorption heat values tend to decrease rapidly with coverage for small molecules, eventually reaching a plateau, while for larger sorbates heats of sorption remain approximately constant[22]. Thirdly, molecular siting may depend on temperature, as illustrated by the case of Xe in zeolite Rho where Xe is exclusively located in the cage-connecting prisms at low temperature[23].

All these facts are easily rationalized within the confinement model which predicts that guest-host relationships prevail, i.e., that molecules will initially sorb in the smallest pores (rings, prisms, side-pockets, channels) where optimal match is achieved, in order to maximize their van der Waals attraction energy. One can also conclude from such considerations that sorbed molecules will firstly achieve the best possible fit between their size (and shape) and that of the intracrystalline environment. This phenomenon explains the apparent physical site heterogeneity sometimes observed in the initial stages of their physisorption. For large and bulky molecules, such a situation occurs only to a limited extent.

Consequently, the molecular trapping efficiency of molecular sieves is also directly related to the increase in physisorption energy resulting from their important surface curvature. This energy is much larger than its thermal counterpart. Thus, the sorption equilibrium is strongly displaced towards adsorption at room temperature or below and physical desorption occurs usually only to a noticeable extent at temperatures higher than ambient. The above facts are implicit manifestations of molecular docking in zeolites, a phenomenon which is sometime evidenced explicitly.

For example, n-hexane and its methyl-pentane isomers have nearly identical sorption heats on ZSM-5 (≈ 65 kJ.mol^{-1}), indicating that all their CH_x segments interact with the zeolite surface. For ALPO-11 the heat of sorption of 2-methyl-pentane is nearly identical to that of n-hexane (≈ 66 kJ.mol^{-1}), whereas the heat of sorption of 3-methyl-pentane is about 9-10 kJ.mol^{-1} smaller (56.8 kJ.mol^{-1}). The latter value however compares well to the measured heat of sorption of n-pentane (56.6 kJ.mol^{-1})[17]. The lower sorption heat of 3-methyl-pentane has been explained by the sticking of the alkyl-C_5 chain on the pore wall with the center 3-methyl group pointing towards the center of the channel along the ALPO-11 elliptical pore long axis, a situation in which the interaction of the 3-methyl group with the surface is minimized[17]. Such guest (molecule) - host (framework) relationships for molecules adsorbed in zeolites are easily visualized using molecular graphics representations from which the amount of van der Waals contact between the sorbed molecules and the pore walls is qualitatively and easily evaluated.

<u>NMR Chemical Shifts of Confined Molecules</u>

At temperatures such that $kT < W(s)$, spherical molecules or atoms will stick and oscillate with respect to the pore wall and hop as a whole over its corrugation barriers, rather than collide with the wall and be reflected through the cavity or pore. This picture is the basis for our interpretation of the structure-dependent NMR chemical shift variation observed for ^{129}Xe and ^{13}C of tetramethylammonium (TMA$^+$) and tetramethylphosphonium (TMP$^+$) cations in zeolites[2,20,24].

Neglecting the role of exchange-cations and noting that the interaction is primarily with the pore wall oxygen anions[13], it can then be conceived using the reaction field concept[15] that the screening by electrons of NMR sensitive nuclei will vary with the degree of confinement and that the position of their NMR resonance will be shifted. Indeed, the adsorbed molecule is submitted to the effect of the surroundings fluctuating electric field whose quadratic value is not averaged out. Thus, a paramagnetic (positive) contribution to the chemical shift should appear[24], which agrees with the experimental observations[20,25,26].

The chemical shift variation δ_s induced by the interaction with the pore wall varies linearly with the molecular interaction energy with the surface: $W(s)$ given by relation (2). That is[2,24]:

$$\delta_s = A + B.W(0).W_r(s) \tag{13}$$

A and B being molecular constants. For a given sorbate of known van der Waals radius, $W_r(s)$ values are readily calculated for a whole series of zeolites on the basis of their known crystallographic structure. The validity of eqn.13 can thus easily be tested.

Correlations between δ_s and $W_r(s)$ according to eqn. 13 are shown in Fig. 2 for the NMR of ^{129}Xe and that of ^{13}C in TMA$^+$ [2,20,24] and TMP$^+$ [20] cations occluded in a variety of zeolites. Extremely and unexpectedly good proportionality is observed in all cases, thereby supporting the physical basis of our model and our interpretation of the chemical shift variation for such spherical species trapped in zeolites.

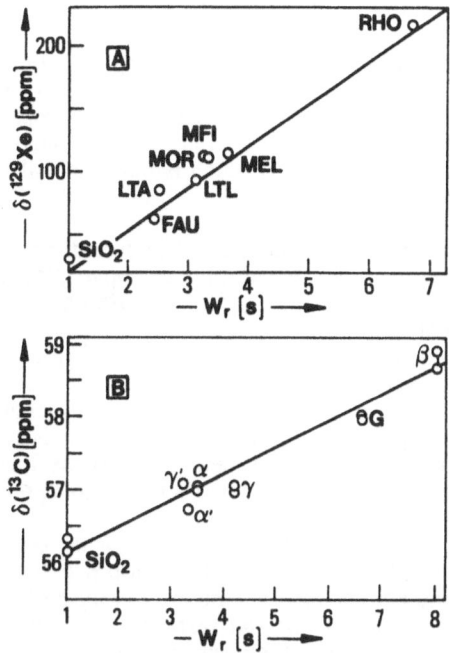

Fig. 2. Variation of the chemical shift (δ_s, ppm) of molecules occluded in the intracrystalline void volume of zeolites as a function of $W_r(s)$. A: ^{129}Xe (structures as indicated); B. ^{13}C of TMA$^+$ (cage type as marked).

The following relationships are operational:

$$\delta_s(^{129}\text{Xe, ppm}) = -5.35 + 33.44\, W_r(s) \tag{14}$$

$$\delta_s(^{13}\text{C,TMA}^+,\text{ppm}) = 55.15 + 0.27\, W_r(s) \tag{15}$$

$$\delta_s(^{13}\text{C,TPA}^+,\text{ppm}) = 8.03 + 0.30\, W_r(s) \tag{16}$$

Provided confinement is the only and main parameter which affects the chemical shift, the model thus predicts limiting (upper) values of δ_s equal to about 262, 58, and 10.5 ppm for ^{129}Xe, ^{13}C in TMA$^+$, and ^{13}C in TMP$^+$, respectively, as the maximum allowed value of $W_r(s)$ is 8.

Because of the above relationships from which the curvature factor $W_r(s)$ is easily derived, NMR becomes a quantitative tool to probe at the atomic scale either the average pore size of zeolites (^{129}Xe) or the size and shape of the direct environment of the ion-exchange sites (^{13}C of TMA$^+$ and TMP$^+$) which is likely to be the locus of the catalytic activity . It has been successfully applied to the determination of the void space in ZSM-20 and to evidence the preferential location of the TMA$^+$ and TMP$^+$ cations in the side-pockets of the mordenite channels[20].

MOLECULAR DIFFUSION IN MICROPORES

We have addressed above the *floating molecule* (s ≈ 1) and the *creeping diffusion* (s < 0.9) concepts for molecular diffusion in the micropores. In the creeping diffusion of hydrocarbons, motion is described as a sequence of segmental rotations of only part of the molecule (around a C-C bond), thereby changing its center of mass but not necessarily implying the simultaneous translation of all its parts[27]. This proposal of Barrer[28] was based on the observation that self-diffusion activation energies (measured by NMR) for linear alkanes approach asymptotically a limit which is only a small fraction of their sorption heat[29,30].

Evidence for the segmental locomotion of hydrocarbons is readily obtained from classical NMR relaxation time measurements. For example, plots of $\ln T_1$ vs. $1/T$ (T_1 is the longitudinal, spin-lattice, relaxation time) for the different CH_x group of the n-hexane molecule adsorbed on zeolite H-Y enable the estimation of activation energies for molecular (segmental) motion(s), equal to 6.6, 7.6, and 8.0 kJ.mol^{-1} (all values ± 1.0 kJ.mol^{-1}) for the CH_3-, alpha-CH_2-, and beta-CH_2- groups, respectively[17]. The terminal methyl groups are thus more mobile (more rotational freedom) than the chain methylene groups. In addition, the activation energy for the motion of -CH_2- groups is comparable to their average heat of sorption, 8.3 kJ/mol[2,3]. Consequently, n-hexane has a indeed a segmental interaction with the zeolite surface, which supports the creeping diffusion model.

Another feature, however subject to controversy, is the presence and the nature of an external surface barrier to uptake sorption. We have proposed that one possible origin for the energy barrier at the pore entrance is the convex curvature of its rim which reduces the adsorption energy as compared to the flat surface value[5]. It has been argued that this proposal would not hold when the atomic corrugation of the surface is considered[31]. However, more theoretical insight is needed at this moment before a firm conclusion can be drawn.

Nevertheless, one can always measure uptake sorption rates of sorbates on materials which possess about identical crystal size and extrapolate these observations to the calculation of diffusion-related parameters. Several problem arise: i. uptake sorption, certainly in its initial stage, is not isothermal because of the adsorption heat release, ii. classical diffusion equations do not describe correctly the intracrystalline diffusion process, iii. one has to distinguish the effects of uptake diffusion (external surface barrier), non-equilibrium intracrystalline diffusion, and equilbrium self-diffusion.

Data exist which characterize by a mass transfer coefficient k (min^{-1}) the rate of uptake of n-hexane in various molecular sieves[17]. For comparable crystal sizes, when there is only a small influence of cation nature and content, a surprising dependence of $\ln k$ on W_r(s) is observed (Fig. 3). In contrast to intuitive reasoning based on pore aperture only, the uptake rate increases as the molecule and pore sizes match each other more intimately, the highest rate being observed for ALPO-11 (small pore, unidimensional) rather than for zeolite Y (large pore, 3-dimensional).

Although the parameter W_r(s) is not related quantitatively to the diffusion rate, it is expected that the higher the value of W_r(s), the lower will be the external surface physical barrier for sorption, as a molecule at the rim of smaller pores will have a higher adsorption energy[2,3]. In addition, the confinement model predicts also that physisorption energy gradients (intracrystalline energy barriers for diffusion) will be enhanced in non-unidimensional structures relative to one-dimensional uniform pore size networks[17]. In the former case indeed, the curvature of the local environment will change along the diffusing molecule trajectory because of the presence of intersections and cages. In the latter case, only the atomic corrugation of the surface will have an effect.

The above remarks are fundamental to rationalize the increase of ln k as a function of $W_r(s)$ (Fig. 3). Clearly, however, such observations need further support and quantitative interpretation, namely as they are most relevant to our understanding of sorption and catalysis by zeolites.

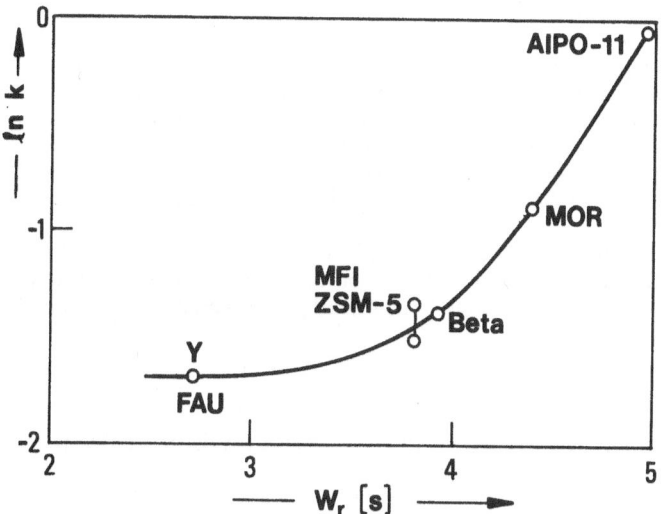

Fig. 3. Variation of ln k with the relative van der Waals interaction energy $W_r(s)$ for the sorption of n-hexane on various zeolites (H-form, as indicated) and ALPO-11. (Adapted from ref. 17).

CONFINEMENT EFFECTS IN CATALYSIS

The Acid Strength of Zeolites

The most utilized catalytic property of zeolites is their carboniogenic activity, e.g., their use as acid catalysts[1]. We will show below that acid catalytic activity is a blend of various parameters, i.e., nature, strength , accessibility, and environment of the acid sites.

Firstly, a general remark. It is customary to characterize the acid strength of solid acids by titration in the presence of indicators and using the Hammett acidity scale[32] (which in theory depends on the nature of the solvent). Several such studies have been reported for zeolites[33] and deserve a word of caution. Confinement predicts indeed that the degree of solvation (by the molecular sieve framework) of both the titrant and the indicator will vary from one zeolite to another. In addition, access of the indicator to the intracrystalline volume of the zeolite may be restricted. As the solvent nature changes with zeolite framework type, it is clear that the comparison of Hammett parameters for zeolites with different topologies has little meaning. The method can however be used to compare isomorphous series of aluminosilicates and some correction should be introduced (see above our discussion of C_z) to handle the case of aluminophosphate-based materials.

Variations in the acid catalytic activity of zeolites of different Al content and/or structure have been attributed customarily to changes in the acid strength of their Brønsted acid sites resulting from collective effects, considering zeolites as ionic solvents[34] or using intermediate electronegativity concepts[35,36]. These models neglect the contributions of geometrical factors and of physisorption. Recently, framework topology has been shown to affect zeolite acid strength below a Si/Al ratio of about 5-10 [37]. There are several indications that acid strength should not vary for materials with Si/Al ratio greater than about 5, for which case two Al structural sites must be separated statisticallly by more than one Si, i.e., when Al sites behave as isolated catalytic entities. Let us mention in particular the results of quantum mechanical calculations[38], characterizations of Y zeolites with different Al contents[39], ^{1}H magic angle spinning NMR studies of various zeolites[40], and catalytic testing of HY, LaY, and HZSM-5 catalysts[41]. Changes in catalytic activity may thus be attributed to "nesting" [4], an effect which we proposed and that can now be quantified in terms of confinement.

The nest effect insists, among other things, on the established fact[4,42,43] that reaction rate constants for intracrystalline zeolite catalysis should be affected by the adsorption equilibria influencing the local (internal) concentrations of all the chemical species present along the reaction pathway. Qualitative support was obtained from a volcano-type correlation between turnover frequency (TOF) and pore diameter[4] for a variety of zeolites effecting n-pentane cracking[44]. The nest effect can also be quantitatively demonstrated.

The following considerations apply to zeolite catalysts which have retained their chemical integrity, if and when diffusion-disguised kinetics does not operate. Paraffin cracking is selected as an example. The rate of paraffin cracking r_A is given by[45]:

$$r_A = k_{HT} \cdot [RH] \tag{17}$$

in which k_{HT} is the rate constant for hydrogen transfer from alkanes to carbenium ions and [RH] the alkane concentration. As catalysis occurs inside the zeolite pores, eqn. 17 must be rewritten as:

$$r_A = k_{HT} \cdot K_a \cdot [RH]_e \tag{18}$$

where K_a is the equilibrium constant for the physisorption of RH, accounting for surface curvature effects, and $[RH]_e$ is the alkane concentration outside the crystallites . The net rate constant k and the turnover frequency TOF are then given by[3]:

$$k = k_{HT} \cdot K_a \tag{19}$$

$$TOF = k_{HT} \cdot K_a / (Al/(Al+Si)) \tag{20}$$

It is easily derived[3] that, at constant temperature, turnover frequencies for the conversion of a given reactant by different zeolites are related by the following equation:

$$\ln(TOF/TOF_r) = \ln R_{TOF} = (Q - Q_r)/RT \tag{21}$$

R_{TOF} is the ratio of the turnover frequency TOF for a given zeolite to the value TOF_r obtained using a reference material (index r). Q and Q_r are the corresponding sorption heats which can be evaluated using eqn.10. Figure 4 shows a correlation according to eqn. 21 for the cracking of n-pentane at 450°C on medium and large pore zeolites (MFI=ZSM-5, MOR, LTL, and Y). Highly siliceous ZSM-5 was taken as reference (low Al content, isolated catalytic sites).

Excellent and unexpected agreement is observed, supporting thereby quantitatively the recent suggestion[3,4] that activity differences arise from nesting rather than variations in acid strength. The slight deviation observed for FAU and LTL is the only indication for a possible effect of "site only" acid strength. It is a factor of about 2 and no more, far different from the two orders of magnitude variation in TOF which is observed experimentally.

Structure-dependent contributions must thus be considered, such as confinement which can act separately or in addition to, and even dominate the effect of acid strength when large enough molecules are considered or when the matching of the molecule and pore conformations becomes ideal. Clearly, an adequate definition and measure of acid strength should not be "static", i.e., involve local, electrostatic, or polarity factors only. The ability of an acidic site to transfer a proton depends on several parameters: the polarity of the Brønsted hydroxyl group, the proton affinity of the accepting molecule, the capacity of the zeolite framework to accept and delocalize the residual negative charge, and the respective configurations and environment of the acid site and the reactant(s). Clearly, this situation is one where the *supermolecule concept* has to be considered in order to rationalize catalytic activities.

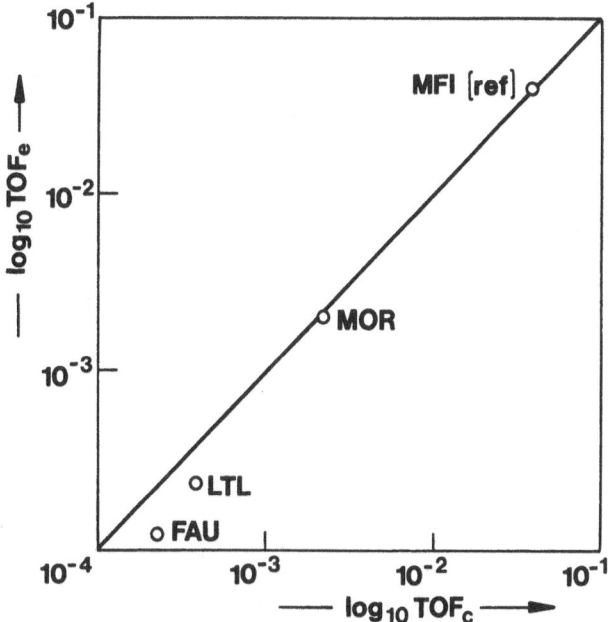

Fig.4. Experimental (TOF_e) vs. calculated (TOF_c) turnover frequency (see eqn. 21) for various zeolites, as indicated. (Adapted from ref. 3; data from ref. 44).

Note finally that confinement predicts that large inhomogeneities in physisorption potential may exist in some zeolites, in particular when they are constituted by interconnected channels, prisms, and cages. This could provide a quantitative basis for the physical understanding of the recently proposed "energy gradient selectivity" (EGS) effect[46].

Receptor Sites and Preorganization in Zeolite Catalysis

It has been demonstrated that solid (microporous) catalysts which confine reactants within their intracrystalline free volume can function as "supermolecular catalysts" resembling enzymes in some of their actions[2,7]. Confinement insists on the role played by structural complementarity for determining the conformation of the sorbed molecule which eventually forms a supermolecule together with its zeolitic environment.

In particular, the unique channel properties of (Ba,K)-exchanged zeolite L may play a role in its unmatched selectivity (>90%) for the dehydrocylization of n-hexane into benzene. The analysis of the aromatization selectivities[47] indicate undubiously (i) that only metal-catalyzed processes are operative and (ii) that direct 1-6 ring closure followed by dehydrogenation is determinant in C_6-C_9 alkane aromatization by zeolite L-based catalysts. The high selectivities of the normal paraffins and 2-methyl-hexane aromatizations are exceptional for heterogeneously catalyzed reactions and tend for an optimized system to those observed for enzyme-catalyzed processes.

Non-binding interaction of n-hexane with the active site environment leads to the recognition of the particular zeolite L cavity shape and the preorganization of n-hexane as a pseudo-cycle (see Fig. 5). The supermolecule is eventually transformed into a metallocycle which can yield cyclic compounds and ultimately benzene. Such stages are conveniently described and visualized by molecular graphics simulations and computations[7].

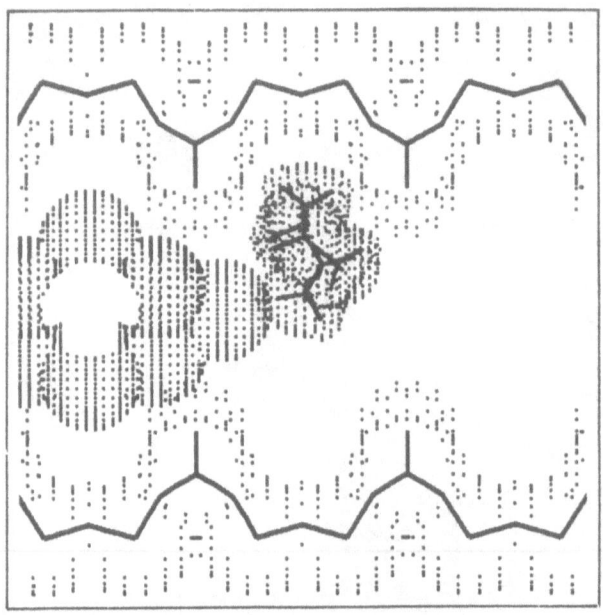

Fig. 5. Preorganization of n-hexane as a pseudo-cycle in the
unidimensional channel structure of zeolite L.

n-Hexane first meets head-on with a Pt atom, either single with free space within the same cage or more probably from a Pt cluster protruding through a window from a nearby cage, and interacts with it by a terminal CH_3 group. Then, the n-hexane molecule or its already partially dehydrogenated product (guest) bound to Pt curves itself in the groove of the zeolite L cage (host) in order to optimize its interaction with the surrounding zeolite framework. The CH_3 group in position 6 is thereby forced back to the vicinity of the Pt active site. Structural recognition and preorganization are thus central determinants for the direct 1-6 ring closure. The higher the binding between guest and host, and the better the supermolecule resembles the transition state configuration, the lower should be the energy of the transition state relative to the reactants and products[48]. Consequently, chain lengths longer than C_6 or branching should congest and destabilize to various degrees the molecule (guest) - zeolite (host) complex and thus decrease catalytic activity, which is indeed observed experimentally. Once a "C-1 - C-6 metallocycle" is formed, the reaction will proceed via a chemistry that has been demonstrated extensively in the literature[49-51].

CONCLUSIONS

The importance of surface curvature, i.e., confinement, effects in pores of atomic size implies that sorption or catalytic sites and their direct environment in zeolites constitute as a whole the entity responsible for the adsorption or catalytic properties, a situation somewhat resembling that encountered in enzymes.

Molecules in zeolites are solvated by the framework and the effect of sorption on their properties can be quantified, as illustrated by the evaluation of sorption heats and the description of NMR chemical shifts variations. The validity of the confinement model is thereby assessed. The solvent action of the framework has other important physical and chemical consequences, e.g., regarding acid strength evaluation, intermolecular interactions, etc...

One area which is likely to witness interesting developments is that of diffusion in zeolite pores. It is indeed likely to be non-Fickian. The proposal of a more realistic diffusion model would greatly enhance our understanding and handling of separation and mass-transfer controlled catalytic processes.

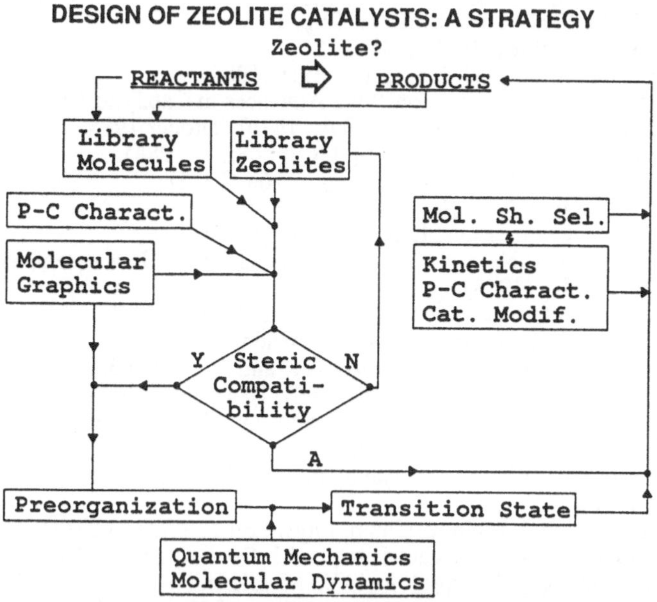

DESIGN OF ZEOLITE CATALYSTS: A STRATEGY

Fig.6. Assisted molecular design of zeolite catalysts: the integration of physico-chemical characterizations, molecular engineering, theoretical computations, and molecular graphics. A novel strategy in molecular sieve-catalyzed organic chemistry[12]. Abbreviations: Y: Yes; N: No; A: Almost; P-C: Physicochemical; Mol. Sh. Sel.: Molecular Shape Selectivity.

Zeolite-catalyzed reactions may involve non-binding interactions. Therefore and in such cases, they must be regarded as supermolecular conversions where guest-host relationships may affect both catalytic activity and selectivity. The demonstration of supermolecular catalysis by zeolites and the advent of reliable molecular graphics simulations open an avenue for new thinking and strategies in the application of zeolite catalysts to selective organic synthesis. Those will integrate the molecular engineering of zeolites with molecular graphics and theoretical computations for the enhanced molecular design of molecular sieve-based catalysts (Fig. 6).

ACKNOWLEDGEMENTS

The author would like to acknowledge the valuable contributions of his colleagues and corworkers whose name appear in his publications cited as references. In particular, thanks are due to Prof. A.A. Lucas, Prof. J.M. André, and Prof. J.P. Vigneron whose input has been essential for the evaluation and development of the confinement model.

REFERENCES

1. P.B. Weisz, Ind. Eng. Chem. Fundam., 25:53 (1986).
2. E.G. Derouane, Mémoire Acad. Roy. Belg., Sciences, 2e Ser., T. LXV, Fasc. 5 (1988).
3. E.G. Derouane, J.M. André, and A.A. Lucas, J. Catal., 110:58 (1988).
4. E.G. Derouane, J. Catal., 100:541 (1986).
5. E.G. Derouane, J.M. André, and A.A. Lucas, Chem. Phys. Lett., 137:336 (1987).
6. D.J. Cram, Science (Washington, D.C.), 240:760 (1988).
7. E.G. Derouane and D. Vanderveken, Appl. Catal., 45:L15 (1988).
8. P.B. Weisz, Pure Appl. Chem., 52:2091 (1980).
9. E.G. Derouane, in "Intercalation Chemistry", M.S. Whittingham and A.J. Jacobson, eds., Academic Press, New York, 1982; p. 101.
10. S.M. Csicsery, Zeolites, 4:202 (1984).
11. E.G. Derouane, Stud. Surf. Sci. Catal., 19:1 (1984).
12. E.G. Derouane, Oral presentation at the Advances in Catalytic Technologies Seminar, Catalytica Inc., Santa Clara, CA., U.S.A., 18-20 October 1988.
13. E.G. Derouane and M.E. Davis, J. Mol. Cat., 48:37 (1988).
14. E.G. Derouane and J.G. Fripiat, Zeolites, 5:165 (1985).
15. L. Onsager, J. Am. Chem. Soc., 58:1486 (1936).
16. E.G. Derouane, Chem. Phys. Lett., 142:200 (1987).
17. E.G. Derouane, J. B.Nagy, C. Fernandez, Z. Gabelica, E. Laurent, and P. Maljean, Appl. Catal., 40:L1 (1988).
18. I. Derijcke, E.G. Derouane, and A.A. Lucas, to be published.
19. J.G. Fripiat and E.G. Derouane, to be published.
20. E.G. Derouane, J. B.Nagy, B. De Roover, and C. Fernandez, in "Proc. 8th Intern. Zeolite Conf., Amsterdam, The Netherlands, 10-14 July 1989.
21. H. Stach, U. Lohse, H. Thamm, and W. Schirmer, Zeolites, 6:74 (1986).
22. L.V.C. Rees, Chem. Ind., 7:252 (1984).
23. I. Gameson, P.A. Wright, T. Rayment, and J.M. Thomas, Chem. Phys. Lett., 123:145 (1986).
24. E.G. Derouane and J. B.Nagy, Chem. Phys. Lett., 137:341 (1987).
25. J. Fraissard, T. Ito, M. Springuel-Huet, and J. Demarquay, in "New Developments in Zeolite Science and Technology", Y. Murakami, A. Iijima, and J. Ward, eds.; Kodansha-Elsevier, Tokyo, Amsterdam, 1986; p. 393, and references therein.
26. S. Hayashi, K. Suzuki, S. Shin, K. Hayamizui, and O. Yamamoto, Chem. Phys. Lett., 113:368 (1985).
27. R.M. Barrer and J.A. Davies, Proc. Roy. Soc. London Ser. A, 322:1 (1971).
28. R.M. Barrer, Symposium on the Characterization of Porous Solids, Neuchatel, 9-13 July 1978.
29. R.M. Barrer, in "Zeolites: Science and Technology", F. Ramoa Ribeiro, D. Rollmann, and C. Naccache, eds., NATO ASI Series E, No. 80, Martinus Nijhoff, The Hague, 1984; p. 227.
30. R.M. Barrer, in "Zeolites: Science and Technology", F. Ramoa Ribeiro, D. Rollmann, and C. Naccache, eds., NATO ASI Series E, No. 80, Martinus Nijhoff, The Hague, 1984; p. 261.
31. F. Vigné-Maeder, J. Catal., in press.
32. K. Tanabe, in " Catalysis Science and Technology", Anderson, J.R. and M. Boudart, eds., Springer-Verlag, Berlin, 1981; 2:242.
33. M.W. Anderson and J. Klinowski, Zeolites, 6:150 (1986).
34. D. Barthomeuf, J. Phys. Chem., 83:249 (1979).
35. W.J. Mortier, J. Catal., 55:138 (1978).
36. P.A. Jacobs, Catal. Rev. Sci. Eng., 24:415 (1982).
37. D. Barthomeuf, Mater. Chem. Phys., 17:49 (1987).
38. E.G. Derouane and J.G. Fripiat, Zeolites, 5:165 (1985).
39. J.R. Sohn, S.J. DeCanio, P.O. Fritz, and J.H. Lunsford, J. Phys. Chem., 90:4847 (1986).

40. D. Freude, M. Hunger, and H. Pfeifer, Chem. Phys. Lett., 128:62 (1986).
41. S. Fukase and B.W. Wojciechowski, J. Catal., 102:452 (1986).
42. J. B.Nagy, M. Guelton, and E.G. Derouane, J. Catal., 55:43 (1978).
43. M.F.M. Post, J. van Amstel, and H.W. Kouwenhoven, in "Proc. 6th Intern. Zeolite Conf.", D.H. Olson and A. Bisio, eds., Butterworths, Guildford, 1984; p. 517.
44. E. Kikuchi, H. Nakano, K. Shimomura, and Y. Morita, Sekiyu Gakkaishi, 28:210 (1985).
45. W.O. Haag and R.M. Dessau, in "Proc. 8th Intern. Congr. Catalysis", Verlag-Chemie, Weinheim, 1985; p. IV-545.
46. C. Mirodatos and D. Barthomeuf, J. Catal., 93:246 (1985).
47. P.W. Tamm, D.H. Mohr, and C.R. Wilson, in "Catalysis 1987", J.W. Ward, ed., Elsevier, Amsterdam, 1987; p. 335.
48. L. Pauling, Am. Sci., 36:51 (1948).
49. L. Nogueira and H. Pines, J. Catal., 70:404 (1981).
50. Z. Paal, Advan. Catal., 29:273 (1980).
51. F.G. Gault, Advan. Catal., 30:1 (1981).

COOPERATIVE AND LOCALISED EFFECTS IN CATALYSIS OVER ZEOLITES

J. Dwyer

UMIST

Chemistry Department, PO Box 88, Manchester M60 1QD, U.K.

ABSTRACT

Several approaches provide explanations for the catalytic properties of zeolites. Some of these emphasise cooperative effects and others focus on influences localised at the reactive sites. Evidence supporting these points of view is considered with a view to clarifying the catalytic processes in zeolites.

INTRODUCTION

The microporous structure of zeolites and related materials means that catalytic sites are mainly intracrystalline. To access the reactive sites molecules must penetrate the pore mouths and enter a pore system having molecular dimensions so that sorbates are always under the influence of pore walls. Powerful electrostatic fields can arise, within the crystals, from charges on cations and framework atoms which can polarise a sorbate leaving it susceptible to attack by strong acid centres. Furthermore zeolites can be effective in stabilising ionic species (1) and may, therefore, facilitate reactions involving ionic intermediates.

Attempts to provide explanations for the properties, particularly the acid catalytic properties of zeolites have considered some or all of the above aspects. These approaches include consideration of zeolites as ionising solvents (2), solid electrolytes (3) or crystalline liquids (4). Explanations, particularly for acid catalysis have also used topological considerations, (5) (6) (7) (8) concepts of electronegativity equilisation (9) (10) and, particularly in theoretical approaches at the ab initio level, relatively small clusters as representative of catalytic sites (11) (12) (13) (14). Extended calculations (15) and more qualitative approaches, for example HSAB theory (61) are also used.

Some of these approaches emphasise cooperative effects, which, in zeolites, can include cooperation within the solid state, to encourage site homogeneity, and cooperation between zeolite and the sorbed phase, to encourage particular configurations for sorbed or diffusing species. Other approaches stress the detailed composition and configuration of specific active sites. This paper examines evidence for cooperative and localised effects in zeolite catalysis.

Guidelines for Mastering the Properties of Molecular Sieves
Edited by D. Barthomeuf *et al.*
Plenum Press, New York, 1990

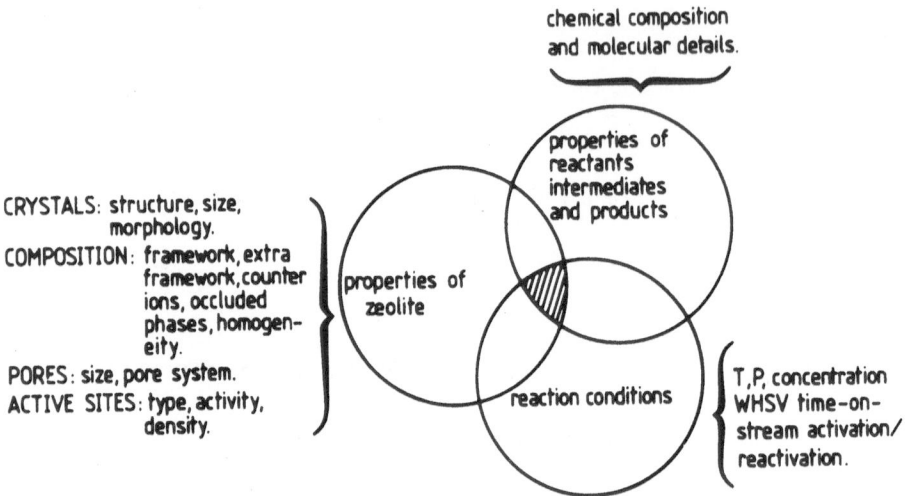

Figure 1. Zeolite catalysis, parameters involved.

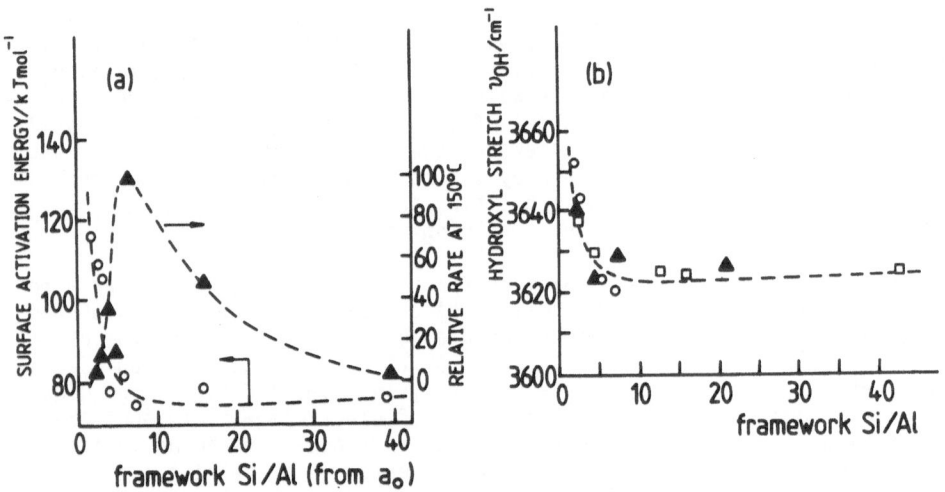

Figure 2. Effect of framework composition on faujasite zeolites.
    a) isomerisation of cyclopropane over steamed/acid leached zeo-
    lite H-forms (15).
    b) shifts in hydroxyl stretch on dealumination by several pro-
    cedures.
       ° steam/acid leach UMIST (1978)
       ▲ SiCl Klinowsky (1986)
       □ SiCl De Canio (1986)

ZEOLITE CATALYSIS: THE PROBLEM

This is illustrated in Fig. 1 where the overlap region of the Venn diagram is of concern. Further complication is common since this region is often time-dependent due to catalyst deactivation. Consequently, it is perhaps suprising that correlations between zeolitic properties and catalytic results are as good as they are. Conversely it is not so suprising that, with so many parameters, limited correlations can be found by selective emphasis of parameters as given in Fig. 1. Within the overlap region we envisage diffusion of reactants to catalytic sites where particular configurations of sorbed species and reactive intermediates may be stabilised by solvation by the zeolite. Transportation to and from the sites may be limited by spatial factors, as may particular reaction intermediates, and overall reaction schemes may have kinetic or thermodynamic limitations. A complete explanation for changes in reaction rates and product distributions, over a range of zeolite structures and compositions, requires a knowledge of the entropic and enthalpic terms associated with each step. Consequently simplification to detect generalisations is required.

In what follows diffusion and sorption are considered very briefly and catalysis is largely confined to hydrogen forms of zeolites although it is abundantly clear that cations and occluded species have a definite role in zeolite catalysis.

DIFFUSION/SORPTION

Diffusion in zeolites is in the configuration regime (16) which has catalytic consequences in both activity and, particularly, in selectivity (17). Diffusivities can be strongly dependent on localised factors, for example amine diffusion in H-forms of zeolites, or can be dominated by cooperative effects over a considerable structural domain (18).

Powerful electrostatic forces arising from charges on framework atoms and compensating cations result in strong specific interactions with sorbates having permanent electrical moments (19). Specific interactions, which tend to be highly localised, can involve considerable charge transfer to produce chemisorbed species. Additionally, non-specific Van der Waals interactions are much stronger in zeolites than in amorphous catalysts due to surface curvature (20) resulting in nesting effects (21) (22). Nesting increases the number of surface atoms interacting with a sorbate molecule but the effect is essentially localised, over a limited domain, unless guest-host interactions result in modification of the framework structure, which is sometimes observed (23), implying cooperative interaction over a much more extended region. On the other hand strong localised (specific) effects can dominate changes in sorption energetics. For example the enthalpy of sorption for olefins is profoundly affected when Ca^{2+} ions replace Na^+ in the large cages of zeolite X (24) but the sorption energy for alkanes of similar size is hardly changed.

For the simple reaction $(A \longrightarrow P)$ when the surface reaction is rate controlling, the surface activation energy (E_s) is related to the heat of sorption $(\lambda = -\Delta H_s)$ by $E_s = E_a + \lambda$

where E_a is the apparent activation energy. Sorption energies can be influenced by "nesting" which can involve; (a) cooperative effects, usually over a limited domain, between the zeolite and the sorbate and (b) specific effects which tend to be more highly localised. Reactive intermediates, frequently, are sorbed species so that both of these

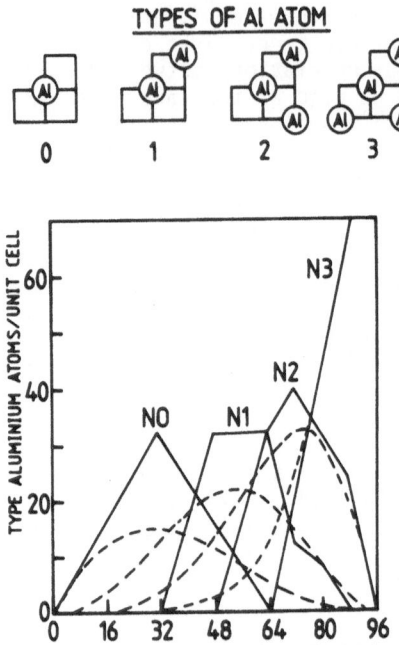

Figure 3. Disribution of types of Al atoms in Faujasites.

——————— unique structures of Cambridge group.
------ disordered unit cell

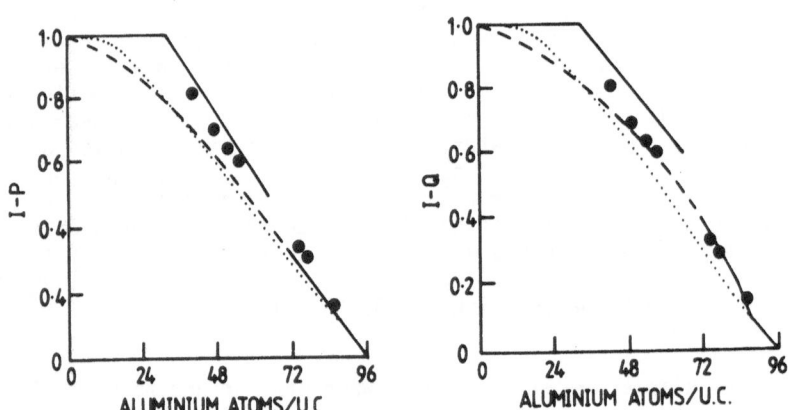

Figure 4. ——————— unique structures of Cambridge Group.
------ disordered unit cell.
..... centre of symmetry maintained.
•• experimental α-values of Barthomeuf and Beaumont.

effects, cooperative and localised, will be involved in the stabilisation of appropriate geometries and delocalisation of charges. These features are important aspects of catalysis. The influence of the two effects ((a) and (b)) is likely to vary considerably in different systems as indicated in Fig 1.

ACID CATALYSIS ON ZEOLITES

As with sorption the properties of the active site, particularly those which stabilise reactive intermediates, will involve some cooperation with neighbouring atoms. Because of their ease of identification, and their direct involvement with the zeolitic framework structure, only the Bronsted sites will be considered here and discussion will focus on hydrogen forms of zeolites for simplicity.

In aluminosilicate zeolites the Bronsted sites are bridged hydroxyls ($\equiv$ Si(OH)Al$\equiv$) giving rise to a hydroxyl stretch ($\bar{\nu}_{OH}$) in the range 3660 cm^{-1} to 3600 cm for hydroxyls in the larger (accessible) pores of zeolites having diverse structures and composition. For aluminium-rich zeolites $\bar{\nu}_{OH}$ decreases as framework Si/Al increases (Fig 2(b)) reflecting a weakening of the O-H bond which is associated [but not directly correlated (10)] with increased acid strength as confirmed by ^{1}HNMR (25) and catalytic studies (26). Typically, it seems that $\bar{\nu}_{OH}$ and acid catalytic activity increase as framework Si/Al increases up to a value of Si/Al of around 6 to 15 depending upon source of data and zeolite structure (8) (Fig 2(a)).

Electronegativity and Acidity

Many results reflecting changes in zeolitic properties with changes in composition, irrespective of structure, can be correlated by a single parameter, the Sanderson intermediate electronegativity (9) (27). This parameter (S), which reflects the equilisation of the electronegativities of hetero-atoms in a compound, has theoretical justification and for the hydrogen zeolites (HZ) is a function of composition (28)

$$S = \left[S_H^x S_{Al}^x S_O^2 S_{Si}^{(1-x)} \right]^{1/(3+x)}$$

where S_H, S_{Al} etc are the (constant) electronegativities of the constituent atoms and x is the framework aluminium fraction. The charge on the hydrogen atom (δH) is then

$$\delta H = (S - S_H)/2.08 \, S_H^{\frac{1}{2}}$$

and is linear in S.

This comprehensively collective approach to zeolite properties implies that all Bronsted sites are modified in the same way as Si/Al changes and all have the same strength, so that the overall rate (R) per unit cell of H-form zeolite may be written*

$$R = N_{Al} r(S) \tag{a}$$

where N_{Al} is the number of available framework aluminiums per unit cell and $r(S)$ is the common specific rate for each site which is determined by

---

* Equations (a) and (b)-(e) can be modified to account, for example, for a spread of acidity associated with a single type of site or for the fractional activity sometimes observed with a particular site but they are here represented simplistically for illustration.

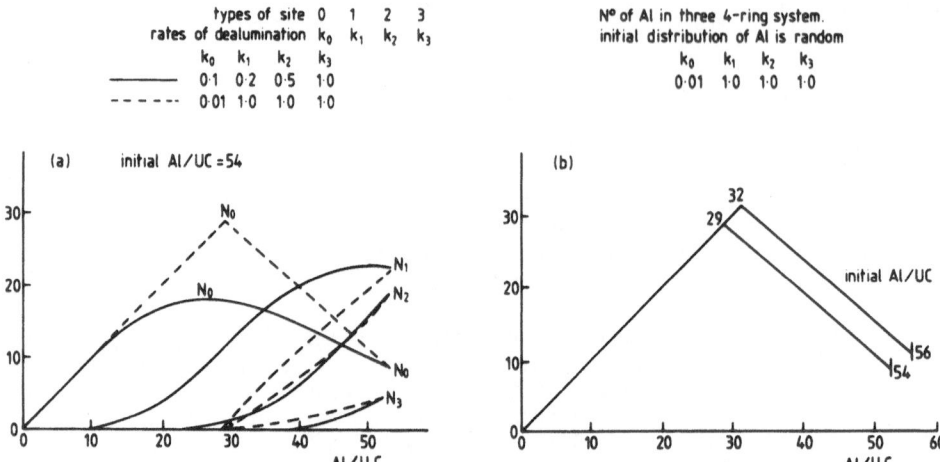

Figure 5. a) Effect of Changes in Rates of Dealumination.
b) Effect of Initial Composition on No (Isolated Sites).

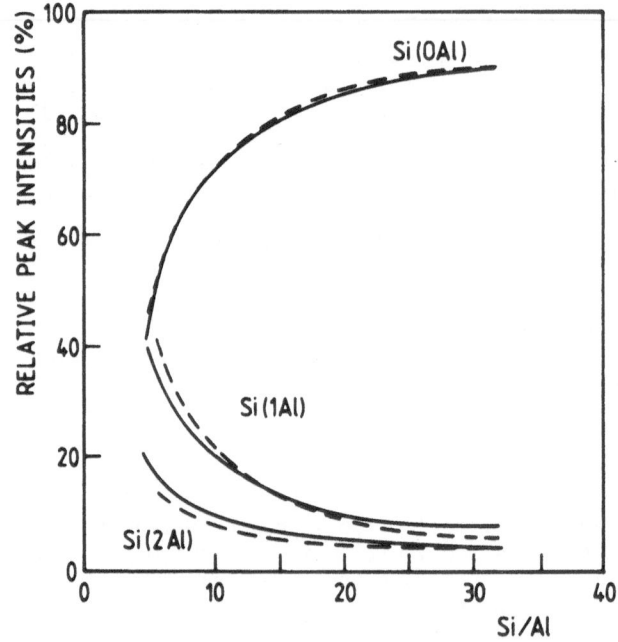

Figure 6. Experimental Si(nAl) intensities-------
Calculated Si(nAl) intensities ―――――
Assuming Al on sites T_3 and T_4 only
with either two or no Al in 4-rings.
(J.B. Nagy and E.G. Derouane 1988).

the intermediate electronegativity (S) and hence by global composition. Excellent correlations are observed over a considerable range of 'S'. Intermediate electronegativity is particularly effective at relatively low values of Si/Al and the effect of charge-balancing (cations) can be incorporated into the correlations. However, this approach cannot account for any structural effects and moreover it incorrectly predicts behaviour at low content of framework aluminium. From the expression above for 'S' it is seen (28) that $\delta S/\delta x$ is maximal, over the range of zeolite compositions ($0 \leq Al \leq 0.5$), as $x \rightarrow 0$ ($Al \rightarrow 0$). This is contrary to much experimental evidence, for example the rate of hexane cracking on highly siliceous H-ZSM-5 is linearly correlated with framework Al content (29), implying a constant turnover number and a zero differential as $x \rightarrow 0$ rather than a maximal value. Further difficulties arise with this collective approach when isomorphous substitution of framework atoms leads to generation of distinctly different Bronsted sites, as discussed subsequently, and infrared results show structural effects in correlations using Sanderson electronegativity (30).

These aspects have recently been addressed (31) in the development of an effective electronegativity which allows for a structural input. Up to date this approach has considered only polymorphs of silica having zeolitic structures but impressive correlations of several physical properties have been demonstrated and, for instance, charges on silicons and oxygens are found to depend upon structure and on particular crystallographic sites so that different 'T' atoms within the same structure do not have the same charge. This is a considerable advance on the Sanderson treatment although the simplicity of the original approach is lost. Importantly, this new approach accepts the role of structure, as well as composition, and allows for local as well as global influences. Moreover, calculated charges on framework atoms vary between structures and can vary between different crystallographic sites within structures so that both electronic and geometrical features differ at specified sites. This is clearly relevant to the siting of aluminium and to the local factors influencing catalytic sites.

Zeolites as Crystalline Liquids

Results demonstrate that zeolites can behave as ionising solvents (2) or as solid electrolytes (3). These cooperative features are likely to play a significant role in stabilisation of ionic intermediates, such as carbocations, and hence in catalysis. This is particularly so at higher content of framework aluminium where these effects are more noticeable. Consideration of these properties leads to the suggestion that zeolites be considered as "crystalline liquids" (4) and that rates of acid catalysed reactions in zeolites are analogous to those in solution;

$$Rate = k[H^+][S]f_H \ f_S/f_x$$

where 'S' refers to the substrate, X the transition state and f_H etc. represent activity coefficients. Titration studies (32) suggest that the activity coefficient of protic hydrogen in zeolites is a function of framework aluminium content. This is a largely cooperative approach to zeolite catalysis but it does allow for $f_H \rightarrow 1.0$ around Si/Al = 6 (faujasites) suggesting 'isolated' sites at Si/Al $\geq$ 6. As with the Sanderson electronegativity, useful correlations and chemically intuitive interpretations of experimental observations can be made using this model (4) (8). An explanation for solution-like behaviour, of which many examples are given, is related to the very large internal surfaces in zeolites, which are not present in typical solids, which allow for contact between all (or most) of the framework atoms with sorbate molecules.

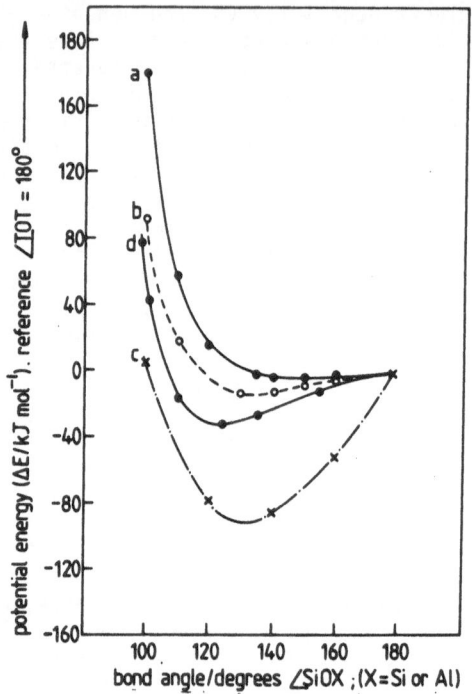

Figure 7. Potential energy curves for (a)SiOSi, (b) SiOAl°, (c) SiOHAl and (d)AlOP units. (STO 3G BASIS).

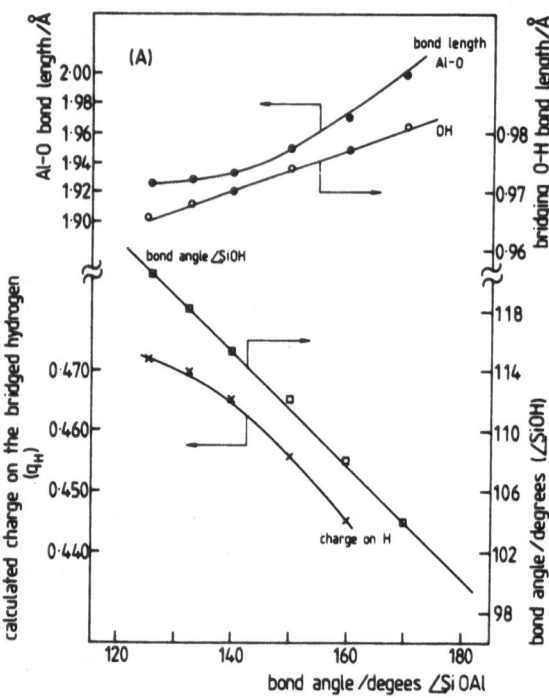

Figure 8a.   Ab-initio MO calculations. Effect of T-O-T angle on $H_3Si(OH)$ AlH_3 unit.

(b) Ab-initio Mo calculation (3-21G; SCF). Symmetrical units X_3 Si(OH) Al X_3 (X=H,F).

Following on from work demonstrating the importance of framework aluminium content (32), explanations for the dependence of zeolite acidity, and catalytic activity, on the distribution and clustering of framework aluminiums appear (5) (6) (7(a)). For example, in the faujasitic framework all 'T' atoms are crystallographically identical having nine second neighbour 'T' atoms three of which are in the three four-ring system and are closer than the other six. This gives rise to four types of Al atom having $i = 0, 1, 2,$ or 3 aluminiums as diagonal neighbours in the four-ring system (6) (7(a)). Weaker acidity, which is observed experimentally at high aluminium content is associated with increased Al site density. The relative numbers of sites N_i ($i = 0, 1, 2, 3$) can, subject to assumptions, be calculated (Fig 3) and strong acidity is usually associated with N_0 and sometimes with both N_0 and N_1 sites. This approach emphasises local composition so that the overall catalytic reaction rate (R) may be represented as

$$R = \sum_i^n n_i r_i \qquad\qquad (b)$$

where n_i are the numbers of sites with specific rate r_i and and $\sum n_i = N_{Al}$ for H-forms of zeolites with all sites (eg per unit cell) available.

Effective acidities may be defined for given compositions and structures, either from experimental (32) or from theoretical considerations (7(a)). Rates may then be defined using a weighted average specific rate ($\bar{r}$) such that,

$$R = N_{Al}\bar{r} \qquad\qquad (c)$$

and $\bar{r}$ is a function of both composition and structure. Effective acidities, which can be related to parameters describing the shielding of aluminiums in the FAU framework by neighbouring aluminiums in either the three four-ring system (p) or in the complete second neighbour tetrahedral system (q), are shown as a function of composition (Fig 4). Both of those parameters are used in correlation of zeolite properties (7(a), 7(b)) (40). Correlation of catalytic rates with these parameters gives

$$R = N_{Al}(1-p)W \qquad or \qquad R = N_{Al}(1-q)W \qquad\qquad (d)$$

where W is a scaling factor. Quite extensive correlations are made using parameter 'q' in recent work (7(b)). More extended coordination spheres may also be considered (8), but recent results (7(a), (b)) suggest that extension beyond the second 'T' atom coordination sphere may be unnecessary. Similarly, ^{29}Si NMR shifts in dealuminated faujasites are interpreted by considering shielding effects of aluminiums in only the first 'T' atom coordination sphere plus the three second coordination 'T' atoms diagonally opposed in the three four-ring system (7(c)). Theoretical (MO) calculations also predict a reduced charge on protic hydrogens in the presence of proximate aluminium 'T' atoms (10).

The dependence of reaction rates on framework composition is reasonably well established for some acid catalysed reactions over faujasitic zeolites. Results for isomerisation of cyclopropane (Fig 2(a)) are fairly typical. Activation energies decrease and reaction rates increase with increase in framework Si/Al until Si/Al $\approx$ 6. Activation energies thereafter remain constant and reaction rates decrease with further increase in Si/Al. These, and similar results, can be interpreted as implying that the strength of the remaining acid sites increases as Si/Al increases to a value of around six, when all the sites are then

249

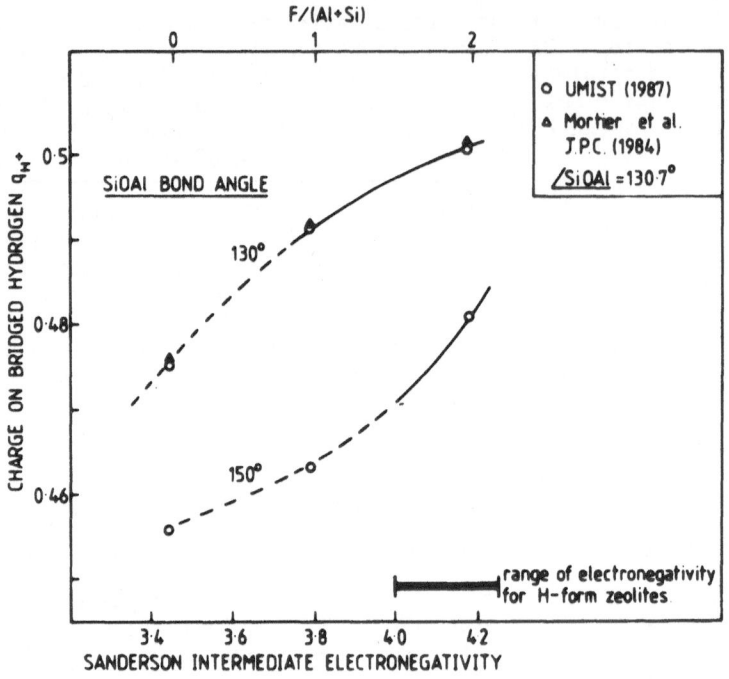

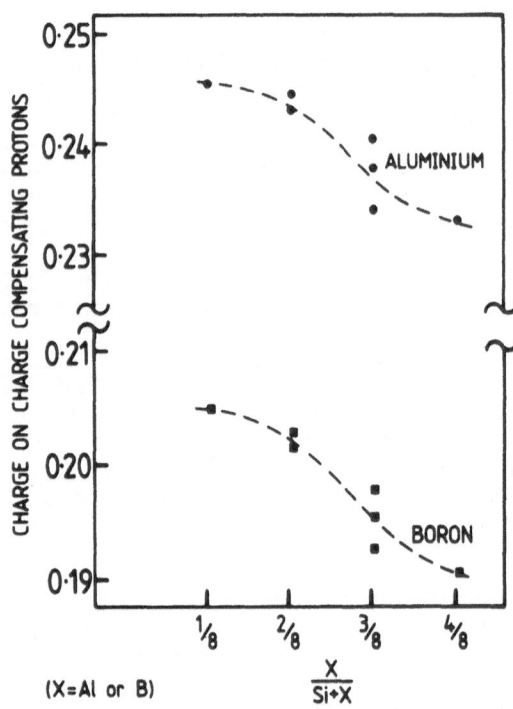

Figure 9. Charges on bridging hydrogens in D4R. $H_xSi_{8-x}Al_xO_{12}(H_8)$.

"strong", resulting in increased reaction rates and reduced activation energies. Further increases in Si/Al result in a loss of strong sites giving reduced reaction rates but cause no further change in activation energy. Assuming that the surface protonation of cyclopropane is rate-determining, the surface activation energy, plotted in Fig 2, is related to the apparent activation (E_a) energy by

$$E_s = E_a + \lambda$$

where λ is the heat of sorption of cyclopropane. In this work both λ and E_a change with Si/Al (Fig 4) but the changes in λ are not sufficient to account for observed changes in E_s so that results in (Fig 2) are not due to sorption effects alone.

These and similar results (8) (29) (33) are frequently interpreted by assuming that, with increase in Si/Al, aluminiums become sufficiently separated to act independently of each other and represent isolated sites. This can be interpreted as implying that cooperative effects between aluminiums are significant when aluminiums are clustered but are negligible when they are separated. A more 'localised' view would emphasise differences in numbers of types of site (N_i). Basically, it is argued that, in the faujasite framework, aluminiums become isolated as Si/Al $\geq$ 6 when sites are all of type N_0. However, this requires implicit assumptions about the dealumination process. For example Fig 5 shows (34) that the dependence of N_0 on Si/Al is influenced by starting composition and relative rates of dealumination of Al clusters (6). Moreover, although catalytic activity per site may become constant at Si/Al > 6, in faujasites, ^{29}Si NMR results suggest that AlOSiOAl units persist to higher levels of Si/Al (26) (35). Thus the influence of local composition is not completely clear in faujasites.

A similar ambiguity holds for H-ZSM-5 and H-MOR. At lower concentrations of framework Al, the rates of acid catalysed reactions over H-ZSM-5 are linearly related to Al content (32). However, at higher levels of aluminium (Al/uc>3) there is evidence both from infrared (30) and catalytic studies (36) for decreased acidity and activity per site. Reduced activity and acidity (Al/uc>3) can be explained by assuming localised cooperative effects which result in mutual screening of aluminiums by neighbouring aluminiums arising from preferential siting. Theoretical (MO) calculations predict preferential siting of Al in T2 and T12 on energetic considerations (37) or in T1, T12 and T9 on the basis of bond angles (38), and the enhanced activity observed on mild steaming of H-ZSM-5 (39) (40) is explained (39) by assuming Al is sited in the four rings (T9, T10). If the latter siting is correct we should expect relative intensities of the ^{29}Si NMR for Si(2Al), Si(1Al) to be as 8:16:64 when there are 8 Al/uc. This pattern is not reported for 8 Al/uc but preferred siting is clearly implied by ^{29}Si NMR at 12 Al/uc (41). The point is that in the presence of preferred siting we cannot be certain that sites are isolated at given values of Si/Al, nor can we be sure that aluminiums in AlOSiOAl units are always less active than isolated aluminiums. Nevertheless there is some evidence that activity per site, in MFI, decreases as aluminium increases and this is reflected in infrared studies (30) but not in catalytic studies of hexane cracking (29). Differences in sample preparation or activation procedures may account for lack of agreement here but a question does remain.

Similar considerations apply to mordenite (MOR). However, structural studies using XRD show, at least for large crystals, that aluminium is (42(a)) preferentially sited in four-rings. Recent work using ^{29}Si NMR suggests that all the aluminiums are paired in synthetic mordenite

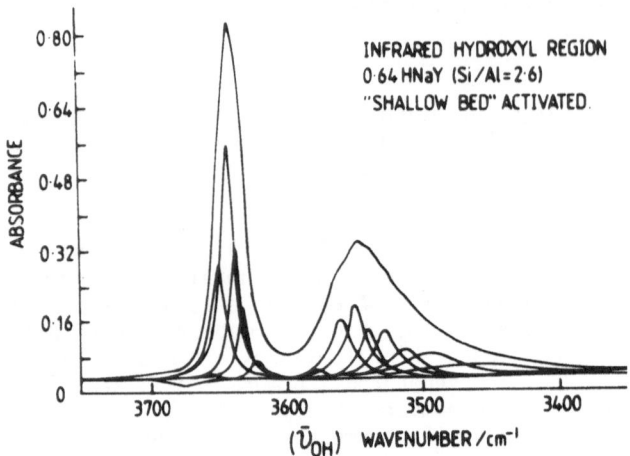

Figure 10. Numerical separation of H.F. and L.F. stretching vibrations
(cauchy shapes). (D. Dombrowsky et al. JCS (F.T.) 1985).

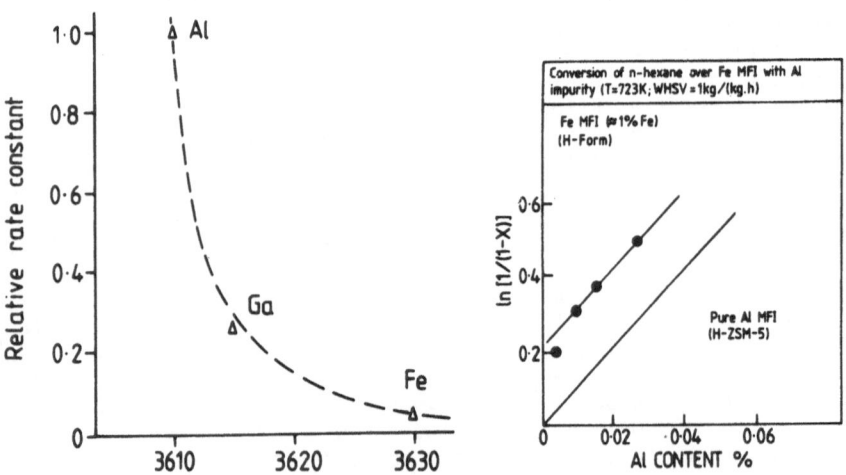

Figure 11.(a) Hydroxyl stretching frequency / cm−1
(b) Effect of Al impurity on catalytic activity of iron
silicates (MFI structure). M.F.M. Post et al. (1988)

(42(b)) and that acid extraction removes either both aluminiums (diagonally paired) or neither (Fig 6). If this interpretation is correct, sites in MOR do not become isolated upon acid extraction so that optimal framework compositions are not explicable in these terms. The optimal rates might then reflect changes in sorption and/or generation of sites of enhanced activity arising from involvement of extra-framework aluminium produced during the extraction. These sites of "enhanced acidity", which are also observed in other zeolites (60) (39) (40), are themselves presumed to arise from highly localised interactions, and theoretical calculations (MO) on simple clusters add some support for this view (43(a)) and (b) as does recent NMR work (43(c)).

Notwithstanding the above it is evident that recent results on dealuminated MOR (43(a)) do show maximal rates at compositions very close to those predicted (8) to have maximum numbers of isolated sites. These results (43(a)) must be accepted as coincidental if the previous considerations, particularly the NMR are correct.

CLUSTER CALCULATIONS: HETEROGENEITY OF BRIDGED HYDROXYLS

Aluminium Siting

Explanations for the dependence of acidity on bulk framework composition necessitates consideration of aluminium siting. Ab-initio MO calculations (37) (38) on simple clusters based on crystallographic coordinates suggest that sites favoured for aluminium are T2 and T12 (37) in MFI, T2 and T4 in FER (37) and T1 and T4 in TON (38). These sites tend to be associated with larger average T-O bond lengths and in FER and TON with smaller average TOT angles. However, in silica-rich zeolites, experimental coordinates are dominated by silicon and oxygen positions and substitution of silicon by aluminium must change coordinates locally to produce more typical Al-O distances. On this basis, and on the results of ab-initio MO calculations for simple units $(HO)_3 Si(OH)Al(OH)_3$, it is likely that low angle sites might show preference for aluminiums which, in the presence of bridged hydrogens, are seen to be particularly stabilised in lower TOT angle sites (125-140 °) as compared to silicons for which SiOSi linkages are stable over a wide range of angles (Fig 7). Consequently it is considered that emphasis be given to low-angle sites when considering aluminium siting and Si(OH)Al links (38).

Acid Strength

Protic hydrogens at lower or moderate temperatures are localised, for example on Si(OH)Al bridges, but have increased mobility at elevated temperature. The properties of the OH bond depend on structure and composition (26). The effect of structure may be simulated by considering the effect of TOT angle on the optimum geometry and charge distribution in units $H_3Si(OH)AlH_3$. Ab initio MO calculations (3-21 G basis with full geometry optimisation) show that the properties of the hydroxyl bond depend on the TOT angle and the Al-O distance (Fig 8a) which may be envisaged, in simple terms, as the polarising effect of AlH_3 on the H_3SiOH unit. As the TOT angle and Al-O distance increase the reduced 's' character in the oxygen hybrid orbital results in weaker OH bonds (increased OH distance). This can clearly influence acidity but the reduced charge separation at high angles may also imply an increased facility for homolytic scission and radical generation.

The importance of local composition is strongly emphasised by ab-initio MO calculations (3-21 G basis) on clusters $H_3Si(OH)TH_3$. Charges on bridged hydrogens and interactions with bases are strongly dependent on

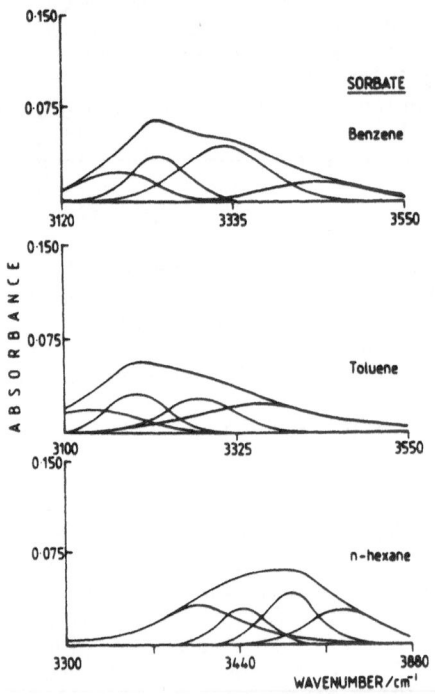

Figure 12. Perturbation of the hydroxy stretch ($\bar{v}_{OH}$) in the I.R. by sorption of weak bases. (J. Datka. et al. J. CAT. 1988).

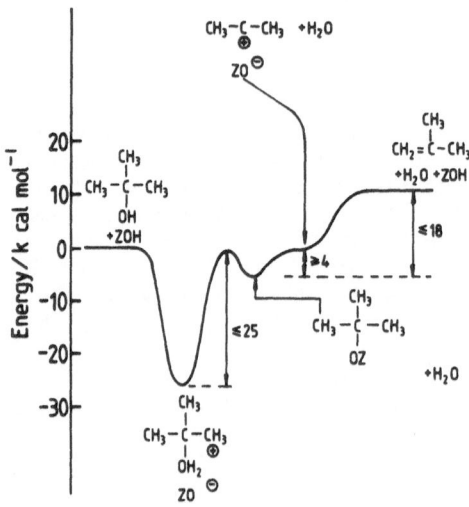

Figure 13. Reaction profile for dehydration of butanol on H–ZSM–5. (Aronson, Gorte, Farneth, White, JACS (1989)).

the substituent 'T' atom (Table 1).

TABLE 1. Effect of isomorphous substitution on Bridged Hydroxyls in unit
(a) $H_3Si(OH)TH_3$ (44(a))

	'T' atoms	Al	B	Ga
Charge on hydrogen	q_H	0.4727	0.4670	0.4710
Bond length	$r_{OH}/Å$	0.967	0.963	0.965
Hydroxyl stretch	$\bar{\nu}_{OH}/cm^{-1}$	3931	3970	3940
Shifts in Hydroxyl Stretch $\Delta\bar{y}/cm^{-1}$ (a) CO		183	82	–
(b) NH_3		1150	548	–

(a) Optimised geometries. C_s symmetry; 3–21G basis.

Similar results are found for MeAPO's (44(b)). Table 2 suggests that strong Bronsted sites may be generated from bridges of Mg or Be to P.

TABLE 2. Characteristics of Bridged Hydroxyls in MeAPO's (44(b))

Unit (a)	$r_e(OH)/Å$	$\bar{\nu}_{OH}/cm^{-1}$	q_H	$|q_0q_H|$
$H_3Be(OH)PH_3$	0.967	3911	0.4910	0.3814
$H_3Mg(OH)PH_3$	0.968	3911	0.5093	0.4229
$H_3Si(OH)AlH_3$	0.967	3931	0.4727	0.4445

(a) Optimised geometries: C_s symmetry; 3–21 G basis.

Compositional effects over larger domains may be simulated in small units by replacing terminal hydrogens by more electronegative atoms such as fluorine (Fig 8(b)). As expected charges on bridged hydrogens increase when the electronegativity of near neighbour atoms is increased (26(b)).

Effects of composition over more extensive regions can be investigated using larger clusters and more empirical MO calculations. For example CNDO/2 calculations on double four-rings (Fig 9) show the weaker acidity associated with clustering of aluminium and confirm the weaker acidity of boron versus aluminium in framework silicates which is suggested by results in Table 1. Similar calculations on double four-rings in SAPO's suggest that the protic charge on bridged Al(OH)Si units can depend upon the position of substitution of 2Si atoms for one Al and one P atom. Substitution in the first 'T' atom coordination results in a higher protic charge than substitution in the second 'T' atom coordination sphere (Fig 14) although both units have the same overall composition $(H_8P_2Si_3Al_3O_{11}(OH)$.

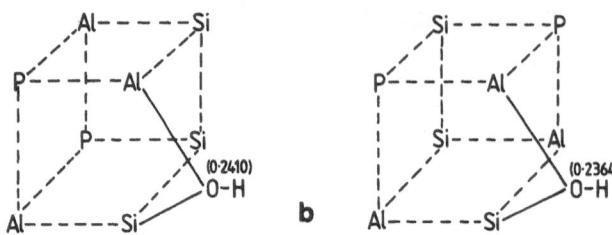

Fig 14 Charges on bridging hydrogens (44(c)) on double four-rings in SAPO's. Units $H_8Al_3Si_3P_2O_{11}(OH)$. Terminal hydrogens and bridging oxygens, other than the protonated oxygen are not shown.

Calculations on both the smaller and larger clusters are in general agreement with experimental results. The double four-ring calculations show that acidity decreases as aluminium content increases and tends to reach a limiting value at low aluminium concentration. The dependence of acidity on isomorphous substitution of Al for B or Ga (Table 1 and Fig 9) is also observed experimentally (62) and is reflected in catalytic activity. Recent results (Fig 11) exemplify this and also suggest (63) that there is no synergism between sites due to Fe and those due to Al (Fig 11(b)). They operate independently with no cooperation suggesting a more general form of equation (b) to allow for 'l' types of active 'T' atoms (Al, Fe, Ga, B etc.) each having n_{ij} sites with specific rates r_{ij} where $\Sigma n_{ij} = n$, the total number of available 'T' sites (eg. per unit cell) for a given type of 'T' site (j). The overall rate/uc (R) is then

$$R = \Sigma \Sigma n_{ij} r_{ij} \qquad\qquad (e)$$

The notion that Bronsted sites function independently is not confined to isomorphously substituted zeolites since three Bronsted sites are reported to exist in H-MOR one of which converts n-octane at seven times the rate of the other two (45). Similarly four types of site are reported, from TPDA studies, in H-Y (6) and IR results (46) are also interpreted, using numerical analysis, by assuming that the high-frequency hydroxyl stretch band in zeolite H-Y is composed of several overlapping bands (Fig 10) so that changes in $\overline{y}_{OH}$ with changes in Si/Al result from changes in the relative proportions of the types of hydroxyl rather than from extensive cooperative influences leading to a change in properties of a single type of hydroxyl (4). Some support for this interpretation appears to come from FTIR resolution enhancement (26) but effects in this early work are complicated by incomplete compensation of water bands. However, similar studies using OD bridges, and OH bridges in completely dry systems, do suggest that the HF band in faujasitic zeolites is not a single band (47). More suprisingly, numerical deconvolution of the hydroxyl bands in Na/H-ZSM-5 which are perturbed by weak bases suggests (Fig 12) that the band around 3610 cm^{-1} is also composed of several overlapping bands (48). It is difficult to explain these results without accepting a localised model for bridged hydoxyls similar to that suggested by theoretical calculations (10).

In practice, of course, there may be averaging effects in reactions at higher temperatures, and reactions involving relatively strong bases requiring only weak acid sites may produce levelling. In this context it should be said that spectroscopic and thermal studies on the sorption and dehydration of alcohols (49) consistently show 1:1 complexes between the alcohol and framework Al sites which tend to decompose at the same rate and yield the same products irrespective of Si/Al ($260 \geq Si/Al \geq 19$) or of structure (ZSM-5, ZSM-12 and MOR). The authors conclude (49) that adsorption leading to reaction takes place at isolated and chemically equivalent sites, associated with framework Al, which are insensitive to extended lattice structure and influenced only by local structure factors.

The Role of Bronsted Sites

Product distributions in hydrocarbon conversions are frequently explained via carbenium ion chemistry. For example it is generally assumed that olefins are directly protonated to generate carbenium ions. However, theoretical calculations predict that the initial interaction involves formation of an alkoxyl specie (10) with the carbenium ion representing an excited state.

$$
\begin{array}{c}
\underset{\equiv Si \quad Al\equiv}{\overset{H}{\underset{|}{\overset{|}{O}}}} \end{array}
\quad \underset{\Longleftarrow}{\overset{C_nH_{2n}}{\Longrightarrow}} \quad
\underset{\equiv Si \quad Al\equiv}{\overset{C_nH_{2n+1}}{\underset{|}{\overset{|}{O}}}}
\quad \underset{\Longleftarrow}{\Longrightarrow} \quad
\underset{\equiv Si \quad Al\equiv}{\overset{\left[C_nH_{2n+1}\right]^+}{O -}}
$$

Although it is possible that such calculations do not make sufficient allowance for the ability of zeolites, particularly when aluminium-rich, to stabilise ions (1), similar conclusions are drawn from studies of alcohol dehydration using ^{13}C NMR, TPD and DTA (49). The alkoxy specie from dehydration of 2Me-2 propanol is of lower energy than the carbenium ion, for which no direct evidence is obtained, but the subsequent chemistry is consistent with carbenium ion intermediates (Fig 13). The positions of the above equilibria are clearly relevant. Free radical species can also be readily generated from olefins or aromatics and the rate of their production can depend upon the acidity of the catalyst (50). In the methanol to gasoline process the involvement of both surface methoxy species and free radicals is proposed (51). There is uncertainty about the location of radicals, which may be defect sites (52), but the alkoxy species are clearly localised and their properties do depend upon the composition and structure of the ≡Si(OH)T≡ bridge (53). The initial activation of hydrocarbons over zeolites is frequently explained, on the basis of product distribution, by direct protonation of C-C or C-H bonds implying superacidity (54) (55) (56). There is growing evidence for these processes but the influences of radicals cannot in all cases be excluded (57).

However, from the present point of view these initial processes can be described as involving a localised active site be it Bronsted, Lewis or radical, although the local environment, geometrically and electronically, may significantly affect the stability of reactive intermediates by cooperative interaction. Similary, secondary reactions, such as bimolecular hydrogen transfer, typically involve interaction between sorbed, localised, carbenium ions and olefins or alkanes. Whilst there may be some mobility of carbenium ions at elevated temperatures it is likely that they are generally stabilised by localisation on strong acid sites although the locality involved in stabilisation may extend over several atoms in a cage depending upon the size of the carbenium ion. These interactions which will be both structure and composition dependent introduce a cooperative effect. Cooperative effects are also seen in the enhanced concentrations of reactive species in micropores (58) which favour the bimolecular hydride transfer over monomolecular cracking processes. Shape-selective effects can also involve the cooperation of more extensive regions of the zeolite either in limiting the formation of bulky transition states, in generating a preferred configuration for sorbed reactants (20) and intermediates, or in cooperatively affecting diffusivities (18). Additionally one can see how cooperative effects could modify the active sites associated with sorbed metallic complexes (or metal clusters) where a cage may force unusual geometry on a complex thus modifying its electronic and geometric properties. If a particular zeolite structure adapts itself to accommodate the complex then we have extensive cooperation between the zeolite and the reactive site in addition to any cooperation between sorbed reactant species and zeolite.

CONCLUSIONS

The activity and selectivity of zeolite catalysts can be relatively well described in terms of both structure and composition. A description of zeolite catalysis in terms of localised versus cooperative effects is less clear not least because these terms require interpretation.

The suggestion that there is extensive cooperation within the solid phase as implied by a simple application of the Sanderson electronegativity concept is not completely satisfactory. This is particularly evident when site concentration is low as in silica-rich zeolites or in isomorphously substituted (siliceous) zeolites where overall activity is additive in terms of site activities. In such cases local structure and local composition largely govern activity. Active sites at low density might almost be considered as defects in the siloxane chains and this is reflected in the band structure (59) for siliceous zeolites. At higher site density (aluminium content) there can be clustering of aluminium and, certainly for clusters involving more than two aluminiums, the activity per aluminium is decreased. Clustering of aluminiums involves cooperative solid-state effects and ESCA studies support this in that the band structure of aluminium-rich zeolites is similar to that of alumina doped with silica (59). However, cooperative effects are, in general, not very significant beyond the second 'T' atom coordination sphere. This is suggested by; (a) explanations for ^{29}Si NMR shifts (7(c)); (b) correlations of catalytic activity, acidic efficiency and hydroxyl frequencies (7(b)) (40) with parameters based on shielding effects; (c) results of molecular orbital calculations on clusters (10) (11). Cooperative solid-state effects operating over a limited region are in keeping with significant covalent bonding in zeolite frameworks. Moreover, the smooth shift in frequency of the HF hydroxyl stretch, on dealumination of FAU, which is taken as evidence for a cooperative effect (4) can also be explained on more localised models for which there is growing evidence.

Long range intermolecular cooperative effects between sorbate molecules forming an intrazeolitic fluid phase can clearly be significant at high loading but are unlikely to be significant in typical catalytic conditions for hydrocarbon transformations.

Sorption and diffusion in zeolites involve interaction between several surface atoms with atoms of the sorbate and there is evidence that energetically favourable sorbate configurations can be stabilised. Similarly the diffusive process can involve guest-host cooperation. All of these effects are, however, present in silica polymorphs which are not catalysts. Catalysis requires sites (localised) where specific interactions culminate in chemical processes. The local structure and composition provide a reactive domain where the interaction between surface atoms and sorbed reactants or intermediates can be very significant. At this time the weight of evidence seems to support the view that catalytic activity is largely determined by local geometry and local composition. However, the capacity of zeolite frameworks to relax over an extended region, in order to optimise geometry at the reactive site may be an important feature. Structural modification by sorbates is known (23) but the significance of extended cooperative relaxation during catalysis is not yet quantified.

Selectivity in zeolite catalysis can involve larger domains, for example when spacial factors limit the size of intermediates in particular cages or pore restrictions reduce access to or migration from active sites. Additionally, there can be selectivity at the reactive site (site selectivity) arising from the generation of preferred reactant or intermediate configurations involving cooperation between the zeolite surface and reactant species over an extended localised domain. This aspect is discussed in this meeting (E.G.D). Reaction over acid zeolites can also be considered in terms of the HSAB theory and orbital versus charge control (61). The "softness" of the zeolite anion is reported to depend on both structure and composition so that cooperation, at least over a reaction domain, is envisaged.

It, therefore, seems reasonable to regard catalysis in zeolites as involving a localised domain in which both structure and composition are of importance. These reactive domains do not have the flexibility found in enzyme catalytic centres but have considerably more site specificity than amorphous acid catalysts. It is, perhaps, because the reaction domain in zeolites is frequently not so extensive that theoretical (MO) calculations on cluster compounds have been so effective in modelling reactive sites.

ACKNOWLEDGEMENTS

I would like to thank my research students and research associates past and present, for their much valued contributions to my education and the Science Research and Engineering Council (SERC) and the Chemicals Industry, particularly BP Research Centre, Shell Research Amsterdam, Crosfield/Unilever and Laporte Industries for financial support.

REFERENCES

1. J.A. Rabo in "Zeolite Chemistry and Catalysis", J.A. Rabo (Editor) A.C.S., 171. 332 (1976).
2. P.H. Kasai and R.J. Bishop, J. Phys. Chem., 77, 2308 (1973).
3. J.A. Rabo, P.H. Kasai, Prog. Solid State Chem., 9, 1 (1975).
4. D. Barthomeuf, J. Phys. Chem., 83. 2, 249 (1979).
5. E. Dempsey, J. Catalysis, 33. 479 (1975).
6. R.J. Mikowsky and J.F. Marshall, J. Catalysis, 44. 170 (1976); R.J. Mikowsky, J.F. Marshall and W.P. Burgess, ibid, 58, 489 (1979).
7. (a) B. Beagley, J. Dwyer, F.R. Fitch, R. Mann and J. Walters, J. Phys. Chem., 88. 1744 (1984).
   (b) U. Lohse, B. Parlitz and V. Patzelova, J. Phys. Chem., 93, 3677 (1989).
   (c) M.T. Melchior and J.M. Newsham, Stud. Sur. Sci. and Catalysis, 49 part B, P.A. Jacobs and R.A. van Santen (Editors), Elsevier (1989) p805.
8. D. Barthomeuf, Mat. Chem. and Phys., 17. 49 (1987).
9. W.J. Mortimer, J. Catalysis, 55, 138 (1978).
10. G.M. Zhidomorov and V.B. Kazansky, Adv. Cat., 14. 131 (1986).
11. J. Sauer, Chem. Rev., 89. 199 (1989); J. Phys. Chem., 91. 2315 (1987).
12. P.J. O'Malley and J. Dwyer, J. Phys. Chem., 92, 3005 (1988).
13. A.G. Pelmenshchikov, V.I. Pavlov, G.M. Zhidomirov and S. Beran, J. Phys. Chem., 91. 3325 (1987).
14. J. Datka, P. Geerlings, W.J. Mortier and P.A. Jacobs, J. Phys. Chem., 89, 3483 (1985); 89, 3489, (1985).
15. R. Vetrivel, C.R.A. Catlow and E.A. Colbourn, in Studies in Surf. Sci. and Catalysis, 37, Editors P.J. Grobet et al., Elsevier (1988) p317.
16. P.B. Weisz, Chem. Tech., 3 498 (1973).

17. (a) P.B. Weisz, Pure Appl. Chem., 52. 2091 (1980).
    (b) E.G. Derouane, Studies in Surf. Sci. and Catalysis, 19.
        S. Kaliaquine and A. Mahay (Editors), Elsevier (1984) pl.
18. R.L. Gorring, J. Catalysis, 31, 13 (1973).
19. R.M. Barrer, "Zeolites and Clay Minerals as Sorbents and Molecular
    Sieves, Academic Press (1978).
20. J.H. de Boer and J.F.H. Custers, Z. Phys. Chem., B25, 225 (1934).
21. E.G. Derouane, J. Catalysis, 100. 41 (1986); 110, 58 (1988); paper in
    the present volume.
22. B. Smit and C.J.J. den Ouden, J. Phys. Chem., 92, 7169 (1988).
23. C. Fyffe et al., J. Chem. Soc. Chem. Commun., 541 (1984).
24. J. Dwyer, Chemistry and Industry, 258 (1984).
25. D. Freude, M. Hunger and H. Pfeifer, Chem. Phys. Lett., 121, 1, 62
    (1986).
26. (a) J. Dwyer, Studies in Surf. Sci. and Catalysis, 37, P.J. Grobet et
        al. (Editors), Elsevier (1988) p383.
    (b) J. Dwyer and P.J. O'Malley, Stud. Surf. Sci. and Catalysis, 35,
        S. Kaliaquine (Editor), Elsevier (1988) p5.
27. P.A. Jacobs, Catal. Rev. Sci. Eng., 24, 415 (1982).
28. J. Dwyer, F.R. Fitch and E.E. Nkang, J. Phys. Chem., 87. 5402 (1983).
29. W.O. Haag, in Proc. 6th Int. Zeolite Conf. D. Olson and A. Bisio
    (Editors), Butterworths (1984) p466.
30. J. Datka and Z. Piwowarska, Zeolites, 8, 30 (1988).
31. W.J. Mortier, Studies in Surf. Sci. Catalysis, 37, P.J. Grobet et al.
    (Editors) Elsevier (1988) p253, K.A. Van Genetchen and W.J. Mortier,
    Zeolites, 8, 273 (1988).
32. R. Beaumont and D. Barthomeuf, J. Catal., 27, 45 (1972).
33. J.R. Sohn, S.J. De Canio, P.O. Fritz and J.H. Lunsford, J. Phys.
    Chem., 90, 4847 (1986).
34. J. Dwyer, F.R. Fitch, unpublished work.
35. A.W. Peters, G.C. Edwards, M.P. Shatlock, T.G. Roberie, K. Rajagopalan
    and E.T. Habib, XI Simposio Iberoamericano de Catalysis, F. Cossio et
    al. (Editors) (1988) p1417.
36. N.R. Meshram, S.G. Hedge and S.B. Kularni, Zeolites, 6, 434 (1986).
37. E.G. Derouane and J.G. Fripiat, Zeolites, 5, 165 (1985).
38. P.J. O'Malley and J Dwyer, Zeolites, 317 (1988).
39. R.M. Lago, W.O. Haag, R.J. Mikowsky, D.H. Olson, S.D. Hellring, K.D.
    Schmitt and G.T. Kerr, Proc. 7th Int. Zeol. Conf., Y. Murokani et al.
    (Editors), Kodansha-Elsevier (1986) p677.
40. A.G. Ashton, S. Batmanian, D.M. Clark, J. Dwyer, F.R. Fitch,
    A. Hinchliffe and F.J. Machado, Studies in Surf. Sci. and Catalysis,
    20, B. Imelik et al. (Editors), Elsevier (1985) p101.
41. C.A. Fyffe, G.C. Gobbi and G.J. Kennedy, Chem. Lett.. 1551 (1983).
42. (a) W.M. Meier, R. Meier and V. Gramlich, Z. F. Krystallog., 147. 329
        (1978).
    (b) J.B. Nagy and E.G. Derouane, A.C.S. Sym. Series, 368, W.H. Flank
        and T.E. Whyte Jr., (Editors) 2 (1988).
43. (a) J. Sauer and W. Schirmer, Stud. Surf Sci. and Catalysis, 37, P.J.
        Grobet et al. (Editors), Elsevier (1988) p39.
    (b) A.G. Pelmenschchikov, G.M. Zhidomorov, and K.I. Zamaraev, Stud.
        Surf. Sci. and Catalysis, 49 part B, P.A. Jacobs and R.A. van
        Santen, (Editors), Elsevier (1989) p741.
    (c) E. Brunner, H. Ernst, D. Freude, T. Frohlich, M. Hunger and H.
        Pfeifer, Stud. Surf. Sci. and Catalysis, 49 part A, P.A. Jacobs
        and R.A. van Santen, (Editors), Elsevier (1989) p623.
44. (a) P.J. O'Malley and J. Dwyer, Chem. Phys. Lett. 143, 97 (1988).
    (b) R. Carson, E.M. Cooke, J. Dwyer, A. Hinchliffe and P.J. O'Malley,
        Stud. Surf. Sci. and Catalysis, 46, A.G. Karge, J. Weitkamp,
        (Editors), Elsevier (1989) 39.
    (c) R. Carson, J. Dwyer and P.J. O'Malley, unpublished work.

45. A.L. Klyachko, G.I. Kapustin, T.R. Bruara and A.M. Rubinstein, Zeolites, 7. 119 (1987).
46. D. Dombrowsky, J. Hoffman, J. Fruwert and T. Stock, J.C.S. Farad. Trans., 1, 81, 2257 (1985).
47. J. Dwyer, J. Dewing, N.E. Thompson, P.J. O'Malley and K.A. Karim, In Press, J.C.S. Chem. Commun. (1989).
48. J. Datka, M. Boczar and P. Rymarowicz, J. Catalysis, 114, 368 (1988).
49. M.T, Aronson, R.J. Gorte, W.E. Farneth and D. White, J.A.C.S., 111. 840 (1986).
50. A.V. Kutcherov and A.A. Slinkin, Studies in Surf. Sci. and Catalysis, 18, P.A. Jacobs et al. (Editors), Elsevier (1984).
51. C.D. Chang, S.D. Hellring and J.A. Pearson, J. Catalysis, 115, 282 (1989).
52. S. Shih, J. Catalysis, 79, 390 (1983).
53. A.M. Breuil, J. Dwyer and P.J. O'Malley, unpublished work.
54. W.O. Haag and R.M. Dessau, Proc. 8th Int. Cong. Catalysis, Berlin, II-305 (1984).
55. A. Corma, J. Planelles, J. Sanchez-Marin and F. Tomas, J. Catalysis, 93, 30 (1985).
56. W.K. Hall, E.A. Lombardo and J. Englehardt, J. Catalysis, 115, 611 (1989).
57. G.M. Kramer and G.B. McVicker, J. Catalysis, 115, 608 (1989).
58. J.A. Rabo, Rev. Sci. Eng., 23. 293 (1981).
59. T.L. Barr, Li Mei Chen, M. Mohsenien and M.A. Lishka.
60. C. Mirodatos and D. Barthomeuf, J.C.S. Chem. Commun., 39 (1983).
61. K.P. Wendlandt and H. Bremer, Proc. 7th Cong. Catal. (Berlin) IV, 507 (1984).
62. C.T-Wu and C.D. Chang, J. Phys. Chem., 89, 1569, (1985).
63. M.F.M. Post, T. Huizinga, C.A. Emeis, J.M. Nanne and W.H.J. Stork. Studies in Surf. Sci. and Catalysis, 46, H.G. Karge and J. Weitkamp (Editors), Elsevier (1989) p365.

LONG RANGE VERSUS SHORT RANGE INTERACTIONS IN ZEOLITES

Wilfried J. Mortier and R. Vetrivel

EXXON Chemical Holland, Basic Chemicals Technology
P.O. Box 7335, 3000 HH Rotterdam
The Netherlands

ABSTRACT

Models based on short range and long range (mainly electrostatic) interaction effects in describing the properties of zeolites are critically examined. Starting point is the Hamiltonian operator. Depending on the property under investigation, the (long range) external potential terms will have to be included in the model. Geometries can be predicted adequately by accounting for short range (repulsive and attractive) interactions only. Energy calculations and molecular interactions are much more sensitive to including the external potential in the calculations because of the predominant influence of the Madelung potential on the charge distribution for inorganic systems. For interactions of molecules with the surface, the absolute magnitude is of no importance, but the potential gradient will determine the properties. The cation distribution is adequately described using short-range interactions only, and is seemingly independent of the Al distribution pattern in the framework. The same could apply to the protons, in which case we might have to revise the current concepts concerning Broensted acidity in zeolites.

INTRODUCTION

Modeling the properties of zeolites increases our scientific understanding of these materials, and our ability for custom- designing catalysts and adsorbents. But since we are dealing with the solid state, traditional computational methods used for isolated molecules do not necessarily apply. The binding in aluminosilicates can be considered to be somewhere intermediate between covalent and ionic, and we can reasonably expect that long range electrostatic effects become more and more important. The question whether we can safely disregard these depends on the property under investigation. We are moreover more often than not limited to empirical models, and extrapolation beyond the observations on which these are based becomes dangerous.

There is no doubt however about the first principles of physics. Solving the Schroedinger equation for our system will guarantee the correct wave function and energy. As soon as we are able to write the Hamiltonian operator, finding the wave function is only a mathematical (although formidable) exercise. In atomic units, the Hamiltonian operator

Guidelines for Mastering the Properties of Molecular Sieves
Edited by D. Barthomeuf *et al.*
Plenum Press, New York, 1990

is written as:

$$\hat{H} = -\frac{1}{2} \sum_{\mu=1}^{N} \nabla_\mu^2 + \sum_{\mu<\nu}^{N} \frac{1}{r_{\mu\nu}} + \sum_{\mu=1}^{N} v(\mu) \qquad (1)$$

$$\hat{T} \qquad + \qquad \hat{V}_{ee} \qquad + \qquad \hat{V}_{ne} \left(v(\mu) = -\sum_\alpha \frac{Z_\alpha}{r_{\mu\alpha}} \right)$$

where the summation indexes run over all electrons (μ,ν) and over all atoms (α) with charge Z_α. It is a characteristic for this operator that for every system of N electrons, the kinetic energy term $\hat{T}$ and the electron-electron repulsion term $\hat{V}_{ee}$ are identical. Two systems can only be different in their external potential V_{ne} (i.e. not only the nuclei, but also all other charges in the surroundings). This forms the basis of the first Hohenberg and Kohn theorem[1], and establishes the foundations of density functional theory[2,3].

We usually have no problem for writing the Hamiltonian operator and solving (or finding approximate solutions to) the Schroedinger equation for a molecule alone in the universe. With more molecules around it, and for the solid state, this becomes however very complicated. Ignoring these on the other hand, means building a bias in our model, and we need to understand the extent of our approximation and the consequences of it.

Bringing two systems together (to form a chemical bond, or allowing a physical interaction only), will usually end up in a lowering of the total energy at first, going through a minimum, and when the electron clouds tend to penetrate, ending in a sharp increase of the energy. Modelling a stable situation then reduces to finding the equilibrium distances and the interaction energy, i.e. trying to find the location of the energy minimum and the depth of the potential well. We will now investigate several situations and try to find some guidelines to whether we may neglect the long range effects or not.

THE GEOMETRY

This basically addresses the position of the energy minimum (distances only) rather than the depth of the potential well. This question has been discussed at length by Catlow and Cormack[4] for the computer modelling of silicates, and the reader is referred to this review paper for a critical survey of the different approximations that have been used.

As long as we are interested in obtaining the geometry of a structure or a complex only, short-range interaction effects suffice, although for silicates, we cannot neglect effects beyond the pair potential of the bonded atoms such as the bond bending terms and the influence of the torsion angles for adequately reproducing the experimental geometries[4].

A recognized difficulty[4] is the inclusion of charge shifts within the framework: a different environment for an atom pair directly influences the size and shape of the electron cloud as well as the electron density. Calculating the charge shifts is most easily done by using the solid state electronegativity equalization method[5,6], and it may be projected that improvements in the geometry calculations can be made. The calculation of the charge distribution however requires taking into account of the long range interactions.

When the long range interactions are included by incorporation of point charges as described below, the possibility of optimizing the geometry of quantum chemical clusters is restricted. It is not possible to optimize the bond lengths of atoms in the boundary of the quantum chemical cluster since they are binding to quantum chemically treated atoms and point charges at the same time. However, this problem can be overcome by using larger clusters so that the atoms for which the geometry needs to be optimized are surrounded by quantum chemically treated atoms only.

We will now no longer focus on the calculation of the geometry and pay more attention to the prediction of the properties.

PREDICTION OF PROPERTIES

If we want to make statements about properties such as the total energy, temperature stability, charge distribution, polarization, electronegativity, etc. ... we obviously have to consider the depth of the potential well for all bonds in the system, which requires a proper consideration of the last term in the Hamiltonian operator (equation 1). For solids such as zeolites, this means that a full summation of $\Sigma Z_\alpha / r_{\mu\alpha}$ will be required.

Charges

For zeolites, the charges matter a great deal in reproducing the magnitude of the total energy. Formal charges Si^{4+}, Al^{3+} and O^{2-} are obviously not applicable in view of the partly covalent bonding. Having access to these charges is of primary importance. The impact of the long range effects on the charge distribution is most easily understood in terms of the recent developments in density functional theory, which allowed an explicit definition of the electronegativity of an atom in a molecule. The external potential explicitly appears in the expression of the electronegativity (χ) for an atom in a molecule[5]:

$$\chi_\alpha = -(\partial E/\partial N_\alpha)_{R_{\alpha\beta}\ldots,N_\beta\ldots} = \chi_\alpha^* + 2\eta_\alpha^* q_\alpha + \sum_{\beta \neq \alpha} \left(\frac{q_\beta}{R_{\alpha\beta}}\right) \tag{2}$$

Defined as in expression (2) (with E referring to the total energy of the system, i.e. molecule or crystal, N_α the number of electrons on atom α, q_α the atomic charge, $R_{\alpha\beta}$ the interatomic distances, χ_α^* and η_α^* the expansion coefficients for the contribution of the intra-atomic energy terms of an atom in a molecule as a function of the atomic charge and using a spherical-atom approximation[5]), the electronegativities for all atoms in the system are equalized to the system's average: $\chi_\alpha = \chi_\beta = \chi_\gamma = \ldots = \bar{\chi}$. This allows the calculation of the charges and the average electronegativity[5].

For the solid state, and certainly for inorganic systems such as zeolites, all interactions have to be taken into account, i.e. it requires a summation of the Madelung potential. This fortunately can be done easily for the crystalline state using the well established methods of Ewald[7] and Bertau[8]. A few points may be stressed here:

(i) The charge distribution is directly influenced by the structure type[6,9,10]. This is obvious from the Hamiltonian as well as from the expression 2 for the electronegativity of an atom in a molecule. Atoms in a different crystallographic site will have different charges. This is thus a direct consequence of the long range interactions. Athough the use of formal charges cannot be correct in evaluating the ionic contribution to the total energy (because these do not depend on the structure type), there is no way of estimating their absolute value either. The atomic charges depend on the definition and for the quantum-mechanical calculations, also on the approximations used. An estimate of the extent of the long range effects on the charge distribution can be made by comparing e.g. the charge on the bridging oxygen in a $(OH)_3Si-O-Si(OH)_3$ cluster with the charges obtained for the same Si-O-Si bridge embedded in an SiO_2 framework. At the same time, a comparison with other clusters for which we know that the chemical properties **must** be drastically different, is instructive[11].

Oxygen charge EEM (ab-initio charge between brackets)[11]

$H_3Si-O-SiH_3$	$(OH)H_2Si-O-SiH_2(OH)$	$(OH)_3Si-O-Si(OH)_3$	FAU	MOR
-0.65039	-0.68765	-0.77565	-0.7991	-0.8426
(-0.66999)	(-0.67983)			

From this table, it is evident that the direct chemical environment will be predominant for the properties (hence the importance of the composition on the chemical properties). A comparison on the other hand with the charge shifts occuring in molecular interactions (of the order of only a few thousands of an electron; see e.g. the charge shifts occuring in water molecules in forming dimeric clusters[12]), we must conclude that embedding the 'active site' in a zeolite will certainly have a significant effect on the properties. Van Santen et. al.[13] estimate the total contribution of the long range effects to the stabilization energy (i.e. necessarily forthcoming from the ionic interactions) to about 10%. With respect to the isolated oxygen in $(OH)_3Si-O-Si(OH)_3$, the average negative charge on the oxygens in FAU increases by 3% and by 8.5% for MOR (all SiO_2 composition), which is of the same order of magnitude as the decrease when **four** hydroxyls are removed from the cluster, i.e. 12%.

(ii) The charge distribution and the average electronegativity refer to two entirely different properties of a system: the charges to an integration of the electron density distribution over the "atomic region", and the electronegativity to the compactness of the electron cloud. We refer to reference 6 for an example: tridymite and crystobalite have about the same electronegativity, but the structure of tridymite is more ionic. The electronegativity inversily correlates with the refractive index and increases with decreasing framework density.

(iii) Unlike for organic chemistry, the electrostatic environment i.e. the long range effects, play a more important role in the description of the properties of a system. Again, this is obvious from equations 1 and 2. This is also the reason why most empirical electronegativity-based formalisms fail to produce good correlations with the properties of inorganic systems: the environment has consistently been neglected[14].

(iv) Adding a constant potential to all atoms does not influence the charge distribution ($\chi_\alpha = \chi_\beta$ is equivalent with $\chi_\alpha + C = \chi_\beta + C$). In terms of quantum chemistry, the energy

$$E' = \int \Psi^* (\hat{H}+C) \Psi \, d\tau = E + C \int \Psi^*\Psi \, d\tau = E + C \tag{3}$$

is minimized for the same wave function and therefore also for the same electron density distribution function. This means that only the potential gradients are really important, in agreement with the experimental findings by Barthomeuf[15].

(v) The EEM method does under no circumstances provide data on the energy of the system, nor about the energy levels. It however supplements the information that can be obtained from quantum chemistry, since it is the natural choice if charge distributions are required. In view of the accurate structure-dependent charges however, calculating the Madelung energy should be directly possible provided that the crystallographic structure can be modeled or used as such.

Binding Energy

Long range effects do directly influence the Hamiltonian operator (equation 1) and therefore also directly the total energy and the binding energy of a system. Equation 2 predicts that any transition from the "isolated" system to the condensed phase will increase the ionicity of the bonds. Embedding an atom in a positive potential will increase its electronegativity and the reverse is true for a negative potential. This is exactly the natural arrangement of the condensed phase. The higher electronegativity for an already negatively charged atom will further enhance its charge, and for the positive atom result in another decrease of the electron density. This effect is certainly not negligible for inorganic crystals such as zeolites.

It is not an easy task to include these long range interactions in the Hamiltonian operator for calculating a wavefunction for a framework moiety. Methods have been demonstrated to account for the long range effects in zeolites[16]. Inclusion of the long range interactions in addition to the short range interactions are practically only possible for small entities of the framework structure. Since we are restricted by the number of atoms that can be treated by quantum mechanics, the long range interactions have to be accounted for by the explicit presence of point charges. Hence, a combination of energy-minimization calculations (for representing the long range interactions) and the quantum mechanical calculations (for the short range interactions) is by far the best possible compromise for simulating a partly ionic material like a zeolite.

Particularly relevant are the attempts in doing so by Vetrivel et. al.[17] and Seiti[18]. Vetrivel et al. analyzed the effect of surrounding charges on the calculated properties of a quantum-cluster. It was realized that it was essential to mimic the crystalline surroundings in that the atoms of the cluster need to be subject to the same Madelung potential and potential gradient. The electron populations are indeed crucially dependent on the geometry and the size of the surrounding point charge cluster (which raises the questions of the appropriate boundary conditions), in agreement of course with equation 2.

The approach followed by Vetrivel et al.[17] involves the interfacing of the quantum chemical calculations with energy minimization calculations. An explicit quantum mechanical cluster of atoms representing the active site is embedded in an array of point charges representing the zeolite framework. The point charges are centres without basis functions and they represent only the long range effects. The presence of point charges does not affect the time needed for the calculation of two-electron integrals but requires the calculation of some one-electron integrals. Fixing the boundary for the quantum chemical cluster by embedding it in an array of ions with their formal charges may be suitable for fully ionic materials, but for a partially covalent material such as zeolites, modifications of the charge and position of the point charge ions in the periphery are necessary for an adequate simulation. These calculations were carried out using the GAMESS (General Atomic and Molecular Electronic Structure System) package developed by Guest and Kendrick[19]. The electronic charges at the boundary of the quantum chemical cluster depend on the formal charge assigned to the point charges.

Another succesful approach was followed by Seiti[18] in evaluating the effect of the environment on the properties of a complex formed between a Broensted acid site and an adsorbed molecule. For the complexation of donor molecules such as NH_3, H_2O, PH_3 and H_2S with the bridging OH groups, two energy minima are consistently found, one for the covalent complex (such as $ZOH...NH_3$) and one for the ionic complex ($ZO^-...NH_4^+$ e.g.). Contrary to experimental observations, the proton transfer seems not to be convincing for these strong donor molecules. Obviously, other effects were not accounted for, i.e. the strong electrostatic fields existing inside the zeolite cavities. These environmental effects were included by accounting for the Madelung potential (using Ewald's method, and EEM charges) in the Hamiltonian operator, for which a method was developed. This electrostatic perturbation stabilises both the ionic and the covalent complexes, but in the case of the interaction of ammonia with the bridging hydroxyls, the ionic complex (formation of ammonium) is further stabilized by about 50 kcal/mole with respect to the covalent complex. In agreement with the previous discussion, the charge on the proton also increases after inclusion of these long range effects in the calculations. It must be stressed that these calculations were performed on the ab-initio level with full geometry optimization.

In view of the drastic effect of the long range interactions for ionic or partly ionic systems, the question should be addressed whether previous quantum-mechanical calculations based on framework moieties are indeed representative. This should affect all properties that are derived from this type of calculations. The prediction of trends caused by the variation of selected parameters is feasible at the most, but this also needs attention.

Zeolites vs. Oxides

An analysis of equation 2 readily reveals that there should be an intrinsic difference between the bulk properties and the surface: the external potential in the bulk will be much more pronounced than at the surface. It is a characteristic of course that for zeolites, the "active surface" is embedded within the crystals, while for oxide catalysts, the molecules can only interact with the surface. This is fundamental, and should therefore lead to specific properties of zeolites which are not found for oxides. Particularly relevant in this respect is the ESCA work by Vinek et al.[20]. Magnesium and oxygen binding energies are compared for a series of acidic and basic oxides with those of Mg-exchanged zeolites. Acidic character (Electron Pair Acceptor Strength) and basic character (Electron Pair Donor Strength) should be associated with high and low electron binding energies respectively. Some of the data is presented in the following table (binding energies of the Mg 2p and O 1s electrons in eV):

	Basic		Acidic				Zeolites	
	La_2O_3	MgO	$MgSO_4$	$MgHPO_4$	AL_2O_3	SiO_2	MgNaX	MgNaY
Mg 2p		48.0	49.1	49.8			50.2	49.4
O 1s	529.0	530.2	531.1	531.8	531.8	533.1	530.6	530.9

The Mg 2p binding energy of Mg-exchanged zeolites ranks among the values found for the acidic catalysts; the oxygen 1s binding energies for the zeolite-framework oxygens rank among those of the basic oxides. This signifies that zeolites do at the same time combine strong acceptor (acid) and strong donor (basic) properties. This certainly is a fundamental distinction between zeolites and other acid or base catalysts, due to their structural arrangement.

Molecular interactions

The above discussions also apply to molecules residing in the zeolite channels: these wil be subject to a different potential gradient than when adsorbed on the surface of the crystals. Also the long range electrostatic interactions will have to be included properly in the Hamiltonian (potential gradients) in order to obtain meaningful results. The polarization of the adsorbed molecules can again accurately be reproduced by the EEM method. Changes in bond ionicity are then indicative for changes in bond strength.

Qualitatively, the EEM method also parallels Gutmann's bond length variation rules[21]. Not all methods will allow the calculation of the correct charge shifts upon molecular interactions: in a study of alkylboranes and alkylamines and their adducts with NH_3 and BH_3, Linert et. al.[22] concluded that CNDO and MINDO MO calculations did not reproduce the correct polarization of the molecules. Also the EHT method often fails to adequately describe ionic systems due to the neglect of the electrostatic forces[23]. It is in this sense contradictory, that for the larger systems the semi-empirical methods would be preferred, while the outcome of the calculations can hardly be trusted.

Traditionally, molecular interactions are approached by looking at the van der Waals interaction energy. The expressions for quantitatively estimating this contain empirical parameters for the electronic description of the molecules which can be adjusted for any type of situation. These parameters refer to the dipole moment and polarizability of the interacting species. These were required however at a time when detailed quantum-mechanical calculations were not yet feasible. It should be possible today to estimate these interaction energies directly starting from ab-initio calculations of the interacting molecules (a warning about the limited applicability of the semi-empirical methods in these matters was already given in the foregoing paragraph). Realizing that these interaction terms are purely electrostatic in nature, and that the EEM-method provides "ab-initio quality" charges and charge shifts, Uytterhoeven[24] came to the conclusion that in principle one should no longer have to fall back on these empirical parametrizations. This was illustrated by calculating the complexation energies of water and ammonia with univalent cations from ΔE(reaction) = E(complex) - n E(isolated molecule). The terms carried in the energy calculations were considered purely electrostatic in nature (E = 1/2 Σ $q_i q_j / R_{ij}$; especially also since only ΔE is important and that it can reasonably be expected that the covalent contributions cancel). This directly addresses at the same time the dipole-dipole interaction energy and polarization of the molecules since not only the extra electrostatic terms arising from bringing the molecules together are accounted for, but also the changed electron density distribution after forming of the complex (polarization energy) is estimated in this way. A single correlation was found for all cation types and for any number of molecules in the complex. The fact that two different lines are found for ammonia and water is just a consequence of the definition of an atomic charge in a molecule: this strongly depends on the method used and it is impossible to attach absolute numbers to it (using formal charges is irrealistic as well). In this case the charges referred to those that would have been obtained after a Mulliken[25] population analysis of STO-3G wave functions. It should also be mentioned that this can also be applied to the solid state since not only charge shifts within the adsorbed molecules, but also changes in the charge distribution for the (zeolite) framework can be calculated by an adequate accounting for the long range effects.

Cation distribution

Long-range effects will also directly influence the potential at the cation-exchange sites (depth of the potential well). The influence of these effects was theoretically studied by Van Dun et. al.[26,27,28] using a statistical thermodynamical approach, and experimentally verified for a series of cation-exchanged Y type zeolites as a function of the temperature (a parameter naturally occuring in a statistical-thermodynamical model)

The following (short-range) interactions were accounted for in the model: repulsive interactions only occur between cations on adjacent sites I and I'; a cation in site II is considered to behave ideally, i.e. it does not interact with any of the cations in other sites. The Bragg-Williams approximation is most transparent and has the advantage that the expressions can be written immediately for any structure type and for any kind of interaction between cation sites and adsorbed molecules (by simply adding the interaction energies to the site energies; this is not so for the exact solutions which essentially give the same picture) The following equations then apply for dehydrated faujasite- type systems: for site I:

$$\frac{P_I}{1-P_I} = j_I \exp (\mu + e_I + 2P_{I'}w)/kT \tag{3}$$

for site I'

$$\frac{P_{I'}}{1-P_{I'}} = j_{I'} \exp (\mu + e_{I'} + P_I w)/kT \tag{4}$$

and for site II

$$\frac{P_{II}}{1-P_{II}} = j_{II} \exp (\mu + e_{II})/kT. \tag{5}$$

Together with a restriction for the total number of cations per unit cell

$$\overline{N} = 16P_I + 32P_{I'} + 32P_{II}. \tag{6}$$

where P_i denotes the populated fraction of the site i; $-e_i$ the stabilization energy of the cation at the site i. (Only energy level differences are important, the energy level of site I was fixed to zero), $-w$ the potential energy for any neighbour pair of cations in sites I and I', and $j_i(T)$ is the partition function for the three vibrational degrees of freedom of a cation at the site i (which we do not have to consider further if these are equal for the three different kinds of cation sites).

For all systems considered, it was found that an accurate prediction of the cation distribution could be made using the same three parameters for all the structures (with the same cation type) and all temperatures, i.e. $e_{II}-e_I$, $e_{I'}-e_I$, and w. These were moreover found to be independent of the number of protons and the details of the cation distribution itself. It was therefore proposed that the cations themselves generate their site energy, primarily determined by short- range interaction with the nearby framework oxygens. The site energy level differences were found to increase in the order $Na^+>Sr^{2+}>Ca^{2+}$. The site preferences are for monovalent $I>II>I'$ and for divalent cations $I>I'>=II$. The energy level differences also vary in a linear way with the number of Al per unit cell.

Apparently, the description of the cation distribution is possible using short-range interaction models only (i.e. irrespective of the details of the cation distribution at other sites). This is because only the energy-level differences are of importance. The cations are located at the energetically most favourable sites (taking the short-range repulsions into account of course). The Al distribution seems also not to matter. The same question can be asked for the location of the protons in the framework: is there any reason why these should be located at the bridging oxygens in a Al-OH-Si configuration, excluding Si-OH-Si cases? What is obvious for isolated framework moieties might no longer apply to the solid state where long range effects dominate the electronegativity of the framework atoms and therefore also their charges. The more negatively charged oxygens (preferred location for the protons, in analogy with the cation distribution) might not necessarily bridge Si and Al. It was indeed found by De Tavernier[29] for different Al distribution scenarios in the mordenite-type framework, that the most negatively charged oxygens are located in the large channels at Si-O-Si bridges. Derouane et al.[30] also came to the conclusion that in view of the insensitivity of the oxygen charge on the isomorphous substitution, a Si-OH-Si should not be ruled out. Do we have to reconsider our concepts about Broensted acidity? Most probably in view of these and previously discussed long range interaction effects.

CONCLUSION

The present "birds-eye view" of the long range interaction effects suggests that we must be willing to revisit several of our concepts. For inorganic systems, the charge distribution, and therefore also the properties, are dominated by the long range electrostatic interactions. We now have several (complementary) tools at our disposal which might make this formidable task possible.

ACKNOWLEGMENT

WJM thanks the Belgian National Science Foundation (NFWO) for funding this research by a former permanent position as Research Director (Onderzoeksdirekteur).

REFERENCES

1. Hohenberg, P. and Kohn, W. Phys. Rev. Sec. B 136: 864 (1964).
2. Parr, R.G. in *Electron Distributions and The Chemical Bond*, Coppens, P. and Hall, M.B. Eds., New York: Plenum Publ. Corp., (1982) p. 95.
3. Parr, R.G. Ann. Rev. Phys. Chem. 34: 631 (1983).
4. Catlow, C.R.A. and Cormack, A.N. Int. Reviews Phys. Chem. 6: 227 (1987).
5. Mortier, W.J., Ghosh, S.K. and Shankar, S. J. Am. Chem. Soc. 108: 4315 (1986).
6. Van Genechten, K., Mortier, W.J. and Geerlings, P. J. Chem. Phys. 86: 5063 (1987).

7. Ewald, P.P. Ann. Phys. **64**: 253 (1923).
8. Bertaut, F. J. Phys. Radium **13**: 499 (1952).
9. Van Genechten, K. *Ph. D. Thesis #160, K.U. Leuven, Fac. Agronomy*, (1987).
10. Van Genechten, K.A., and Mortier, W.J. Zeolites **8**: 273 (1988).
11. Uytterhoeven, L. *Ph. D. Thesis #000, K.U. Leuven, Fac. Agronomy*, in preparation.
12. Kollmann, P.A., and Allen, L.C. J. Chem. Phys. **51**: 3286 (1969).
13. van Santen, R.A., van Beest, B.W.H. and de Man, A.J.M. *this conference*.
14. Mortier, W.J. Structure and Bonding **66**: 125 (1987).
15. Mirodatos, C. and Barthomeuf, D. J. Catal. **114**: 212 (1988).
16. Jackson, R.A. and Catlow, C.R.A. Mol. Simulation **1**: 207 (1988).
17. Vetrivel, R., Catlow, C.R.A., and Colbourn, E.A. Proc. R. Soc. Lond. **A 417**: 81 (1988).
18. Seiti, K. *Doct. Thesis, Univ. PARIS VI, Chimie*, Paris (1988).
19. Guest. M.F. and Kendrick, J. *GAMESS user manual, University of Manchester Computer Centre*, Manchester (1986).
20. Vinek, H., Noller, H., Ebel, M. and Schwartz, K. J. Chem. Soc. Faraday Transactions I **73**: 734 (1977).
21. Gutmann, V. *The Donor-Acceptor Approach to Molecular Interactions* New York: Plenum Press (1978).
22. Linert, W., Gutmann, V., and Perkins, P.G. Inorg. Chim. Acta **69**: 61 (1983).
23. Wang, Y., Nordlander, P. and Tolk, N.H. J. Chem. Phys. **89**: 4163 (1988).
24. Uytterhoeven, L. *Proceedings of the 8th Int. Zeolite Conference*, Amsterdam (1989), in press.
25. Mulliken, R.S. J. Chem. Phys. **23**: 1833 (1955).
26. Van Dun, J.J. and Mortier, W.J. J. Phys. Chem. **92**: 6740 (1988).
27. Van Dun, J.J., Dhaeze, K. and Mortier, W.J. J. Phys. Chem. **92**: 6747 (1988).
28. Van Dun, J.J., Dhaeze, K., Mortier, W.J. and Vaughan, D.E.W J. Phys. Chem. Solids **50**: 469 (1989).
29. De Tavernier, personal communication.
30. Derouane, E.G., Fripiat, J.G. and André, J.-M. *personal communication*.

ACID FUNCTION IN ZEOLITES: RECENT PROGRESS

Jule A. Rabo and Gregory J. Gajda

UOP RESEARCH AND MOLECULAR SIEVE TECHNOLOGY
Tarrytown Technical Center
Tarrytown, New York 10591

INTRODUCTION

The purpose of this paper is to review the chemical and structural characteristics of zeolites deemed relevant to acidity and to acid catalysis. Several tentative conclusions are also drawn, in order to interpret some of the important chemical phenomena. In view of the availability of several detailed reviews on catalytic data, these will not be reviewed here.

ACID-BASE CHEMISTRY

An acid is generally defined as a species that can donate a proton, and a base as a species that can accept it (Bronsted). The dissociation of an acid HA in solvents is described as an acid-base equilibrium: $HA+S \Longleftrightarrow A^- + SH^+$. In solution, the ionization of acid HA leads to a new acid with solvent S, forming HS^+. This acid-base concept is applied in a wide chemical scope from neutral to positively and negatively charged systems. The extent to which an acid ionizes depends on the acid itself and on the basicity of the solvent applied. Therefore, the solvent plays a major role in the ionization process[1]. Hammett introduced a method to determine acid strength by measuring the degree of protonation of basic indicators in acid solutions. The Hammett acidity function, H_o, is derived from $B + SH^+ \Longleftrightarrow BH^+ + S$ (SH^+ is solvated proton, B is the base, S is solvent)

$$H_o = pk_{BH+} - \log \frac{[BH^+]}{[B]}$$

The Hammett acidity concept postulates that the H_o function was unique for any particular series of acid solutions of varying acid concentration. This postulate has broad applications but it also has limitations with bases which do not give a straight plot between log $[BH^+]/[B]$ and H_o. In solution system, the solvent may have major effect upon both the proton donor and acceptor, depending on character-istics such as polarity, polarizability, etc. best represented by the dielectric constant. Thus, the acid function in solution can be adequately characterized, at least for systems for which the properties of constituent molecules are known[1].

Guidelines for Mastering the Properties of Molecular Sieves
Edited by D. Barthomeuf *et al.*
Plenum Press, New York, 1990

With the advent of recent progress in organometallic chemistry, producing a large variety of complex molecules with acidic or basic character, many new observations were made requiring refinements of the rules governing acid-base interaction. The refinements advanced by Chatt and Pearson categorize systems as hard and soft acids and bases(2,3,4,5). While the concept was developed mainly to explain the interaction between Lewis bases and Lewis acids they are also relevant to protic systems. It was found that in nucleophilic displacement reactions the rate of displacement was in most cases influenced by the proton affinity of the base. However, in other cases the rate depended mainly on the polarizability of the base. To explain these rate phenomena, acids and bases were grouped into "hard" versus "soft" species on the basis of low versus high polarizability, respectively. In equilibrium measurements a similar characterization was found useful for acids and bases, also allowing for a base to be both soft and strongly proton bonding. In general, hard acids prefer to bind with hard bases and soft acids with soft bases. Importantly, among the hard acids H^+ is found to be the most typical(3,4). It should be considered that in the formation of covalent bond the two-center coulomb integral is maximized using bonding orbitals of similar size and similar electronegativity. This explains that hard acids prefer hard bases. For soft acids and bases, a significant or even a major part of the interaction between the acid-base system is derived from various aspects of electron correlation. This interaction may rest on the polarizability interaction (London) but it may also involve electronic correlation between high-energy, exited state orbitals of the acid and base molecules.

Recently, Pearson(4) defined the terms "absolute electronegativity" and "absolute hardness" based on molecular orbital considerations. Accordingly, the absolute hardness is twice the energy gap between the highest occupied valence orbital and the lowest unoccupied orbital. With this definition in mind he suggests the following characteristics in the chemistry and interaction of soft and hard acids and bases

- Hard acids prefer bonding with hard bases and soft acids with soft bases.

- Hard molecules have a large "energy gap" while soft molecules have small "energy gap".

- High polarizability is characteristic of soft acids and bases.

- While the electron chemical potential is constant in a molecule, the hardness varies from atom to atom.

- Soft molecules undergo unimolecular rearrangements more readily than hard molecules.

These generalized considerations are of particular importance because they are consistent with the results of recent theoretical work on the chemical bonding in large electron deficient (cationic) molecules such as carbocations, carboranes and their metal complexes.(6,7).

In spite of obvious differences between solution chemistry and zeolite acids, the basic aspects of the chemistry between the solid proton donor and the acceptor molecules are the same for both systems. Therefore, the acid catalytic activity of H-zeolites depends on the following factors

- Acid strength and concentration of zeolite hydroxyl groups.
- Media (solvent) and diffusion effects in zeolites.
- Base strength of reactant molecules.
- Chemistry of the carbocation-like reaction intermediates.

274

It is clear that in zeolites the interaction between acid sites and reactant molecules involves not only the zeolite's protic site but also its media/solvent effect: the contribution from the surrounding zeolite crystal(8). The following sections will discuss zeolite characteristics relevant to the acid function.

CHEMISTRY AND CHARACTERIZATION OF ACID SITES IN ZEOLITES

Formation of Acidic O-H Groups

Contemplating the chemistry of introducing protons into zeolites one should keep in mind that the protonation is carried out after crystallization, usually by thermolyzing NH_4^+ or other proton precursor cations(9,10,11). Following removal of the base one frequently observes some loss of zeolite crystallinity as well as a tendency toward framework aluminum hydrolysis, especially in the presence of steam. This is particularly true for aluminum rich zeolites (A,X) where conversion of the NH_4^+ form to the H^+ form by thermal treatment results in total loss in crystallinity(10). Thus, our experience shows that with zeolites the proton is not an ordinary cation like alkali or other metal cations. Clearly, the interaction between proton and zeolite framework oxygen can be fairly called a "proton attack" since it causes substantial changes in the bonds surrounding the newly formed hydroxyl group(12). This, of course, should be expected considering the very high electron affinity of the proton (13.5 eV) relative to even the smallest alkali cation (Li = 5.3 eV). Thus, it is expected that the formation of acidic O-H groups will cause substantial changes in adjacent T-O-T bonds in the crystal framework.

One of the most important questions is, whether the change brought about by the "proton attack" on zeolite oxygens remains a local phenomenon limited to atoms adjacent to the O-H group or the zeolite crystal responds by a concerted readjustment of the whole crystal lattice, in order to minimize loss of lattice resonance and crystallinity. See the discussion in the following sections.

Measurements of Acid Strength and Distribution

The measurement of the number, type and strength of the acid sites is the key experimental data regarding zeolite acidity. Our discussion will focus on Bronsted acidity, although methods to distinguish Lewis acidity also have been developed. See the discussion by Aboul-Gheit, et al(13), for example. The traditional methods for measuring the strength of strong acids involve the use of colorimetric indicators, such as the Hammett indicators, which are titrated against the acid. An example using amorphous aluminosilicate is given by Hashimoto, et al(14) who find a distribution of both Bronsted and Lewis sites with maximum Hammett acidities of about -13. Unfortunately, these indicators are too large to fit into the pore system of most zeolites. Other methods, such as thermometric titration, differential scanning calorimetry (DSC), temperature programmed desorption (TPD) and catalytic rate measurements (e.g. cumene cracking, olefin oligomerization) have been proposed to measure some function of the acid strength and number.

Calorimetric Methods: Thermometric Titration and DSC

The use of calorimetric methods of measuring acid strength distributions is well established. These are based on the assumption that the acid strength is directly related to the heat of adsorption (or desorption) of a standard base. Examples include the work of Tsutsumi et al (15) and Auroux et al (16). Two specific methods will be discussed: thermometric titration and DSC. Thermometric titration using n-butylamine

275

in benzene was used by Bezman[17] to measure the acid strength distribution
in HY and by Pellet et al [12] for various treated H-Y zeolites and
amorphous aluminosilicates. This method works well for large pore
materials with 3-dimensional channel systems. But, as discussed by Bezman
and Pellet et al, it is unsuccessful even with mordenite and also with
smaller pore zeolites. They propose that the problem is the blockage of
the channels in mordenite by strongly adsorbed n-butylamine. Thus this
method is of limited utility, but could possibly be extended by use of
more competitive solvents or higher reaction temperatures to provide
greater mobility of the titrant. Extension of the experimental technique
to permit the use of thermometric titration with a wider variety of
zeolites would be quite valuable. An alternative microcalorimetric
measurement is discussed by Macedo et al[18], using ammonia to study the
acid strength in HY zeolites after various dealumination treatments. The
thermometric titration with n-butylamine shows that the initial heat of
sorption is much greater for HY than for amorphous aluminosilicate
(~35 kcal/mol versus ~20 kcal/mol, respectively). This implies that HY
is a stronger acid. The same thermometric titration also shows that the
amorphous alumina-silica debris phase formed in H-Y zeolite upon extensive
steaming closely resembles the acidity of the amorphous silica-alumina gel.

An alternative to thermometric titration is differential scanning
calorimetry (DSC) which measures the size of the endotherm associated with
desorption (usually of ammonia in tests of zeolite acidity) from a solid
material. This technique thus provides the temperature and heat of
desorption of standard bases. DSC has been applied by Aboul-Gheit
et al[19] to the study of aluminosilicates and H-mordenite. Aboul-Gheit
et al have developed techniques[20] for eliminating endothermic peaks due
to water desorption from the zeolite, thus simplifying the spectrum. This
technique combines features of TPD and thermometric titration. Care must
be taken to insure that the measurements are not limited by heat or mass
transfer, and that desorption from weak sites is not blocked by adsorbates
at strong acid sites. The former can be checked by varying the rate of
heating the sample, the sample size or the flow rate, while the latter can
be controlled by preloading the sample with adsorbate at various temperatures and comparing the distributions. These procedures are similar to
those discussed by Kapustin, et al[21] for controlling similar effects in
TPD spectra. In addition, Vannice et al[22] discuss errors that may
result from changes in the thermal conductivity of the gases surrounding
the sample.

Temperature Programmed Desorption

Recent reviews of TPD are given by several authors[23], while Forni
and Magni[24] discuss TPD as applied to acid strength distributions in
zeolites. TPD consists of heating a sample at a constant rate and
measuring the quantity of material desorbed at each temperature. As
discussed by Hashimoto et al[25], this data can be converted into a
desorption energy distribution which is related to the acid strength
distribution of the material. The assumption is that the desorption
energy is primarily the reprotonization energy of the acid. Thus a
stronger acid requires a higher energy to reprotonate, but this direct
relationship may be modified by the type of acid site. This technique can
be extended, by suitable choice of adsorbates to differentiate intrapore
acidity from surface acidity, as demonstrated by Take et al[26].

Typically it is assumed is that the calculated energy is primarily that
due to the desorption of the ammonia or other adsorbate. As discussed by
Forni and Magni[24] and others[23], the apparent energy may be dominated
by diffusional (or heat transfer) effects. Forni et al[27] have tested
the effects and propose that in the case of ammonia the intracrystalline

diffusion is controlling in zeolites. The measured activation energy for diffusion is so large ($E_a \sim 30$ Kcal/mole) that they suggest a major interaction between the desorbing ammonia and available acid sites (e.g. multiple adsorption/desorption steps). These effects, and others, are discussed. The effects of heat and mass transfer can be quantified by varying the catalyst quantity and flowrate during TPD measurements. When these effects are properly taken into account, very good agreement between TPD and calorimetric data is obtained, although the TPD data provides a partially averaged value for acid strength, rather than a full distribution.

Summary

- H-zeolites have larger heats of adsorption for nitrogen bases compared to amorphous aluminosilicates. This implies that H-zeolites are stronger acids.

- Calorimetric methods provide the most quantitative data on the strength of the acid/base interaction.

- Thermometric titration has the advantage of being essentially isothermal, but is subject to channel blockage effects due to strongly adsorbed bases.

- DSC provides data similar to thermometric titration but is subject to additional complexities in the calculation of acid strength distribution.

- TPD is useful, simple and versatile, but is strongly affected by heat and mass transfer. These effects must be carefully taken into account for high accuracy.

- Forni et al, and others, have suggested the strong influence of diffusion (probably multiple adsorption/desorption phenomena) during TPD on zeolites.

X-ray Photoelectron Spectroscopy and X-ray Emission Spectroscopy

X-ray emission spectroscopy uses a monochromatic x-ray source to excite secondary x-ray emission with frequencies characteristic of the various elements in the sample. This technique probes the full bulk of the crystal. For details of the experiment see Dowell(28). X-ray photoelectron spectroscopy (XPS) is a surface sensitive technique which uses a monochromatic x-ray source to excite photoelectrons from the sample. The energy of these electrons is characteristic of elements in the sample for "core-level" electrons, but may also represent hybridized molecular orbitals for "valence level" electrons. A description of an XPS system is given by Barr(29). As discussed by Okamoto et al(30), the surface sensitivity of the XPS techniques requires appropriate precautions to insure that the surface of the material is representative of the bulk.

Both spectroscopic techniques show changes in the binding energy (BE) of electrons on Si, Al and O with changes in the zeolite framework Si/Al ratio. In addition, there is a large shift in BE of Al on changing from tetrahedrally coordinated to octahedrally coordinated(31). This provides a method of detecting non-framework Al in steamed zeolites, as does solid-state-NMR (see below).

Greater detail is achieved in the XPS data, and more detailed analyses have been undertaken. Taniguchi and Takaishi(32) have correlated the shifts in the Si(1s) and Si (KLL) lines with the oxygen polarizability.

West and Castle(33) determined a correlation of Auger parameter
[BE(SI$_{1s}$-BE(SI$_{KLL}$)-hv(incident x-ray energy)] and silicate polarizability.
Taniguchi and Takaishi argue that in zeolites, the major polarizability
will be due to the oxygens. By extending the correlation to the measured
values for Na-mordenite and NaA zeolites, they derived oxygen polarizabilities [α(O)] of 2.2Å^3 and 2.5Å^3, respectively. This compares
to a calculated value for cesium mordenite of 2.45$\pm$0.15Å^3 by Takaishi
and Hosoi(34). They also calculate the α(O) value for A zeolite for a
range of ionicities [q=8 is the total negative charge on the four oxygens
attached to a T atom, thus q=8 is fully ionic] and derive α(O)=3.88Å^3
for q=8 and α(O)=2.0Å^3 for q=2 in NaA. The q value is related to the
ionicity of the M-O bond and gives a partial charge on Na of ~0.38 e$^-$.
The trend toward higher polarizabilities with decreasing Si/Al ratio is in
accordance with theoretical predictions, such as Derouane and Fripiat(35),
based on the assumption that polarizability of O is directly related to
the effective negative charge on O. They calculate an effective charge on
O of -0.695 for SiO$_4$ and -0.700 for HAlO$_4$.

Okamoto, et al(30) performed a series of experiments to determine the
effects of Si/Al ratio and cation type. They determined that all XPS lines
[O(1s), Si(2s), Si (2p) and Al(2p)] show increases in binding energy with
increases in the Si/Al ratio. These results were relatively independent
of crystal structure for Na zeolites, suggesting that the primary cause of
the shift is common to all zeolites examined (A, faujasite and mordenite).
They also investigated the effect of changing cation with the result that
the O(1s) binding energies decrease in the order H$^+$>Li$^+$>Na$^+$>K$^+$>Cs$^+$. The
effect is expected from electronegativity considerations (see, for example,
Pauling(36)) and the larger binding energies imply increased electron
density in the O---cation bond. Thus the ionicity of the O---cation bond
is H$^+$<Li$^+$<Na$^+$<K$^+$<Cs$^+$, as expected. Beran and Dubsky(37) have calculated
an approximate effective charge on Na$^+$ in faujasite with a range of
0.23-0.29 e$^-$. Interestingly, their results indicate that the effective
charge declines with decreasing Si/Al ratio (from 6.0 to 1.0). This
suggests that, within the limits of their calculation, the dipolar
repulsion outweighs the increase in polarizability in determining the
ionicity of the bond. Mortier(38) calculates the effective charge on H$^+$
in faujasite and computes 0.12 e$^-$ in X zeolite (Si/Al=1.25) and 0.14 in
Y zeolite (Si/Al=2.5). His results are consistent with Beran, et al and
also reproduce the experimental XPS data that H$^+$ is less ionic than Na$^+$
in faujasite.

Barr and Liska(39) have used high resolution valence band XPS to study
the surfaces of several zeolites. They found evidence for a number of
"impurity" phases, such as metal aluminates, on the surface of several
zeolite crystal samples and emphasize the need for careful sample preparation and analysis. Their data, however, indicates the presence of "group"
rather than "elemental" shifts in the binding energies. This proposal is
extended by Barr, et al(40) to imply the existence of

$$(SiO_2)_x\bullet(M_{y/p}^{+p} \ AlO_2^-)_y\bullet ZH_2O$$

units in zeolites that can be resolved into broad structural groups.
For zeolites with Si/Al<5, they propose an effective structure of
SiO$_2$-doped NaAlO$_2$, while for Si/Al>5, the effective structure is
NaAlO$_2$-doped SiO$_2$. These results differ from those expected from
simple solid solutions of SiO$_2$ and NaAlO$_2$ (a possible model for
amorphous aluminosilicates) especially in the low Si/Al region. However,
amorphous aluminosilicates were not examined. These data suggest the
existence of "global" electronic effects in the crystalline zeolites.
Barr, et al further distinguish between these two groups as cage-like (low
Si/Al) and chain-like (high-Si/Al), according to the types of zeolites

they examined in each category (A, faujasite, L and mordenite, ZSM-5, silicalite, respectively), but have no examples of high silica cage structures or low silica chain structures to establish this division of structure type versus Si/Al ratio effects.

Summary

- XPS data imply the existence of "group" rather than "elemental" binding energy shifts. This suggests the existence of "global" electronic effects responding to changes in Si/Al ratios and changes in electronegativity of the exchangable cation.

- The electron binding energies for Si, Al and O all increase with increasing Si/Al ratio.

- Shifts in the Si(1s) and Si(KLL) lines correlate with changes in the oxygen polarizability.

- The O(1s) binding energy shifts with cation electronegativity. The largest shift, by far, is observed with H^+ cation.

- XPS data suggests only partial charge, even for Na cation (~$0.25e^-$), and even smaller charge on O-H hydrogen ($0.1-0.2e^-$).

Solid State NMR Spectroscopy

Solid state, magic-angle spinning NMR spectroscopy utilizes small samples of powdered materials spun at high (~ few kHz) frequencies at an angle of 54°44' to average out chemical shift anisotropy. Recent reviews of the technique and recent experiments are: Nagy and Derouane(41), Thomas and Klinowski(42) and Pfeifer, Freude and Hunger(43).

As discussed in the review by Nagy and Derouane(41), solid state magic angle spinning NMR of several nuclei has been applied to the study of zeolites. The ^{29}Si spectrum provides a series of peaks corresponding to silicon atoms with 0 to 4 Al nearest neighbors. From this NMR data the Si/Al framework ratio may be calculated. The exact frequencies of a given peak (e.g. Si with 1 Al nearest neighbor) vary slightly with zeolite type (see, for example,(44) comparing Y and ZSM-20) possibly reflecting differing framework electronic polarizabilities. Also ^{29}Si NMR has demonstrated single sharp peaks for zeolite systems of varying Si/Al ratio(45). This is consistent with a "global" electronic effect with the electronegativity of the entire framework being averaged. Individual Si atoms still adopt geometries consistent with the crystal structure and the number of Al nearest neighbors, which produces the typical five peak pattern of ^{29}Si NMR with Si/Al ratios greater than one. In addition, ^{29}Si MNR has shown single sharp peaks with partial cation exchange(45). This may be due to hydrated cations having mobility and giving a "rapid-exchange-limit" spectrum. The ^{29}Si NMR of carefully dehydrated partial-exchange systems would provide useful additional information on cation effects. ^{27}Al NMR has been applied primarily to observe the tetrahedral (framework) or octahedral (non-framework) form of Al in zeolites.

Solid-state NMR of amines (using ^{15}N) and phosphines (using ^{31}P) has been used as a probe of the Bronsted and Lewis acid sites in zeolites(46). Chemical shift differences indicate that formation of protonated (Bronsted) or non-protonated (Lewis) complexes with the acid site. In addition, chemical shift differences between protonated species indicate differences in the strength of complexation: tightly held or weakly bound. Finally, different molecules sample different subsets of acid sites, depending on the accessability, type or acid strength of the site.

The most direct evidence for the nature of zeolite acidity is expected to come from studies involving ^{1}H. Early work by Stevenson(47) used broadline ^{1}H NMR to locate the protons in dehydrated H-Y and established an approximate geometry for the bridging hydroxyls. His results indicated that the protons are attached to O_1 and O_3 with an O-H bond length of ~1.00 to 1.03Å. More recently, direct measurement of the chemical shift of ^{1}H in zeolites and amorphous aluminosilicates has provided further data. Pfeifer, et al(43,48) obtained ^{1}H MAS NMR spectra of HY and amorphous aluminosilicate. The HY spectrum shows four peaks, one at ~2 ppm (a), two partially overlapping peaks at ~4.5 ppm (b&c) and one peak at ~7 ppm (d). The amorphous aluminosilicate shows one peak at ~2 ppm with a broad peak at ~4.5 ppm. They assign line (a) to non-acidic structural hydrogens, noting that adsorption of pyridine shifts this line to lower field, suggesting hydrogen bonded molecules. They argue that protonation of pyridine would lead to considerable line broadening which they did not observe. They assign line (d) to the residual NH_4^+ ions left after activation. This agrees both with the reduction in intensity at increased activation temperature and the assignment of NH_4^+ compounds(49). They assign the two remaining lines to bridging hydroxyl groups. A ^{1}H MAS-NMR spectrum of H-mordenite produces only one, relatively broad, line in the bridging hydroxyl region. Pfeifer, et al use this as evidence to correlate the ^{1}H-MAS NMR peaks with observed IR hydroxyl peaks. By plotting only the amount of protons indicated by the intensity of the (c) peak, they suggest that a linear relationship of cumene cracking activity versus area of the (c) bridging hydroxyl peak exists. They then conclude that this demonstrates that the acidity of zeolite and amorphous aluminosilicate acidity are qualitatively similar and differ only in quantity.

More recently, Pfeifer(50) has modified this conclusion, by using the area of the (b) peak to generate a rate vs intensity plot that gives significant differences for H-Y and H-mordenite indicating that differences between zeolites exist. The residual correlation between H-Y and amorphous aluminosilicate may be an artifact of amorphous material produced by the thermal treatment to reduce the bridging hydroxyl content in the HY. The H-Y and H-mordenite also have different Si/Al ratios (2.6 and 7, respectively) which may be expected to influence the acidity of the hydroxyl protons. Pfeifer(50) also changes the assignments of (b) & (c) with the IR hydroxyl peaks to provide a direct relationship of the larger NMR chemical shift correlating with the lower IR hydroxyl stretching frequency. Since the assignment for the lower frequency IR hydroxyl band in HY is as an inaccessible hydroxyl in the small cages, this leads to the change in correlating cumene cracking with NMR peak area discussed above. Uncertainty in the current state of knowledge regarding assignments of ^{1}H MAS-NMR frequencies to specific types of protons in the zeolite structure makes any correlations with IR hydroxyl stretching frequencies quite tentative.

Briefly discussed by Pfeifer, et al(43) is the significance of the absolute chemical shift magnitude and the effects of solvation on proton shifts. They note, for example, that liquid water gives a proton signal at 4.8 ppm(51a) while gas phase water gives a signal at 0.31 ppm(51b). Even in the gas phase, some hydrogen bonding may occur, with a consequent down-field shift. Pfeifer(50) provides examples of ^{1}H chemical shift versus gas phase acidity for several "hydroxylic" compounds (MeOH, phenol, etc). Pfeifer also discusses the problems of interpreting the NMR experiments because of complications due to observing only the residence time of a proton on a given oxygen atom. He notes that the mean lifetime can be greatly reduced by the presence of even very small quantities of bases.

Finally, Freude, et al(52) report that the chemical shift of the bridging hydroxyl line increases as the Si/Al ratio in faujasite increases in the range 1.4 to 7, but remains constant for Si/Al > 10. This probably represents the "isolated" aluminum site with "maximum O-H polarity", discussed in Section (C-6).

Recent work by Haw, et al(53) using solid state ^{13}C NMR has demonstrated the existence of "complex" type reaction intermediates for propene adsorbed on HY at low temperatures (~200 K). They identify the complex as an isopropoxy species by its chemical shift. They also note that the species spontaneously oligomerizes as the system is warmed to room temperature and forms a new species, tentatively identified as an alkyl substituted cyclopentenyl carbocation. This new species has much greater cation stability and is representative of a "free" carbocation as discussed below.

Summary

- ^1H MAS NMR experiments have, so far, not provided definitive information on the differences in acidity between H-zeolites and amorphous aluminosilicalites.

- ^{29}Si NMR provided information on the framework Si/Al ratio.

- ^{27}Al NMR provided information on framework and non-framework Al species.

- ^1H NMR indicates the existence of "bridged" and "terminal" hydroxyl groups in zeolites and amorphous aluminosilicates.

- The ratio of "bridged" to "terminal" hydroxyls is much higher for zeolites compared to amorphous aluminosilicate.

- ^1H NMR suggests that Al sites become "isolated" at a Si/Al ratio of ~7 in faujasites.

- ^{31}P and ^{15}N NMR of probe molecules can distinguish Lewis and Bronsted acid sites. ^{13}C NMR can distinguish the type of complex formation (strongly or weakly complexed to the acid site) between Bronsted acids and probe molecules.

Infrared Spectroscopy

Several articles summarize various aspects of the study of zeolites with IR spectroscopy. These include: Flanigen(54) and Ward(55) who discuss transmission IR studies in the framework and hydroxyl regions, Maroni, et al(56), and Janin, et al(57) who discuss the use of probe molecules to study the acidity of zeolites by diffuse reflectance and transmission techniques, and Baker, et al(58) who discuss the study of extra-framework cations with far IR spectroscopy. Each of these articles summarizes the general state of the respective arts as of the date of the articles.

Jacobs and Mortier(59) summarize studies of zeolite acidity by hydroxyl IR spectroscopy. They rationalize the observed hydroxyl stretching frequencies with two general principles: Sanderson intermediate electronegativity (which depends on chemical composition) and the presence of small (6 or 8 member) rings in certain zeolites. They thus propose that the primary effect is an electronic effect due to zeolite composition (e.g. Si/Al ratio), and cite a linear correlation between hydroxyl stretching frequency and Sanderson intermediate electro- negativity as evidence. This explanation accounts for the "high-frequency" hydroxyl band in a considerable number of zeolites. The second effect, presence of restricted rings is proposed to explain the "low-frequency" hydroxyl band by suggesting that hydroxyl groups located in such rings will undergo a bathochromic (low frequency) shift, due to electrostatic interactions with the nearest oxygen atoms, which is inversely propor- tional to the average squared distance. Thus the effect is only observed with the smallest rings.

More recent work has focused on using IR to probe the types of acid sites and the nature of the local structure in zeolites. Datka, et al(60) have investigated the hydroxyl spectrum of H-Na-ZMS-5 and concluded that there is evidence for heterogenity of hydroxyl groups in the zeolite. They propose either: an effect due to differing numbers of Al NNN in the immediate vicinity of the hydroxyl or (their preferred explanation) a dependence on the local Al-O-Si bond lengths and angles. They calculate proton affinities for five distinct subbands ranging from ~1320 to ~1200 KJ/mole. These differences represent a range roughly comparable to the difference between acetic and trifluoroacetic acids, yet the five subbonds all have the same IR frequency (~3609 cm^{-1}) in the absence of probe molecules.

Multi-technique studies, including IR, have been used to address the questions raised by various treatments to dealuminate zeolites. Zi and Yi(61) have compared several techniques including hydrothermal treatment, EDTA extraction and ammonium hexafluorosilicate (AFS) treatment. They find that the asymetric framework stretching frequency is a linear function of the framework Al/(Al + Si) molar ratio.

Summary

- IR spectroscopy can distinguish several types of hydroxyl groups in zeolites. As discussed by Jacobs and Mortier, the stretching frequency of the "high-frequency" band is primarily determined by the electronegativity of the framework (Si/Al ratio). In zeolites with small ring structures, hydroxyl groups located in such rings will undergo a bathochromic (low frequency) shift, esulting in an additional "low-frequency" band.

- Datka, et al, propose a method of determining proton affinity using IR spectroscopy of probe molecules. They demonstrate that a single IR band can give rise to several subbands with different proton affinities.

Theoretical Studies and Modeling

Theoretical calculations and modeling studies of zeolites fall into two broad categories: semi-quantitative (those which use quantitative measures such as Si/Al ratio to give rank orderings of some property such as acidity) and quantitative (studies such as ab initio calculations that attempt to predict quantitative results of experimental zeolite properties). The first category encompasses many theories which attempt to identify the key parameters influencing zeolite properties, such as acidity, often based on direct experimental data. The second category encompasses both ab initio calculations which use model "clusters" to mimic zeolite structures and statistical mechanics calculations which attempt to predict dynamic distributions (such as cation site populations during "titration").

The semi-quantitative theories primarily consider Al site distributions as the primary determinant of acid strength. This was discussed by Pine, et al(62) to explain variations in acid activity in faujasite-based cracking catalysts. Their argument is essentially that the structure of faujasite requires a given Al atom to have 4 Si atoms in the 1st surrounding layer (nearest neighbors) and 9 Al or Si atoms in the 2nd layer (next-nearest-neighbors or NNN). In the case of X, all of these 9 atoms will be Al, by Lowenstein's rule(63), while with increasing Si/Al ratio, some Al will be replaced by Si. At a sufficiently large ratio, all 9 will be Si and the original Al site will be isolated. The strength of the original Al site will depend on the number of Al NNN, reaching a maximum at O NNN. This argument was extended by Wachter(64) who provided detailed statistical calculations of Al NNN population at various Si/Al ratios, and determined the Si/Al "limit" when essentially all Al sites were "isolated" to be ~5 (not including Lowenstein's rule constraints) or ~7 (including Lowenstein's rule contraints). This level of the theory predicts that all zeolites with the same number of Al NNN (e.g. faujasite, A, chabazite), will show a maximum acid strength at similar Si/Al ratios. Barthomeuf(65) has extended this idea by using topological densities to include the effects of layers 1 through 5 surrounding the Al atom. For a discussion of topological density see Meier and Moeck(66) or Barthomeuf(65). From these numbers and the experimental data regarding the "isolated Al" limit for faujasite activity, Barthomeuf predicts the limiting Si/Al ratios for isolated Al in these zeolites. As several zeolites give different predictions for the limiting Si/Al value by the two theories (e.g. Al NNN theory predicts the same limit for faujasite and zeolite A of ~7, while topological density theory predicts 6.8 for faujasite and 7.6 for A), tests of the theories are possible by preparing faujasite and zeolite A with varying Si/Al ratios. These theories predict the effect of changing Si/Al ratio on relative acidity with any zeolite crystal type, but cannot predict relative acidities between different zeolites or amorphous alumino silicates.

Models for the acid site in amorphous aluminosilicates have been proposed by Hansford(67) and Peri(68). Hansford proposed a surface silanol rendered more acidic by the presence of a neighboring aluminum ion (similar to structure II in Figure 2. Peri has proposed a series of small (2-4 T atom) ring structures and especially favors Al-O-Al bridged structures. The lack of direct structural probes greatly hinders the testing of these models, although most available data favors framework Al Lewis acid-base interactions with silanols, rather than Al-O-Al bond structures.

The proper description of the primary electronic units is an important part of model construction. Pelmenshchikov, et al(69) have used $Si(OH)_4$, $Al(OH)_3$, NaOH and H_2O as primary building units and constructed simple models to determine the stability of various small linear or cyclic structures. They emphasize the short range aspects of structure, as discussed by Zhidomirov and Kazansky(70). Their results demonstrate the strong tendency to form Al-O-Al bonds in small clusters. However, their small structures lack any stabilization energies due to long range order (crystallinity) and may be applicable primarily to amorphous materials. Barr et al (40) emphasize the global nature of the electronic structure in zeolites, based on his XPS data.

A general theory regarding differences in acid strength between zeolites and aluminosilicates has been discussed by Mortier(71). Using bond valence, Gutmann's rules for interatomic interactions and principles of electronegativity equalization, Mortier proposed an enhanced electron donor-acceptor interaction in zeolites as shown in Figure 1. This emphasizes the reactive aspect of acidity (a molecule is not an acid until a base is present for it to react with) and the electron delocalization into both Al-O and Si-O (oxygen bridge bonds). A "resonance" depiction of this idea is shown in Figure 2. Structure I is a fully bridged oxygen with a weakly bonded proton while structure II is a silanol with a weak Lewis acid interaction of the hydroxyl oxygen with the Al. The actual situation is a resonance hybrid.

These arguments lead us to propose that zeolite crystals respond to "proton attack" by a global readjustment of the bond structure of the whole zeolite matrix, in order to prevent major local distortions in bonding, leading to loss of long range symmetry, and thus, crystallinity. In addition, energetics would favor equalization of topologically equivalent electronic interactions. Thus, there would be a significant stabilization resulting from the electronic structure of the Al-O and Si-O bonds becoming more equivalent. This implies a dominance of structure I in the resonance. For an amorphous material, there is no long range stabilization for structure I and structure II can dominate. This interpretation is consistent with the available NMR evidence which shows a large proportion of H in "terminal" hydroxyls in amorphous alumino-silicates. As Mortier notes, the interaction of O with Al will weaken the O-H bond and increase acidity. This then may represent the fundamental reason for the difference in acidity between crystalline zeolites and amorphous aluminosilicates.

Quantitative calculations attempt to provide direct predictions of magnitude differences in the physical properties of zeolites. They are generally less successful in dealing with amorphous materials due to the lack of structural models. Early calculations provided estimates of partial charge densities of framework elements and cations. Examples include Mortier(38) who calculated the partial charge on H in a faujasite hydroxyl with values of 0.12 (Si/Al=1), 0.14 (Si/Al=2.5), 0.18 (Si/Al=infinity). The increase in ionic character for hydrogen can be explained by the higher electronegativity of Si versus Al causing an increased electron transfer from oxygen to Si with increased electron transfer from H to oxygen. Thus, the OH bond becomes more ionic. A similar effect is seen in data for Na by Beran and Dubsky(37).

Theoretical considerations of the affect of T-O-T bond angles on the bond structure and specifically on the electronegativity of the bridging oxygen is of particular importance in view of the fact that zeolites of higher protic acidity have a range of T-O-T bond angles higher than those of somewhat lower protic acidity. Compare the bond angle range of 137-177° for ZSM-5(72a) or 143-180° for mordenite(72b) with the range of 138-147° for HY(73). Quantitative calculations yield estimates of the effect of T-(OH)-T bond length and angle on the acidity of the bridging hydroxyl. Dwyer and O'Malley(74) calculated an optimized angle of ~144° for both protonated and unprotonated forms. Carson, et al(75) have calculated the energy for protonated and unprotonated forms, and noted that the unprotonated forms show little change from ~130°-175° while the protonated forms become markedly less stable for larger angles. This suggests that the deprotonization energy decreases with increasing T-O-T angle and thus the proton becomes more acidic. Mortier and Geerlings(76), however, calculated the hydroxyl stretching frequency for a $(HO)_3Si(OH)Al(OH)_3$ cluster for Si-O-Al angles of 140°-160° and find an increase of only 13 cm^{-1}, suggesting little change in bond strength. Kasseb, et al(77) have used a similar cluster to calculate minimum energies, for at T_1-O-T_2 angles of 134.2 (protonated) and 179.3 (deprotonated). They also calculated a deprotonation energy of ~1320

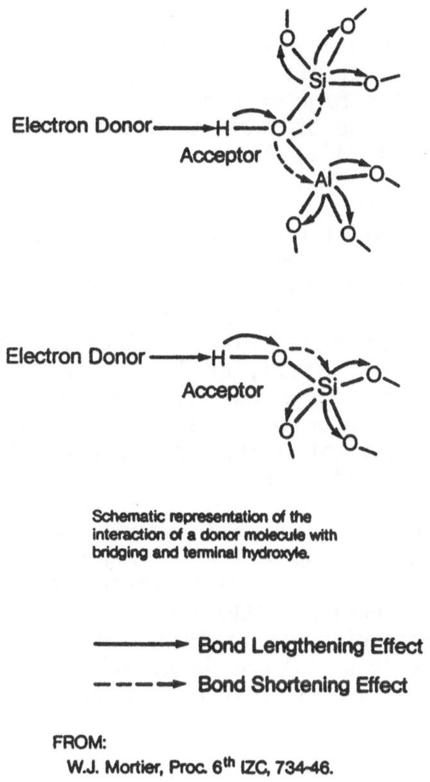

Schematic representation of the
interaction of a donor molecule with
bridging and terminal hydroxyle.

→ Bond Lengthening Effect

---→ Bond Shortening Effect

FROM:
W.J. Mortier, Proc. 6th IZC, 734-46.

Figure 1
Schematic Representation of the Interaction of a Donor Molecule
with Bridging and Terminal Hydroxyls.
From Mortier (71).

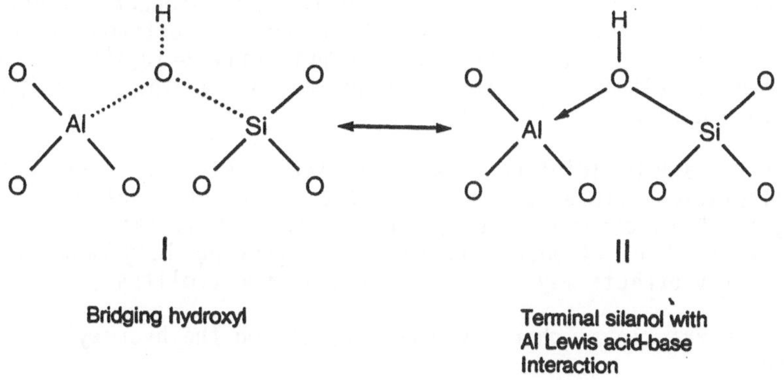

I

Bridging hydroxyl

II

Terminal silanol with
Al Lewis acid-base
Interaction

Figure 2
RESONANCE MODEL OF T–(OH)–T
BOND STRUCTURE

285

KJ/mole with a H$^+$ charge of ~0.05 for the hydroxyl proton. In addition, they calculated deprotonization energies of ~1320 (Si/Al), ~1380 (Si/B) and ~1300 KJ/mole (Si/Ge) for the respective T_1/T_2 pairs.

Derouane and Fripiat(78) have calculated that the stability of the T-O-T bridge increases with increasing electropositive character of the cation. Thus the protic form of a zeolite is the least stable. They calculated a partial charge of 0.276 on the hydrogen. They also found that the substitution of Si by Al occurs with local deformation of the Si-O-Al bridge. This maximizes the ionicity of the O-H bond. They do not indicate any possible effect of this deformation on the crystal stabilization energy. Gourset, et al,(79) used extended Huckel calculations for the acidity in offretite considering the two non-equivalent sites. They find that T_2 is more acidic than T_1, and that less acidic Al are less stable (more easily removed), in accordance with experiment (see, for example, Fernandez, et al(80)). Finally, Derouane and Fripiat(35) have calculated preferential Al siting at T_2 and T_{12} in ZSM-5. They note that this implies possible Al in NNN sites, (although the relatively high Si/Al in ZSM-5 makes this unlikely, see Barthomeuf(65)), and that the framework is highly covalent with the negative charge extensively delocalized upon proton abstraction.

A series of statistical mechanics calculations have been carried out by Mortier and co-workers(81,82,83) regarding cation site preferences in Y. Mortier, et al(81) calculated site preferences of K in hydrated KY, while Van Dun and Mortier(82) calculated site preferences for Na in Y, noting the predominant influence of short range cation-framework interactions and the influence of even small quantities of water. In a welcome development, Van Dun, et al(83) test their model with experimental data on Na HY, CaHY and SrY.

<u>Summary</u>

- Semi-quantitative theories by Pine, et al and Barthomeuf propose that the number of neighboring Al atoms determine the acidity of a given acid site.

- Mortier proposed the existence of an enhanced donor/acceptor interaction in zeolite to explain the stronger acidity compared to amorphous aluminosilicate. This interaction can be extended into a model of the Al-(OH)-Si bond structure as a resonance structure with "bridging hydroxyl" and "terminal silanol" as extreme limits. For zeolites, the actual OH group is primarily bridging, while in amorphous aluminosilicate it is primarily terminal, in agreement with NMR data.

- The primary building units and the type of electronic structure calculations (local or "global") are important. XPS data suggests "global" electronic effects in zeolites. Calculations optimizing only local electronic structure and ignoring long-range crystal symmetry effects may not be good models for zeolites.

- Only a fractional charge (~0.2e$^-$) exists on the hydroxyl proton.

- Bond angle in the T-(OH)-T bond strongly influences deprotonation energy. High T-OH-T bond angles tend to increase H acidity by increasing the "s" character in the T-O bonds and reducing the "s" character in the O-H bond. This increases the positive charge on the H, resulting in stronger protic acidity. High T-(OH)-T bond angles also weaken the T-O-T bond strength.

286

- Theoretical calculations indicate the existence of preferred T-atom sites for Al in zeolite structures with non-equivalent T-atom sites. This is in agreement with experimental data.

- Statistical mechanics has been used by Mortier, et al, to predict cation distribution in zeolites.

MEDIA (SOLVENT) AND DIFFUSION EFFECTS IN ZEOLITES

Zeolites as Media (Solvent)

In case of solutions the effect of a solvent can be readily calculated by taking into account the properties of the molecules involved such as polarity, polarizability or dielectric constant. The characterization of the zeolite media is much less tractable in a quantitative sense.

Aluminum Rich Zeolites

Adsorption measurements, theoretical modeling, and extensive reaction chemistry using both Na-Y and H-Y zeolites indicate that the intracrystal-line void space displays high polarity. Very large electrostatic field gradients, of the order of 1 volt/Angstrom, may exist in the α cages of Y zeolite at a distance of 2.5Å from the center of site-2 bivalent cations. Similarly, oxygen rings unoccupied by cations provide large areas of negative electrostatic fields(84).

The large electrostatic fields derived by ab initio calculations are consistent with reports on the reaction chemistry of zeolites with probe molecules. The results of these experiments show that Na-Y zeolite at SiO_2/Al_2O_3 ratio of 5 can polarize and even ionize a variety of atoms and molecules, forming remarkably stable occluded ions or ionic clusters in the zeolite crystal. These phenomena include the irreversible occlusion of salt molecules(85), the ionization of alkali metals forming a variety of subvalent alkali clusters (Na_4^{3+}, K_6^{5+}, etc) in the cavities(84) and the ionization of $NO + NO_2$ radicals, forming [NO^+ x NO_2^-] ionic complex in the zeolite cavity(86). These reactions are endothermic by up to ~140 kcal/mole. Thus, the energy required for these processes is provided by the zeolite. Such interaction between occluded molecules and zeolite crystals displays the characteristics of a strong electrolyte or solvent(8,84).

In addition to theoretical modeling and reactions with probe molecules, the media (solvent) strength in zeolites can be estimated by xenon NMR spectroscopy(87). This non-destructive method gives information on the electrostatic field in the vicinity of accessible cations. In addition, it gives information on the short and long range geometry of the intracrystalline pores, and on the nature of occluded metal clusters.

One of the important practical results of the solvent effect in catalysis with Y zeolite is the concentration enhancement of hydrocarbon reactants within the zeolite crystal relative to the vapor phase. The result of this phenomenon is a great enhancement of the rates of bimolecular reaction steps such as the hydride transfer reactions in catalytic cracking, relative to the rates of monomolecular reaction steps. This effect is significant even at very low Na and framework Al concentration (88). Significantly, this selective, reaction rate enhancing phenomenon has been demonstrated for both non-protic (K exchanged) and for H-Y zeolites(8,89). This shows, that directionally the media effect is similar in both H-Y and in alkali-cation Y.

The media effect in zeolites is centered on the polarity of O-Na or O-H groups and on the polarity and polarizability of Al-O linkages. Therefore, high Al and cation concentrations contribute to increased media effect. At very high concentrations, however, there is a reversal in the media effect, due to dipole relaxation between adjacent polar groups, and to competitive, counteracting effects of adjacent polar sites with a given adsorbed molecule. The deleterious effect of very high Al concentrations on media effect is similar to its effect on O-H acid strength, since both are based on dipole relaxation processes. The precise value for the optimum Si/Al ratio for maximum media effect depends both on the zeolite structure and on the size and chemistry of the reactant molecule. It is likely that in general the decline in media effect becomes large with Si/Al ratios below 2.5.

On the basis of model calculations and chemical probing of the intra-crystalline reaction environment discussed above, it can be concluded that high concentrations of framework aluminum and surface metal cations (including protic sites) result in very strong solvent effects. Direct-ionally, this is similar to the effect of polar solvents in solutions, with specific features due to the micro environment in the intracrystal-line zeolite pores and cavities.

High-Silica Zeolites

The most dramatic difference in media (solvent) effect between aluminum rich and silicon rich zeolites is displayed by the interaction with water. Aluminum rich zeolites are all strongly hydrophillic, and this property is similar for both metal cation exchanged and H-zeolites. They adsorb water at low pressures, filling up the zeolite cavities with water. In contrast, silica rich molecular sieves, particularly at high Si/Al ratios, are distinctly hydrophobic. These, when tested with water containing a few percent butylalcohol, selectively adsorb the butylalcohol. Such strong hydrophobic character shows that polarity based interactions are minimal. The interaction between the microporous crystal and adsorbed molecules rests on London-type interactions, based on the relatively smaller polarizability of the silica-rich zeolite framework. These findings are fully consistent with similar conclusions deduced from mechanistic studies in the cracking of n-hexane over aprotic silicalite. This study showed very strong n-hexane concentration enhancement for K-Y zeolite while it found no concentration enhancement for silicalite(8).

In respect to media (solvent) effects in zeolite acid catalysed hydrocarbon reactions, the following conclusions are offered

- Very high aluminum and cation/H^+ concentrations (Si/Al <2.5) are expected to reduce media (solvent) effects by competing and counteracting polarization between adjacent polar sites.

- Aluminum and cation (including H^+) rich zeolites enhance the concentration of reactants in the intracrystalline pores/cavities, according to the polarity and polarizability of the zeolite-reactant system.

- Aluminum and cation (including H^+) rich zeolites contribute to the ionization of adsorbed atoms and molecules.

- The effect of media declines at low aluminum and cation concentrations.

Diffusivity in Zeolites

The effects of diffusion on the interpretation of temperature programmed desorption (TPD) data are discussed by Forni and Magni(24), who find that diffusion effects may dominate the TPD data. Forni, et al,(27) analyzed experimental TPD data and found that intracrystalline diffusion controlled the desorption rate. From the large apparent activation energy (D_a~30 Kcal/mole framework) they suggest that interactions of the desorbing species with available acid sites (multiple desorption/adsorption steps) was occurring. Eic and Ruthven(90,91,92) used zero length chromatography (ZLC) to measure diffusion rates for desorption of material from zeolites. Studies of xylene and benzene in NaX gave results much lower than NMR self-diffusivities, but they were in line with gravimetric measurements. Eic and Ruthven(90) also obtained similar results for n-butane in NaX, but achieved agreement with NMR data for CaA. NMR self-diffusion measurements determine intracrystalline diffusion coefficients. For a discussion see Ruthven,(93) and experimental data by Karger and Ruthven(94), Karger and Pfeifer(95,96) and Karger, et al(97). ZLC primarily measures similar effects, unless a significant surface to pore barrier exists. The data of Eic and Ruthven suggest that a barrier does exist, as discussed below.

Diffusion effect can also be used to advantage, as in the study by Caro et al(98). They used diffusion measurements and models, to determine the location of N-bases in H-ZSM-5 by the effect on methane self-diffusion. Their results confirm previous suggestions that Bronsted acid sites are sited in the channel intersections.

Derouane, et al,(99) have used sorption kinetics, heats of sorption, NMR relaxation times and molecular graphics simulations to study the diffusion of alkanes in zeolites. They find that confinement effects regulate the diffusion, which is consistent with a "segmental diffusion" mode for the alkanes as proposed by Barrer and Davies(100). They also considered the "floating" and "creeping" molecule concepts, as discussed below. For the zeolites considered, n-hexane has a "creeping" diffusion in large pore zeolites but may change to "floating" in the case of AlPO-11.

There has been a large effort to apply theoretical models and related calculations to study diffusivity in zeolites. A brief review of Monte Carlo models for diffusion in zeolites is given by Aust, et al(101). They assess the influence at different model assumptions on the concentration dependence of transport and self diffusion in a two dimensional zeolite network.

Derouane, et al(102) discuss the effects of surface curvature on diffusion of molecules in zeolites. This is an extension of the "nest effect", proposed by Barrer and modified by Derouane(103), in which the adsorbed molecule and zeolite reciprocally optimize their van der Waals interactions.

In zeolites, this can lead to electronic structure changes, which may be manifested as changes in global crystal symmetry, upon adsorption of molecules such as benzene in ZSM-5. This also leads to the concept of "supermolecular" interactions, in which the total system (zeolite plus adsorbate) must be considered. Observations of large intracrystalline diffusion rates for alkanes has lead to the concept of configurational diffusion, as proposed by Weisz and Haag, et al(104). Derouane, et al extend this concept to discuss two potential energy situations: "floating" molecules, when the channel diameter is approximately equal to the molecular diameter, and "creeping" molecules, when the channel diameter is "large" (approximately two or more times the molecular diameter). Using these concepts, Derouane et al can rationalize semi-quantitatively a number of experimental observations including: the role of zeolites as molecular traps, the origin of the surface diffusional barrier and rapid diffusion of molecules in "tight-fitting" pores. Derouane et al provides an excellent conceptual framework for describing the physical determinants of diffusional effects in zeolites.

We wish to emphasize two points from the work of Derouane et al (102) to discuss the difference in the diffusional characteristics of zeolites and amorphous materials. Derouane et al show the approximate transition from "floating" to "creeping" diffusion occurs over the range of pore diameters from ~1 to ~2 times the molecular diameter for spheres. Importantly, the tendency to maximize attractive potentials for non-spherical molecules will tend to make this transition occur at ~2 times the mean molecular diameter, although the total attractive energy will be maximized by smaller pore diameters. For typical molecules of interest, the mean molecular diameter is <~10Å, thus "floating" diffusion is largely restricted to pores <20Å in size. The effects also occur only if the pore is of relatively constant curvature, thus long, crystallographically regular pores maximize the effect, while constantly varying pore diameters minimize the effect.

The surface diffusional barrier is a result of the sudden change in curvature upon entering a small, molecule sized pore. For a gradual curvature change, such as entering a large pore in an amorphous material, no barrier is expected. Derouane et al use this barrier to explain the much smaller uptake relative to NMR self-diffusion co-efficients and note that rate differences of ~10^5 at 273K are possible due the magnitude of the potential barrier. As discussed above, Eic and Ruthven(90) see differences in rate even for desorption processes. This can be explained by noting the reciprocal effect of the transition from pore to surface also has the same change in curvature, thus inducing potential energy barriers to molecular desorption from the pores. The effect should be greatest for "floating" molecules, which already have enhanced intra-crystalline diffusion, and should decrease with the transition to "creeping" diffusion and with more gradual pore size changes toward the surface (e.g. funnel-shaped pores). Thus zeolites possess potential energy barriers to desorption from the pore system that are not present in amorphous materials. This barrier can greatly increase the residence time of molecules in zeolites due to physical adsorption alone. Chemisorption on acid sites will have addition effects, as seen in the TPD analysis by Forni et al (27).

Summary

- The high frequency of readsorption in zeolite pores can give rise to large diffusion barriers, amounting to ~30 Kcal/mole. This may substantially affect the practical, acidic catalytic activity of H-zeolites.

- Diffusion in well defined molecule-sized pores formed in certain zeolites produces unusual effects.

- For pores approximately equal in diameter to the size of the molecule, "floating" diffusion, which is very rapid, may occur.

- There is a surface potential barrier that can hinder the <u>outward</u> diffusion of molecules from small, well-defined, pores of zeolite crystals.

290

• These diffusion effects can be readily identified and measured in zeolites, while they are negligible in large-pored materials such as amorphous aluminosilicates.

FORMATION OF CARBOCATIONS

The direct characterization of carbocations formed in H-zeolites has been attempted by NMR spectroscopy in several studies, [see recent review by Lombardo et al(105)]. The only hydrocarbon cation unequivocally identified so far is the triphenylcarbenium ion, formed on both H-Y and on amorphous silica-alumina catalysts. The chemical shift observed with this stable carbocation is almost identical to the one measured for this carbocation in a superacid. Therefore, it can be concluded that this species is a "free" cation. Unfortunately, formation of this cation is prevented in other zeolites by its large size. Other, smaller carbocations have been also tentatively identified in zeolites (106). However, the high reactivity of these reactive carbocations to olefins, forming oligomeric species, cast doubt on the interpretation of some of the NMR assignments.

Considering the important role of carbocations in acid catalysis, those zeolite acid characteristics are of particular interest which influence the formation, stability and reaction chemistry of these hydrocarbon reaction intermediates. In strong acid solutions, particularly in superacids, this reaction intermediate is represented as a free solvated carbocation(1,5). It is formed either by protonation of olefins or aromatics, by hydride abstraction from saturated hydrocarbons or by forming five-coordinated C^+ centers(1,5,107). In strong acid solutions a large variety of stable carbocations have been prepared and characterized by NMR and other techniques.

Theoretical modeling of the polarizability of carbocations shows that the molecular bonds in these species utilize a variety of orbitals, including excited state orbitals, which tend to increase the polarizability(6,7). This characteristic of carbocations is important in acid catalysis, because of the relationship between softness (polarizability) and reactivity of these species for intramolecular rearrangements.

In some cases, such as in catalytic cracking, the acid catalysis proceeds by a chain reaction, involving hydride transfer from reactant molecules to chemically transformed carbocations. Thus, one assumes that catalysis is initiated when the reaction conditions permit rapid hydride transfer. The initial cations, are generally produced by protonated olefins. In bifunctional catalysis (using a combination of a dehydrogenation catalyst and a solid acid) the olefin produced over the metal is isomerized over the acid sites and the isoolefins are presumably hydrogenated to isoparaffins over the metal catalyst. Therefore, there is no need to invoke a chain reaction mechanism. Whatever the sequence of steps in the H-zeolite catalysed reaction mechanism, it seems relevant to consider the implication of the effect of both O-H acid strength and media effect upon carbocation formation and structure.

The review of the chemistry of zeolite acid sites demonstrated that in the absence of strong added solvents such as water, the zeolite O-H groups carry hydrogen atoms with only a fractional positive electronic charge (0.12-0.25e^-). It was also discussed that cation and aluminum rich zeolites readily ionize a variety of absorbed molecules(8,84,85). Therefore, these zeolites are expected to contribute to the polarization of the hydrocarbon reactants and to the stabilization of carbocations as well. On the other hand, silica-rich zeolites demonstrated no media effect, and therefore, they are not expected to contribute via media effect to the formation of carbocations.

On the basis of this information it can be expected that in strong-electrolyte type zeolites the zeolite media (crystal field) contributes to the formation of weakly bound carbocations, to the extent that they may be linked mainly by ionic interaction to zeolite framework oxygens. Such "free" carbocations would be relatively mobile, depending upon the degree of covalent linkage with the zeolite lattice, and free to assume the thermodynamically most stable configuration in the zeolite.

In contrast to strong electrolyte type zeolites, the media effect is very small in high-silica zeolites, vide infra. However, the lack of solvation effect in these molecular sieves may be counterbalanced by a stronger O-H polarity. Here, it is conceivable that for lack of media contribution to the stabilization of a "free" carbocation, the positively charged hydrocarbon reaction intermediate formed by proton transfer from the H-zeolite to an olefin may be held as a hydrocarbon ligand, complexed through the acidic zeolite hydrogen (H^+) to the zeolite framework.

The formation of a positively charged, hydrocarbon complex strongly linked to a zeolite framework oxygen should influence the effective charge, the life time (stability) and the steric hindrance imposed on this species. It can be also expected that these characteristics would also affect the reaction rates and the reaction mechanism. Therefore, one may expect substantial differences in the reaction chemistry between positively charged hydrocarbon "complexes" and "free" carbocations. Since the formation of "free" carbocations requires higher activation energy, involving the deprotonization energy of the zeolite, it is likely that complex type intermediates may be formed at relatively low temperatures while "free" carbocations are generated at relatively higher temperatures. The reaction temperature used in catalytic processes such as FCC is high enough (~500°C) to provide sufficient energy to break even stable C-C or C-H bonds. Thus, under these conditions the formation of free carbo-cations is likely, particularly when the positively charged carbon has strong electron donor ligands.

The mechanistic aspects of carbocation chemistry over protic zeolites have been extensively studied by many investigators, and the work has been reviewed by Poutsma(89), with enlightening discussion on the greatly diverse chemical and physical effects of H-zeolites on hydrocarbon reaction mechanisms. Some of the effects are derived from geometric (shape selectivity) considerations while others are derived from solvation and protic strength, giving rise to mechanisms involving proton donors, hydride abstractions or pentacoordinated C^+ species(89). More recently, the formation of unusual product patterns obtained over strong acid H-zeolites has been discussed on the basis of pentacoordinated C^+(107a) and radical cation reaction intermediates(107b). In recent discussions the "superacid" reaction pattern has been ascribed to very strong acid sites. We believe, however, that some of these mechanistic effects may be related to "complex" versus "free" carbocation type reaction intermediates.

CONCLUSIONS

Reviewing the literature on the acid function in zeolites provides an excellent demonstration of the need for interdisciplinary research in surface chemistry and catalysis. The results show that in this complex field a single experimental technique or theoretical calculation uncovers only a small segment of the relevant chemistry. Past studies on zeolite acidity and acid catalysis cover a vast scientific effort, applying an extraordinary wide array of techniques and theories. In spite of this effort, some of the key questions including the cause of differences in acid strength between H-zeolites and amorphous silica alumina, and between zeolites of different structures, remained unanswered. Nevertheless, the most recently available data base is of high quality and it provides a multifaceted view of the chemistry. On this basis, we suggest several tentative conclusions on zeolite acidity, subject to experimental verification in the future.

1. The strong acidity of H-zeolites versus amorphous aluminosilicates of similar compositions seems to be based on a "global" readjustment of the zeolite lattice bond structure to "proton attack", in order to minimize lattice deformations leading to loss of resonance energy and crystallinity (long-range symmetry). This is in agreement with experimental XPS and NMR data. It is proposed that in order to maximize resonance energy, the zeolite crystal tends to equalize T-O-T bonds by enhancing the bonding in Al-O linkages to become similar to Si-O linkages. This trend would render the oxygens linked to aluminum more electronegative. Both proposed effects tend to enhance the positive charge on the hydrogen attached to oxygens bridging between Al and Si atoms. On the other hand, amorphous aluminosilicates, with chain or, more likely, ring type Si-O-Al-O-Si-linkages, probably respond to proton attack by reducing the bond strength of the coordinative linkage between Al and the bridging oxygen. This would tend to reduce both the interaction between oxygen and aluminum as well as the net positive electronic charge on the hydrogen, resulting in more covalent and consequently less acidic, terminal type Si-O-H groups.

2. We propose that differences in O-H acid strength between different zeolites rest on differences in crystal structure and related bond structure, T-O-T bond angles and lengths, and crystal resonance energy. Large T-O-T bond angles such as those found in ZSM-5 enhance the "s" character of the T-O bonds, and consequently, the electronegativity of the oxygen. The result is larger positive charge on the hydrogen and stronger protic acidity.

3. In general, the strength of zeolite media (solvent) effect is expected to grow with increasing Si/Al ratios, from 1 to about 2.5, and to reach maximum in the range of Si/Al of 2.5 to about 5. Beyond a Si/Al ratio of about 5 it should decline with lower aluminum content. In contrast to the media effect, the intrinsic O-H acid strength of individual protic sites increases with increasing Si/Al ratios, reaching a near maximum value at a ratio of about 7 (depending on the zeolite structure type).

4. All H-zeolites, even the strongest acids, carry O-H hydrogen with only a fractional positive charge ($\sim 0.2e^-$). Therefore, the strength of the zeolite media (solvent) is expected to strongly contribute to the formation of "free" carbocations. It can be also expected that in cases when H^- abstraction is not dominant in the mechanism, the reaction intermediate in hydrocarbon catalysis may be a hydrocarbon complex, linked through the acidic zeolite hydrogen to framework oxygens. Such complex-like reaction intermediate should have different reactivity and reaction chemistry from "free" carbocations. The tendency to form such complex like intermediates should be inversly related to the strength of the intracrystalline zeolite media (solvation). Carbocation mechanisms involving five-coordinated C^+ centers in other strong acid zeolites may, in part, result from "complex" type reaction intermediates. Furthermore, low temperatures should favor "complex" like species while high temperatures should favor "free" carbocations.

293

5. Diffusion of molecules in zeolites with approximately molecule-sized
    channels differs greatly from diffusion in amorphous materials.
    Multiple adsorption/desorption reactions may dominate the diffusion in
    zeolites. Molecules in zeolites may also exhibit enhanced ("floating")
    diffusion and a channel-mouth energy barrier to diffusion.

6. The effect of hard versus soft character of the carbocation and the
    zeolite lattice in acid catalysis is not well understood. It is
    expected, however, that larger polarizability for the reaction
    intermediate and for the lattice oxygen should affect the reaction
    chemistry of "complex" type hydrocarbon reaction intermediates.
    Substitution of zeolite framework cations by transition metal ions is
    expected to increase the softness of the zeolite bond structure by
    providing polarizable "d" orbitals. Such phenomena may be the cause
    of new, not strictly acid strength related catalytic phenomena
    reported for transition metal substituted molecular sieves(108).

REFERENCES

 1 G.A. Olah, G.K.S. Prakash, J. Sommer, Superacids,, John Wiley & Sons,
   (1985)
 2 R.G. Pearson, J. Am. Chem. Soc., 85:3533 (1963)
 3 R.G. Pearson, J. Am. Chem. Soc., 107:6801 (1985)
 4 R.G. Pearson, Proc. Natl. Acad. Sci., 83:8440 (1986)
 5 G.A. Olah, G.K.S. Prakash, R.E. Williams, J.D. Field, D. Wade,
   Hypercarbon Chemistry, John Wiley & Sons, (1987)
 6 a) J.F. Stanton, W.N. Lipscomb, R.J. Bartlett, M.L. McKee, Inorg. Chem.,
      28(1):109 (1989
   b) J.F. Stanton, R.J. Bartlett, W.N. Lipscomb, Chem. Phys. Lett.,
      138(6):525, (1987)
   c) K.L. Krause, K.W. Volz, W.N. William, J. Mol. Biol., 193(3):527,
      (1987)
 7 a) M. Bremer, P. Schleyer, U. Fleischer, J. Am. Chem. Soc., 111(3):1147
   b) A.E. Reed, C. Schade, P. Schleyer, P.V. Kamath, J. Chandrasekhar,
      J. Chem. Soc., Chem. Commun., (1):67, (1988)
   c) M. Saunders, K.E. Laidig, K.B. Wiberg, P. Schleyer, J. Am. chem. Soc.
      110(23):7652, (1988)
   d) W. Bauer, M. Feigel, G. Mueller, P. Schleyer, J. Am. Chem. Soc.,
      110(18):7652, (1988)
   e) P. Schleyer, Pure Appl. Chem., 59(2):1647, (1987)
   f) P. Schleyer, T.W. Williams, W. Koch, A.J. Kos, H. Schwarz, J. Am.
      Chem. Soc., 109(23):6953, (1987)
   g) P. Schleyer, B.T. Luke, J.A. Pople, Organometallics, 6(9):1997,
      (1987)
 8 J.A. Rabo, Catal. Rev.-Sci. Eng., 23(1&2):293, (1981)
 9 D.W. Breck, Zeolite Molecular Sieves,, Wiley-Interscience, New York,
   (1974)
10 J.A. Rabo, P.E. Pickert, D.N. Stamires and J.E. Boyle, in Actes du
   Deuxième Congrès International de Catalyse, Paris, (1960) p. 2055
11 J.A. Rabo, in Zeolite Chemistry and Catalysis (ACS Monograph 171),
   Am. Chem. Soc., 5 (1976)
12 R.J. Pellet, C.S. Blackwell and J.A. Rabo, J. Catal., 114:71, (1988)
13 A.K. Aboul-Gheit, M.A. Al-Hajjaji, M.F. Menougy and S.M. Abdel-Hamid,
   Anal. Lett., 19:529, (1986)
14 K. Hashimoto, T. Masuda and H. Sasaki, Ind. Eng. Chem. Res., 27:1792,
   (1988)
15 K. Tsutsumi, H.Q. Koh, S. Hagiwara, H. Takahasi, Bull. Chem. Soc. Jap.,
   48:3576, (1975)
16 A. Auroux, V. Bolis, P. Wierzchowski, P. Gravelle and J.C. Vedrine,
   J. Chem. Soc. Faraday Trans. I, 75:2544, (1979)
17 R. Bezman, J. Catal., 68:242 (1981)

18 A. Macedo, A. Auroux, F. Raatz, E. Jacquinot and R. Boulet, <u>ACS Symp. Ser.</u>, 368:98 (1968)

19 a) A.K. Aboul-Gheit and M.A. Al-Hajjaji, <u>Anal. Lett.</u>, 20:553 (1987)
   b) A.K. Aboul-Gheit, <u>J. Catal.</u>, 113:490 (1988)

20 A.K. Aboul-Gheit, M.A. Al-Hajjaji and A.M. Summan, <u>Thermochim Acta</u>, 118:9 (1987)

21 G.I. Kapustin, T.R. Brueva, A.L. Klyachko, S. Beran and B. Wichterlova, <u>Appl. Catal.</u>, 42:239 (1988)

22 B. Sen, P. Chou, M.A. Vannice, J. Catal. 101:517 (1986)

23 a) J.L. Falconer and J.A. Schwarz, <u>Cat. Rev. Sci. Eng.</u>, 25:141 (1983)
   b) R.J. Gorre, <u>J. Catal.</u>, 75:164 (1982)
   c) R.A. Demmin, R.J. Gorre, J. Catal., 90:32 (1984)
   d) Y.-J. Huang, J.A. Schwarz, <u>J. Catal.</u>, 99:249 (1986)
   e) J.M. Criado, P. Malet, G. Munuera, <u>Langmuir</u>, 3:973 (1987)

24 L. Forni and E. Magni, <u>J. Catal.</u>, 112:437 (1988)

25 K. Hashimoto, T. Masuda and T. Mori <u>in</u> "Proc. 7th Intern. Zeolite Conf., 503 (1986)

26 J. Take, H. Yoshioka and M. Misono, <u>in</u> "Proc. 9th Intern. Conf. Catal., 372 (1988)

27 L. Forni, E. Magni, E. Ortoleva, R. Monaci and V. Solinas, <u>J. Catal.</u>, 112:444 (1988)

28 L.G. Dowel, J.M. Bennett and D.E. Passoja, 25th Ann. Denver X-ray Conf. (1976)

29 a) T.L. Barr, <u>Am. Lab.</u> 10:40 (1978)
   b) T.L. Barr, Am. Lab. 10:65 (1978)

30 Y. Okamoto, M. Ogawa, A. Maezawa, T. Imanaka, <u>J. Catal.</u>, 112:427 (1988)

31 R.L. Patton, E.M. Flanigen, L.G. Dowell, D.E. Passoja, <u>ACS Symp. Ser.</u>, 40:64 (1977)

32 K. Taniguchi and T. Takaishi, <u>in</u> Proc. 7th Intern. Zeolite Conf., 155 (1986)

33 R.H. West and J.E. Castle, <u>Surf. Interf. Anal.</u>, 4:68 (1982)

34 T. Takaishi and H. Hosoi, <u>J. Phys. Chem.</u>, 86:2089 (1982)

35 E.G. Derouane and J.G. Fripiat, <u>Zeolites</u>, 5:165 (1985)

36 e.g. L. Pauling, "The Nature of the Chemical Bond", Cornell, NY, (1960)

37 S. Beren and J. Dubsky, <u>J. Phys. Chem.</u>, 83:2538 (1979)

38 W.J. Mortier, octoral Thesis, Catholic University of Leuven (1978)

39 T.L. Barr and M.A. Lishka, <u>J. Am. Chem. Soc.</u>, 108:3178 (1986)

40 T.L. Barr, L.M. Chen, M. Mohsenian and M.A. Lishka, <u>J. Am. Chem. Soc.</u>, 110:7962 (1988)

41 J.B. Nagy and E.G. Derouane, <u>ACS Symp. Series.</u>, 368:2 (1988)

42 J.M. Thomas and J. Klinowski, <u>Adv. Cat.</u>, 33:199 (1985)

43 H. Pfeifer, D. Freude, M. Hunger, <u>Zeolites</u> 5:274 (1985)

44 E.G. Derouane, N. Dewaele, Z. Gabelica and J.B. Nagy, <u>Appl. Cat.</u>, 28:285 (1986)

45 a) G. Engelhardt and D. Michel, "High Resolution Solid-State NMR of Silicates and Zeolites", Wiley, New York (1987) p.228,257,260
   b) J.M. Bennett, C.S. Blackwell and D.E. Cox, <u>J. Phys. Chem.</u>, 87:3783 (1983)

46 a) J.H. Lunsford, W.P. Rothwell and W. Shen, <u>J. Am. Chem. Soc.</u>, 107:1540 (1985)
   b) W.L. Earl, P.O. Fritz, A.A.V. Gibson and J.H. Lunsford, <u>J. Phys. Chem,</u>, 91:2091 (1987)
   c) L. Baltusis, J.S. Frye and G.E. Maciel, <u>J. Am. Chem. Soc.</u>, 108:7119 (1986)
   d) L. Baltusis, J.S. Frye and G.E. Maciel, <u>J. Am. Chem. Soc.</u>, 109:40 (1987)

47 R.L. Stevensen, <u>J. Catal.</u>, 21:113 (1971)

48 M. Hunger, D. Freude, H. Pfeifer, H. Bremer, M. Jank and K.P. Wendlandt, <u>Chem. Phys. Lett.</u>, 100:29 (1983)

49 B.B. Whipple, P.J. Green, M. Ruta, R.L. Bujalski, <u>J. Phys. Chem.</u>, 80:1350 (1976)

50 H. Pfeifer, J. Chem. Soc. Faraday Trans. I, 84:3777 (1988)
51 a) J.W. Emsley, J. Feeney, and L.M. Sutcliffe, "High Resolution NMR
    Spectroscopy" Pergamon, Oxford (1966)
    b) R.K. Harris and B.E. Mann, "NMR and the Periodic Table" Academic
    Press, London (1978)
52 D. Freude, H. Pfeifer, M. Hunger, in "Proc. 7th Intern. Zeolite Conf.,
    107 (1986)
53 J.F. Haw, B.R. Richardson, I.S. Ohsiro, N.D. Lazo and J.A. Speed,
    J. Am. Chem. Soc., 111:2052 (1989)
54 E.M. Flanigen, in "Zeolite Chemistry and Catalysis", ACS Monograph 171
    (Ed. J.A. Rabo) (1976) p.80
55 J.W. Ward, in "Zeolite Chemistry and Catalysis" ACS Monograph 171,
    (Ed. J.A. Rabo) (1976) p.118
56 V.A. Maroni, K.A. Martin and S.A. Johnson, ACS. Symp. Ser., 368:85,
    (1987)
57 A. Janin, J.C. Lavalley, A. Macedo and F. Raatz, ACS. Symp. Ser.,
    368:117 (1988)
58 M.D. Baker, J. Godber and G.A. Ozin, ACS Symp. Ser., 368:136 (1988)
59 P.A. Jacobs and W.J. Mortier, Zeolites.,2:226 (1982)
60 J. Datka, M. Boczen and P. Rymarowicz, J. Catal., 114:368(1988)
61 G. Zi and T.Yi, Zeolites., 8:232 (1988)
62 L.A. Pine, P.J. Maher and W.A. Wachter, J. Catal., 85:466 (1984)
63 W. Lowenstein, Am. Mineral., 39:92 (1954)
64 W.A. Wachter, in Proc. 6th Intern. Zeolite Conf.,(1984) p. 141
65 D. Barthomeuf, Mat. Chem. Phys., 17:49 (1987)
66 W.M. Meier and H.J. Moeck, J. Solid State Chem., 27:349 (1979)
67 R.S. Hansford, Ind. Eng. Chem., 39:849 (1947)
68 J.B. Peri, J. Catal., 41:227 (1976)
69 A.G. Pelmenshchikov, E.A. Paukshtis, V.S. Stepanov, K.G. Ione,
    G.M. Zhidomirov and K. I. Zamaraev in "Proc. 9th Intern. Cong. Catal."
    (1988) p.404
70 G.M. Zhidomirov and V. Kazansky, Adv. Catal., 34:131 (1986)
71 W.J. Mortier, in "Proc. 6th Intern. Zeolite Conf.", (1984) p.734
72 a) K.J. Chou, J.C. Lin, Y. Wang and G.H. Lee, Zeolites., 6:35 (1986)
    b) J.L. Schlenker, J.J. Pluth and J. V. Smith, Mat. Res. Bull., 14:849
    (1979)
73 D.H. Olson and E. Dempsey, J. Catal., 13:221 (1969)
74 J. Dwyer and P.J. O'Malley, Stud. Surf. Sci. Cat.,35:5 (1988)
75 R. Carson, E.M. Cooke, J. Dwyer, A. Hinchliffe and P.J. O'Malley,
    in "Zeolites as Catalysts, Sorbents and Detergent Builders", Extended
    abstracts,
76 W.J. Mortier and P. Geerlings, J. Phys. Chem., 84:1982 (1980)
77 E. Kassab, K. Seiti, and M. Allavena, J. Phys. Chem., 92:6705 (1988)
78 E.G. Derouane and J.G. Fripiat, J. Phys. Chem., 91:145 (1987)
79 A. Goursot, F. Fajula, C. Daul and J. Weber, J. Phys. Chem., 92:4456
    (1988)
80 C. Fernandez, A. Auroux, J.C. Vedrine, J. Grosmangin and G. Szabo,
    in "Proc. 6th Intern. Zeolite Conf." (1986) p.345
81 W.J. Mortier, D.E.W. Vaughan and J.M. Newsam, ACS Symp. Ser. 368:194
    (1980)
82 J.J. Van Dun and W.J. Mortier, J. Phys. Chem., 92:6740 (1988)
83 J.J. Van Dun, K. Dhaeze, W.J. Mortier, J. Phys. Chem., 92:6747 (1988)
84 J.A. Rabo, C.L. Angell, P.H. Kasai, and V. Schomaker, Discuss. Faraday
    Soc., 41:328 (1966)
85 J.A. Rabo and P.H. Kasai, Progress in Solid State Chemistry., 9:1 (1975)
86 P.H. Kasai and R.J. Bishop, ACS Monograph., 171:350 (1976)
87 a) T. Ito and J. Fraissard, J. Chem. Soc. Faraday Trans. I., 83:451
    (1987)
    b) J. Fraissard, T. Ito, M. Springuel-Huet and J. Demarquay, in "Proc.
    7th Intern. Zeolites Conf.," (1986) p.393
    c) L. Petrakis, T. Ito, M.A. Springuel-Huet, T. Hughes, I.Y. Chen and

J. Fraissard, in "Proc. 9th Intern. Congress on Catal.", (1988) p.348
d) T. Ito, M.S. Springuel-Huet and J. Fraissard, Zeolites., 9:68 (1989)
e) M.A. Springuel-Huet, J. Demarquay, T. Ito and J. Fraissard,
"Innovation in Zeolite Material Science" (Ed. P.J. Grobet et al.)
Elsevier (188) p.183

88 S.M. Brown, W.J. Reagan and G.M. Wolterman, U.S. Patent 4.325,813.

89 M.L. Poutsma, ACS Monograph., 171:437 (1976)

90 M. Eic and D.M. Ruthven, Zeolites., 8:472 (1988)

91 D.M. Ruthven and M. Eic, ACS Symp. Ser., 368:362 (1988)

92 M. Eic and D.M. Ruthven, Zeolites., 8:40 (1988)

93 D.M. Ruthven, "Principles of Adsorption and Adsorption Processes",
Wiley, New York, (1984)

94 J. Karger and D.M. Ruthven, J. Chem. Sóc. Faraday Trans.I., 77:1485
(1981)

95 J. Karger and H. Pfeifer, Zeolites., 7:90 (1987

96 J. Karger and H. Pfeifer, ACS Symp. Ser., 368:376 (1988)

97 J. Karger, J. Chem. Soc. Faraday Trans.I., 76:717 (1980)

98 J. Caro, M. Bulow, J. Karger and H. Pfeifer, J. Catal., 114:186 (1988)

99 E.G. Derouane, J. B. Nagy, C. Fernandez, Z. Gabelica, E. Laurent and
P. Maljean, Appl. Cat., 40:L1 (1988)

100 R.M. Barrer and J.A. Davies, Proc. Royal Soc. London, Ser. A., 322:1
(1971)

101 E. Aust, K. Dahlke and G. Emig., J. Catal., 115:86 (1989)

102 E.G. Derouane, J.M. Andre and A.A. Lucas, J. Catal., 110:58 (1988)

103 a) R.M. Barrer, "Zeolites and Clay Minerals as Sorbents and Molecular
Sieves", Academic Press, London, (1978)
b) E.G. Derouane, J. Catal.,100:541 (1986

104 a) P.B. Weisz, Chemtech., 3:498 (1973)
b) W.O. Haag, R. Lago and P.B. Weisz, Farad. Disc. Chem. Soc., 72:317
(1981)

105 E.A. Lombardo, J.M. Dereppe, G. Marcelin and W.K. Hall, Catal.,114:167
(1988)

106 a) M.C. Grady and R.S. Gorte, J. Phys. Chem., 89:1305 (1985)
b) M.T. Aronson, R.J. Gorte, and W.E. Farneth, J. Catal., 98:434 (1986)
105:455 (1987)
c) A. Webb, in "Actes Deuxième Int. Congr. Catal.," Technip, 1:1289
(1960) : W.K. Hall comment, "Actes Deuxième Int. Congr. Catal., Technip,
1:1307 (1960)
d) M. Zardkoohi, J.F. Haw, and J.H. Lunsford, J. Amer. Chem. Soc.,
109:5278 (1987)
e) J.P. van den Berg, J.P. Wolthuizen, A.D.H. Clague, G.R. Hays, R.
Huis and J.H. Van Hooff, J. Catal., 80:130 (1983)

107 a) W.O. Haag and R.M. Dessau, in "Proc. 8th International Congress
on Catalysis" Berlin (1984) p.305 Dechema Frankfurt-am-Main (1984)
b) J.A. Rabo, in "Proc. Sixth Int. Zeolite Conf. (1985) p.41

108 R.J. Pellet. P.K. Coughlin, E. Shamshoum and J.A. Rabo, ACS Symp. Ser.,
368:512 (1988)

DIRECTING ACTIVITY AND SELECTIVITY OF LARGE PORE ACID ZEOLITES THROUGH THE CONTROL OF THEIR PHYSICOCHEMICAL PROPERTIES

Avelino Corma

Instituto de Catálisis y Petroleoquímica
C.S.I.C.
28006 Madrid, Spain

INTRODUCTION

The catalytic activity and selectivity of zeolites are related to both crystalline structure and chemical composition. Crystalline structure is responsible not only for the so-called shape selectivity, but also will determine the number, accesibility and strength of the active sites due to changes in the T-O-T angles. If one refers to acid zeolites, the deprotonation energy of a bridged hydroxyl group depends on its local geometry, which is determined by the framework structure, and by the possibilities to relax this geometry within the framework subjected to deprotonation[1]. In this way it has been found that the deprotonation energy decreases with increasing Si-O-Al angles[2]. Then, when calculations have been made using observed atomic coordinates in HZSM-5, acidity differences of NH_3 adsorption over 33 Kcal.mol^{-1} can be inferred[1], for hydroxyl sites on different crystallographic positions[3]. Besides framework geometry, the chemical composition of the zeolite has also an important influence on the number and stability of the anions formed in the deprotonation process, and therefore on the number and strength of the acid sites formed.

In the present paper we will discuss the influence of the chemical composition of some large pore acid zeolites on their adsorption and catalytic properties.

INFLUENCE OF ZEOLITE COMPOSITION ON ACTIVITY

The Brönsted acid sites in zeolites are associated to protonic hydrogens from bridged structures Al(OH)Si. In principle, the potential number of acid sites is equal to the number of framework Al, the actual number being smaller due to cation exchange and dehydroxylation occuring during activation. If the total number of acid sites is important for acid catalyzed reaction what becomes crucial is the number of acid sites strong enough to carry out a given reaction. Therefore, in order to design active and selective acid catalysts based on zeolites, one needs to know what are

Guidelines for Mastering the Properties of Molecular Sieves
Edited by D. Barthomeuf *et al.*
Plenum Press, New York, 1990

the parameters which control the acid strength. It was proposed[4] that the chemical environment of the sites should affect their acidic and, therefore, their catalytic properties. In this way it has been shown[2] that the density of positive charge increases when decreasing the number of Al in the second coordination sphere of a given Al. This has been explained considering the zeolites as solid electrolytes[5]. In this sense, and taking into account the high concentration of protons in zeolites, activity coefficients should be considered, as it is done in aqueous solution[5]. An efficiency coefficient α_0, which behaves as an activity coefficient, has been determined in faujasites[5]. This factor increases from $\alpha_0=0$ for a faujasite with Si/Al=1, where every Si atom is surrounded by four Al atoms, to $\alpha_0=1$ for Si/Al close to 6 where each hexagonal prism contains 2 Al atoms one in each face[6]. Taking into account the Loewenstein rule and assuming a random distribution of Al one can find, in a faujasite structure, up to 9 different types of Al. The strongest acid sites will be associated to those Al with zero Al in the next nearest neighbours. The relative population of them will change with the number of Al per unit cell. Then, if a random distribution occurs, one should found a maximum in activity for a given reaction at a certain framework Si/Al ratio. The particular value will depend on the acid strength needed for that reaction.

In Figures 1 and 2 the activity of two 12 MR zeolites versus framework Si/Al ratio have been plotted for several reactions[7-14], all of them requiring strong acid sites. From a close observation, we can see that there is not a general trend as the theory would predict, i.e. for a Y zeolite, when going from 2.5 to ~6.0 Si/Al by de-alumination, the activity for acid demanding reactions should increase reaching a maximum around Si/Al=6 and then decrease. In the case of offretite and mordenite, the maximum should occur at a Si/Al ~ 7 and 9 respectively[5]. These results indicate that other compositional factor such as the extraframework (EFA1) may play an important role.

It has been reported for steamed mordenite[15], that there are acid sites which retain NH_3 at temperatures as

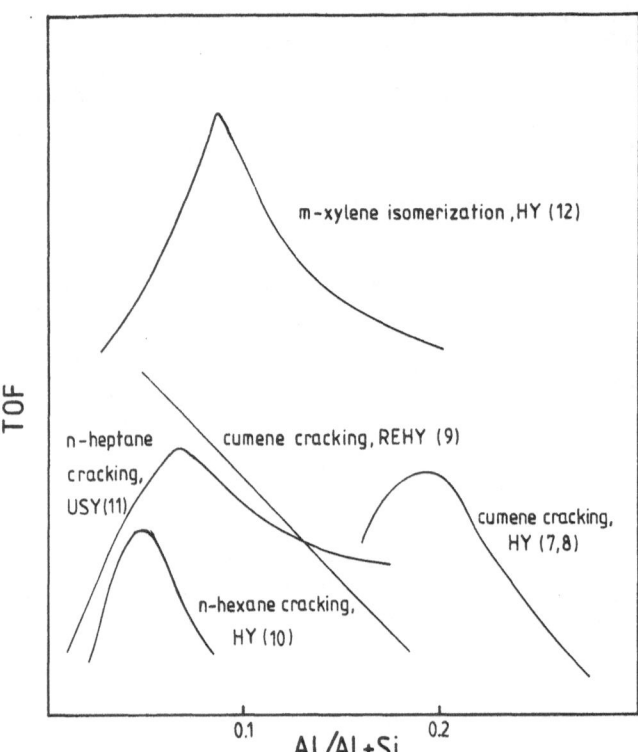

Fig. 1. Influence of framework Al in TOF

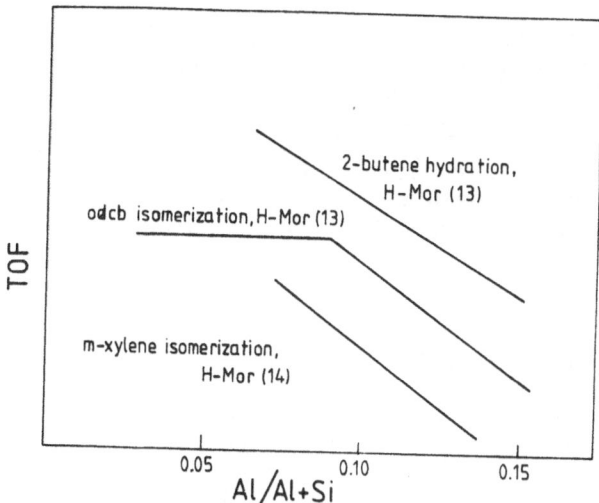

Fig. 2. Influence of framework Al in TOF

high as 600°C. It is there postulated that an enhancement of the acidity of some framework Brönsted sites can occur by interacting with EFAl charged species of the type:

$$\begin{array}{c} H \\ | \\ O \text{----} (AlO^+)_p \\ / \ \backslash \\ Al \quad Si \end{array}$$

If this is so, when steaming the zeolites in adequate conditions to form species of the type $[Al(OH)_2]^+$, $[Al(OH)]^{2+}$ or $(AlO^+)_p$, the surface of the zeolite would be quite different from that corresponding to an ideal HY. Indeed, under these conditions, one would have on the surface species of the type:

This means that normal framework acid sites, cationic Lewis sites, and "superacids" Brönsted sites generated by interaction of a normal Brönsted and a Lewis EFAl species, or a partially dislodged framework Al, are present.

Those "superacid sites" has been claimed to exist in partially dealuminated Mordenite[15] and in mild steamed ZSM-5 [16,17] and in steamed HY[18], in order to explain an increase in the turnover frequency (TOF) when the zeolites were partially dealuminated by steam. In HZSM-5, it was suggested a one to one correspondence of Brönsted and Lewis sites for optimum activity in cracking of n-hexane[16]. The enhanced activity sites for n-hexane cracking in H-ZSM-5 which appear after steaming has been related to the presence in zeolite H-ZSM-5 of paired Al regions. One of the two Al is partially hydrolyzed and acts as a strong electron withdrawing center for the remaining tetrahedral Al, thus creating a stronger Brönsted site in agreement with the general hypothesis presented previously by Lunsford[19]. On the other hand, it has been shown[20] that in well prepared and carefully treated H-ZSM-5 zeolites, a direct correlation between activity and Al/UC exists, at least in the region 0 to 10 Al/UC. If one assumes, however, that in H-ZSM-5 there is preferential siting of aluminium in the four-ring system, then either the activity of a hydroxyl group related to one Al with zero and one Al in the second coordination sphere is the same[21], or it becomes difficult to explain the direct correlation between n-hexane cracking and the number of Al/UC in unsteamed samples.

Recently[22], the number of aluminiums in the second coordination sphere in HZSM-5 with Si/Al ratios between 24.6 and 623 has been calculated from the ^{29}Si NMR results. It has been found that in this range all Al are completely isolated, i.e. no Al are present in the second coordination sphere. Moreover, the Si/Al calculated from NMR and from chemical analysis are the same, indicating that dealumination has not occured under their activation conditions.

On the other hand, in the case of H-Mordenites with Si/Al ratios up to 10.2 there are still pairing Al sites[22]. In the case of faujasites, dislodged aluminium is distributed between both small and large cages, and the proportion depends on the steaming conditions. Therefore, in the case of faujasite type structures it becomes difficult to establish an optimum ratio of EFAl to FAl[16].

Recently we have shown[23] that, in USY samples, a new i.r. band at 3600 cm^{-1} associated to very strong Brönsted sites (i.e. sites retaining pyridine at 400°C and 10^{-4} Torr) is present when the samples were steamed at mild temperatures (up to 700°C). At the same time an amorphous silica-alumina phase was detected[24]. There, the hydroxyls associated to the 3600 cm^{-1} band were related with the amorphous silica-alumina phase. However, there is another possibility to explain the 3600 cm^{-1} hydroxyl groups which can not be ruled out and this is the enhancement of the acidity of some framework HF hydroxyl groups by cationic EFAl species. This interaction will produce a shift in the wavenumber of about 30 cm^{-1}.

In Figure 3 the i.r. spectra of a USY sample steamed at 650°C is given. It can be seen (Fig. 3b) that after adsorbing pyridine and desorbing it at 250°C and 10^{-4} Torr., there are hydroxyl groups retaining pyridine at 3630 (HF), 3560 (LF), 3600 and 3525 cm^{-1}.

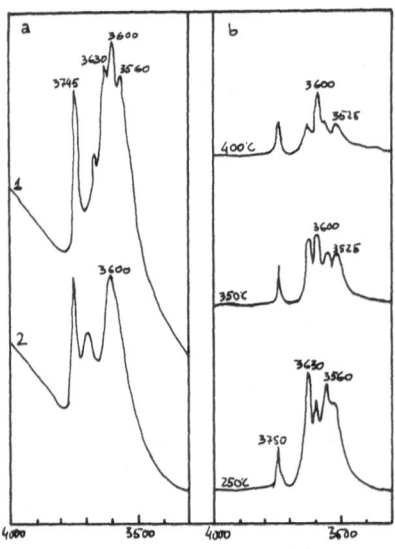

Fig. 3. a) I.R. spectra of the OH groups in a USY (24.43Å before (a1) and after (a2) adsorption of pyridine and desorption at 250° C
b) Differnce spectra after adsorption of py and desorption at different temperatures

302

Table 1.   Temperature of maxima (T_{max}), activation energy of desorption (E) and amount of NH_3 desorbed, for the dealuminated HY.

HY Samples			Peak 1			Peak 2		
Al/uc	a_0 A	Si/Al	T_{max} °C	E Kcal/mol	$n.10^4$ mol/g	T_{max} °C	E Kcal/mol	$n.10^4$ mol/g
22	24.43	7.7	234	29.2	5.84	331	34.8	5.63
18	24.39	9.9	209	27.8	6.25	381	37.7	5.94
13	24.35	13.6	204	27.5	4.34	379	37.6	5.48
8	24.30	24.5	199	27.2	2.75	367	36.9	3.95

The presence of acid hydroxyls at 3525 cm^{-1} would be in favour of the interaction of EFAl with Brönsted sites, if one assumes that in the case of the LF hydroxyls a similar shift than in the case of HF can be produced, due to the interaction with cationic EFAl. Moreover, after desorption at 400°C and 10^{-4} Torr, only the hydroxyls at 3600 and 3525 cm^{-1} are acid enough to retain pyridine, while at 350°C desorption temperature, all Brönsted acid sites HF, LF and the 3600, and 3525 cm^{-1} do still retain.   When NH_3 t.p.d. has been carried out[25] (Fig. 4) on LaHY samples, the results indicate (see Table 1) that in peak 2 there are sites with energies of adsorption ranging from 130 to 158 $KJ.mol^{-1}$.   It should be expected that sites adsorbing NH_3 with heat of adsorption equal to 130-140 KJ mol^{-1}, will have a TOF lower than those with heats > 150 $KJ.mol^{-1}$.   Indeed, it has been observed that in H-Mordenites there are two types of Brönsted sites with different specific activity.   One type, with heat of NH_3 adsorption equal to 150 $\pm$ 5 $KJ.mol^{-1}$, show octane cracking activities almost one order of magnitud higher than the other type, with NH_3 heats of adsorption in the range of 120-140 $KJ.mol^{-1}$ [26].

In our case, if we plot the acid sites corresponding to the HF + LF hydroxyl and the number of acid sites corresponding to 3600 + 3525 cm^{-1} hydroxyls from pyridine adsorption, for a series of samples steam dealuminated at increasing steaming temperature (Fig. 5), one can see that very strong Brönsted sites (3600 and 3525 cm^{-1}) go through a maximum for 15-25 Al/UC.   In other words, the amount of very strong acid sites

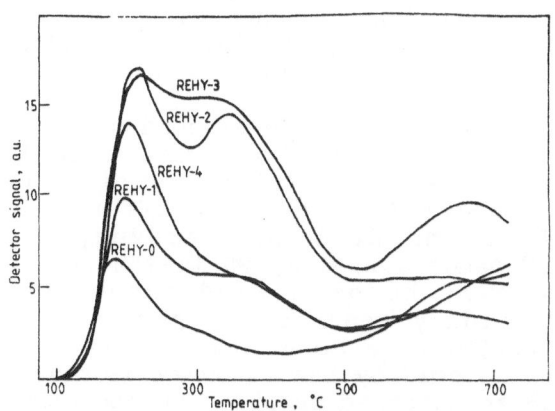

Fig. 4.   TPD of NH_3 on REHY zeolites. REHY-0, 0%wt La_2O_3; REHY-1, 3.3%; REHY-2, 8.3%; REHY-3, 12.8%; REHY-4, 17.3%

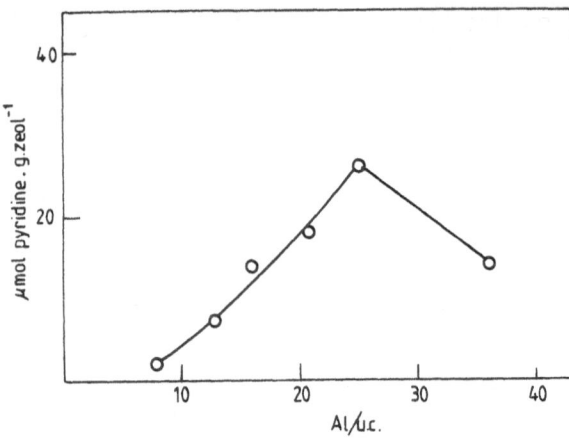

Fig. 5. Pyridine retained on hydroxyls at 3600 and 3525 cm^{-1} after 400°C and 10^{-4} torr desorption for USY zeolites.

is maximum for mild steaming conditions and dissapears for higher steaming temperatures. If the TOF for n-heptane cracking, and for m-xylene isomerization are plotted versus the Al per unit cell, a constant value is not observed for samples with Si/Al ratios higher than ~ 6 (Fig. 1). The TOF goes through a maximum for 15-25 Al/UC and then decreases, something which could be explained considering that the strongest Brönsted acid sites appearing at 3600 and 3625 cm^{-1} show a higher TOF than the classical HF and LF hydroxyls. A simple calculation from the results of Figures 1 and 5, indicates that for xylene isomerization the stronger sites have a TOF ~ 20 times higher than the others. This is consistent with the behaviour of the activation energy for the isomerization of xylenes, with dealuminated zeolites[27].

If the high TOF sites were generated through an enhancement of their acidity by cationic (Lewis acid type) EFAl species, a chemical treatment of the zeolite, selective enough to remove them without dealuminating extensively the framework, should have a direct impact on the i.r. spectra of the USY in the OH region, as well as on catalytic activity.

Nevertheless one should take into account that during steam dealumination, aluminium cationic species of the type $[Al(OH)_2]^+$, $[Al(OH)]^{2+}$, $(AlO^+)_p$ are formed and they can also occupy cation exchange positions neutralizing, therefore, Brönsted hydroxyl groups, while stabilizing the zeolite structure in a similar way as $[La(OH)_2]^+$ or $[La(OH)]^{2+}$ species do. Macedo et al.[28] have shown that in mild steamed HY zeolites, an adequate acid leaching produces an increase of the HF OH band in the i.r. spectra. On the other hand, if the HY samples were steamed at higher temperatures (> 700°C), either a much smaller or no increase in acidity was observed, after acid leaching.

In our case we have taken a USY sample steamed at 650°C in which all hydroxyl groups described above were clearly visible (Fig. 3) and then it was treated with a solution of $(NH_4)_2SiF_6$. Figures 6a and 6b[29] show that after this treatment, by which 60% of the EFAl was removed, the intensity of the HF and LF bands increase, while the acidic 3600 and 3525 cm^{-1} hydroxyl bands have disappeared. Nevertheless the Brönsted acidity measured by pyridine at 250°C and 10^{-4} Torr. desorption conditions (Fig. 6c and 6d) has increased, while the Lewis acidity has strongly decreased with respect to the $(NH_4)_2SiF_6$ unwashed sample. On the other hand, the chemical treatment has produced a small dealumination (from 24.43 to 24.39 Å u.c. size) of the framework. Despite this framework dealumination occurring

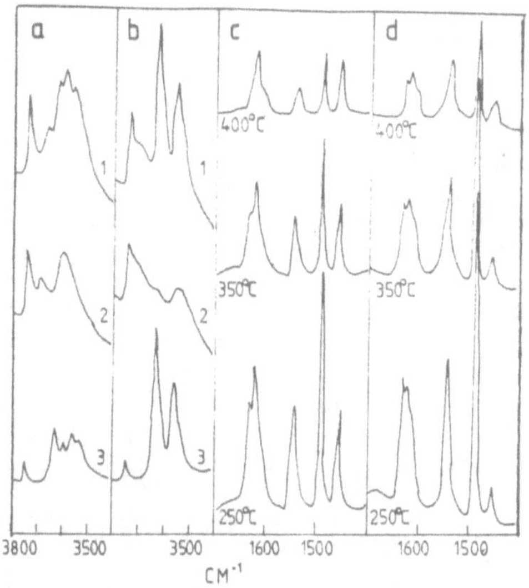

Fig. 6. Hydroxyl and pyridine i r spectra of USY sample, before a and c and after b and d treatment with $NH_4)_2SiF_6$.
1) sample degassed at 400°C, 10^{-4} torr overnight.
2) after adsorbtion of pyridine and desorption at 250°C.
3) difference (1-2).

during the chemical treatment, the Brönsted acidity related to FAl has increased. However the TOF for cracking n-heptane, isomerizing m-xylene and cracking gasoil were reduced 60, 60 and 50% of the original. Notice that from these results and from those presented by Macedo et al.[28] the concept of TOF in zeolites should be revised, since part of the framework Al are neutralized by cationic EFAl species.

The results obtained with the USY samples before and after treatment with $(NH_4)_2SiF_6$ indicate that one should not necessarily expect a direct correlation between total acidity and catalytic activity in USY samples, since sites of enhanced acidity and consequently of higher activity than the classical HF and LF are present. These sites are certainly related to the presence of EFAl, either forming part of an amorphous silicoaluminate, or enhancing the acidity of conventional Brönsted sites. What becomes apparent is that if these EFAl species are either removed by chemical treatments or polymerized by hard steaming conditions, the activity of the zeolite for carbonium ion reactions strongly decreases.

In the case of another large pore zeolite such as Beta, a different behaviour has been observed. Indeed, Beta zeolite is synthesized with Si/Al ratios > 7 [30], and while it becomes dealuminated by the activation treatments[31], we have not seen another acid hydroxyl band in the Al(OH)Si region than the 3615 cm^{-1} corresponding to Beta. Nevertheless, and due to the geometry of this zeolite, a higher TOF for n-heptane cracking[31] and m-xylene isomerization[32] has been observed on Beta than in HY zeolite.

In conclusion, it could be said that from the point of view of the active acid sites, the activity of a zeolite depends not only on the framework Si/Al ratio, but also on the type and amount of EFAl. At the same time these parameters are controlled by the level and procedure of dealumination.

INFLUENCE OF ZEOLITE COMPOSITION ON SELECTIVITY

Acid strength

Changes in product selectivity for acid catalyzed reactions due to different zeolite composition have been related to changes in acid strength and site density. The influence of acid strength can easily be visualized, if we take into account that different acid catalyzed reactions need sites with different acid strength. On the other hand,

we said above that by dealumination, one changes the number
of Al in the next nearest neigbours (NNN) position and,
therefore, the population of sites with different acid
strength. In this way it is possible, for instance, to
design zeolite catalysts which can carry out branching
isomerization with minimum cracking[33].

It has been found in electrophilic alkylation of
aromatics on zeolites that depending on acidity the
selectivity to the alkylation on para or ortho changes. For
instance, in the methylation of toluene[34] the ratio of para
to ortho xylene increases in the following series of
zeolites: NaY < $Ca_{0.5}Y$ < $Ca_{0.3} Na_{0.4}Y$ < $H_{0.3}Na_{0.7}Y$ <
$H_{0.7}Na_{0.5}Y$ < HY (Si/Al=17).

In another case[35], toluene has been alkylated with
methanol on Y zeolite exchanged with mono, di, and trivalent
cations. There the para to ortho xylene in the products
increases in the order mono < di < trivalent.

During the alkylation of o-xylene on faujasite
zeolites[36], it has been found that BaNaX and MgNaY direct to
the formation of 1,2,3-trimethylbenzene, while methylation of
o-xylene on HY samples gives preferentially 1,2,4-
trimethylbenzene.

Another parameter on the zeolite which has been changed
is the level of Na^+ exchanged by NH_4^+ [37]. In that case it
has been observed that during the alkylation of toluene by
methanol, the ratio of para to ortho xylene increases when
the level of Na^+ exchanged increases.

Changes in the framework Si/Al ratio of a given zeolite
also produces changes in the para to ortho xylene during the
methylation of toluene. Then, increasing the Si/Al ratio of
a faujasite increases the para/ortho ratio. However,
extremely silica rich acid faujasite and mordenite do not
follow such sequence, and relatively low para/ortho ratios
are found[34].

In all these cases the zeolite behaviour has been
correlated, but not explained, with acid strength. Indeed,
from the point of view of charge density and acid strength,
it will become very difficult to explain the para selectivity
induced by certain composition of large pore zeolites, in
which no geometrical constrains could be expected to occur.
However, Wendlant and Bremer have related the para
selectivity with the hardness of the acid sites on zeolites.

Therefore, if coulombic interactions can not explain
the selectivities observed, another factor should be
considered, as for instance the frontier electron density on
the different positions of the nucleophilic molecule, and the
possibilities of orbital controlled reactions on zeolites.

ESTIMATION OF CHEMICAL REACTIVITY

The energy gained and lost (ΔE) when the orbitals of
one reactant overlap with those of another, was derived by
Klopman[38,39] and Salem[40] using the perturbation theory:

$$\Delta E = - \sum_{ab} (q_a + q_b) \beta_{ab} S_{ab} + \sum_{K<1} \frac{Q_K Q_1}{\epsilon R_{K1}} +$$

$$\underbrace{\hspace{4cm}}_{\text{first term}} \qquad \underbrace{\hspace{3cm}}_{\text{second term}}$$

$$+ \sum_{r}^{occ} \sum_{s}^{unocc} - \sum_{s}^{occ} \sum_{r}^{unocc} \frac{2(\sum_{ab} C_{ra} C_{sb} \beta_{ab})^2}{E_r - E_s} \quad [1]$$

---

third term

q_a and q_b are the electronic populations in the atomic orbitals $\underline{a}$ and $\underline{b}$. β, the resonance integral, is the energy associated with having the electron shared by the atoms in the form of the covalent bond. S is the overlap integral, which gives a measure of how effective the overlap is in lowering the energy. Q_K and Q_1 are the total charges on atoms K and l. ϵ is the local dielectric constant, and R_{Kl} is the distance between the atoms K and l. C_{ra} is the coefficient of atomic orbital $\underline{a}$ in molecular orbital $\underline{r}$, where $\underline{r}$ refers to the molecular orbitals on one molecule and s refers to those on the other. Finally, E_r is the energy of molecular orbital $\underline{r}$.

The first term in equation [1] is the first order closed-shell repulsion term, and comes from the interaction of the filled orbitals of one molecule with the filled orbital of the other. Its main effect on chemical reactivity can probably be identified with the fact that the smaller the number of bonds to be made or broken at a time, in a chemical reaction, the better. Concerning to its importance on influencing selectivity, it can be said that it is usually very similar for each of two possible pathways. In this way, if a molecule can be attacked at two possible sites, the first term will be very similar for attack at each site.

The second term is the coulombic repulsion or attraction, and becomes specially important in reactions involving ions or polar molecules.

The third term, "the orbital controlled" represents the interaction of all the filled orbitals with all the unfilled of correct symmetry. In other words, the occupied orbitals, specially the HOMOs, of each interact with the unoccupied orbitals, specially LUMOs, of the other while an attraction between the molecules is produced.

When discussing chemical reactivity on the basis of equation [1], it should be taken into account that factors which affect the entropy of activation and steric effects are not included.

Concerning acid-base reactions, it is known that the facility with which an acid-base reaction takes place depends on the strength of the acid and base, but it also depends on another quality called the hardness or softness of the acid or base (HSAB)[41,42]. This quality can not be precisely measured, but qualitatively estimated[43], and the HSAB principle says that hard acids prefer to bind to hard bases, and soft acids prefer to bind to soft bases.

Pearson[41] has defined hard and soft acids and bases in the following way:

<u>Soft bases</u> : The donor atoms are of low electronegativity and high polarizability and are easy to oxidize. They hold their valence electrons loosely.

Hard bases : The donor atoms are of high electronegativity and low polarizability and are hard to oxidize. They hold their valence electrons tightly.

Soft acids : The acceptor atoms are large, have low positive charge and contain unshared pairs of electrons p or d, in their valence shells. They have high polarizability and low electronegativity.

Hard acids : The acceptor atoms are small have high positive charge, and do not contain unshared pairs in their valence shells. They have low polarizability and high electronegativity.

It is possible to explain the principle that hards acids form stronger bonds and react faster with hard bases, and soft acids form stronger bonds and react faster with soft bases by means the above mentioned perturbation theory. We start from the principle that the hard acids have high-energy LUMOs and the hard bases have low-energy HOMOs[44]. In other words, the higher the energy of the LUMO of an acid the harder it is as an acid, while the lower the energy of the HOMO of a base the harder is a base and, on the other hand, we explain the reactions mainly through the HOMO-LUMO interaction.

These principles can be used to explain not only the position of attack (ortho, meta, para) of electrophiles to aromatics, reaction of big importance in catalysis, but also could explain the ortho/para ratio obtained, when electrophiles (carbocations) attack substituted benzenes.

In the case of substituted benzenes of the type 1 for instance:

the total electron population is the same on the ortho and the para position, but the frontier electron density is higher in the para position. Therefore, one should expect that the softer electrophiles will give more substitution in the para position. Nevertheless, and this should be considered along these discussions, the steric effects are not taken into account, since here we are more interested in explaining how, with the same electrophile and nucleophile, the ortho/para ratio will change when the hardness of the acid site is being changed.

For substituted benzene molecules of the type 2 for instance:

the total electron population is meta >> ortho > para. The frontier electron population is meta > ortho > para. Then, since all three molecular contributions are in the same direction high ortho/para ratios are found for these compounds.

Finally in the case of alkylation of benzene derivatives of type 3, for instance:

OCH   OH   NR$_2$ He

for which heterogeneous catalysts and more specifically zeolites are used, the total charge is larger on ortho. However, the frontier electron population is larger on the para position. Therefore it should be expected softer electrophiles to give more para substitution.

HSAB : CHARGE AND ORBITAL CONTROLED REACTIONS IN ZEOLITES

It has been said above that the harder the acid the higher the energy of LUMOs. When a zeolite acts as a catalyst and attacks to a nucleophile to form a "carbocation," it has to be taken into account that very seldom there is a total proton transfer from the zeolite to the nucleophile, i.e. a complete ionic interaction, but what exists is the formation of a complex of the type:

$$Zeo^- .... [H.....Reactant]^+ \qquad [2]$$

There, a partial electron transfer from the reactant to the zeolite occurs. We can relate the HOMO and LUMO of the complex to the LUMO of the zeolite and the HOMO of the nucleophile combination. The energy level of the LUMO of the complex [2], and therefore its hardness, will depend on the LUMO energy of the zeolite itself but also on the energy difference between the LUMO of the zeolite and the HOMO of the nucleophilic reagent (Fig. 7).

Notice that it has been considered here a close energy between the LUMO of the "carbocation" and the HOMO of the zeolite in order to produce a sensible amount of covalent bonding.

The energy of the HOMO of the zeolite can be modified by changing the average electronegativity of the structure, then the energy of the HOMO will depend on factors such as framework composition, level of exchange, type of cation, structure etc.

If we take into account a series of complexes [2] formed on zeolites with different framework Si/Al ratios (Fig. 8). It can be expected that when increasing the Si/Al ratio, i.e. the average electronegativity of the zeolite, the energy of the LUMO of the complex decreases. This indicates that the higher the electronegativity of a given zeolite, the softer the acid-base complex, which in turns means that for a given nucleophile the higher the framework Si/Al of an acid zeolite, the softer is the acidity of the zeolite. Notice that this is a clear case for which stronger acidity (higher Si/Al ratio) does not correspond to harder acidity.

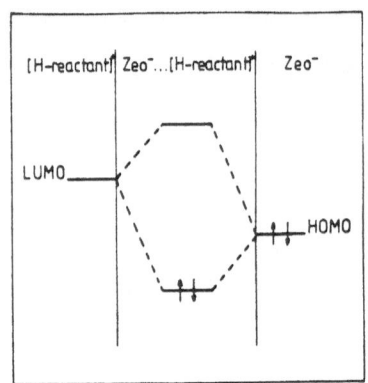

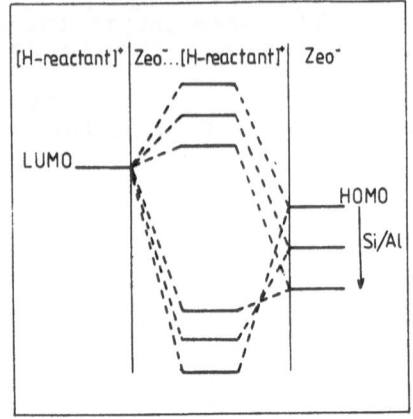

Fig. 7. Qualitative LUMO and HOMO MO energy
levels of the complex [Zeo⁻...electro-
phile] related to the MOs of the zeolite and the electrophile.

Fig. 8. Influence of the framework Si/Al of the
zeolite on the LUMO and HOMO energy level of the complex [Zeo⁻...electrophile]

We can discuss from this point of view chemical orientation effects of zeolites in an important type of reactions: electrophilic alkylation of aromatics. If we take as an example the alkylation of toluene by methanol on an acid zeolite, the reaction can be written as:

1- Protonation of methanol.

$$Zeo^- - H^+ + CH_3OH \longrightarrow Zeo^- \cdots \left[CH_3 - O \begin{matrix} H \\ H \end{matrix} \right]^+$$

2- Alkylation of the electrophile (Toluene).

$$Toluene + Zeo^- \cdots \left[CH_3 - O \begin{matrix} H \\ H \end{matrix} \right]^+ \longrightarrow xylenes + Zeo^- - H^+$$

Notice that during the alkylation process on zeolites, the alkylating agent has to be considered as the complex carbocation-zeolite and not only the carbocation.

In the case of toluene, the coefficients of the HOMO are high on the ortho and para position and zero on the meta 44. Thus, the third term of equation [1], will very strongly favour reaction at ortho and para. However, the frontier electron density is higher in the para position. One should therefore expect that the lower the energy LUMO of the

electrophile (Zeo⁻...$\left[CH_3 - O \begin{matrix} H \\ H \end{matrix} \right]^+$), i.e. the softer the

electrophile, the smaller will be the energy difference between that and the HOMO of the nucleophile (toluene), i.e. the smaller (E_r-E_s) in the third term of equation [1]. Then, the smaller (E_r-E_s), the more "orbital controlled" will be the reaction, and higher selectivity to the para product should be expected. If this is so, and taking into account Fig. 8, one should expect that the higher the framework Si/Al ratio of a H-zeolite and therefore the higher the average electronegativity, i.e. the softer the zeolite, the softer would be the alkylating complex, and therefore more para

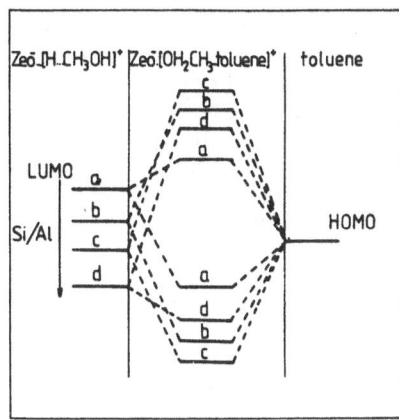

Fig. 9.
Qualitative LUMO and HOMO of the complex of alkylating and nucleophile, related to the MOs of the nucleophile, and the alkylating agent formed on zeolites of different Si/Al.

substitution during the electrophilic activation of alkylaromatics should be expected.

The same reasoning could be applied to explain the selectivity for any electrophilic or nucleophilic type of reaction, on any zeolite and on any zeolite composition.

However, we would like to discuss also here the reported observation[34] that some highly dealuminated zeolites produce a para/ortho ratio lower than expected. To look at this we are going to present qualitative energy levels for the MOs corresponding to the interaction of a series of electrophiles

$$(Zeo^- \ldots \left[CH_3 - O \overset{+}{\underset{H}{\overset{H}{\diagup}}} \right])$$ in which

the original zeolite has different framework Si/Al ratios (Fig. 9).

It can be seen, that depending at the relative level of the different LUMOs of the zeolites and the HOMO of the toluene, there may exist one inversion in the increasing covalent character, i.e. "orbital control," achieved by decreasing the LUMO energy. Indeed, in Fig. 8, it becomes apparent, that by increasing the framework Si/Al ratio from zeolite A to C, a decrease in the (E_r-E_s) is obtained and therefore the para to ortho ratio should increase from A to C. However when going from C to D, (E_r-E_s) increases and therefore the para to ortho selectivity will be smaller for D than for C.

It can be deduced as a whole that it would be better to relate the average electronegativity of an acid zeolite with the acid softness instead than with the acid strength. In this way, it becomes quite clear that by dealumination, the softness of the zeolite will keep increasing continously, even though the acid strength will reach a maximum and then will stay constant.

Following this, it becomes apparent that even if the acid strength remains the same for several zeolites and the turnover number is also the same, differences in selectivity could still exist since the hardness of these zeolites (the average electronegativity) could be different.

It is now the moment to question if the selectivity changes observed during alkylation of aromatics and substituted aromatics in medium pore zeolites with different Al contents[45,46] or even with different heteroatoms[47], are always due to a "shape selective effect" or to a chemical effect related with the frontier orbitals and chemical reactivity. We think that the "orbital control" effect should be taken into account in these cases, since in H-ZSM-5 zeolites with similar crystal sizes but with decreasing amounts of Al, a higher para to ortho selectivity is observed during aromatic alkylation[45]. This could be predicted from our model since, by decreasing the Al content in H-ZSM-5 zeolites, even if the acid strength of the sites does not change, the intermediate electronegativity of the zeolite is still increasing and so does the softness of the zeolite, and therefore, the para directing effect.

Then from the intermediate electronegativity calculated by Mortier[48] for a series of zeolites, one could predict not only the softness of those zeolites but also the para/ortho evolution during alkylation, or more generally the selectivity evolution for "orbital controlled" reactions catalyzed by zeolites or by any other solid acid.

ACID SITE DENSITY

The density effect has been used in the literature to explain differences in selectivity between uni and bimolecular reactions. In this way it has been claimed that molecular reactions occurring in zeolites, such as alkylaromatics disproportionation and hydrogen transfer need the presence of two close acid sites[48,49]. This was implicitly assuming that the reaction was taking place through a bimolecular Langmuir-Hinselwood type mechanism, for which the two reactant molecules are adsorbed in two adjacent acid sites. This type of mechanism is difficult to imagine, specially on highly dealuminated zeolites in which most of the Al have 0 and 1 Al in the NNN and therefore the acid sites are strong enough to protonate benzene rings or olefins. Then in this case, it seems unlikely that two positively charged molecules would encounter.

We have carried out a series of experiments to check this hypothesis looking at the influence of Brönsted acid site density on hydrogen transfer and disproportionation of xylenes.

In our case we have started from the hypothesis that both type of reactions are bimolecular following a Eley-Rideal. This implies that the reaction takes place between one molecule adsorbed on a Brönsted site:

$$A + H^+ \xrightleftharpoons{K_a} AH^+ \qquad \text{(Adsorbed)}$$

$$AH^+ + A \text{ or } B \xrightarrow{K'_a} C + DH^+ \qquad \text{(Adsorbed)}$$

$$DH^+ \longrightarrow D + H^+$$

and another in the "gas phase". In this case molecules in "gas phase" means molecules weakly adsorbed, which behave more like fluids in the intrazeolitic environment. There, the molecule does not have to be adsorbed necessarily on another Brönsted site. If this is so, the kinetic rate expression for a bimolecular reaction, as for instance the disproportionation of m-xylene, will be:

$$r_0 = \frac{K_0 [S_0] K_a [A]^2}{1 + K_a [A]} \qquad [3]$$

In which K_0 is the kinetic rate constant for the controlling step. $[S_0]$ is the initial concentration of active sites. K_a is the adsorption constant for reactant A, and [A] is the concentration of reactant inside the zeolite pores.

On the other hand, the isomerization of m-xylene to o- and p-xylene is a first order reaction which fits well[50] the following model:

$$r_i = \frac{K_o \ [S_o] \ K_a \ [A]}{1 + K_a \ [A]} \qquad [4]$$

From [3] and [4] it appears that the selectivity for isomerization will depend, among other structural confinement factors and the possibility that both reactions need sites of different strength, on the "effective" concentration of reactants inside the pores of the zeolite. In other words, a zeolite with a higher adsorption capacity for a given reactant should, in principle give a lower ratio isomerization to disproportionation than another with a lower adsorption capacity. The same would be true for the ratio cracking (unimolecular) to hydrogen transfer (bimolecular).

In Figure 10 it can be seen that when decreasing the number of Al per unit cell in a HY zeolite the isomerization to disproportionation ratio for m-xylene increases. Meanwhile, the adsorption isotherms for m-xylene adsorption[51] on two HY zeolites with different number of Al/UC indicate that the adsorption decreases when the Al/UC decreases.

In the case of hydrogen transfer the paraffin/olefin ratio in the n-heptane cracked products, and therefore the hydrogen transfer decreases when decreasing the Al/UC. Meanwhile n-butene adsorption isotherms clearly show that the adsorption of olefins decreases when decreasing the number of Al/UC (Fig. 11). From all these results one may conclude, that by changing the framework composition, besides changing acid strength, one may change the adsorption capacity for a given reactant and changes in adsorption can produce changes in selectivity. Moreover results in Figure 12, when comparing n-butene and n-butane adsorption[52], shows that not only the adsorption capacity changes but also the adsorption selectivity. For instance, olefins are proportionally less selectively adsorbed on USY zeolites than paraffins when decreasing the number of Al/UC, this being specially clear for samples with less than 10 Al/UC, or what is equivalent in this case, samples which were steamed above 700 °C. Notice, that the adsorption capacity is going to depend

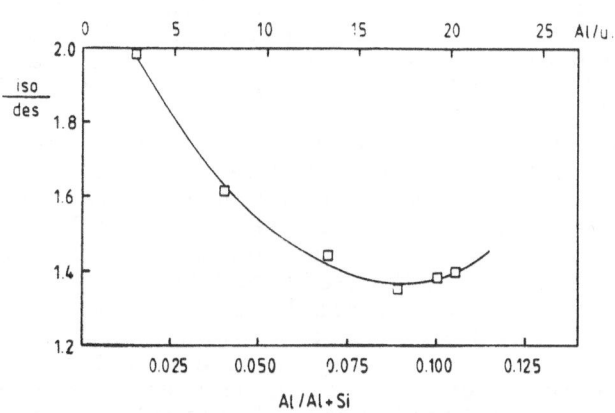

Fig. 10. Isomerization/ Disproportionation ratio vs. Al/u.c. in m-xylene isom. on SiCl₄ treated Y zeolites

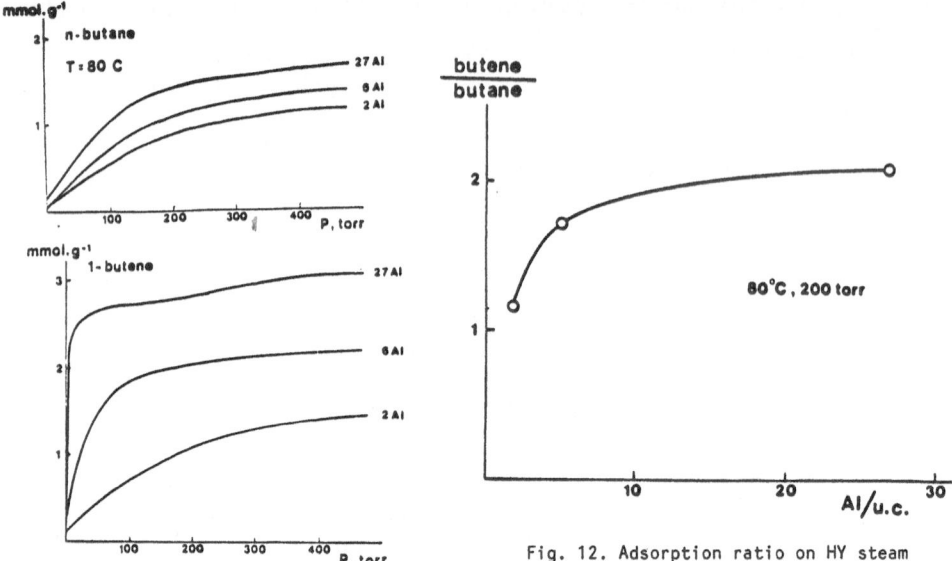

Fig. 11. Adsorption isotherms on HY steam

Fig. 12. Adsorption ratio on HY steam

on the electric field inside the zeolite, and therefore on chemical factor such as a presence of cationic species $[La(OH)_2]^+$ and type of them, framework Si/Al of zeolite, presence of framework atoms other than Al, etc.

CONCLUSIONS

In conclusion, the activity of a given large pore zeolites depends not only on the framework Al content but also on the amount and type of EFAl. At mild dealumination conditions (either by steam or $SiCl_4$), EFAl is mainly present as cationic species, a part of them neutralizing acid sites and increasing the steam stability of the zeolite. Another part will be in the form of amorphous silica-alumina and/or interacting with Brönsted sites enhancing its acidity. Two band of hydroxyls at ~ 3600 and 3525 cm^{-1}, which are strongly acidic, and give a higher TOF than the HF and LF hydroxyls, are related with that EFAl. At harder dealumination conditions the EFAl polimerizes and migrates to the surface, while the 3600 and 3525 cm^{-1} decrease and finally disappear. This goes parallel with the TOF of the zeolite for acid demanding reactions.

There are reactions for which the coulombic interactions can not explain the observed selectivities. For instance in the case of electrophilic alkylations, the distribution of the frontier orbital of the nucleophile is going to direct the attack. Moreover the orbital controlled effect and therefore the para/ortho selectivity will also depend on the level of the LUMO orbital of the alkylating reactant. It has to be taken into account that the alkylating reactant is not only the electrophile reactant but the complex zeolite-electrophile. Then the softer the zeolite acidity the more orbital controlled will be the reaction. This depends on the electronegativity of the zeolite. The higher the electronegativity of the zeolite the softer will be the acidity of the zeolite, the lower the LUMO and therefore the more orbital controlled will be the reactions.

Finally, when there are competing uni- and bimolecular reactions, changes in selectivity with dealumination can be explained on the bases of changes in the adsorption capacity and selectivity of the zeolites.

ACKNOWLEDGEMENT

The author thanks Drs. J. Planelles, F. Melo and V. Fornés for helpfull discussions, and to CAICYT MAT 85/0284.

REFERENCES

1. J. Sauer and W. Schirmer, "Innovation in Zeolite Materials Science", P.A. Grobet et al., ed., Elsevier, Studies in Surf. Sci. and Catal., 37, 323 (1988).
2. I.N. Senchenya, V.B. Kazansky and S. Beran, J. Phys. Chem., 90:4857 (1986).
3. J.G. Fripiat, F. Berger-André, J.M. André and E.G. Derouane, Zeolites, 3:306 (1983).
4. D. Barthomeuf and R. Beaumont, J. Catal., 30:288 (1973).
5. D. Barthomeuf, Materials Chem. and Phys., 17:49 (1987).
6. D. Barthomeuf, "Catalysis by Zeolite", Elsevier, ed., Studies in Surf. Sci. and Catal., 5, 55 (1980).
7. K. Tsutsumi and H. Tokahashi, J. Catal, 24:1 (1972).
8. K.V. Topchieva and H.S. Thnoang, Kin. i. Katal., 11:490 (1979).
9. A. Corma, V. Fornés and F.V. Melo, 10th Iberoamerican Congress of Catalysis, 647 (1986).
10. D. Barthomeuf, "Zeolites Science and Technology" NATO ASI Series, ed., F. Ramoa, A.E. Rodrigues, L. Deane, C. Naccache, 80:317 (1984).
11. A. Corma, V. Fornés, A. Martínez and A.V. Orchillés in "Perspectives in Molecular Sieve Science" W.H. Flank and T.E. Whyte, Jr., ed., ACS Symp. Ser., 368, Chap. 35, 542 (1988).
12. A. Corma, V. Fornés, J. Pérez-Pariente, E. Sastre, J.A. Martens and P.A. Jacobs, ibid chap. 36, 555 (1988).
13. B. Coq, R. Durand, F. Fajula, C. Moreau, A. Finiels, B. Chiche, F. Figueras and P. Geneste, "Heterogeneous Catalysis and Fine Chemicals", Stud. in Surf. Sci. and Catal., 41, M. Guisnet et al, ed., 241 (1988).
14. C. Mirodatos, B.H. Ha, K. Otsuka and D. Barthomeuf, Proc. Fifth Int. Conf. Zeol., L.V.C. Rees, ed., Naples 382 (1980).
15. C. Mirodatos and D. Barthomeuf, J. Chem. Soc. Chem. Commun., 39 (1981).
16. A.G. Ashton, S. Batmanian, D.M. Clark, J. Dwyer, F.R. Fitch, A. Hinchcliffe and F.J. Machado, "Catalysis in Acid and Bases" B. Imelik et al., ed., Elsevier, Studies in Surf. Sci. and Catal., 101 (1985).
17. R.M. Lago, W.O. Haag, R.J. Mikousky, D.H. Olson, S.D. Hellring, K.D. Schmitt and G.T. Kerr, Proc. 7th Int. Zeolite Conf., Y. Murokani et al., ed., Kodansha-Elsevier, 677 (1986).
18. R.A. Beyerlein, G.B. McVicker, L.N. Yacullo and J.J. Ziemak, J. Phys. Chem., 92:1967, (1988).
19. J.H. Lunsford, J. Phys. Chem., 72:4163 (1968).

20. W.O. Haag, R.M. Lago and P.B. Weisz, _Nature_, (London), 309:589 (1984).
21. E.G. Derouane and J.G. Fripiat, _Zeolites_, 5:165 (1985).
22. S. Segawa, M. Sakaguchi and Y. Kurusu, "Methane Conversion", D.M. Bibby et al., ed., _in_ Surf. Sci. and Catal, 42:579 (1988).
23. A. Corma, V. Fornés, A. Martinez, F.V. Melo and O. Pallota, "Innovation in Zeolite Materials Science", Stud. in Surf. Sci. and Catal., 37, P.J. Grobet et al., ed., 495 (1988).
24. J. Sanz, V. Fornés and A. Corma, _J. Chem. Soc. Faraday Trans. I_, 84:3113 (1988).
25. A. Corma, V. Fornés, F.V. Melo and J. Herrero, _Zeolites_, 7:559 (1987).
26. A.L. Klyachko, G.I. Kapustin, T.R. Brueva and A.M. Rubinstein, _Zeolites_, 7:119 (1987).
27. A. Corma, E. Sastre, F. Llopis and J. Montón, to be published.
28. A. Macedo, F. Raatz, R. Boulet, A. Janin and J.C. Lavalley, "Innovation in Zeolite Materials Science", P.J. Grobet, ed., Elsevier, Stud. in Surf. Sci. and Catal., 37, 375 (1988).
29. A. Corma, V. Fornés, F. Mocholí and F. Rey, to be published.
30. J. Pérez-Pariente, J.A. Martens and P.A. Jacobs, _Appl. Catal._, 31:35 (1987).
31. A. Corma, V. Fornés, J. Pérez-Pariente, A.V. Orchilles in "Fluid Catalytic Cracking Role in Modern Refining", ed., M.L. Occelli, A.C.S. Symp. Ser., Chapt 3:24 (1988).
32. A. Corma, J. Pérez-Pariente, E. Sastre, J. Martens and P.A. Jacobs, _Appl. Catal._, in press.
33. P.A. Jacobs, _Catal. Rev. Sci. Eng._, 24:415 (1982).
34. K.P. Wendlandt, H. Bremer, Proceeding 8th Int. Congr. Catal., Verlag-Chemie, Vol. IV, 507 (1988).
35. T. Yashima, H. Ahmad, K. Yamazaki, M. Katsuka and N. Hara, _J. Catal._, 16:273 (1970).
36. Ju. N. Sidosenko, _Avt. Svid._, 59:1889 (1985).
37. T. Yashima, H. Ahmad, K. Yamazaki, M. Katsuka and N. Hara, _J. Catal._, 17:151 (1970).
38. G. Klopman, _J. Am. Chem. Soc._, 105:7512 (1983).
39. G. Klopman, ed., "Chemical Reactivity and Reaction Paths", Willey N.Y (1974).
40. L. Salem, _J. Am. Chem. Soc._, 90:543 (1968).
41. R.G. Pearson, _J. Am. Chem. Soc._, 85:3533 (1963).
42. H. Ho, in Hard and Soft Acid and Bases, Principle in Organic Chemistry, Academic Press, NY. (1977).
43. Parr and R.G. Pearson, _J. Am. Chem. Soc._, 105:7512 (1983).
44. I. Fleming, "Frontier Orbitals and Organic Chemical Reactions", John Wiley & Sons, N.Y (1976).
45. Cv. Bezouhanova, Chr. Dimitrov, V. Nenova, B. Spassov and H. Lechert, _Appl. Catal._, 21:149 (1986).
46. J.M. Jacobs, R.F. Parton, A.M. Boden and P. Jacobs, in "Heterogeneous Catalysis and Fine Chemicals", Stud. Surf. Sci. and Catal., 41, M. Guisnet et al., ed., 221 (1988).
47. R.B. Borade, A.b. Halgeri, and T.S.R. Prasada Rao, 7th Int. Zeol. Conf., Stud. in Surf. Sci. and Catal., Y. Murakami et al., ed., 28:851 (1986).

48.   M.L. Martin de Armando, N.S. Gnep and G. Guisnet, _J._ _Chem. Res._ (M), 1:241 (1981).
49.   L.A. Pine, P.J. Maher and W.A. Watcher, _J. Catal._, 85:466 (1984).
50.   H. Matsumoto, K. Yasui and Y. Morita, _Bull Japan_ _Petrol.Inst._, 10:8 (1968).
51.   A. Corma, A. Mifsud, E. Sastre, to be published.
52.   A. Corma, M. Farraldos and A. Mifsud, _Appl. Catal._, 47:125 (1989).

IS THE SHAPE SELECTIVITY ALWAYS RESPONSIBLE FOR THE ORIENTATION OF THE REACTION PATH IN ZEOLITE CATALYSIS?

Wolfgang F. Hoelderich

BASF, Ammonia Laboratory

6700 Ludwigshafen, Federal Republic of Germany

INTRODUCTION

Zeolites are of great significance as catalysts in the field of petro-chemistry, and are gaining steadily in importance for use in the synthesis of organic intermediates and fine chemicals[1-22].

The wide range of applications of zeolites may be seen from the fact that the catalytic properties can be varied in almost limitless fashion, namely by appropriate choice of the mode of preparation including isomorphous substitution[23-25] or by subsequent modifications of the zeolite (e.g. ion exchange, steaming, acid leaching etc.). Acidic, neutral and basic zeolite catalysts can be prepared, and in addition it is possible to employ zeolites as carriers for active components. Despite these manipulations, however, the zeolitic pore framework is retained. Accordingly, these catalysts can, in principle, still display the shape selectivity characteristic of zeolites.

The course of a chemical reaction can be determined by the different types of shape selectivity[26-35]. Reactant selectivity allows only part of the reacting molecules to pass through the catalyst pores. In product selectivity, only products with the correct dimensions can diffuse out of the pores. When only those reactions can take place in which the transition state of the reactants fits the internal pores or cavities, then one speaks of restricted transition state selectivity. Related to this type of shape selectivity is the so-called restricted growth type selectivity, in which the formation of higher addition, substitution or oligomerisa-

Guidelines for Mastering the Properties of Molecular Sieves
Edited by D. Barthomeuf *et al.*
Plenum Press, New York, 1990

319

tion products is suppressed due to the geometric dimensions of these products.

Shape selectivity can only operate when the reaction occurs within the zeolite pore framework. There are, however, cases in which it is uncertain whether reaction takes place on both the inner and the outer surfaces or, on the other hand, only on the outer surface of the zeolite. Furthermore, examples are known in which the molecular dimensions would certainly permit reaction within the zeolite framework, but where the course of the reaction is apparently not determined by shape selectivity but instead by other parameters such as thermodynamic considerations. It is my aim at this NATO workshop to consider such examples from the field of organic synthesis, and to discuss them together, so as to decide which factors determine the course of the reaction and to what extent, if any, shape selectivity plays a role.

ALKYLATION REACTIONS

The alkylation of the aromatic nucleus[2,16,19,36-40] or of the side-chain[16,41-44] of alkylbenzenes are competing reactions which have been extensively studied. A very important feature of the pentasil zeolites is their capacity to direct the alkylation of alkylbenzenes selectively into the para-position[45]. In most cases pure pentasil zeolites still do not have an adequate p-selectivity. However, the thermodynamically favored m- and o-isomers can be suppressed in favour of the p-isomers by modification, e.g. by means of metals or by adding N-heterocyclic compounds[46-50]. For example, in the methylation of toluene, 99 % p-selectivity at 86 % conversion is obtained using Mg-ZSM-5. The part played by doping in increasing p-selectivity is discussed comprehensively in several articles[51-53].

Alkylation of phenol

Alkylation reactions of arenes containing functional groups can also be carried out on zeolitic catalysts. These conversions, however, are more complex than the alkylation of alkylbenzenes, because attack at the nucleus as well as at the functional group can take place i.e. in the case of phenol not only carbon (C-) but also oxygen (O-) alkylation is possible[19,54-57]. An excellent review on this topic has been written by P.A. Jacobs et al.[19]. These authors reported that most of such reactions are carried out in the vapor phase at atmospheric pressure and at temperatures between 200 and 350 °C using pentasil and faujasite-type zeolites. A broad spectrum of products, consisting essentially of cresols, xylenols, anisoles, methylanisoles and diphenyl ether, is obtained when phenol is methylated using zeolites according to equ. 1.

$$
\begin{array}{c}
\text{OH} \\
\text{(phenol ring)} + CH_3OH \longrightarrow
\end{array}
\quad
\begin{array}{l}
\text{anisole} \\
\text{o,p,m - cresols} \\
\text{methylanisoles} \\
\text{xylenols} \\
\text{diphenylether}
\end{array}
\qquad (1)
$$

Strong acid sites favour C-alkylation over O-alkylation. An increase in temperature and residence time causes a decrease of anisole in favour of cresol. Therefore it is assumed that the cresol fraction is not only of a primary but also of a secondary nature, whereas anisole is only a primary product. The formation of the m-isomer, which is the most thermo-dynamically stable isomer and is formed especially at high conversion in the presence of the acidic zeolite and which is difficult to separate, can be reduced by doping the zeolites with alkali or rare earth metals. The hydroxyl group of phenol directs the alkylation preferentially in the o- and p-position. o-Cresol is always obtained with high selectivity - mostly higher using pentasil than using faujasite zeolites. The expected p-selectivity, as is found in the methylation of toluene, is, however, not observed. p-Cresol is only formed preferentially at a yield of 22 % using K-Y-zeolite at 250 °C[55]. In this reaction the Y-zeolite affects the shape selective formation of p-cresol more than the pentasil zeolite.

In contrast, the isomerization of m- as well as of o-cresol over HZSM-5 yields mostly the p-isomer especially at the beginning of the conversion, although this isomer is thermodynamically less stable than the o- and m-isomers[19]. Product diffusion and restricted transition state selectivity are responsible for this effect. Therefore the reason for high o-cresol selectivity in the alkylation of phenol with methanol cannot be shape-selectivity.

Three hypotheses have been put forward to explain the favoured formation of o-cresol. However two of them, proposed by K.Tanabe et al.[58-60] (figures 1 and 2) and by P.Beltrame et al.[61] (figures 3) can not explain the higher p-selectivity using Y-zeolite compared to pentasil zeolites. The third one suggested by P.A.Jacobs et al.[62] (figure 4) is based on the idea that intramolecular transformation of anisole, which is formed as an intermediate, to o-cresol occurs, and a bimolecular reaction of methanol and phenol is suppressed in the pentasil network.

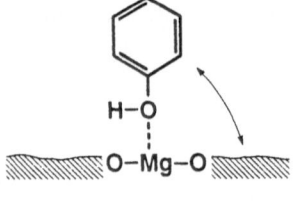

Figure 1. Alkylation of phenol using basic cata-lysts[58-60]

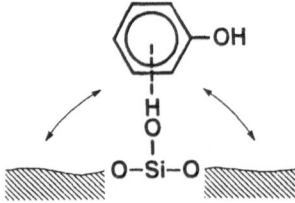

Figure 2. Alkylation of phenol using acid cata-lysts[58-60]

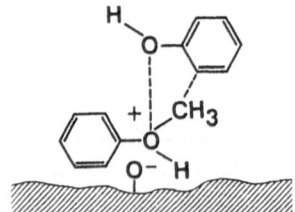

Figure 3. Possible transition state of the alkyla-tion of phenol by an-isole[61]

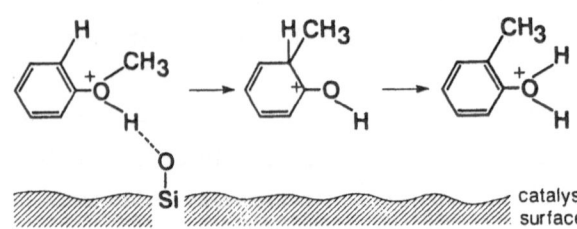

Figure 4. Intramolecular rearrangement of anisole to o-cresol[62]

Nevertheless this hypotheses is also unsatisfactory because, although it explains the high proportion of o-cresol, it fails to explain why shape selectivity does not lead predominantly to the p-isomer. The pentasil zeolites should provide sufficient space in the intracrystalline framework for the attack of methanol in the p-position of phenol as well as for the intermolecular transalkylation by the methyl group from one anisole to another anisole (methylanisole is observed) or to phenol in the p-position as favoured by pentasil zeolites. In other words, there are still open questions, particularly in view of the fact that even in homogeneous medium the o-/p- ratios are much higher than the statistical 2/1 ratio, and on amorphous catalysts without shape selective features o-cresol selectivities as high as 100 % can be obtained[19].

On the other hand, in the case of larger molecules[63-66], such as amylphenol or octylphenol, the reaction in the liquid phase on large-pore zeolites results preferentially in a high proportion of the p-isomer (figure 5).

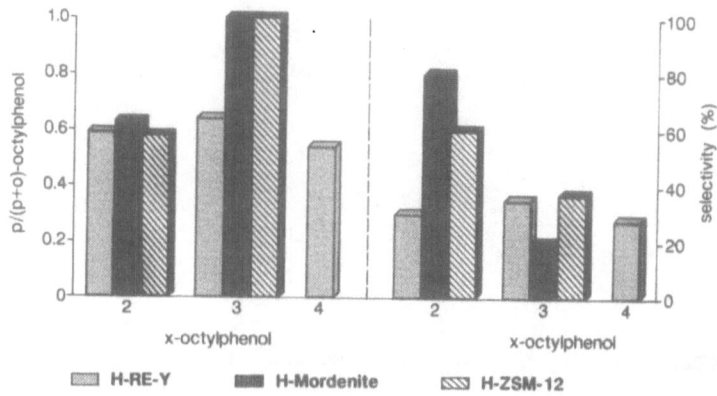

Figure 5. Selectivity and p/(p+o) ratio in the alkylation of phenol with octanol[65]

Even at high conversion, the thermodynamically most stable m-isomers are not present. Here the deviation from the thermodynamic equilibrium in the distribution of the product reflects the transition state shape selectivity. Why does this not apply to the conversion of phenol with methanol?

Alkylation of heteroaromatics

Only few references dealing with alkylation of N-containing heteroaromatics are known[67], perhaps because such systems are apparently intrinsically more complex.

The alkylation of pyridine according to equ. 2 takes place at the aromatic nucleus. In subsequent reactions the picoline which is formed can either undergo further ring alkylation or side chain alkylation to yield ethyl or vinyl pyridine. When HY, LiY, SrY or BaY are employed, ring alkylation occurs almost exclusively to give picolines and lutidines (Table 1), with the alkaline earth doped Y-zeolites leading to higher yields than HY and LiY.

$$\text{(2)}$$

1.) H-Y, Li-Y, Sr-Y, Ba-Y

1.+ 2.) Na-, K-, Cs-Y   Na-, K-, Cs-X

TABLE 1

Productspectrum of the reaction pyridine and methanol depending onto the catalyst[67]

catalysts	H-Y	Li-Y	Sr-Y	Ba-Y
conversion (%)	32	31	62	63
yield (%):				
pyridines	18	13	31	35
lutidines	6	6	22	22
ethylpyridines	1	2	/	/
vinylpyridines	/	/	/	/

Reaction conditions: 400° C, LHSV = 1.3 h^{-1}, methanol/pyridine = 8 molare

Pyridines can be alkylated in the nucleus using faujasites, with the ß-position being preferentially attacked when H-Y or Li-Y-zeolites are used, and the α- and γ-positions in the case of alkali-earth Y-zeolite catalysts. On H-Y 3 % α-, 12 % ß- and 3 % γ-picoline are obtained at 32 % conversion, whereas on Ba-Y 23 % α-, 4 % ß- and 8 % γ-picoline are produced at 63 % conversion (reaction conditions: 400 °C, methanol:pyridine = 8 mol, LHSV = 1.3 hr^{-1}, N_2). Isomers of lutidine are formed as by-products. An explanation why one or the other position is favoured in the alkylation of pyridine with methanol is not obvious. There is no clear correlation between shape selectivity and product distribution.

On the other hand, when pyridine is treated with methanol in the presence of X- and Y-zeolites ion-exchanged with alkali metals (not Li), side chain alkylation is also observed as a consecutive reaction (Table 2). The principal products are ethylpyridine and vinylpyridine as well as isomers of picoline and lutidine.

TABLE 2

Productspectrum of the reaction pyridine and methanol depending onto the catalyst[67]

catalysts	Na-Y	Cs-Y	Na-X	Cs-X
conversion (%)	65	82	45	58
yield (%):				
pyridines	13	15	8	10
lutidines	3	2	1	1
ethylpyridines	13	27	11	21
vinylpyridines	5	7	6	6

Reaction Conditions: 450°C, LHSV = 1.3 H^{-1}, methanol/pyridine = 8 molare

When, for example, Cs-Y-zeolite is used as catalyst (reaction conditions: 450 °C, LHSV = 1.3 hr^{-1}, methanol/pyridine = 8 mol) the yield of ethylpyridine is approx. 27 % at 82 % conversion. Apparently, in this conversion the side chain alkylation as a consecutive reaction is due to the basicity of the zeolite catalyst. Shape selectivity is not involved.

In the side-chain methylation of picolines the following order of reactivity is observed: $\alpha > \gamma > \beta$, with ß-picoline being almost inert. Thus, the maximum yield of 2-ethylpyridine (approx. 30 %, together with 3 % 2-vinylpyridine at 83 % conversion and 450 °C) is obtained by treating α-picoline with methanol on Cs-Y-zeolite. Conversely the yield of 2-vinylpyridine is highest (10 %, 425 °C) when Cs-X-zeolite is used. If H_2 instead of N_2 is used as the carrier gas, the content of ethylpyridine can be increased at the expense of vinylpyridine. Why is vinylpyridine preferentially formed on X- and ethylpyridine on Y-zeolite?

The above examples indicate how it is possible to suppress or to induce consecutive reactions during the alkylation of pyridine merely by means of the appropriate choice of dopant in the zeolite catalyst. In contrast to the side-chain alkylation of toluene using alkali-earth-doped faujasites, it is characteristic of the side-chain alkylation of picolines that alkylation of the nucleus also occurs [e.g.16]. Li-X or Li-Y-zeolite only lead to nuclear alkylation.

Acylation Reactions

It is possible to carry out not only alkylations, but also Friedel-Crafts acylations on zeolitic catalysts. Aromatics such as toluene[67-71] and phenol[72-74] can be acylated using carboxylic acids, acid anhydrides or acid chlorides.

$$CH_3-\underset{}{\bigcirc} \;+\; CH_3-(CH_2)_n-COOH \xrightarrow[-H_2O]{} CH_3-\underset{}{\bigcirc}-\overset{O}{\underset{}{C}}-(CH_2)_n-CH_3 \qquad (3)$$

n = 0 - 19

The liquid phase reaction of toluene with C_1-C_{20} carboxylic acids on Ce-Y-zeolite according to equ. 3 at 150 °C proceeds with increasing yield as the chain length of the acylating agent increases, and reaches a maximum of 96 % with dodecanoic acid[68]. In the case of all acylated toluenes more than 93 % of the para-isomer is obtained. The high proportion of p-isomer is not due or is only to a limited extent due to the shape selectivity of the zeolite, since the reaction on Al-montmorillonite or on the homogeneous catalyst $AlCl_3$ also leads to a preponderance of the p-isomer[19].

$$\underset{}{\overset{OH}{\bigcirc}} \;+\; CH_3COOH \xrightarrow[-H_2O]{} \underset{}{\overset{O-\overset{O}{\overset{\|}{C}}-CH_3}{\bigcirc}} \;+\; \underset{}{\overset{OH}{\bigcirc}}-\overset{O}{\underset{}{C}}-CH_3 \qquad (4)$$

	silicalite 300°C	H-ZSM-5 250°C
conversion of phenol % :	67	15
selectivity % :		
2-hydroxyphenylmethylketone	47	99
phenylacetate	32	–

In contrast, high proportions of ortho-isomer are obtained from the acylation of phenol according to equ 4. On a S-115 (silicalite, UCC) at 300 °C and with a residence time of 8 sec., phenol reacts with acetic acid to yield 2-hydroxyphenylmethylketone with 47 % selectivity and 67 % phenol conversion[73]. The para-isomer is only formed to an extent of 2 %. The remainder of the reaction product consists largely of phenyl acetate (32 %). In fact, on HZSM-5[74] at approx. 250 °C the ortho-isomer is actually obtained with almost 99 % selectivity at 15 % phenol conversion. As in the case of the alkylation with methanol the acylation of phenol yields almost exclusively the o-isomer. However, the intramolecular migration of an acetyl group from phenol acetate to form 2-hydroxyphenyl-methylketone, as suggested by P.A.Jacobs et al.[19, 62] for the formation of o-cresol from anisole, might be more difficult because the acetyl group requires in the zeolite pore more space than the methyl group. This finding contrasts with the results of the Fries-rearrangement of phenol acetate[72] at 400 °C on HZSM-5, in which an o/p-product ratio of approx. 1 is observed at only 4 % conversion and 20 % selectivity. This product ratio is attributed to the shape selectivity of the HZSM-5.

Nevertheless, on the basis of the available results, it must be pointed out that principally electronic effects and thermodynamics determine the reaction path and not the shape selectivity of the catalyst.

Halogenation Reactions

The nucleophilic substitution of aliphatics and the alkylation and acylation of aromatics as electrophilic substitution reactions certainly proceed via a carbonium ion mechanism as a consequence of the acidity of the zeolites.

$$(5)$$

	RE-Y	H-ZSM-5
conversion % :	100	86
selectivity % :		
chlorobenzene	97	13
hexachlorocyclohexane	3	87

Surprisingly, however, in the chlorination according to equ. 5 of benzene one observes not only ionic substitution but also an addition reaction indicative of a radical mechanism. The ratio of substitution to addition product is dependent upon the zeolite catalyst employed; on rare earth-doped Y-zeolites at 175 °C chlorobenzene (CB) is formed in 97 % and the addition product hexachlorocyclohexane (HCCH) in 3 % yield. Conversely, on HZSM-5 87 % HCCH and 13 % CB are obtained at 86 % conversion[75]. By further narrowing the pore width using alkali doping, the yield of the addition product can be increased up to 90 %. On the basis of the smaller pore diameter of ZSM-5 compared with Y-zeolite one would anticipate the opposite result. However, the HZSM-5 catalyst evidently induces a radical reaction, which proceeds to a large extent on the external surface of the zeolite or at the pore mouths. The formation of radicals during the adsorption of aromatics and olefins on H-mordenite and HZSM-5 has been demonstrated by means of ESR spectroscopy[76-78]. Nevertheless, the results obtained when benzene is chlorinated on HZSM-5 are remarkable and are not readily understandable because there are no spatial restrictions preventing the penetration of the reactants into the pentasil zeolite. Particularly in view of the gas phase hydroxylation of benzene on HZSM-5 using N_2O as a radical oxidant - for which reaction a radical mechanism has been ruled out[79,80] - the suggested formation of Cl-radicals on the external surface is not self evident.

ISOMERIZATION REACTIONS

Isomerization reactions of aliphatic compounds (carbon skeleton- and double bond isomerization reactions) can be directed by shape selectivity of the zeolites [e.g.11,15,16]. In particular, the conversion of substituted arenes into the desired isomers can also be brought about with high yields using shape selective zeolites. This reaction is therefore frequently employed as a model reaction for shape selectivity [e.g.3,4,16, 26,81-83]. Zeolite catalysed rearrangements represent a special category of skeletal isomerization reaction; Beckmann-, pinacolone-, Wagner-Meerwein-, epoxide- and dicyclopentadiene-/adamantane rearrangements have been the subject of many investigations [e.g.8,11,15,16,19,20,84-87].

Epoxide Rearrangements

$$\text{(6)}$$

R=alkyl-, alkoxyl-, halogen-, halogenalkyl-

Ring opening of epoxides to aldehydes in either the liquid or gas phase can be carried out very efficiently using zeolites. Styrene oxides can be converted to phenylacetaldehydes on Ti-zeolites in acetone at 30 - 100 °C or on pentasil zeolites in the gas phase according to equ. 6. The yields frequently exceed 90 %[16].

$$(7)$$

Recent publications describe the isomerization of limonene-1,2-epoxide (equ. 7) on Ca-A-zeolite at 150 - 210 °C to carvenone, a valuable inter-mediate for perfume manufacture. The reaction involves the simultaneous shift of a double bond and proceeds with 90 % yield[86,87].

$$(8)$$

Epoxide rearrangement with simultaneous ring contraction occurs when 2-pinene oxide is converted to the perfume intermediate camphorene alde-hyde (equ. 8). The reaction is carried out in ethylene dichloride at 35 °C in the presence of Na-mordenite and formic acid and provides a yield of 90.5 %[86,87].

At present it is not possible to estimate to what extent the shape se-lectivity of the zeolites plays a role in these reactions. In the former isomerization the reaction takes place at the outer surface of Ca-A-zeo-lite. It seems that only a few weakly acidic sites are necessary for this epoxide rearrangement and a high yield is obtained without involving shape selectivity. In the latter case, the conversion of 2-pinene oxide probably occurs inside the zeolite. However, higher yields are not ob-tained. The ring contraction might represent evidence for a constrained environment in the Na-mordenite.

Dicyclopentadiene/adamantane rearrangement

One example of skeletal rearrangement is the complex reaction of te-trahydrodicyclopentadiene in the presence of a multifunctional zeolitic catalyst to yield adamantane (AdH)[88,92] according to equ.9.

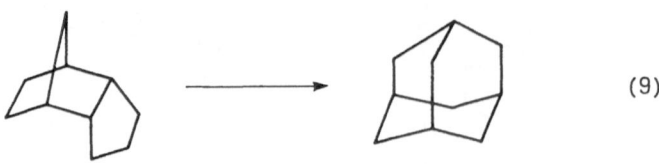

(9)

This rearrangement is carried out in the presence of hydrogen and hydrogen chloride at 200 - 250 °C. Multifunctional catalysts such as Co or
Ni on RE-Y-zeolites (RE = rare earth) are more active than RE-Y-zeolites
alone (TABLE 3). These catalysts deactivate very rapidly within 20 hours.
However, by additional doping with Pt and Re it is possible to increase
considerably both the activity and effective lifetime (approx. 200 hr) of
the catalysts, but at the cost of diminished selectivity (figure 6).

TABLE 3

Productspectrum of the reaction pyridine and methanol depending onto the
catalyst[88]

Catalyst	Conversion	AdH-Yield %	Selectivity %
Re-Y	43.6	18.0	41.2
Co-RE-Y	54.7	26.0	47.5
Pt-Re-RE-Y	93.5	23.0	24.6
Pt-Re-Co-RE-Y	91.5	31.0	33.8
Pt-Re-Ni-RE-Y	68.7	30.0	43.0

Conditions: 250 °C, 2 h, p (HCl) = 1 atm, p (H$_2$) = 15 atm

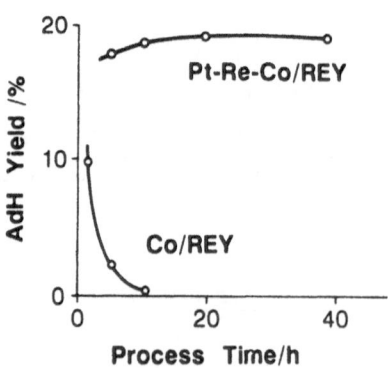

Figure 6. Lifetime of the catalysts in the reaction of tetra-
hydrodicyclopentadien to adamantane

The functionality of the individual components of this highly complicated multifunctional catalyst are not yet understood. Nevertheless, this

example illustrates how it is possible, even in the case of complex reactions, to increase the yield and effective lifetime by means of the appropriate choice and combination of several active components on zeolitic supports. If the internal pores and cavities are filled with several metals migration into the zeolitic framework may be prevented and thus the shape selectivity does not influence the reaction path at all.

Beckmann rearrangement

The most important industrial example of the Beckmann rearrangement is the reaction of cyclohexanone oxime to ε-caprolactam (equ. 10).

$$\text{(structure)} =NOH \quad \xrightarrow[\text{mordenite}]{\text{X-, Y- zeolite}} \quad \text{(structure)} =O \qquad (10)$$

The classical synthesic route involves the oximation of cyclohexanone with hydroxylamine-sulphate and the subsequent rearrangement of the oxime in concentrated sulfuric acid. Approx. 4 - 5 t $(NH_4)_2SO_4$ per t caprolactam are inevitably obtained as co-product[93]. Futher problems encountered include handling a large amount of fuming sulfuric acid and corrosion of the apparatus by the acid. In order to eliminate these problems and also the formation of the co-product, which can only be taken up by the fertilizer industry to a limited extent, continuing efforts are being made to switch from a homogeneous to a heterogeneous catalytic process. Attempts have been made to use zeolites for this purpose[93-102].

As long ago as the 1960's P.S.Landis and P.B.Venuto attempted to use X- and Y-zeolites as well as mordenite in the H-form or doped with rare earth or transition metals[8,94]. Thus cyclohexanone oxime (30 wt% dissolved in benzene) is converted at 380 °C and WHSV = 1.2 hr^{-1} to ε-caprolactam with 76 % selectivity and 85 % conversion during the first two hours. The principal by-product is 5-cyanopent-1-ene together with traces of cyclohexanone and cyclohexanol. As the reaction is continued, the overall conversion decreases to about 30 % after 20 hrs with a drop to 50 % selectivity for caprolactam. As far as the mechanistic considerations are concerned, a protonic catalysed reaction is assumed, whereby initial adsorption of the ketoxime at a catalyst acid site is probably involved.

The greatest limitations to the rearrangement of the ketoxime using Y-and X-zeolites or mordenite were the rapid catalyst aging and the low

331

selectivity for caprolactam. It is not possible to avoid these disadvantages even by employing the strongly acidic, hydrophobic HZSM-5 with SiO_2/Al_2O_3 = 78. A 14 % solution of cyclohexanone oxime in benzene at 350 °C, 1 atm and LHSV = 1.7 hr^{-1} is converted nearly quantitatively for a period of up to 15 hr. Afterwards the conversion drops rapidly to about 40 % at 21 hr on stream[97].

In recent years, therefore, considerable interest has focussed upon zeolite catalysts of reduced acidity, in particular on the outer surface and upon other weakly acidic microporous materials[98-102]. H.Sato and co-workers demonstrated[99,100] that both the catalytic activity and the selectivity of lactam formation increase with increasing Si/Al ratio in HZSM-5 catalysts.

Influence of Si / Al of
HZSM - 5 at 350°C and
WHSV = 38.5 h^{-1}

Influence of acid
amount on the outer
surface of HZSM - 5

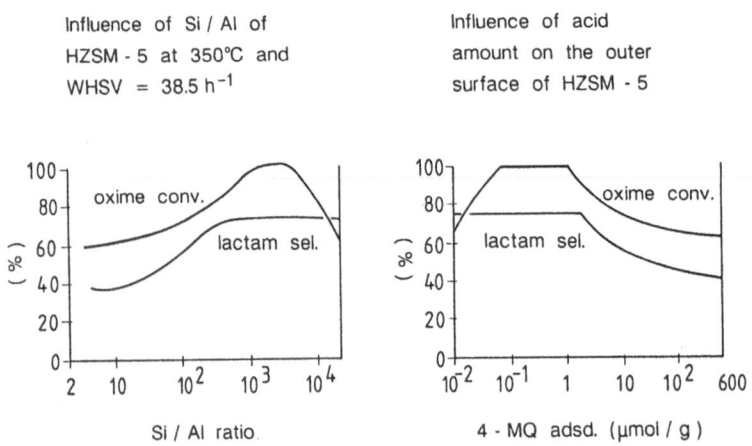

Figure 7. Beckmann rearrangement of 8 % solution of caprolactam-oxime in benzene[99,100]

By measuring the amount of 4-methyl-quinoline (4-MQ) adsorbed on HZSM-5, the number of acidic sites on the outer surface of samples with various Si/Al ratios can be determined (figure 7). It was found that both the activity and the selectivity increase as the quantity of 4-MQ adsorbed on the outer surface decreases. This, and other quoted results indicate that the rearrangement takes place on the outer surface or at the pore mouth of the pentasil zeolites. The relatively large molecular diameter of the cyclohexanone oxime and the $\mathcal{E}$-caprolactam are probably responsible. Even HZSM-5 catalysts with a Si/Al ratio of more than 2000 show high activity. Furthermore the service life of the catalysts increases with increasing Si/Al ratio. But even these catalysts lose 50 %

of their activity in the course of 12 hr. Generally speaking the weaker the acidic centers on the external zeolite surface the better are the results.

The same research group[101] reduced the acidity of the external surface of HZSM-5 catalysts by treatment with organometallic compounds such as chlorotrimethylsilane. The comparison, given in Table 2, of silanated with nonsilanated HZSM-5 shows the advantageous effects of this silanation treatment on the service life of the catalyst and the selectivity for ε-caprolactam.

TABLE 4

Comparison of silanated with nonsilanated HZSM-5 in the reaction[a] of cyclohexanone oxime to ε-caprolactam[101]

catalyst	time on stream [hr]	conversion [%]	selectivity [%]
silanated[b]	3,3	100	95,0
	31,0	98.2	95,0
nonsilanated	3,3	100	79,7
	27,0	95,8	89,4

[a]reaction conditions: 8 wt% solution of oxime in benzene, 350 °C, WHSV = 11.7 hr^{-1}, 1 atm, CO_2 as carrier gas, oxime/CO_2/benzene = 1/5.6/18.3 mol
[b]HZSM-5 with Si/Al = 1600, treated with chlorotrimethylsilane at 350 °C for 4 hr

Pursuing the idea of reducing the acidity of zeolite in order to achieve high selectivity and long catalyst lifetime, micro-porous materials such as the weakly acidic non-zeolitic molecular sieves (e.g. the middle-pore-sized SAPO-11 or SAPO-41) have been used for the Beckmann rearrangement[102]. Using SAPO-11, a 5 % by wt solution of cyclohexanone oxime in acetonitrile is converted at 350 °C, atmospheric pressure and WHSV = 10.8 hr^{-1}, to ε-caprolactam with 95 % selectivity at 98 % conversion rate.

However, also in the latter case, it must be pointed out that the Beckmann rearrangement of cyclohexanone oxime by means of zeolites or other non-zeolitic molecular sieves does not at present, for reasons of service life, constitute an alternative to the homogeneous process now practiced. The development of heterogeneous zeolitic and non-zeolitic catalysts indicate that the reaction does not necessarily take place at the acidic centers as active sites, as was generally believed in the past.

Only rather weak acidity, or no acidity, is required in order to achieve high selectivity at high conversion and long catalyst service life. Furthermore the reaction probably occurs at the outer surface of the zeolite particularly of pentasil zeolites. The shape selectivity does not, therefore, direct the course of the rearrangement, rather 5-cyano-pent-1-ene may be formed preferentially inside the zeolitic pores because of the product selectivity.

RING TRANSFORMATIONS

Ring transformations are to be understood as the exchange of oxygen atoms in cyclic compounds for sulphur[8], nitrogen[8,103-112] or selenium[113]. Such exchange reactions can , in some cases, be more effectively catalysed by bases, in others by acids. The ring size, the type of ring system (saturated or unsaturated) and other structural parameters can play a significant role in this connection.

Reaction of Caprolactone with Ammonia

It has been indicated in the patent literature that one can obtain caprolactam by a route analogous to the above reactions, i.e. by reaction of caprolactone with NH_3 on HY-zeolite[114] or on HZSM-5[115]. However, no precise data have been disclosed. On the contrary, it has been found[116] that in this reaction no O/N exchange takes place, but instead ring cleavage and simultaneous dehydration to hexenenitriles occur according to equ. 11.

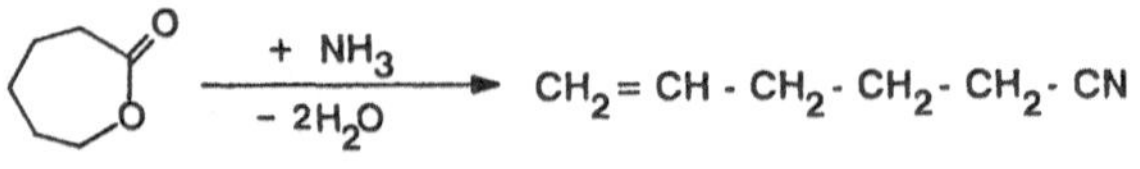

$$\text{(structure)} \xrightarrow[- 2H_2O]{+ NH_3} CH_2 = CH \cdot CH_2 \cdot CH_2 \cdot CH_2 \cdot CN \qquad (11)$$

For example, caprolactone reacts with NH_3 (mole ratio 1:7) at 400 °C on weakly acidic boron zeolites to yield hexenenitrile with 86 % selectivity and on strongly acidic aluminium zeolites of the pentasil and faujasite type with selectivities of 82 and 75 % respectively. Conversely, if the reaction is performed in the presence of a conventional catalyst such as Al_2O_3, then only low yields of hexenenitrile are obtained, since the reaction ceases at the intermediate α-hydroxycapronitrile stage.

The question arises as to why we obtain hexenenitrile instead of caprolactam in the conversion of caprolactone with NH_3. If this result were in agreement with the idea of shape selectivity, i.e. the less bulky hexenenitriles preferentially formed rather than the bulky caprolactam, then caprolactone would have to migrate in the pores of the pentasil type zeolite. On the other hand, it is known from the investigation of the Beckmann rearrangement[99-101] that caprolactam formation takes place at the outer surface, because the compounds are bulky. There may be two explanations for the formation of hexenenitrile. Firstly, the ring opening of the caprolactone occurs at the pore mouth and then the stretched fragment migrates inside the zeolite pore to react with NH_3, and the product diffusion and restricted transition state shape selectivity are responsible for the product distribution. Secondly it may just be a question of the reaction rates, i.e. after the ring opening and addition of NH_3 the dehydration to hexenenitrile occurs faster than the ring closure to form the caprolactam, and shape selectivity is not involved at all.

MEERWEIN-PONNDORF-VERLEY-REDUCTION

In the Meerwein-Ponndorf-Verley reduction, aldehydes and ketones are reduced to the corresponding alcohols by means of Lewis acids such as Al- or Mg-alcoholate in combination with an alcohol such as isopropanol.

$$R-C\overset{O}{\underset{H}{\diagup}} + \overset{H_3C}{\underset{H_3C}{\diagdown}}CH-OH \xrightarrow{NaX} R-CH_2OH + H_3C-\overset{O}{\overset{\|}{C}}-CH_3 \qquad (12)$$

R = Alkyl C_5H_{11} - $C_{12}H_{25}$

A Na-X-zeolite can also take over the function of the alcoholate[117]. It is possible to reduce C_5 to C_{12}-aldehydes to the corresponding 1-alcohols using a Na-X-zeolite/isopropanol system at 180 °C, with conversions of approx. 98 % and selectivities of over 95 % (equ. 12).

$$R-C\overset{O}{\underset{H}{\diagup}} + 2 \overset{H_3C}{\underset{H_3C}{\diagdown}}CHOH \xrightarrow{CaX} R-CH\overset{O-\diagdown}{\underset{O-\diagdown}{\diagup}} + H_2O \qquad (13)$$

R = Alkyl C_5H_{11} - $C_{12}H_{25}$

On the other hand, if the aldehydes are allowed to react on an Ca-X-zeolite/isopropanol system, then acetalization according to equ. 13 takes place almost exclusively. The observed drastic change in the reaction can be tentatively attributed to the higher acidity of the Ca-X-zeolite,

since the acetalization process requires protonic acidity. It is a surprising result, because the acetals formed are bulkier than the alcohols
and the pore size is narrowed by Ca-doping compared to Na-X-zeolite.

conversion: 87 %   conversion: 77 %

selectivity: 86 %   selectivity: 92 %

Futhermore, non-conjugated aldehydes can be easily reduced. Citronellol, for example, can react with isopropanol in a Meerwein-Ponndorf-Verley reduction. On alkali-X-zeolites at 150 °C and with a LHSV of 4.0 h^{-1}
the reaction yields either citronellol or, as a result of ring closure,
isopulegol, depending upon the metal used for doping (equ. 14).

Using Na-X-zeolite, 86 % isopulegol and 14 % citronellol are obtained
at a conversion of 87 %. When, on the other hand, ion-exchanged Cs-X-zeolite is used, the reaction yields 92 % citronellol at 77 % conversion.
The authors attribute these results to the difference in the steric requirements for the two competing reactions. Smaller pore sizes would be
expected to force the aldehyde to adopt a stretched conformation within
the channel, thereby preventing cyclisation and favouring the formation
of citronellol. This observation is attributed to the constrained environment inside the zeolite framework. In addition to this shape selective
behaviour, one must certainly also take into account the changes in the
acidity-basicity relationship in passing from Li to Na to Cs.

CONDENSATION REACTIONS

Condensation reactions are of great industrial importance and can be
catalysed by either acids or bases. Accordingly such reactions represent
a suitable area of application for zeolites, since their acidity or basicity can be adjusted, while the shape selectivity is maintained. In
this way such catalysts can be tailored to the desired reaction.

Condensation Reactions Yielding N-Heterocycles

One route for the synthesis of pyridines is the condensation of alde-

hydes and ketones with ammonia on zeolite catalysts [e.g.8,14,16,20].
ß-Substituted pyridines can be synthesised by the reaction of acrolein
and an alkanal with ammonia according to equ. 15[20].

R = alkyl-, aryl-, aralkyl-, alkylaryl-

Reaction of butanal on a boron zeolite which has been treated with HF
yields ß-ethylpyridine with 72 % selectivity. Analogously ß-butylpyridine
is obtained with 78 % selectivity from hexanal and ß-hexylpyridine with
90 % selectivity from octanal. In all cases conversion is complete and
catalyst lifetime > 48 hs. The increasing selectivity with increasing
chain length of the ß-substituent is surprising, because more extensive
cracking of the long-chain alkanals would have been expected. The transition state shape selectivity of the pentasil zeolite is probably responsible for this effect, since the long-chain alkanals are restricted in
their movements within the zeolite channels and adopt definite positions
in relation to the other reaction components. This effect is reminiscent
of the acylation of toluene using zeolites, in which the selectivity of
the para-isomer increases with the chain length to very high values (see
capter "acylation reactions").

OLIGOMERIZATION REACTIONS

Oligomerization reactions of olefins on zeolites, which have been the
subject of numerous studies, generally yield mixtures of dimers, trimers
and tetramers [e.g.16,118]. On the other hand, oligomerization reactions in
the presence of NH_3 have received little attention. One of the infrequent
examples is the synthesis of dimethylphenylpyridine by cotrimerization of
phenylacetylene and acetonitrile according to equ. 16 in the presence of
a Co-doped Y-zeolite, which had been specially reduced with $NaBH_4$ in the
reaction mixture[119].

Ph = C_6H_5.

At 100 °C, an 80 % yield of diphenylmethylpyridine is obtained. In this case, the heterogeneous catalyst is superior to the homogeneous monovalent Co complexes for reasons of separation and product work-up. The question arises as to whether such a bulky pyridine can be formed inside the large cavity of the Y-zeolite and can diffuse out of the narrow pore.

CONCLUSION

The examples which have been chosen demonstrate that in zeolite catalysed reactions the principle of shape selectivity must not necessarily always apply. In some cases the reaction takes place, unexpectedly only on the external surface of the zeolite catalyst. In other cases the conversion does indeed occur within the pore system, but electronic as well as thermodynamic effects dominate over the shape selectivity influence. The questions which arise in this connection will be discussed at this NATO-workshop and possibly also answered.

Nevertheless the various types of shape selectivity exhibited by medium and large pore zeolites play an extremely important part in acid, neutral and basic catalysed reactions in the field of organic chemistry. It is vital that further developments be made in the areas of new zeolitic and non-zeolitic microporous materials such as SAPO's[120,121], VPI-5 with > 10A pore openings[122], artificially expanded clay minerals such as the pillared clays[e.g.123], modified carbon-molecular sieves[e.g.124] and last but not least enzyme mimics[19,125-135]. With the aid of the tailored catalytic features and the shape selectivity properties which may be anticipated, it should be possible to achieve progress in carrying out organic reactions with improved conversion, selectivity, effective working time and energy requirements, as well as reduced environmental impact.

REFERENCES

1 P.B. Weisz, Pure Appl.Chem. 52:2091 (1980)
2 W.F. Hoelderich and E. Gallei, Ger. Chem. Eng. 8:337 (1985)
   Chem.-Ing.-Tech., 56:908 (1984)
3 N.Y. Chen and W.E. Garwood, Catal. Rev.-Sci. Eng. 28:185 (1986)
4 W.O. Haag and N.Y. Chen, in "Catalyst Design, Progress and Perspectives" (Ed. L.L. Hegedus), John Wiley, New York (1987) p. 163
5 J.E. Maxwell, Catal. Today 1:385 (1987)
6 N.Y. Chen, Stud. Surf. Sci. Catal. 38:153 (1988)
7 P.S. Landis and P.B. Venuto, J. Catal., 6:245 (1966)
8 P.B. Venuto and P.S. Landis, Adv. Catal., 18:259 (1968)
9 P.B. Venuto, Chem. Tech., 215 (1971)
10 W.F. Hoelderich, in "Proc. 7th Int. Zeolite Conference, Tokyo, Japan 1986" (Ed. Y. Murakami, A. Iijima and J.W.Ward)Kodanska,Tokyo(1986)p.827
11 W.F. Hoelderich, Pure Appl. Chem., 58:1383 (1986)

12 F.J. van der Gaag, F. Louter and H. van Bekkum, in "Proc. 7th Int. Zeolite Conference, Tokyo, Japan 1986" (Ed. Y. Murakami, A. Iijima and J.W. Ward) Kodansha, Tokyo, (1986), p.763

13 D. Arntz and M. Baacke, in "Proc. 7th Int. Zeolite Conference, Tokyo, Japan 1986" (Ed. Y. Murakami, A. Iijima and J.W. Ward) Kodansha,Tokyo, (1986) Poster 2 D-13

14 H. van Bekkum and H.W. Kouwenhoven, Stud. Surf. Sci. Catal., 41:45 (1988)

15 W.F. Hoelderich, Stud. Surf. Sci. Catal., 41:83 (1988)

16 W.F. Hoelderich, M. Hesse and F. Naeumann, Angew. Int. Edit., 27:226 (1988)

17 Y. Ono, in "Catalysis by Zeolites" (Ed. B. Imelik et al.) Elsevier, Amsterdam (1980) p.19

18 J.A. Rabo, in "Proc. Int. Symp. on Zeolites as Catalysts , Sorbents and Detergent Builders", Würzburg, FRG, 1988, Elsevier, Amsterdam (1989) Stud. Surf. Sci. Catal., 46:1 (1989)

19 P.A. Jacobs, J.M. Jacobs and R. Parton, in "Proc. Int. Symp. on Zeolites as Catalysts , Sorbents and Detergent Builders", Würburg, FRG, 1988, Elsevier, Amsterdam (1989) p.163

20 W.F. Hoelderich, in "Proc. Int. Symp. on Zeolites as Catalysts , Sorbents and Detergent Builders", Würzburg, FRG, 1988, Elsevier, Amsterdam (1989) p. 193

21 B. Notari, Stud. Surf. Sci. Catal., 37:413 (1987)

22 Y.I. Isakov and K.M. Minachev, Russ. Chem. Rev., 51:1188 (1982)

23 M. Tielen, M. Geelen and P.A. Jacobs, in "Proc. ZEOCAT Symp., Siofok, Hungary, 1985, Acta Physica et Chemica, 31:1 (1985)

24 J. Weitkamp, H. Beyer, G. Borbely, V. Cortes-Corberan and S. Ernst, Chem.-Ing.-Tech., 58:969 (1986)

25 J. Jansen, E. Biron and H. van Bekkum, Stud. Surf. Sci. Catal., 37:133 (1988)

26 J. Weitkamp, S. Ernst, H. Dauns and E. Gallei, Chem.-Ing.-Tech., 58:623 (1986)

27 S.M. Csiscery, ACS Monograph, 171:680 (1976)

28 S.M. Csiscery, Zeolites, 4:202 (1984)

29 S.M. Csiscery in "4th Symp. on Zeolite Science and Technology of Japan", Tokyo, Nov. 20-22 (1988)

30 W.O. Haag, R.M. Lago and P.B. Weisz, Faraday Disc., 72:317 (1982)

31 N.Y. Chen, ACS Symposium Series 368 : 468 (1988)

32 S.M. Csiscery, J. Catal., 108:433 (1987)

33 P.A. Jacobs and J.A. Martens, Stud. Surf. Sci. Catal., 28:23 (1986)

34 E.G. Derouane, J. Catal., 100:541 (1986)

35 P. Ratnasamy, Indian J. Technol., 25:653 (1987)

36 P.J. Lewis and F.G. Dwyer, Oil Gas J., 75:55 (1977)

37 F.G. Dwyer in "Chemical Industries, Vol.5" (Ed. W.R. Moser), Marcel Dekker, New York (1981) p.39

38 W.W. Keading, G.C. Barile and M.M. Wu, Catal. Rev.-Sci. Eng., 26:597 (1984)

39 F.R. Chen, G. Coudurier and C. Naccache, Stud. Surf. Sci. Catal., 28:733 (1986)

40 D. Fraenkel, M. Cherniavsky, B. Ittah and M. Levy, J. Catal., 101:273 (1986)

41 T. Yashima, K. Sato, T. Hayasaka and N. Hara, J. Catal., 26:303 (1972)

42 M.L. Unland and G.E. Barker, in "Chemical Industries, Vol.5" (Ed. W.R. Moser), Marcel Dekker, New York (1981) p.39

43 J.M. Garces, G.E. Vrieland, S.I. Bates and F.M. Scheidt, Stud. Surf. Sci. Catal., 20:67 (1985)

44 C.Lacroix, A. Deluzarche, A. Kiennemann and A. Boyer, Zeolites, 4:109 (1984)

45 D.G. Parker, Appl. Catal., 9:53 (1984)

46 L.D. Rollmann, EP 39 205, Mobil Oil Corp (Nov.4,1981)

47 Q.H.Xu, X.Qinhua and J. Zhu, Acta Phys. Chem., 31:181 (1985)
48 J.P. McWilliams, US 4 447 666, Mobil Oil Corp. (May 8, 1984)
49 C.C. Chu, US 4 420 418 , Mobil Oil Corp. (Dec.13, 1983)
50 N.B. Forbus, W.W. Keading, EP 89 787, Mobil Oil Corp. (Sept.28, 1983)
51 W.W. Keading, G.C. Barile and M.M. Wu, Catal. Rev.-Sci. Eng., 26:597 (1984)
52 W.W. Keading, L.B. Young, C.C. Chu, J. Catal., 89:267 (1984)
53 K.H. Chandavar, S.G. Hedge, S.B. Kulkarni, P. Ratnasamy, G. Chitlangia, A. Singh and A.V. Deo, in "Proc. 6th Int. Zeolite Conf., Reno, (USA), 1983" (Ed. D.H. Olson and A. Bisio), Butterworth, Guildford (1984) p.325
54 P.D. Chantal, S. Kaliaguine and J.L. Grandmaison, Appl. Catal., 18:133 (1985) and Stud. Surf. Sci. Catal., 19:93 (1984)
55 S. Namba, T. Yashima, Y. Itaba and N. Hara, Stud. Surf. Sci. Catal. 5:105 (1980)
56 S. Balsama, P. Beltrame, P.L. Beltrame, P. Carniti, L. Forni and G. Zuretti, Appl. Catal., 13:161 (1984)
57 T.G. Dakuchaeva, S.A. Kozhevnikov, D.A. Sibarov and V.F. Timofeer, Neftekhimiya 1985
58 K. Tanabe, Stud. Surf. Sci. Catal., 20:1 (1985)
59 K. Tanabe and T. Nishizaki, in "Proc. 6th Int. Congr. Catal., London" 2:863 (1976)
60 H. Hattori, K. Shimazu, N. Yoshii and K. Tanabe, Bull. Chem. Soc. Japan, 49:969 (1976)
61 P. Beltrame, P.L. Beltrame, P. Carnitti, A. Castelli and L. Forni, Appl. Catal., 29:327 (1987)
62 R.F. Parton, J.M. Jacobs, H. van Ooteghem and P.A. Jacobs, in "Proc. Int. Symp. on Zeolites as Catalysts , Sorbents and Detergent Builders," Würzburg, FRG, 1988, Elsevier, Amsterdam (1989) p. 211
63 M.M. Wu, UA 4 391 998, Mobil Oil Corp (1983)
64 H. Miki, T. Shivahata and Y. Iwase, Jap. Pat. 60 181 042, Mitsui Petrochem. Ind. (Sept.24, 1985)
65 L.B. Young, EP 29 333, Mobil Oil Corp, (May 27, 1981)
66 A. Corma, H. Garcia and J. Primo, J. Chem. Res.(S) 40 (1988)
67 H. Kashiwagi, Y. Fujiki and S. Enomoto, Chem. Pharm. Bull.30:404 and 30:2575 (1982)
68 B.H. Chiche, A. Finiels, C. Gauthier, P. Geneste, J. Graille and D. Pioch, J. Org. Chem.51:2128 (1986)
69 B.H. Chiche, C. Gauthier and P. Geneste, FR 2.592.039, CNRS (Jun. 26, 1987)
70 B.B.G. Gupta, EP 239.383, Celanese Corp. (Sept. 30, 1987)
71 M. Mueller, DE-OS 3.704.810, Boehringer Ingelheim K.-G. (Aug. 25, 1988)
72 Y. Pouilloux, N. Gnep, P. Magnoux and G. Perot, J. Mol. Catal., 40:231 (1987)
73 I. Nicolau and A. Aguilo, US 4.652.683, Celanese Corp (Mar. 24, 1987)
74 B.B.G. Gupta, US 4 668 826, Celanese Corp (May 26, 1987)
75 T. Huizinga, J.J.F. Scholten, T.M. Wortel and H. van Bekkum, Tetrahedron lett., 21:3809 (1980)
76 A.A. Slinkin, A.V. Kucherov, D.A. Kondratev, T.N. Bondarenko, A.M. Rubinshtein and K. Minachev, in "Proc. Int. Symp. on Zeolites as Catalysts , Sorbents and Detergent Builders", Würzburg, FRG, 1988, Elsevier, Amsterdam, 819 (1989) and J. Mol. Catal.35:97 (1986)
77 T. Ichikawa, M. Yamaguchi and H. Yoshida, J. Phys. Chem.91:6400 (1987)
78 S.A. Surin, L.A. Federova, C. Gullyev, G.D. Chukin, B.K. Nefedov and E.D. Radchenko, Kinet. Katal., 27:1174 (1986)
79 E. Suzuki, K. Nakashiro and Y. Ono, Chem. Lett. 6:953 (1988)
80 Y. Ono, K. Tohmori, S. Suzuki, K. Nakashiro and E. Suzuki, Stud. Surf. Sci. Catal., 41:75 (1988)
81 J.A. Martens, M. Thielen, P. Jacobs and J. Weitkamp, Zeolites 4:98 (1984)

82 N.Y. Chen, ACS Symposium Series 368:468 (1988)

83 N.Y. Chen, Stud. Surf. Sci. Catal., 38:153 (1988)

84 G. Paparetto and G. Gregoria, Tetrahedron Letters, 29:1471 (1988)

85 K. Honna, H.Ichikawa and M. Sugimoto, Nippon Kagaku Kaishi, 10:1298
    and 10:1310 (1986)

86 Y. Fujiwara, M. Nomura and K. Igawa, JP 62 114 926 (May 26, 1987) and
    JP 62 19 549 (Jan. 28, 1987), Toyo Soda MfG

87 M. Nomura and Y. Fujiwara, Nippon Kagaku Kaishi., 5:883 (1987)

88 K. Honna, M. Sugimoto, N. Shimizu and K. Kurisaki, Chem. Lett., 315
    (1986)

89 H. Iida and K. Honna, JP 60 246 333, Iaemitsu Kosan Co.(Dec. 6 1985)

90 K. Honna, H. Iida, N. Shimo, T. Ishibashi and K. Osada, Nippon Kagaku
    Kaishi, 1103 (1986)

91 K. Honna and H. Iida, Nippon Kagaku Kaishi, 1297 (1986)

92 G.C. Lau and W.F. Maier, Langmuir., 3:164 (1987)

93 O. Immel, H.H. Schwarz, H. Starke and W. Swodenk, Chem.-Ing.-Techn.,
    56:621 (1984)

94 P.S. Landis and P.B. Venuto, J. Catal., 6:245 (1966)

95 M. Burguet, A. Aucejo and M. Sancho-Tello, An. Quim. Ser.A, 81:259
    (1985)

96 A.Aucejo M. Burguet, A. Corma and V. Fornes, Appl. Catal., 22:187
    (1986)

97 W.K. Bell and C.D. Chang, EP 056 698, Mobil Oil Corp.,(June 5, 1985)

98 P.M.G. Frenken, EP 086 543, Stamicarbon B.V., (May 7, 1986)

99 H. Sato, N. Ishii, K. Hirose and S. Nakamura, Stud. Surf.Sci. Catal.
    28:755 (1986)

100 H. Sato, K. Hirose, N. Ishii and Y. Umada, EP 234 088, Sumitomo Chem.
    Co. (Sept. 2, 1987)

101 H. Sato, K. Hirose, M. Kitamura, H. Tojima and N. Ishii, EP 236 092,
    Sumitomo Chem. Co. (Sept. 9, 1987)

102 K.D. Olson, EP 251 168, UCC, (Jan. 7, 1988)

103 D. Barthomeuf, G. Coudurier and J.C. Vedrine, Mat. Chem. Phys.,
    18 : 553 (1988)

104 Y. Ono, Heterocycles, 16:1755 (1981)

105 Y. Ono, Stud. Surf. Sci. Catal., 5:19 (1980)

106 Y. Ono, K. Hatada, K. Fujita, A. Halgeri and T. Keii, J. Catal., 41:322
    (1976)

107 Y. Ono, T. Mori and K. Hatada, Acta Phys. Chem., 24:233 (1978)

108 Y. Ono,A. Halgeri, M. Kaneko and K. Hatada, ACS Symp. Series., 49:596
    (1977)

109 K. Hatada, M. Shimada, K. Fujita, Y. Ono and T. Keii, Chem. Lett.,
    439 (1974)

110 K. Hatada, M. Shimada, Y. Ono and T. Keii, J. Catal., 37:166 (1975)

111 K. Hatada and Y. Ono, Bull. Chem. Soc Japan., 50:2517 (1977)

112 Y. Ono, Y. Takeyama, K. Hatada and T. Keii, Ind. Eng. Chem. Prod. Res.
    Dev. 15:180 (1976)

113 E.S. Mamedov, R.A. Babakhanov, R.Y. Akhverdieva, A.K. Veinberg, R.D.
    Mishiev, S.S. Nabisov and M.Y. Lidak, Khim. Geterosikl, Soedin.,
    11:1478 (1986)

114 G.T. Kerr, Ch.J. Plank and E.J. Rosinski, US 3 442 795, Mobil Oil Corp.
    (May 6, 1969)

115 Ch.J. Plank, E.J. Rosinski and G.T. Kerr, US 4.011 278, Mobil Oil Corp.
    (March 8, 1977)

116 W.F. Hoelderich and M. Schwarzmann, EP 267 438, BASF AG, (May 18, 1988)

117 J. Shabtai, R. Lazar and E. Biron, J. Mol. Catal., 27:35 (1984)

118 M.L. Occelli, I.T. Hsu and L.G. Galya, J. Mol. Catal., 32:377 (1985)

119 Y. Ben Taarit, Y. Diab, B. Elleuch, M. Kerkani and M. Chihaoui,
    J. Chem. Soc. Chem. Commun., 5:402 (1986)

120 S.T. Wilson, B.M. Lók, C.A. Messina and E.M. Flanigen, in "Proc. 6th
    Int. Zeolite Conf., Reno, (USA), 1983" (Ed. D.H. Olson and A. Bisio),
    Butterworth, Guildford (1984) p.97

121 E.M. Flanigen, B.M. Lok, R.L. Patton and S. T. Wilson, in "Proc. 7th Int. Zeolite Conference, Tokyo, Japan 1986" (Ed. Y. Murakami, A. Iijima and J.W. Ward) Kodansha, Tokyo (1986) p. 103

122 M.E. Davis, C. Saldarriaga, C. Montes, J.M. Garces and C. Crowder, Zeolites., 8:362 (1988)

123 T.J. Pinnavaia, Science., 220:365 (1983)

124 J.M. Garces, G.E. Vrieland, S.I. Bates and F.M. Scheidt, Stud. Surf. Sci. Catal., 20:67 (1985)

125 E.S. Shpiro, G.VN Antoshin, O.P. Tkachenko, S.V. Gudkov, B.V. Romanovskii and Kh.M. Minachev, Stud. Surf. Sci. Catal., 18:31 (1984)

126 B.V. Romanovsky, in "Proc. 8th Int. Congr. Catal." Verlag Chemie, Weinheim, 4:657 (1984)

127 B.V. Romanovsky in "137" 215-222

128 T.V. Korolkova and B.V. Romanovskii, Neftekhimiya., 26:546 (1986)

129 B. De Vismes, F. Bedioui, J. Devynck, C. Bied-Charreton and M. Perree-Tauvet, Nouv. J. Chim., 10:81 (1986)

130 Y.W. Chan and R.B. Wilson Jr., Prepr. Pap-Am. Chem. Soc., Div. Fuel Chem., 33:453 (1988)

131 F.L. Pettit and M.A. Fox, J. Phys. Chem., 90:1353 (1986)

132 N. Herron, A.D. Stucky and C.A. Tolman, J. Chem. Soc. Chem. Commun., 1521 (1986)

133 C.A. Tolman and N. Herron, ACS Prep. Div. Petr. Chem.,32(3):798 (1987)

134 N. Herron and C.A. Tolman, ACS Prep. Div. Petr. Chem., 32(1):200 (1987)

135 N. Herron and C.A. Tolman, J. Am. Chem.Soc., 109:2837 (1987)

FACTORS INFLUENCING THE SELECTIVITY OF HYDROCRACKING IN ZEOLITES

Jens Weitkamp[1] and Stefan Ernst[2]

[1]Institute of Chemical Technology I, University of
Stuttgart, Pfaffenwaldring 55, D-7000 Stuttgart 80, FRG
[2]University of Oldenburg, Department of Chemistry, Chemical Technology, Ammerlaender Heerstrasse 114-118,
D-2900 Oldenburg, FRG

INTRODUCTION

Hydrocracking is an industrial refinery process which is widely used to convert vacuum gas oil (VGO) to gasoline [1-3]. In addition, there are process variants which convert a variety of other feedstocks, including residues or waxy distillates, to produce valuable hydrocarbon fuels, such as middle distillates or steamcracker feedstocks [4-6]. To a large extent, this process versatility has its origin in the availability of a broad spectrum of hydrocracking catalysts, tailored for specific purposes. Usually, these catalysts consist of an acidic carrier loaded with a hydrogenation/dehydrogenation component, in other words they are bifunctional. Typical acidic components are zeolites, amorphous silica-alumina and alumina. Typical hydrogenation components include palladium and platinum, or non-noble metals such as cobalt, nickel, molybdenum and tungsten [2]. The latter metals are usually in a sulfided form.

The present paper focusses on hydrocracking in large pore zeolites. The reaction mechanisms which are operative in this type of catalyst have been investigated in considerable detail and are now fairly well understood [7-10]. In this article, typical hydrocracking selectivities will be described and rationalized. Mechanisms and hence selectivities of hydrocracking mainly depend on (i) the catalyst composition, (ii) the nature of the feed hydrocarbons and, (iii) the presence or absence of catalyst poisons. The influence of these factors during hydrocracking in zeolite catalysts will be discussed.

TYPES OF HYDROCRACKING

The essential feature of hydrocracking is the cleavage of organic molecules in a hydrogen containing atmosphere. At least three different routes for hydrocracking can be discerned: (i) Thermally induced cleavage via hydrocarbon radicals ("hydropyrolysis"), (ii) monofunctional hydrocracking on hydrogenation components like platinum, palladium or nickel ("hydrogenolysis") and, (iii) scission of carbon-carbon bonds on bifunctional catalysts consisting of a hydrogenation component dispersed on a porous carrier with acidic properties. In commercial petroleum re-

Guidelines for Mastering the Properties of Molecular Sieves
Edited by D. Barthomeuf *et al.*
Plenum Press, New York, 1990

343

fining, neither hydropyrolysis nor hydrogenolysis play an important role; they mostly occur as side reactions in some processes. Therefore, this article will be restricted to bifunctional hydrocracking with particular emphasis on zeolite based catalysts. Depending on the chemical composition and pretreatment of such catalysts, either the acidic or the metal function may predominate. The relative strength of both functions has a large influence on the selectivity of hydrocracking.

It is a useful concept [8,11] to look at hydrocracking over bifunctional catalysts with a strong hydrogenation component and catalytic cracking over acid catalysts lacking a hydrogenation component as limiting cases or pure cracking mechanisms. Other pure cracking mechanisms with entirely different product patterns are hydrogenolysis over metals, oxides or sulfides, and non-catalytic thermal cracking or hydrocracking via radicals. Depending on the properties (e.g., the chemical composition) of the catalyst, two or more of these pure mechanisms may be operative simultaneously and independently, and contribute to the actual product distribution. Likewise, a single cracking mechanism may be operative which combines features of at least two of the pure mechanisms. This complex situation has been visualized by the so-called cracking tetrahedron (cf. Fig. 1) [8,11]. It has at its corners the pure or ideal cracking mechanisms. Also described in Fig. 1 are the catalysts on which the pure mechanisms can be realized.

As indicated in Fig. 1 (right-hand side, top corner), ideal hydrocracking is closely related to skeletal isomerization. On the bifunctional catalysts required for this pure mechanism, both skeletal rearrangement and carbon-carbon bond rupture occur in carbocations formed from the feed molecules. The carbocations are chemisorbed at the acidic sites of the catalyst. One of the roles of the metal site is to open a fast

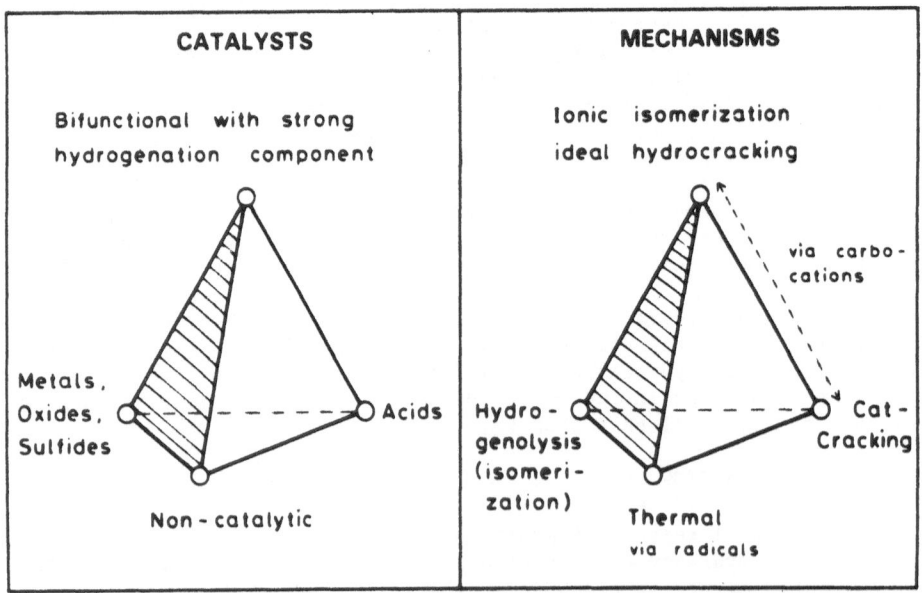

Fig. 1. The cracking tetrahedron (right-hand side) is to symbolize the large variety of cracking mechanisms with four limiting cases arranged at its corners. The catalysts over which the pure mechanisms can be operative are indicated at the corners of the left-hand tetrahedron.

route for interconversion of saturated hydrocarbons, e.g., alkanes and carbocations via alkenes. Today, the bifunctional conversion of hydrocarbons is considered to proceed via the following steps [12-14]: The saturated feed molecule is first converted into an alkene at the dehydrogenation/hydrogenation component of the catalyst. The olefinic intermediate moves to a Brönsted acid site where it is chemisorbed. During chemisorption, the acid proton adds to the double bond of the alkene which results in the formation of a carbenium ion, i.e., a hydrocarbon species with a positively charged trivalent carbon atom. While chemisorbed, the carbenium ion can undergo a bunch of reactions, such as cleavage, skeletal rearrangement or ring closure. The products of these steps are again carbocations or, in the case of ionic cleavage, an alkene and a carbenium ion. They are desorbed from the acid sites as alkenes, leaving behind the proton. The alkene now moves to the hydrogenation component where it is again chemisorbed and takes up hydrogen. Hence, the final products in this sequence of events are again saturated, e.g. lower alkanes, skeletal isomers or cycloalkanes.

The relative strengths of the hydrogenation and the acid component of the catalyst can have a strong effect on hydrocracking selectivities as shown in Fig. 2, in which typical carbon number distributions of the hydrocracked products from n-hexadecane are depicted. For comparison, data for catalytic cracking over $SiO_2-Al_2O_3-ZrO_2$ (from ref. 15) are also included.

Both hydrocracking catalysts have a moderate to strong acid component, viz. amorphous $SiO_2-Al_2O_3$ and CaY zeolite, respectively. The significant difference is in their hydrogenation/dehydrogenation component. It is much weaker for the sulfided CoMo catalyst than for the noble metal catalyst. Note that with each catalyst, the reaction conditions were chosen in such a manner that about 50 % of the hexadecane was hydrocracked.

On Pt/CaY, the molar distribution of the cracked products is fully symmetrical. This is indicative of a pure primary cracking, i.e. one carbon-carbon bond in the feed hydrocarbon is ruptured whereupon the

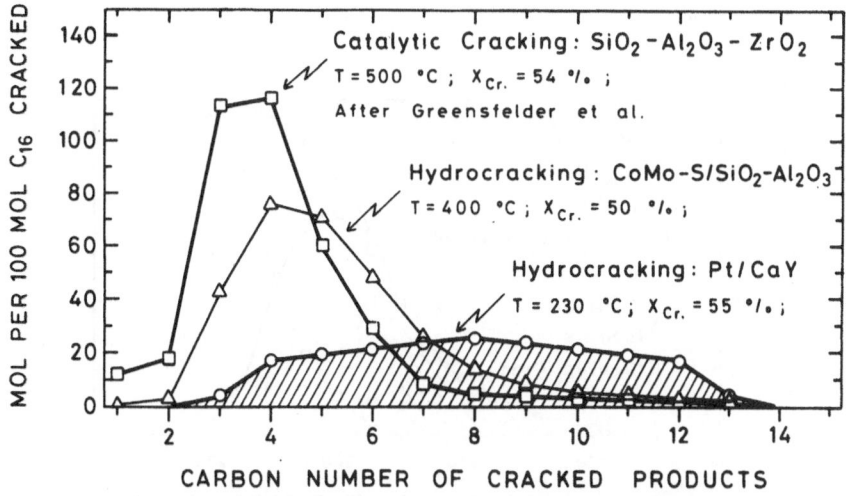

Fig. 2. Molar carbon number distributions in catalytic cracking and catalytic hydrocracking of n-hexadecane at ca. 50 % conversion. Data for catalytic cracking after ref. 15, for hydrocracking after ref. 8.

cracked products are desorbed from the catalyst surface before a second carbon-carbon bond rupture can occur. In other words, an efficient and fast desorption mechanism must be operative on Pt/CaY. The data for Pt/CaY further reveal that the probability for a bond to be broken is highest in the center of the feed molecule (formation of 2 C_8, $C_7 + C_9$, ...). Neither C_1 nor C_2 and only little C_3 are formed. These are salient features for a cracking mechanism via carbocations.

A completely different selectivity is encountered in catalytic cracking, where the catalyst does not contain any hydrogenation component: The vast majority of product molecules is in the range between C_3 and C_6, in other words, extensive secondary cracking must have taken place. Obviously, there is no efficient desorption mechanism in catalytic cracking. An intermediate behaviour is observed in hydrocracking over the catalyst with a weak hydrogenation activity.

For hydrocracking of the type observed on Pt/CaY (see Fig. 2) the term "ideal hydrocracking" has been widely used in the scientific literature [7-11,16]. It has been demonstrated [17] that an ideal hydrocracking catalyst can be produced by loading a strongly acidic zeolite with small amounts of a strong hydrogenation component (e.g., Pt/CeY). Even a Pt-loading of 0.05 wt.-% can be sufficient to open the ideal bifunctional reaction pathway [17]. Hydrocracking on catalysts consisting of a strong hydrogenation component (e.g., Pt) on a weakly acidic or non-acidic form of a zeolite tends to proceed via the hydrogenolysis mechanism on the metal. If this mechanism is operative, it is easily recognized by the nature of the hydrocracked products from, e.g., a long-chain n-alkane: Large amounts of methane and ethane are formed along with higher n-alkanes while iso-alkanes are absent or nearly absent.

THE MECHANISM OF BIFUNCTIONAL HYDROCRACKING

The mechanism of bifunctional hydrocracking in large pore zeolites has been investigated in detail [7-12,16-19]. In the majority of these studies, pure n-alkanes were used as feed hydrocarbons. Typical results for n-tridecane conversion on a Pt/CaY-zeolite are depicted in Fig. 3. Under mild reaction conditions (shaded area) skeletal isomerization is

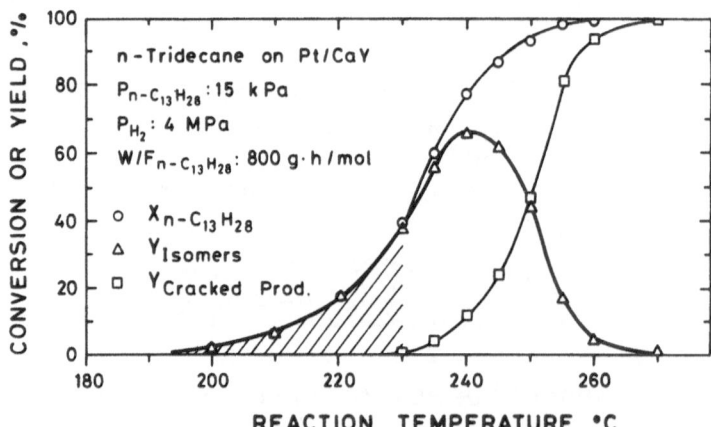

Fig. 3. Isomerization and hydrocracking in model compound studdies with n-tridecane and a Pt/CaY zeolite catalyst (after ref. 21).

346

the only reaction. With increasing severity, the yield of iso-tridecanes passes through a maximum, because more and more feed molecules undergo hydrocracking. Mechanistically, this observation strongly suggests that carbon-carbon bond cleavage is preceded by skeletal rearrangement.

A model which accounts for the consecutive occurrence of isomerization and hydrocracking has been advanced by Coonradt and Garwood [20]. A simplified version of their reaction scheme is shown in Fig. 4.

A long-chain n-alkane is first dehydrogenated at the metallic sites to the mixture of n-alkenes. These diffuse to the Brönsted acid sites where they are protonated to give n-alkyl cations. While chemisorbed, they are branched via so-called type B rearrangements. This type of isomerization is nowadays believed to proceed via protonated cyclopropanes [21]. The resulting iso-alkyl cations can either be desorbed or undergo ionic β-scission. Desorption results in long-chain i-alkenes which are readily hydrogenated into i-alkanes with the same carbon number as the feed. It is clear from this scheme, that the overall selectivity is determined by the relative rates of desorption and β-scission of the iso-alkyl cations. According to Coonradt and Garwood, their rate of desorption is influenced by the steady state concentration of n-alkenes, i.e. the postulated desorption mechanism is a displacement of iso-alkyl cations by n-alkenes through competitive sorption/desorption. This explains how the strength of the dehydrogenation/hydrogenation component of the catalyst can influence the rate of desorption of carbocations from the acid sites.

Exactly the same model is used to explain the observed carbon number distributions of the cracked products. The moieties of the primary β-scission steps are rapidly desorbed from the acid sites by long-chain alkenes through competitive sorption/desorption. If the steady state concentration of long-chain alkenes is low (weak or no dehydrogenation component in the catalyst), the moieties formed in the primary β-scission remain chemisorbed and undergo secondary cracking reactions, even at low conversions of the feed hydrocarbon.

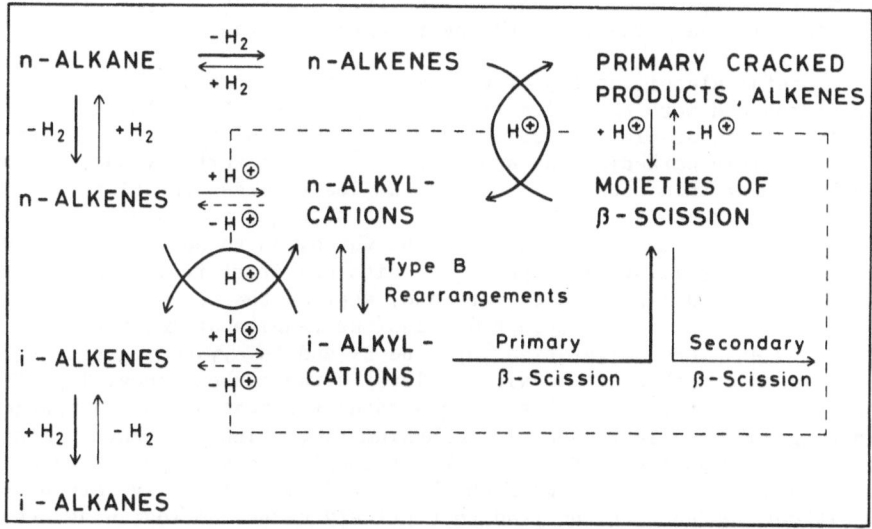

Fig. 4. Schematic representation of the role of competitive adsorption/desorption on hydrocracking catalysts with a strong hydrogenation/dehydrogenation component, after Coonradt and Garwood (ref. 20).

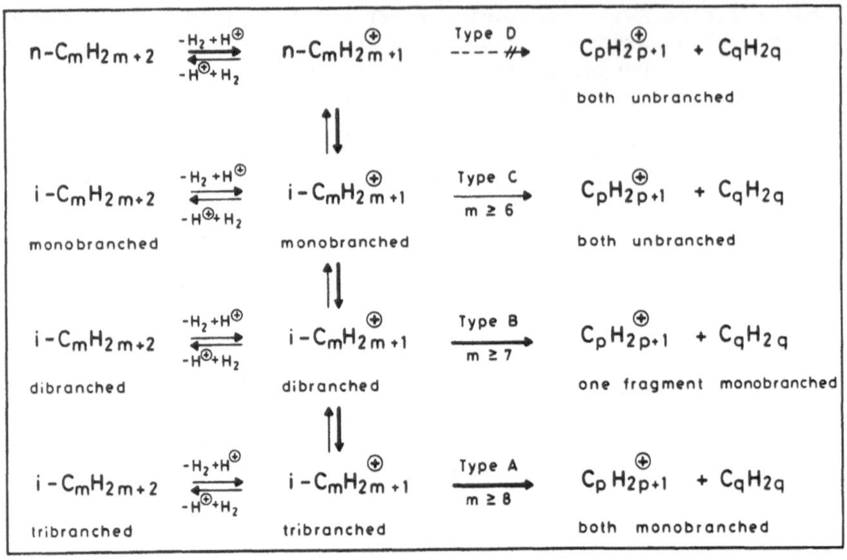

Fig. 5. Reaction network for isomerization and hydrocracking of alkanes on bifunctional catalysts.

The actual reaction scheme for hydrocracking of paraffins in large pore zeolites is somewhat more complex than that proposed by Coonradt and Garwood. A fairly complete network is depicted in Fig. 5: In a first step, a normal alkane is isomerized into its monobranched isomers. These are again isomerized in a consecutive step. The resulting dibranched isomers rearrange again into tribranched species which, however, cannot be observed as such in the product because they undergo a very rapid β-scission. This so-called type A β-scission (vide infra) of tri-branched precursor ions is the main pathway leading to the cracked products. Besides, some β-scission of di-branched species occurs as well. The ultimate reason for this route has to do with the facts that β-scission starts from alkylcarbenium ions and the rate constants for β-scission of these ions increase sharply with their degree of branching.

A useful classification of the different modes of β-scission has been introduced more than five years ago [14,22] (cf. Fig. 6). The energetically most favorable mode of β-scission is that of type A. It starts from a tertiary carbenium ion and leads again to a tertiary one, beside an alkene. For this type of β-scission at least three branchings in an α,α,γ-configuration and consequently at least eight carbon atoms are required. There are two type B β-scissions which, up to now, cannot be distinguished experimentally. Type B_1 β-scissions start from a secondary carbenium ion and lead to a teriary one, whereas the converse holds for type B_2 β-scissions. Both type B β-scissions require at least two branchings (in an α,α-configuration for type B_1 and an α,γ-configuration for type B_2) and a carbenium ion with at least seven carbon atoms. Type C β-scissions start from and lead to a secondary carbenium ion. A single branching suffices in the parent carbenium ion, which must have a minimum of six carbon atoms. For the sake of completeness, type D β-scissions may be considered as well which start from a secondary carbenium ion without a branching and lead to a primary carbenium ion; under the reaction conditions usually applied in catalytic hydrocracking, type D β-scissions probably play a negligible role, if they take place at all.

The rate constants decrease markedly from type A to type D β-scissions. Therefore, a carbocation derived from a long-chain n-alkane will

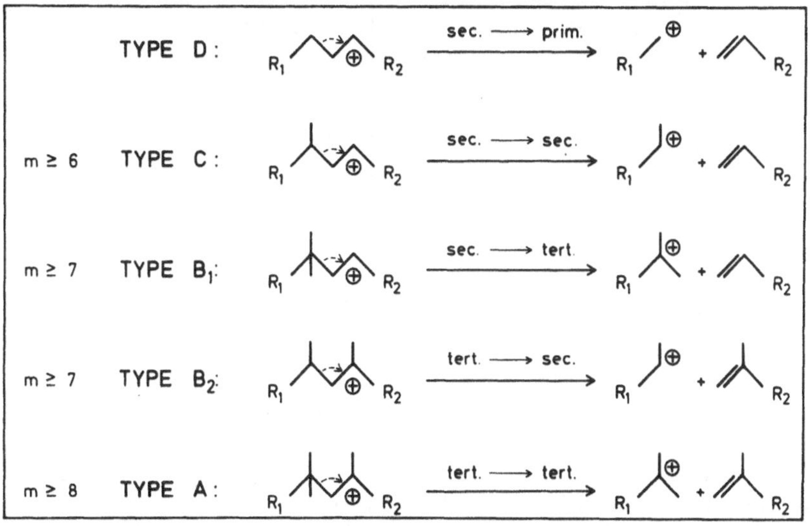

Fig. 6. Modes of β-scission of alkylcarbenium ions (after refs. 14 and 22).

preferentially rearrange several times until a skeleton is arrived at which can undergo the favorable type A β-scission. And this has indeed been observed experimentally with a variety of Y-type zeolites: Provided that the carbon number of the feed is sufficiently large (viz. ≥ 8) and the highly branched feed isomers can be formed in the pores of the zeolite, the predominant hydrocracking pathway is via type A β-scissions (in addition, there is some contribution of type B β-scissions) [9,10,16,19]. As an important consequence, the hydrocracked products contain large amounts of branched isomers. This is of importance for the manufacture of gasoline in a hydrocracker: Branched hydrocarbons have considerably higher octane numbers than unbranched ones. However, most alkanes in hydrocracker gasoline contain one or, at the best, two branchings, the very desirable tribranched alkanes are virtually absent. If they were formed, they would themselves undergo the rapid type A β-scission.

Essentially the same mechanistic rules as in hydrocracking of alkanes can be applied to the hydroconversion of naphthenes on bifunctional catalysts [14]. However, there are some peculiarities which originate in the sluggishness of carbon-carbon bond cleavage inside the naphthenic ring. Two possible reasons for this low rate of ring rupture have been discussed: Chevron researchers [23] argued that the alkenyl cation formed by β-scission of a cycloalkyl cation has a high tendency to recyclize, in other words the thermodynamics of ionic ring opening are considered to be unfavorable. An alternative explanation has been proposed by Brouwer and Hogeveen [24]. According to these authors the low rate of β-scission in cycloalkylcarbenium ions has to do with the unfavorable orientation of the vacant p-orbital at the positively charged carbon atom and the β-bond which is to be broken (cf. Fig. 7): In an aliphatic carbenium ion, there is a free rotation around the α-bond, and in the most stable conformation the β-bond to be broken and the vacant p-orbital are ideally coplanar. This results in a maximum orbital overlap in the transition state of β-scission. Conversely, in an alicyclic carbenium ion, especially if the positively charged carbon atom forms part of a five-membered ring, the β-bond is fixed in a position which is perpendicular or near-perpendicular with respect to the vacant p-orbital. These are very unfavorable conditions for orbital overlap in the transition state.

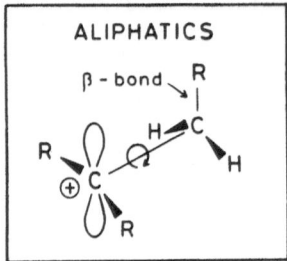
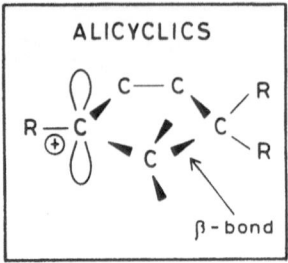

Fig. 7. The role of orbital orientation for the easiness of
β-scission (after ref. 24).

Due to the sluggishness of carbon-carbon bond cleavage inside the naphthenic ring, cycloalkylcarbenium ions will isomerize until a carbon skeleton is attained which allows exocyclic β-scission, especially type A
β-scission in the alkyl side chain. As a direct repercussion, a large
portion of the naphthenic rings initially present in the hydrocracker
feedstock will survive in the product. This explains why gasoline from
a hydrocracker is so rich in naphthenes and hence an excellent feedstock
to catalytic reforming.

The sluggishness of endocyclic carbon-carbon bond rupture has been
repeatedly observed in hydrocracking of naphthenes with a sufficiently
large carbon number on large pore zeolites, e.g., with C_9- through C_{11}-
naphthenes in 0.27 wt.-% Pd/LaY [25]. In shape-selective zeolites, C_{10}-
naphthenes were discovered to hydrocrack with significantly different
selectivities [26,27], and this effect can be exploited for probing the
effective pore width of molecular sieve materials.

INFLUENCE OF COMPETITIVE ADSORPTION ON HYDROCRACKING SELECTIVITIES

Almost all of the publications dealing with the mechanism of bifunctional hydrocracking are restricted to the conversion of pure hydrocarbons. However, there are numerous indications that the selectivity of
hydrocracking is strongly influenced by competitive adsorption/desorp-
tion (viz. preferential adsorption of species with higher molecular
weight or enhanced polarity) at the acidic sites of the zeolite [7,8,10,
28]. Only recently, the influence of the presence of a second model hydrocarbon was explored. Chen et al. investigated isomerization and hydrocracking of n-hexane, either pure or in admixture with aromatics, on
a 0.6 wt.-% Pt/H-Mordenite catalyst [29,30]. In principle, they observed
that the reaction rate of n-hexane decreases considerably if increasing
amounts of benzene are added to the feed [29]. Two possible reasons have
been suggested to explain this observation: (i) A stronger and hence
preferred adsorption of the aromatic molecules at the acid sites (due to
their higher polarity) and, (ii) a strongly hindered diffusion of n-
hexane through the unimensional pore system of mordenite due to a reduction of the pore cross-section by preferentially adsorbed aromatic molecules.

The reaction rate of benzene is influenced in the reverse manner:
Upon dilution of the aromatic feed with 20 wt.-% of n-hexane, the benzene conversion is considerably enhanced. To explain this unexpected behaviour, Chen et al. suggest a certain self-induced blockage of the catalytic sites by benzene molecules, which leads to a self-inhibition of

the benzene conversion. According to the authors, this inhibition is mitigated by dilution with n-hexane [30]. The speculative nature of this explanation reflects that there is still much knowledge lacking on hydrocracking of mixed feeds.

The influence of competitive adsorption on activity and selectivity in hydrocracking long-chain n-alkanes was studied by Dauns and Weitkamp [31]. Fig. 8 shows that n-dodecane is converted twice as fast as n-decane, if the pure paraffins are fed to the reactor. However, if an equimolar mixture of both hydrocarbons is used as feedstock, n-dodecane is converted approximately five times as fast as n-decane. These findings are consistent with the assumption that competitive adsorption governs the reaction rate of the paraffins, and that the species with the longer carbon chain is preferentially adsorbed.

The selectivities for individual products of isomerization and hydrocracking were found to be identical for the pure and the mixed feed, if they are compared at similar conversions.

The preferred conversion of the longer chain alkanes due to competitive adsorption does by no means exclude completely that lower molecular weight hydrocarbons, either initially present in the feed or produced in-situ by hydrocracking reactions, can be converted to some extent. This has been demonstrated with a dual component feed mixture consisting of 96 mole-% n-dodecane and 4 mole-% of 2-methyloctane [31]. As a pertinent result of this study, 2-methyloctane starts to isomerize into 3- and even 4-methyloctane, long before the higher molecular weight n-dodecane is completely consumed by hydrocracking.

This finding has considerable bearing on the mechanism of hydrocracking: It indicates that some secondary isomerization of the hydrocracked products from a long-chain alkane inevitably occurs after the cracking steps, so it would be premature to draw quantitative conclusions on the primary hydrocracking modes from the isomer distributions in the hydrocracked products.

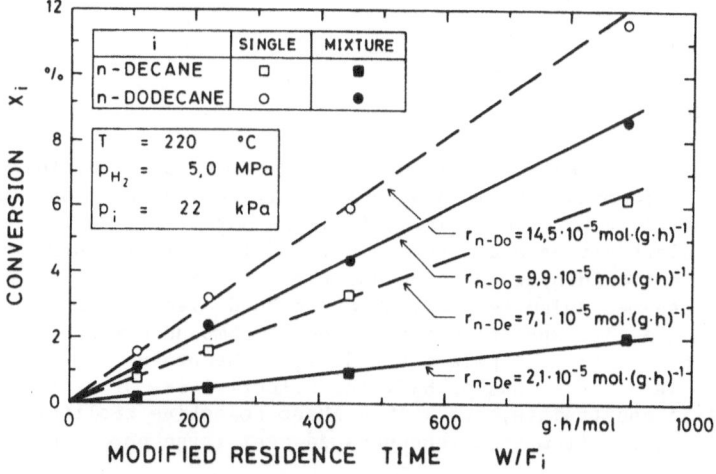

Fig. 8. Hydroconversion of pure n-decane and n-dodecane and their equimolar mixture on Pd/LaY-zeolite.

351

INFLUENCE OF H₂S AND NH₃ ON HYDROCRACKING ACTIVITY AND SELECTIVITY

Zeolite based hydrocracking catalysts are less sensitive against substantial concentrations of ammonia in the feed than silica-alumina based catalyst. This has been attributed to the higher acidity (enhanced density and/or strength of acid sites) in zeolites [2]. Nevertheless, a lower <u>activity</u> of the catalyst is usually observed in the presence of NH_3 as compared to the ammonia-free system [2]. Up to now, systematic investigations on the influence of NH_3 on hydrocracking <u>selectivities</u> are lacking.

The influence of H_2S on the activity of the hydrogenation and acidic functions of bifunctional zeolites was investigated recently by appropriate model reactions of ethylbenzene, viz. its hydrogenation to ethylcyclohexane and its disproportionation to benzene and diethylbenzenes, respectively [32,33]. It has been concluded that the influence of H_2S is twofold: (i) The number and/or strength of the acid sites is reduced. This could happen through hydrogen bonding between the acid OH groups and H_2S, in agreement with spectroscopic data by Karge and Raskó [34]. Alternatively, H_2S could lower the rate of hydrocarbon conversion due to competitive adsorption at the acid sites. Such a mechanism has been envisaged to explain the inhibiting effect of water in hydrocracking over Pt/HY [35]. (ii) The presence of H_2S lowers the hydrogenation activity of the noble metal. This is an expected result since it is known, that metal sulfides are considerably less active for hydrogenation than the noble metals [36].

As a consequence of the strongly reduced hydrogenation activity, it is no longer possible to obtain high yields of feed isomers on a Pd/LaY zeolite catalyst. Rather, hydrocracking predominates already at low conversions. This effect is consistent with the reaction scheme proposed by Coonradt and Garwood for bifunctional hydrocracking [20].

As shown in Fig. 2, one of the salient features of ideal bifunctional hydrocracking is a pure primary cracking selectivity, i.e., a symmetrical molar carbon number distribution of the cracked products up to high conversions. In the presence of H_2S, by contrast, secondary hydrocracking always occurs (i.e., the largest fragments from the primary hydrocracking steps undergo some further degradation) leading to low carbon number products. In the bifunctional reaction scheme, secondary hydrocracking can be readily explained by a reduced hydrogenation activity of the catalyst [8,20,28].

CONCLUSIONS

The mechanism of bifunctional hydrocracking has been widely studied using model compounds and is nowadays fairly well understood. In the absence of steric constraints, i.e., in large pore zeolites, the hydrocracking selectivities are governed by: (i) The relative strengths of the hydrogenation/dehydrogenation and the Brönsted acid component, (ii) the preferred mechanism for carbon-carbon bond cleavage via ionic type A β-scission, (iii) competitive adsorption of compounds with higher molecular weight or stronger polarity and, (iv) the presence or absence of catalyst poisons, viz. NH_3 or H_2S. The selectivities of hydrocracking under steric constraints, i.e., in a shape selective zeolite, can differ significantly from those encountered over large pore zeolite catalysts.

ACKNOWLEDGEMENTS

Financial support by Deutsche Forschungsgemeinschaft, Fonds der Chemischen Industrie and Max Buchner-Forschungsstiftung is gratefully acknowledged.

REFERENCES

1. A.P. Bolton, Hydrocracking, Isomerization and Other Industrial Processes, in: "Zeolite Chemistry and Catalysis", J.A. Rabo, ed., p. 714/779, ACS Monograph 171, American Chemical Society, Washington, D.C. (1976).
2. J.W. Ward, Design and Preparation of Hydrocracking Catalysts, in: "Preparation of Catalysts III", G. Poncelet, P. Grange and P.A. Jacobs, eds., p. 587/616, Studies in Surface Science and Catalysis, Vol. 16, Elsevier Science Publishers, Amsterdam, Oxford, New York (1983).
3. I.E. Maxwell, Zeolite Catalysis in Hydroprocessing Technology, Catalysis Today 1: 385 (1987).
4. K. Hedden and J. Weitkamp, Das Hydrocracken schwerer Erdölfraktionen, Chem.-Ing.-Tech. 47: 505 (1975).
5. J.G. Sikonia, W.L. Jacobs and S.A. Gembicki, Hydrocrack for More Distillates, Hydrocarbon Process. 57 (No. 5): 117 (1978).
6. S.D. Light, R.V. Bertram and J.W. Ward, Hydrocrack Heavier Feeds, Hydrocarbon Process. 60 (No. 5): 93 (1981).
7. J. Weitkamp, The Influence of Chain Length in Hydrocracking and Hydroisomerization of n-Alkanes, in: "Hydrocracking and Hydrotreating", J.W. Ward and S.A. Qader, eds., p. 1/27, ACS Symposium Series, Vol. 20, American Chemical Society, Washington, D.C. (1975).
8. J. Weitkamp, Hydrocracken, Cracken und Isomerisieren von Kohlenwasserstoffen, Erdöl, Kohle - Erdgas - Petrochem. 31: 13 (1978).
9. J. Weitkamp and P.A. Jacobs, Isomerization and Hydrocracking of Long Chain Alkanes: New Insight into Carbenium Ion Chemistry, Preprints, Div. Petr. Chem., Am. Chem. Soc. 26: 9 (1981).
10. M. Steijns, G. Froment, P.A. Jacobs, J. Uytterhoeven and J. Weitkamp, Hydroisomerization and Hydrocracking. 2. Product Distributions from n-Decane and n-Dodecane, Ind. Eng. Chem., Prod. Res. Dev. 20: 654 (1981).
11. H. Pichler, H. Schulz, H.O. Reitemeyer and J. Weitkamp, Über das Hydrocracken gesättigter Kohlenwasserstoffe, Erdöl, Kohle - Erdgas - Petrochem. 25: 494 (1972).
12. G.A. Mills, H. Heinemann, T.H. Milliken and A.G. Oblad, Catalytic Mechanisms, Ind. Eng. Chem. 45: 134 (1953).
13. P.B. Weisz, Polyfunctional Heterogeneous Catalysis, Adv. Catal. 13: 137 (1962).
14. J. Weitkamp, S. Ernst and H.G. Karge, Peculiarities in the Conversion of Naphthenes on Bifunctional Catalysts, Erdöl, Kohle - Erdgas - Petrochem. 37: 457 (1984).
15. B.S. Greensfelder, H.H. Voge and G.M. Good, Catalytic and Thermal Cracking of Pure Hydrocarbons: Mechanisms of Reactions, Ind. Eng. Chem. 41: 2573 (1949).
16. J.A. Martens, P.A. Jacobs and J. Weitkamp, Attempts to Rationalize the Distribution of Hydrocracked Products. I Qualitative Description of the Primary Hydrocracking Modes of Long Chain Paraffins in Open Zeolites, Appl. Catal. 20: 239 (1986).
17. J. Weitkamp, W. Gerhardt and P.A. Jacobs, Isomerization and Hydrocracking of Alkanes on Pt/CeY, Pt/LaY and Pd/LaY Zeolites - Bifunctional or Metallic Catalysis?, in: "Proc. Intern. Symp. on Zeolite Catalysis", Siófok, Hungary, May 13 - 16, 1985, p. 261.

18. J.A. Martens, P.A. Jacobs and J. Weitkamp, Attempts to Rationalize the Distribution of Hydrocracked Products. II. Relative Rates of Primary Hydrocracking Modes of Long Chain Paraffins in Open Zeolites, Appl. Catal. 20: 283 (1986).

19. P.A. Jacobs, J.A. Martens, J. Weitkamp and H.K. Beyer, Shape Selectivity Changes in High Silica Zeolites, Farad. Discuss. Chem. Soc. 72: 353 (1982).

20. H.L. Coonradt and W.E. Garwood, Mechanisms of Hydrocracking: Reactions of Paraffins and Olefins, Ind. Eng. Chem., Proc. Des. Dev. 3: 38 (1964).

21. J. Weitkamp, Isomerization of Long Chain n-Alkanes on a Pt/CaY-Zeolite Catalyst, Ind. Eng. Chem., Prod. Res. Dev. 21: 550 (1982).

22. J. Weitkamp, P.A. Jacobs and J.A. Martens, Isomerization and Hydrocracking of C_9 Through C_{16} n-Alkanes on Pt/HZSM-5 Zeolite, Appl. Catal. 8: 123 (1983).

23. C.J. Egan, G.E. Langlois and R.J. White, Selective Hydrocracking of C_9- to C_{12}-Alkylcyclohexanes on Acidic Catalysts. Evidence for the Paring Reaction, J. Am. Chem. Soc. 84: 1204 (1962).

24. D.M. Brouwer and H. Hogeveen, The Importance of Orbital Orientation as a Rate - Controlling Factor in Intramolecular Reactions of Carbonium Ions, Rec. Trav. Chim. 89: 211 (1970).

25. S. Ernst and J. Weitkamp, Hydrocracking of C_9 Through C_{11} Naphthenes on Pd/LaY and Pd/HZSM-5 Zeolites, in: "Proc. Intern. Symp. on Zeolite Catalysis", Siófok, Hungary, May 13 - 16, 1985, p. 457.

26. J. Weitkamp, S. Ernst and R. Kumar, The Spaciousness Index: A Novel Test Reaction for Characterizing the Effective Pore Width of Bifunctional Zeolite Catalysts", Appl. Catal. 27: 207 (1986).

27. J. Weitkamp and S. Ernst, Shape Selective Hydroconversion of Hydrocarbons: A Powerful Tool for Characterizing the Effective Pore Width of Zeolites and Related Materials, in: "Catalysis 1987", J.W. Ward, ed., p. 367/382, Studies in Surface Science and Catalysis, Vol. 38, Elsevier, Amsterdam, Oxford, New York (1988).

28. H. Schulz and J. Weitkamp, Zeolite Catalysts: Hydrocracking and Hydroisomerization of n-Dodecane, Ind. Eng. Chem., Prod. Res. Dev. 11: 46 (1972).

29. J.-K. Chen, A.M. Martin, Y.G. Kim and V.T. John, Competitive Reaction in Intrazeolitic Media, Ind. Eng. Chem. Res. 27: 401 (1988).

30. J.-K. Chen, A.M. Martin and V.T. John, Modifications of n-Hexane Hydroisomerization over Pt/Mordenite as Induced by Aromatic Cofeeds, J. Catal. 111: 425 (1988).

31. H. Dauns and J. Weitkamp, Modelluntersuchungen zum Isomerisieren und Hydrocracken von Alkan-Gemischen an einem Pd/LaY-Zeolith-Katalysator, Chem.-Ing.-Tech. 58: 900 (1986).

32. H. Dauns, S. Ernst and J. Weitkamp, The Influence of Hydrogen Sulfide in Hydrocracking of n-Dodecane over Palladium/Faujasite Catalysts, in: "New Developments in Zeolite Science and Technology", Y. Murakami, A. Iijima and J.W. Ward, eds., p. 787/794, Kodansha, Tokyo and Elsevier, Amsterdam (1986).

33. J. Weitkamp and H. Dauns, Hydrieraktivität und Acidität bifunktioneller Katalysatoren: In-situ-Charakterisierung mittels Umsetzung von Ethylbenzol, Erdöl, Kohle - Erdgas - Petrochem. 40: 111 (1987).

34. H.G. Karge and J. Raskó, Hydrogen Sulfide Adsorption on Faujasite-Type Zeolites with Systematically Varied Si-Al Ratios, J. Colloid Interface Sci. 64: 552 (1978).

35. T.Y. Yan, The Promotonial Effect of Water in Hydrocracking, J. Catal. 25: 204 (1972).

36. O. Weisser and S. Landa, "Sulphide Catalysts, Their Properties and Applications", p. 18/22, Pergamon Press, Oxford, New York and Vieweg-Verlag, Braunschweig (1973).

HYDROISOMERIZATION AND HYDROCRACKING OF ALKANES. 6. INFLUENCE OF THE PORE

STRUCTURE ON THE SELECTIVITY OF Pt-ZEOLITE

Giuseppe Giannetto[1], Fernanda Alvarez[2], Fernando Ramoa
Ribeiro[2], Guy Pérot[3] and Michel Guisnet[3*]

1. Universidad Central de Venezuela, Escuela de Ingenieria
de Petroleo, Caracas 47100, Vénézuela
2. Instituto Superior Tecnico, 1096 Lisboa Codex, Portugal
3. URA CNRS DO350, Catalyse en Chimie Organique, Université
de Poitiers, 40, avenue du Recteur Pineau, 86022 Poitiers
Cedex, France

INTRODUCTION

Bifunctional metal-acid zeolites are used in numerous industrial processes in petroleum refining and in petrochemical industries[1-3] : hydrocracking, hydroisomerization of light alkanes or of C_8 aromatics... On these catalysts alkane hydrocracking requires i) chemical steps on the metallic sites (dehydrogenation of alkanes and hydrogenation of olefinic intermediates) and on the acid sites (isomerization and cracking of olefins) and ii) diffusion steps of the intermediates from the metallic sites to the acid sites and vice versa[3]. Therefore the activity, the stability and the selectivity of these bifunctional catalysts depend first on the acid and on the metallic functions[3,4]. We have recently shown in n-heptane transformation on PtHY catalysts that a definite correlation exists between the catalytic properties and the balance between the metallic and the acid functions characterized by nPt/nA, the ratio of the number of accessible platinum atoms to the number of strong acid sites (sites on which the heat of ammonia adsorption is greater than 100 kJ per mol). Thus for low values of this ratio the activity is low, the deactivation rapid and apparently n-heptane leads directly to all the isomerization and cracking products. For high values the activity is great, the deactivation very slow and n-heptane transforms successively into monobranched isomers, dibranched isomers and cracking products.

Obviously, the catalytic properties of metal/acid zeolites depend also i) on the characteristics of the diffusion path of reactants, of products and of intermediates (pore dimensions, pore structure (tortuosity...)) and ii) on the size and on the shape of the space available near the active sites for the formation of transition states[6,7]. This effect of the pore structure is shown here in n-heptane transformation on 5 series of platinum zeolites in which the catalysts differ by the nPt/nA ratio. We have chosen only zeolites with large pores (apertures delimited by 12-membered rings of oxygen atoms : 12 MR) or with intermediate-size pores (10 MR). We did not study zeolites with

Guidelines for Mastering the Properties of Molecular Sieves
Edited by D. Barthomeuf *et al.*
Plenum Press, New York, 1990

355

small pores (8 MR) but we have used offretite, a zeolite with a double pore system : small pores (8 MR), accessible only to linear alkanes or olefins and large pores (12 MR). We shall show that n-heptane hydro-conversion is, like n-decane hydroconversion[7], a good model reaction for determining the pore structure of zeolites.

EXPERIMENTAL

Catalysts

The zeolites used were : USHY (Si/Al framework = 5) obtained by calcination under dry air flow at 550°C of USNH$_4$Y (LZY82 from Union Carbide, Si/Al = 3), HMOR (900 H Zeolon from Norton, Si/Al = 7), HOFF (from Grace Davison, Si/Al = 3.4), HBETA (Si/Al = 9) prepared by Guth and coll., UA CNRS Matériaux Minéraux, Mulhouse, HZSM5 (Si/Al = 45) synthetized according to Mobil Patents[8]. Their acidity was characterized by ammonia adsorption followed by calorimetry. The number of strong acid sites (sites on which the heat of adsorption is greater than 100 kJ per mol.), nA (10^{20} sites g^{-1}) and the average value of the heat of adsorption on the A sites (kJ per mol) were the following : USHY (3.4 ; 115), HMOR (7.5 ; 135), HOFF (6.8 ; 130), HBETA (4.2 ; 130), HZSM5 (1.6 ; 120).

Platinum zeolites (wt % platinum from 0.05 to 1.5) were prepared through exchange of zeolites by $[Pt(NH_3)_4]^{2+}$. The conditions of exchange and of activation have been described[9]. The dispersion of platinum was estimated by electron microscopy and/or by H_2-O_2 titration.

n-Heptane transformation

The reaction was carried out at 250°C in a flow reactor with a hydrogen/n-heptane molar ratio equal to 9. The activities ((molar flow rate x conversion)/weight of catalyst) were measured at a conversion below 10 %. The product distribution as a function of the conversion (2-80 %) was obtained using different weights of catalysts (50-200 mg) and different n-heptane flow rates (0.1-12 ml/h).

RESULTS

1. Initial activity and stability

Figure 1 shows the change of the initial activity (Ao extrapolated to reaction time zero) as a function of nPt/nA. With PtUSHY, PtHZSM5, Pt HBETA, Ao first increases with nPt/nA and then above a certain value of nPt/nA (about 0.03) remains constant. PtHMOR and PtHOFF behave differently : with these catalysts, Ao passes through a maximum.

Except with PtHZSM5 catalysts (which are perfectly stable), the activity decreases rapidly during the first hour of reaction, then more slowly. The ratio of the final activity (Af, measured after 5 hours' reaction) to the initial activity, Ao, increases with nPt/nA. It depends very much on the zeolite pore structure. With PtHOFF no deactivation affects the isomerization activity ; the cracking activity decreases with time-on-stream and the product distribution changes : decrease in the percentage of C_3, increase in the percentage of C_4 and increase in the isoC$_4$/nC$_4$ ratio.

2. Selectivity

The distribution of the products was determined on samples deactivated during 5 hours. On all the catalysts n-heptane leads to isomers (I) and to cracking products (C). The isomers were classified

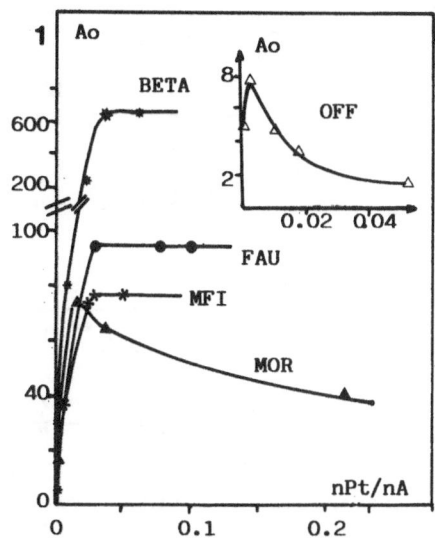 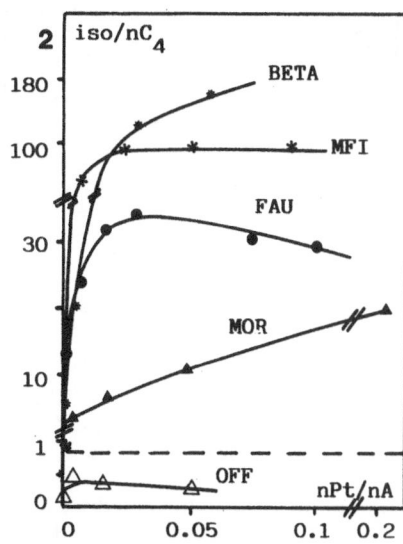

Fig. 1. and 2. Initial activities of Pt zeolites (Ao, 10^{-3} mol $h^{-1}g^{-1}$) and isobutane/n-butane ratio for a conversion of n-heptane into cracking products equal to 5 % (iso/nC_4) versus the ratio of platinum sites/acid sites (nPt/nA). --- Equilibrium.

into monobranched isomers (M : methylhexanes and ethylpentane) and multibranched isomers (B : dimethylpentanes and trimethylbutane). Except on PtHOFF catalysts cracking products C are practically only isobutane and propane in equimolar amounts. On PtHOFF catalysts other light products, mainly n-butane but also pentanes and hexanes (essentially linear), appear and the molar C_3/C_4 ratio is greater than 1.

With all the Pt zeolites, I/C, the ratio of isomerized to cracked n-heptane increases with nPt/nA. I/C is very dependent on the zeolite pore structure : very great on PtUSHY, medium on PtHBETA and PtHOFF and small on PtHMOR and on PtHZSM5. It must be noted that cracking is consecutive to isomerization on PtUSHY for nPt/nA $\geqslant$ 0.03, on PtHBETA for nPt/nA $\geqslant$ 0.06, on PtHMOR for nPt/nA $\geqslant$ 0.2. On PtHOFF and on PtHZSM5 cracking products appear always as primary products.

The ratio of isobutane to n-butane in the cracking products (iso/nC_4) was determined with all the catalysts for a conversion of n-heptane into cracking products equal to 5 %. Iso/nC_4 first increases with nPt/nA then becomes constant. Iso/nC_4 depends very much on the zeolite pore structure : this ratio is greater than 100 on PtHBETA and on PtHZSM5, equal to about 30 on PtUSHY to 10 on PtHMOR and only to 0.3 on PtHOFF (Figure 2).

The ratio of monobranched to bibranched isomers M/B was estimated for conversion close to zero and at maximum isomerization conversion (Figure 3). In both cases monobranched isomers are highly favoured on PtHZSM5 : M/B is equal to ∞ at zero conversion and to about 12 at maximum isomerization conversion. At zero conversion, M/B is great on PtUSHY (about 10) which have high nPt/nA ratios and on all PtHOFF catalysts (about 12). On these catalysts the formation of bibranched isomers B is consecutive to that of monobranched isomers M. It is not the case with PtHBETA and PtHMOR even for high nPt/nA values. Whatever the zeolite there is no equilibrium between M and B even at maximum isomerization conversion. At this maximum, on PtUSHY and on PtHMOR, M/B is 3-4 times higher than at equilibrium, on PtHBETA 4-5 times, on PtHOFF 4-6 times. M/B increases with nPt/nA on all the catalysts except with PtHOFF.

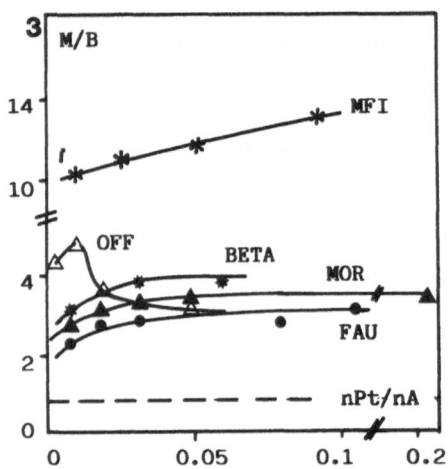 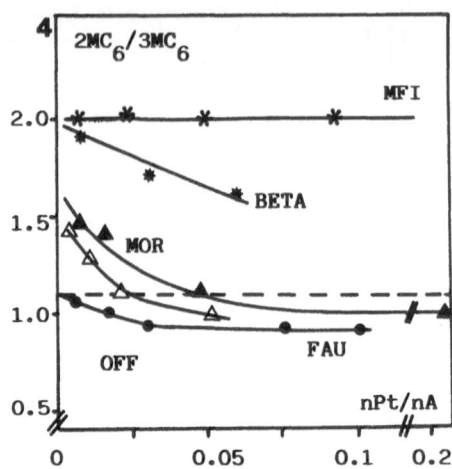

Fig. 3. and 4. Monobranched/multibranched isomer ratio at maximum isomerization conversion (M/B) and 2-methylhexane/3-methylhexane ratio at zero conversion ($2MC_6/3MC_6$) versus the ratio of platinum sites/acid sites (nPt/nA). --- Equilibrium.

The ratio of the initial rates of formation of 2-methylhexane and of 3-methylhexane ($2MC_6/3MC_6$) is plotted on Figure 4 as a function of nPt/nA. With all the catalysts, $2MC_6/3MC_6$ decreases when nPt/nA increases. For high nPt/nA values, $2MC_6/3MC_6$ is greater than the equilibrium value on PtHZSM5 and on PtHBETA and equal to or smaller than the equilibrium value on the other catalysts.

The distribution of the multibranched isomers determined at maximum isomerization conversion depends little on nPt/nA . Table 1 shows this distribution for a nPt/nA value equal to about 0.05. On PtUSHY and PtHMOR the only difference with the equilibrium value concerns the percentages of 2,3 and 2,4 dimethylbutanes. With PtHOFF and PtHBETA 2,2,3-trimethylbutane is unfavored and on PtHZSM5 3,3-dimethylpentane and 2,2,3-trimethylbutane are not formed.

DISCUSSION

The activity, the stability and the selectivity of platinum zeolites in n-heptane hydroconversion depend on the characteristics of the active sites (in particular on the balance between the acid and the hydrogenating activities represented here by nPt/nA) and on the characteristics of the zeolite pore structure : space available near the active sites and diffusion path of reactants and of products. Qualitatively the effect of nPt/nA on the catalytic characteristics is practically the same for all the series of Pt zeolites. Thus, whatever the zeolite, the activity first increases with nPt/nA and above a certain value remains constant or decreases ; the stability increases with nPt/nA ; the ratios of isomerized to cracked n-heptane (I/C), of isobutane to n-butane (iso/nC_4), of monobranched to bibranched isomers (M/B) increase with nPt/nA while the initial ratio of 2-methylhexane to 3-methylhexane ($2MC_6/3MC_6$) decreases. The distribution of multibranched isomers at the maximum isomerization conversion is practically independent of nPt/nA. However big differences in the significance of the changes with nPt/nA are observed between the zeolites. Therefore the activity, the stability and the selectivity must be compared for a given value of nPt/nA. Since our objective is to discriminate between the effects due to the cata-

DMP dimethylpentanes, TMB trimethylbutane.

Table 1. Distribution of dibranched isomers.

Zeolite	2,2 DMP	3,3 DMP	2,4 DMP	2,3 DMP	223 TMB
FAU	25	16	24	31	4
MOR	22	12	27	35	4
OFF	26.5	12	27	33.2	1.3
BETA	21	8	32.5	37	1.5
MFI	8	2	43	47	0
Equilibrium[10]	19	15	11.5	50	4.5

lytic sites and those due to the zeolite pore structure, we shall choose the highest possible value of nPt/nA which can be obtained with all the zeolites. Indeed the greater nPt/nA the greater the similarity between the apparent reaction scheme of n-heptane transformation and that of transformation of the olefinic intermediates on the acid sites[6]. A value of about 0.05 which corresponds to a platinum content (if the platinum dispersion is close to 100 %) between 0.3 % on PtHZSM5 (which has the lower acid site density) and 1 % on PtHMOR and PtHOFF (which have the greater density) has been chosen. The catalytic characteristics obtained for this nPt/nA value are given in table 2. For the discussion, PtUSHY for which configurational limitations and steric constraints are not very significant will be taken as reference.

1. Turnover number of the acid sites

The initial activity of PtUSHY catalysts (Ao) remains constant for nPt/nA values greater than 0.03 (Figure 1). This means that there are enough hydrogenating sites to feed all the acid sites with olefinic intermediates[6]. The reaction rate is therefore proportional to the rate of transformation of olefinic intermediates on the acid sites. Ao in absence of configurational limitations and of steric constraints (as is the case with PtUSHY catalysts) is a characteristic of the zeolite acidity. The differences in the Ao/nA values can be due to differences between the proportions of the Lewis and of the Brönsted strong acid sites, to differences in acid strength or in pore structure. The samll differences between Ao/nA values for USHY, ZSM5 and BETA can be explained by differences in strength of the acid sites. The very low Ao/nA values found for MOR and OFF which have very strong acid sites are probably due to limitation or to blockage of the access of olefinic intermediates to the acid sites. With HMOR, a monodimensional zeolite this limitation is probably due for a large part to a blockage of the channels by coke deposited during the first moment of the reaction (before the first measurement of activity)[6]. The pore system of HOFF, which is monodimensional for branched molecules and which presents cavities with small apertures is also very sensitive to pore blockage[12]. Platinum crystallites are also responsible for a blockage of the large channels of this zeolite[13].

Table 2. Pore system and catalytic characteristics

Zeolite	Pore system[a]			Catalytic characteristics						
	Structure	Pore	Dimensions (nm)	$\frac{Ao}{nA}$ (h^{-1})	$\frac{Af}{Ao}$	$\frac{I}{C}$	$\frac{iso}{nC_4}$	$\frac{M}{B}$	$\frac{2MC_6}{3MC_6}$	TMB (%)
FAU (USY)	***	12 MR cage	0.74 1.2	180	0.7	20	35	3	0.95	4
MOR	* *	12 MR 8 MR	0.65x0.70 0.26x0.57	55	0.3	0.6	10	3.4	1.1	4
OFF	* **	12 MR 8 MR	0.67 0.36x0.49	1.5	0.55	1.1	0.3	3.4	1.0	1.3
BETA	***	12 MR	0.66x0.81 0.56x0.65	900	0.7	2.5	150	3.8	1.7	1.5
MFI (ZSM5)	***	10 MR	0.53x0.56 0.56x0.65	300	1	0.3	100	12	2.0	0

[a] From Ref. 11 ; *, **, *** one, two or three-dimensional ;
TMB : trimethylbutane.

2. Stability

It is now well known that the lower the acid site density, the lower the coking rate and therefore the slower the deactivation[14]. It is probably why PtHZSM5 catalysts do not deactivate whatever the value of nPt/nA. Indeed Pt/dealuminated Y zeolite does not deactivate either[15].

The main parameter of the pore structure which determines the zeolite stability is the mode of diffusion. If the diffusion occurs monodimensionally the deactivation is very rapid[14]. Indeed a single coke molecule is enough to limit or to block the access of the reactants to all the acid sites of a channel. This is not obviously the case when the diffusion occurs tridimensionally. The low stability of MOR, a monodimensional zeolite was therefore expected, as well as the quasi similarity of the stabilities of USHY and of BETA, tridimensional zeolites having similar acid site densities. The very different deactivation rates found for isomerization and for cracking on OFF catalysts show that these reactions do not occur on the same acid sites. This confirms that there exists acid sites not only in the large channels but also in the gmelinite cages[16].

3. Isomerization/cracking selectivity (I/C)

With PtUSHY catalysts with great nPt/nA values, cracking is consecutive to isomerization and therefore I/C measured at low conversion is very high. A low value of I/C means that the diffusion of olefinic intermediates between 2 metallic sites is slower than their reaction on the acid sites and that consequently these intermediates undergo several successive reactions[6]. This must occur with zeolites having narrow pores (the diffusion of dibranched and even of monobranched olefinic C_7 intermediates is then limited or inhibited). Thus I/C is low with PtHZSM5 catalysts which have narrow pores (10 MR) and with PtHOFF catalysts which have, besides large pores (12 MR), gmelinite cages with small apertures (8 MR) from which only linear products (therefore cracking products) can desorb. With PtHOFF catalysts the formation of linear cracking products is practically not inhibited by pyridine poisoning at the reaction temperature while that of isomers is completely suppressed[17]. With PtHMOR catalysts which have however large pores, low values of I/C are also obtained. This can be explained by the diffusional limitations caused by coke deposited in the large channels. It must be noted that the great strength of the HMOR acid sites could also be responsible for the low I/C value. The intermediate value found for PtHBETA shows that the diffusional limitations are not very pronounced. It can be noticed that on PtHBETA with nPt/nA = 0.06, cracking is consecutive to isomerization. This observation is characteristic of a zeolite with large pore apertures (12 MR).

4. Iso/nC_4 ratio

The iso/nC_4 ratio found with PtUSHY catalysts is characteristic of cracking in complete absence of steric constraints of dibranched olefinic intermediates. A greater value is found with PtHZSM5 catalysts and also with PtHBETA. To explain this it must be assumed that particular transition states are highly favored by the shape and the size of the space available near the acid sites[18]. The characteristics of this space in HZSM5 and in HBETA should therefore be similar.

The relatively low value found with PtHMOR could be attributed to limitations, caused by coke deposits, in the desorption of iC_4 from large channels while the low value found with PtHOFF would be due to the fact that cracking occurs for a large part in the gmelinite cages from which only linear compounds can desorb. This latter part is confirmed by the value of the C_3/C_4 ratio greater than 1 and by the formation of nC_5 and nC_6 without the complementary C_2 and C_1 fragments[17].

5. Monobranched/multibranched isomers (M/B)

A value of 3-4 for M/B measured at maximum isomerization conversion is characteristic of large pore zeolites (12 MR). A much greater value is found on PtHZSM5 catalysts because the diffusion of dibranched isomers in the narrow channels of this zeolite is very limited.

6. Ratio of 2-methyl to 3-methylhexane ($2MC_6/3MC_6$)

It is generally admitted that branching occurs through protonated cyclopropanes (PCP)[19,20]. Because of the lower number of PCP structures leading to $2MC_6$ compared to $3MC_6$, $2MC_6/3MC_6$ should be much smaller than 1 (about 0.4). With PtUSHY catalysts, this ratio is equal to 0.9 that is to say slightly smaller than the equilibrium value. With PtHBETA and more so with PtHZSM5, values much greater than the equilibrium value are found. These values can only be explained by transition state selective effects[21,22] : the size and/or the shape of the space available near the acid sites would favour the formation of the transition state leading to $2MC_6$. It must be noted that Martens et al[7] found in n-decane hydroconversion that on PtHZSM5, $2MC_9$ was highly favoured compared to the other isomers but that on PtHBETA it was the contrary.

7. Distribution of multibranched isomers

At maximum isomerization conversion, multibranched isomers are practically in equilibrium with PtUSHY and PtHMOR. On the contrary, the bulkiest isomers : 3,3-dimethylpentane and trimethylbutane are highly unfavoured with PtHZSM5 catalysts which have narrow pores (10 MR). Trimethylbutane is also unfavoured with PtHOFF and PtHBETA catalysts. At least in the case of PtHOFF this can be explained by a limited diffusion of this bulky molecule in the channels of the zeolite.

CONCLUSION

In heptane hydroconversion, the activity, the stability and the selectivity of bifunctional zeolitic catalysts depend not only on the characteristics of the active sites but also on the pore structure. The size of the pores, the mode of diffusion (mono or tridimensional) are responsible for differences that exist between the activities of the acid sites of the various zeolites, the stability and the selectivity (isomer/cracking rate ratio, monobranched/bibranched isomer ratio, isobutane/n-butane, distribution of multibranched isomers). The distribution of monobranched isomers ($2MC_6/3MC_6$) and of cracking products (iso/nC_4) depends also on the size and on the shape of the space available near the acid sites. Therefore this model reaction is quite suitable for determining the pore structure of zeolites. This reaction, simpler than n-decane hydroconversion, gives practically the same information.

REFERENCES

1. A.P. Bolton, in: "Zeolite Chemistry and Catalysis", J. Rabo Ed., ACS Monograph 171, Washington D.C. (1976) p. 714.
2. Kh. M. Minachev and Ya. Y. Isakov, ibid, p. 552.
3. M. Guisnet and G. Pérot, in: "Zeolites : Science and Technology", F.R. Ribeiro et al. Eds., NATO ASI Series E 80, Martinus Nijhoff Publishers, The Hague (1984) p. 387.
4. P.A. Jacobs, J.B. Uytterhoeven, M. Steyns, G. Froment and J. Weitkamp, in: "Proceedings of the 5th International Conference on Zeolites", L.V.C.Rees Ed., Heyden, London (1980) p. 607.
5. G.E. Giannetto, G.R. Pérot and Michel R. Guisnet, Ind. Eng. Chem. Prod. Res. Dev., 25:481 (1986).
6. M. Guisnet, F. Alvarez, G. Giannetto and G. Pérot, Catalysis Today, 1:415 (1987).
7. J.A. Martens, M. Tielen, P.A. Jacobs and J. Weitkamp, Zeolites, 4:98 (1984).
8. R.J. Argauer and R.G. Landolt, U.S. Patent 3.702.886 (1972).
9. F. Alvarez, G. Giannetto, A. Montes, F. Ribeiro, G. Pérot and M. Guisnet, in: "Innovations in Zeolite Materials Science", P.J. Grobet et al Eds., Studies in Surface Science and Catalysis Vol. 37, Elsevier, Amsterdam (1988) p. 479.
10. D.R. Stull, E.F. Westrum and G.C. Sinke, "The Chemical Thermodynamics of Organic Compounds", Wiley, New York (1969).
11a. W.M. Meier and D.H. Olson, "Atlas of Zeolite Structure Types", Butterworths, London (1987).
  b. J.B. Higgins, R.B. La Pierre, J.L. Schlenker, A.C. Rohrman, J.D. Wood, G.T. Kerr and W.J. Rohrbaugh, Zeolites, 8:446 (1988).
12. S. Mignard, P. Cartraud, P. Magnoux and M. Guisnet, J. Catal., to be published.
13. G. Giannetto, P. Cartraud, F. Alvarez and M. Guisnet, React. Kinet. Catal. Lett., 36(2):251 (1988).
14. E.G. Derouane, in: "Catalysis by Acids and Bases", Imelik et al Eds., Studies in Surface Science and Catalysis, Vol. 20, Elsevier, Amsterdam (1985) p. 221.
15. F. Alvarez, G. Giannetto, M. Guisnet and G. Pérot, Applied Catalysis, 34:353 (1987).

16. G. Bourdillon, M. Guisnet and C. Gueguen, Zeolites, 6:221 (1986).
17. G.E. Giannetto, F.B. Alvarez and M.R. Guisnet, Ind. Eng. Chem. Res. 27:1174 (1988).
18. G. Pérot, P. Hilaireau and M. Guisnet, in: "Proceedings of the 6th International Zeolite Conference", D. Olson and A. Bisio Eds., Butterworths, London (1984) p. 427.
19. F. Chevalier, M. Guisnet and R. Maurel, in: "Proceedings of the 6th International Congress on Catalysis", G.C. Bond et al Eds., The Chemical Society, London (1977) p. 478.
20. D.M. Brouwer and J.M. Oelderik, Recl. Trav. Chim., Pays-Bas, 87:721 (1968).
21. P. Jacobs, J. Martens, J. Weitkamp and H. Beyer, Faraday Discussion Chem. Soc., 72:354 (1982).
22. G. Giannetto, G. Pérot and M. Guisnet, in: "Zeolites : Synthesis, Structure, Technology and Application", B. Drzaj et al Eds., Studies in Surface Science and Catalysis, Vol. 24, Elsevier, Amsterdam (1985) p. 631.

The Structural Elements of Faujasite and their Impact on Cracking Selectivity

A. W. Peters, W. C. Cheng, M. Shatlock, R. F. Wormsbecher, and
E. T. Habib, Jr.

W. R. Grace – 7379 Route 32, Columbia, MD 21044 USA

ABSTRACT

The location of alumina in the dealuminated faujasite framework is described. The selectivities of dealuminated and rare earth exchanged faujasite, including selectivities for octane, coke, hydrogen transfer, and isomerization, are shown to depend on the relative amounts of paired and isolated sites present. Pentacoordinated and octahedral nonframework alumina do not contribute to selectivities in cracking. Selectivity and activity is controlled by the framework aluminum sites.

INTRODUCTION

While the framework structure of faujasite is well known largely from xrd analysis (1), the structural details concerning the locations and chemistry of the aluminum atoms are less well known. This is especially true of the partially exchanged and hydrothermally dealuminated zeolite used in an operating hydrocracking or FCC catalyst. These details may be responsible for important differences in activity and selectivity since it is believed that the acidity of the zeolite is provided by aluminum atoms in and perhaps outside of the framework.

The placement of the aluminum atoms within the framework of the dealuminated zeolite cannot be determined from the xrd results and has only recently been inferred from [29]Si masnmr results. Some of these results will be discussed here.

Aluminum also occurs outside the framework. Hydrothermal dealumination removes part of the aluminum from the framework. If this aluminum is not removed from the system, it becomes a nonframework part of the zeolite. Very little is known about the chemistry, the coordination and the placement of the nonframework aluminum in the zeolite or about the effect of this material on catalyst activity and selectivity. Again, what is known is largely inferred from [27]Al masnmr results.

There has been a continuing effort to correlate activity and selectivity with structure. Barthomeuf and Beaumont (2) showed that there are weak and strong acid sites in faujasite, that dealumination produces a relative increase in strong sites and few or no weak acid sites, and that the strong sites are more active for cracking than the weak ones. Dempsey and coworkers (3) suggested that site strength results from site isolation. Isolated aluminum sites provide stronger acidity and are more catalytically active than aluminum sites containing one or more second nearest aluminum neighbors. Finally Pine and coworkers (4) showed that dealumination of the zeolite in a cracking catalyst was associated with a high olefin content and research octane number in the gasoline product. Rajagopalan and Peters showed a somewhat different relationship between coke selectivity and dealumination (5). New work by Cheng and Rajagopalan shows that other selectivities including hydrogen transfer and isomerization show a dependence on framework alumina content (6).

Guidelines for Mastering the Properties of Molecular Sieves
Edited by D. Barthomeuf *et al.*
Plenum Press, New York, 1990

EXPERIMENTAL

Faujasite was prepared at various Si/Al ratios from seeded slurries using conventional methods (7). The HZSM-5 was prepared using conventional methods (8) at a Si/Al ratio of 13. Catalytic materials were prepared for testing in a low soda form, 0.3% Na2O or less, and were characterized by XRD and by BET surface area. ^{29}Si MASNMR spectra were obtained on a Bruker AM-400 spectrometer using a 10 degree pulse at 0.1s intervals. About 10,000 scans were collected per spectrum. ^{27}Al MASNMR spectra were obtained at 104.2 MHz and at spinning rates of 4-8KHz using a similar delay and pulse angle, and at 130.321 and 156.376 MHz at spinning rates of 3-15 KHz. Alumina free Vespel spinners were used. At the higher fields 40000 scans were collected per spectrum. Spectra at the higher fields were obtained by Colorado State Regional NMR Center, National Science Foundation Grant No. CHE-8616437. The sample of andalusite was obtained from the National Museum of Natural History, Smithsonian Institution, catalog number 120399. Hydrothermal treatments were carried out at 760°C and 100% steam for 6 hours unless otherwise noted.

FRAMEWORK ALUMINUM

Lippmaa and coworkers (9,10) were the first to probe atomic environments in zeolites using magic angle spinning nmr (masnmr) experiments. With this technique Melchior, Vaughan, and Jacobson (11) were able to show that Lowenstein's rule is strictly obeyed in faujasite. Based on Pauling's valence bond theory, Lowenstein had proposed that Al-O-Al bonds involving tetrahedral aluminum atoms are unstable and will not occur in aluminosilicates (12). Although counter examples have been found, the rule is generally obeyed.

Dempsey had proposed that second nearest neighbors tend to avoid each other for the same reasons. Although masnmr results have not been consistent with this particular proposal, the results do indicate the existence of an avoidance principle (13,14).

The distribution of silicon atoms bonded through oxygen to zero, one, two, three, or four aluminum neighbors can be determined by ^{29}Si masnmr. Results collected from the literature and determined in this laboratory are presented in Figure 1. The intensities of the five different silicon species are reported in the five graphs (a) through (e) normalized to the total Si per unit cell. These masnmr measurements and correlations have been previously reported (15). The solid curve is the calculated most probable distribution using the analytical expression obtained by Peters (14) in combination with a bookkeeping rule intended to guarantee that Lowenstein's rule is assumed in connecting the prisms.

The results in Figure 1 show that the intensity of the Si (4Al) peak (a) falls to zero more rapidly than expected relative to the most probable distribution curve. The zero intercept occurs at about 138 Si/uc corresponding to Si/Al=2.46 and a unit cell of 2.468 nm. This means that faujasite prepared with higher silicon content will not contain Si (4Al) structures. Compensating intensity changes in the region between 96 and 128 Si/uc relative to the most probable distribution are evident in the distributions of the other species. The intensity of the Si(0Al) peak is well below the solid line while the intensities of the peaks corresponding to Si (2Al) and Si (3Al) are above the line resulting in an intensity distribution that is more sharply peaked than would be observed for the most probable distribution.

While the intensity of the Si (3Al) peak (b) is above the line in the region between 96 and 128 Si/uc, beyond 128 Si/uc the intensity of this peak drops rapidly below the line and approaches zero between 150 Si/uc and 163 Si/uc. Thus the nmr evidence is that faujasite structures containing about 160 or more silicon atoms per unit cell will not contain any Si (3Al) types of structures. The practical consequence of this result is that at lower unit cell sizes every aluminum has zero (isolated sites) or one (paired sites) next nearest aluminum neighbors. Nests of three aluminum atoms attached to one silicon cannot occur. Chains of paired structures involving alternating silicon and aluminum atoms can occur and will result in aluminum atoms with two next nearest neighbor aluminum atoms. These structures are less likely at higher silicon content above 170 silicon atoms per unit cell and at lower unit cell sizes below 24.40.

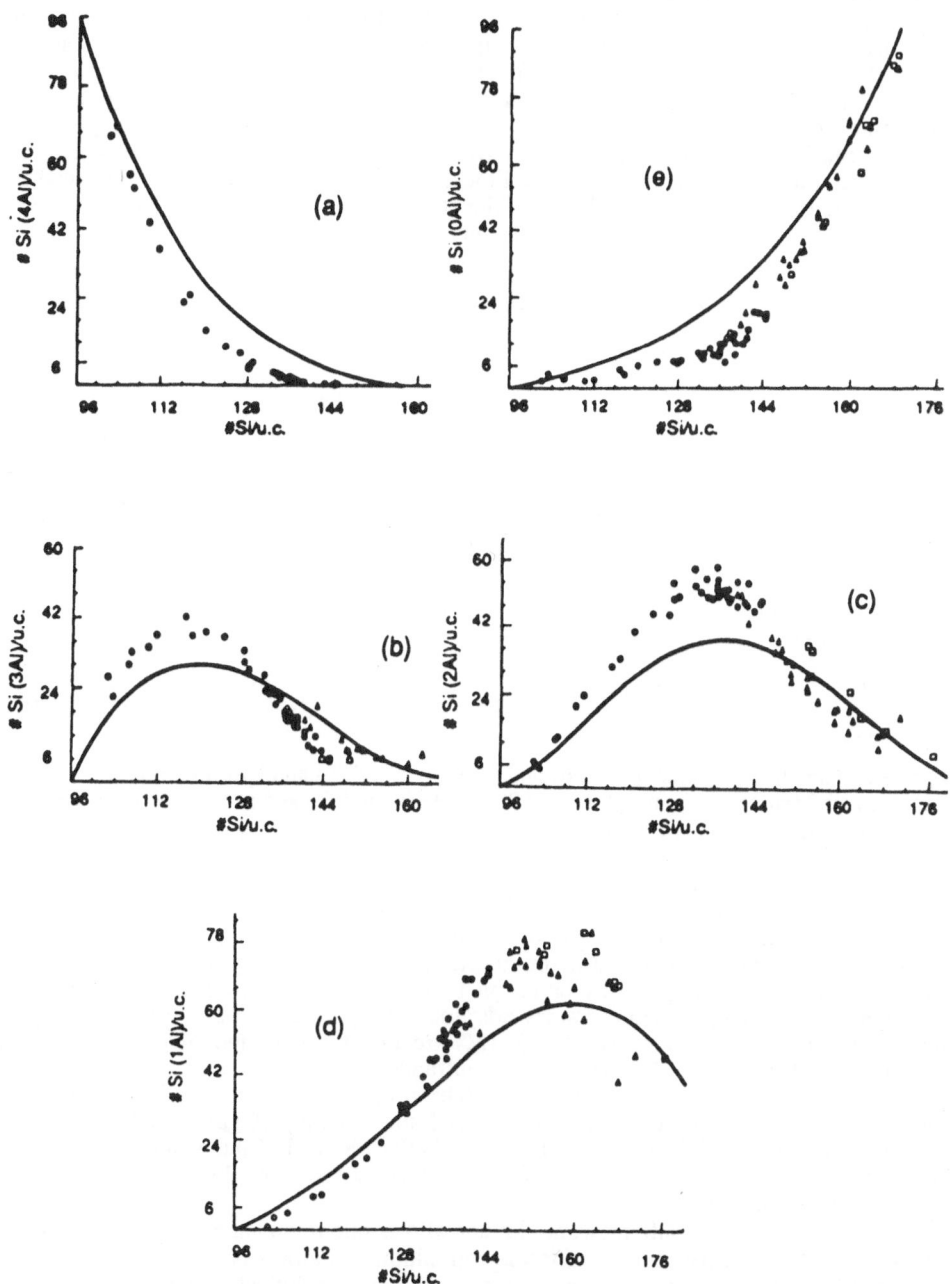

Figure 1. The number of silicon atoms per unit cell of faujasite bonded through oxygen to four (a), three (b), two (c), one (d), or zero (e) aluminum neighbors as determined by [29]Si masnmr. Experimental results include as synthesized (O) as well as hydrothermally (Δ) and chemically (□) (silicon tetrachloride, hexafluorosilicate) dealuminated samples. The solid curve is the most probable distribution estimated according to reference (14) with a modification discussed in the text.

REMOVAL OF NON-FRAMEWORK ALUMINA

Hydrothermally treated faujasite contains tetrahedral framework alumina occurring as a peak at about 60 ppm along with pentacoordinate alumina at about 30 ppm and an octahedral form at about 0 ppm (16,17). The identification of the pentacoordinate peak (18) is based on the spectrum of andalusite, known to contain pentacoordinate alumina (19), and shown in Figure 2 at different spinning rates and at different field strengths. The central peak in Andalusite is shifted from 18.9 ppm to 22.7 ppm as the field varies from 130.321 to 156.376 MHz, while there is little or no shift in the three peaks observed for hydrothermally treated faujasite shown in the same Figure. In the case of the faujasite an increase in the spinning rate significantly increases the intensity of the pentacoordinate peak at about 30ppm. A similar result was observed for hydrothermally treated HZSM-5, Si/Al = 23.

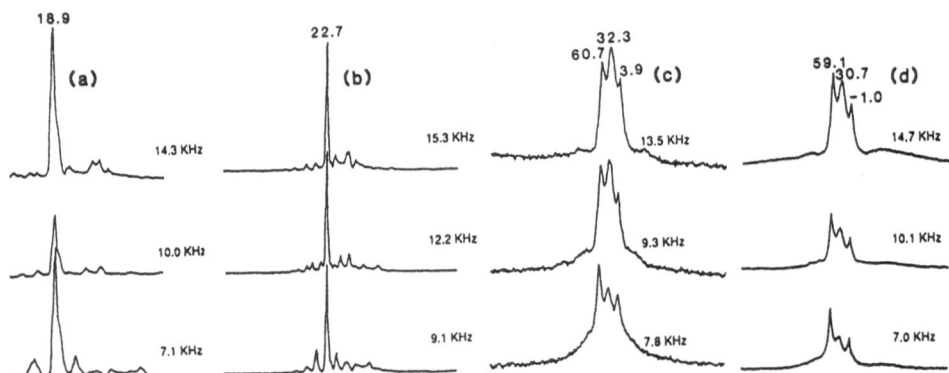

Figure 2. The ^{27}Al masnmr spectra of andalusite (a) and (b) and hydrothermally dealuminated faujasite (c) and (d) at 130.321 MHz (a) and (c) and at 156.376 MHz (b) and (d) at spinning rates of 7-15 KHz.

Most of the nonframework alumina can be removed by washing with ≈pH 3 acid. Spectra of two different samples of USY prepared at the same unit cell of 2.450 nm by either ammonium hexafluorosilicate or by hydrothermal dealumination are shown in Figure 3 after a severe additional hydrothermal treatment to a very low unit cell. In the low unit cell material prepared by either method a nonframework tetrahedral peak at 55 ppm occurs as a shoulder on the framework aluminum peak at 60 ppm (20). This peak is not removed by an acid or caustic (0.1M) treatment, while the treatment removes pentacoordinate and most of the octahedral nonframework alumina.

Previous work (15) has shown that octahedral and pentacoordinated nonframework alumina does not have any effect on the catalytic cracking of gas oils. In that work catalysts were prepared from zeolites before and after removal of the pentacoordinate and octahedral alumina. These results do not preclude the possible involvement of tetrahedral nonframework alumina.

The removal of nonframework alumina allows one to obtain a correlation between the unit cell size and the number of framework aluminum atoms per unit cell at low unit cell size. The results obtained in Figure 4 are given by the analytical expression

$$\#Al = 95\pm10 \, (a_0 - 24.17\pm.04).$$

This relation agrees better with the older Breck and Flanigen equation (21),

$$\#Al = 115.2 \, (a_0 - 24.191)$$

than with the more recent Lunsford equation (22),

$$\#Al = 107.1 \, (a_0 - 24.238).$$

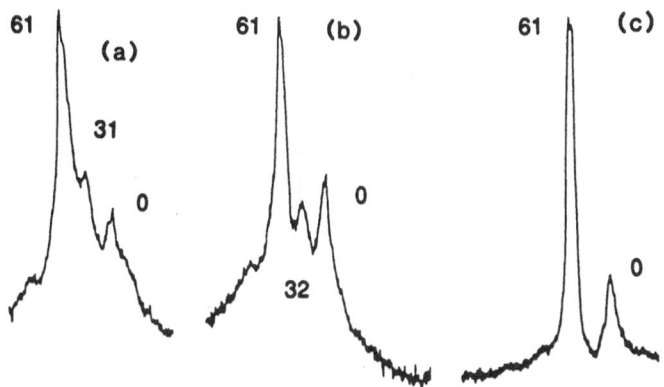

Figure 3. [27] Al MASNMR spectra of severely dealuminated low unit cell USY (a) hexafluorosilicate / hydrothermal dealumination, 2.427nm unit cell, (b) hydrothermal dealumination, and (c) after pH 3 acid wash of (b) showing framework (61 ppm) and tetrahedral (a shoulder at 57 ppm), pentacoordinate (31 ppm), and octahedral (0 ppm) non-framework alumina.

One might think that a relationship based on [29]Si masnmr (the Lunsford equation) would be the most reliable. However non-framework alumina may occupy exchange sites or coordinate to the framework and produce changes in the masnmr spectra that can lead to underestimation of the aluminum content. Hydrothermal dealumination can also result in the formation of mesoporosity and in an increase in the number of SiOH groups which may appear with a shift similar to aluminum coordination in the nmr spectrum.

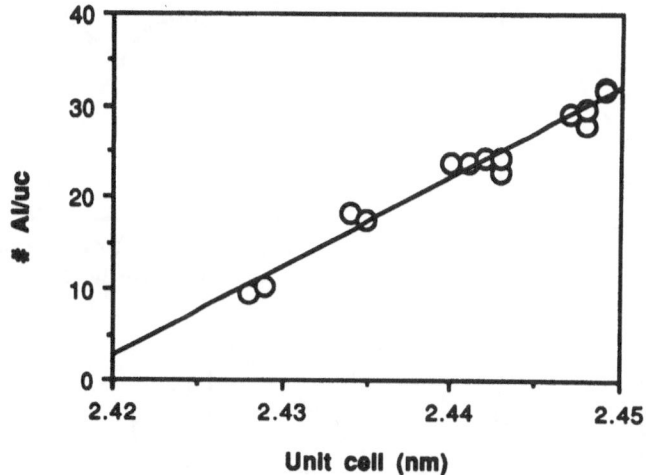

Figure 4. The number of aluminum atoms per unit cell for low unit cell dealuminated and decationated faujasite determined by XRD measurements and chemical analysis.

CORRELATIONS BETWEEN STRUCTURE AND CATALYTIC ACTIVITY AND SELECTIVITY

It has been observed that coke (5) and octane (4) selectivities are strongly dependent on Si/Al ratio, and it has been argued that there is a correlation with site isolation. Beagley, Dwyer and coworkers (23) give a quantitative measure of site isolation for various proposed models and at various aluminum contents. The results of this work show that even for a random distribution a significant degree of site isolation occurs at relatively high cell sizes of about 2.440nm (≈23 Si/u.c.) and in the case of models proposing specific ordered structures isolation occurs even earlier in the dealumination process. Thus there is an inconsistency between the site isolation model and the selectivity results.

The results from Figure 1 show that below 2.440 nm unit cell size there are only two kinds of sites, isolated and paired, where two aluminum atoms are attached to the same silicon. An estimate of the relative amounts of sites formed from isolated and paired aluminum atoms at a silicon content above 160 Si/u.c. is given using the methods of reference (14) where structures involving silicon attached to three or four aluminum atoms are forbidden. The number of paired aluminum sites is estimated to be 1.33 times the number of silicon atoms with two aluminum neighbors. The number of isolated sites is the remainder of the aluminum atoms. As a check, four times the number of isolated aluminum sites plus 2.33 times the number of paired sites should give approximately the number of silicon atoms connected to one aluminum.

Figure 5 shows how the densities of these sites vary with unit cell size using the correlation from Figure 4. The density of binary sites decreases rapidly and the number of isolated sites increases to a maximum at about 180 silicon atoms/u.c. or a unit cell of 2.43 nm. The research octane number of the gasoline obtained by cracking gas oil over a catalyst containing the appropriate zeolite has also been shown to vary with unit cell size (4). Since research octane number is flat down to about 2.430 and increases rapidly with further unit cell decreases, octane does not correlate well with changes in acid strength resulting from site isolation.

Coke selectivity has a different dependence on unit cell. Coke selectivity improves as the unit cell decreases to about 2.430 nm and then flattens out. This behavior correlates well with the decrease in binary sites with unit cell, also Figure 5.

The behavior of the octane selectivity curve with unit cell is suggestive of a structure sensitivity. If one assumes that cracking can take place on all acid sites, but that hydrogen transfer only occurs on binary sites, then the ratio of cracking activity to hydrogen transfer should be given approximately by the ratio of total sites, isolated plus binary, to binary sites. This ratio correlates well with research octane results obtained from the literature (24), Figure 6.

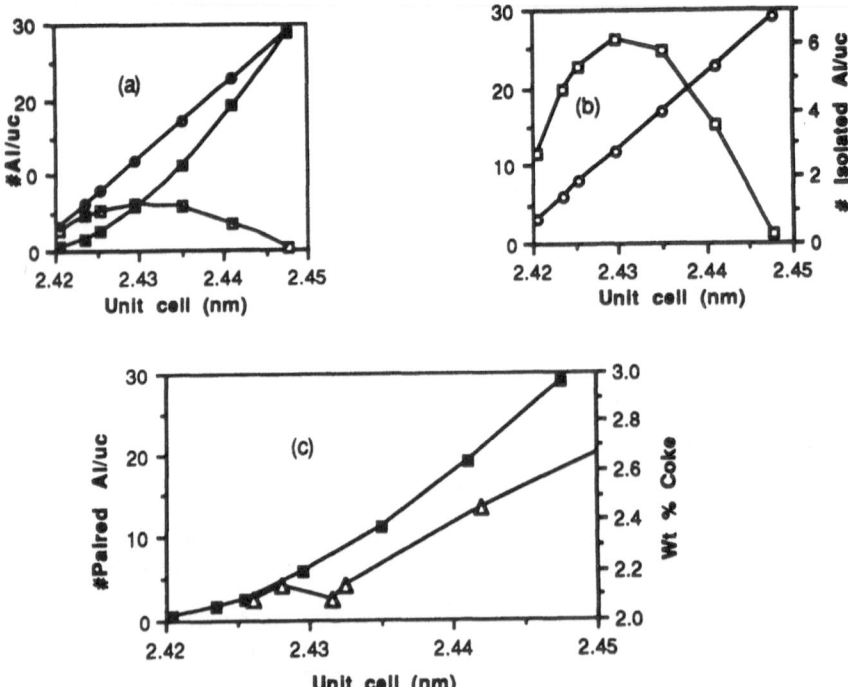

Figure 5. The relationships between the unit cell size and the total site density (O), isolated site density (□), paired site density (■), and coke selectivity (Δ) in low unit cell faujasite. Site densities are the number of Al atoms/unit cell. The correlation with the total number of aluminum atoms per unit cell is obtained from Figure 4 and the related equation in the text connecting the unit cell size and the aluminum content per unit cell.

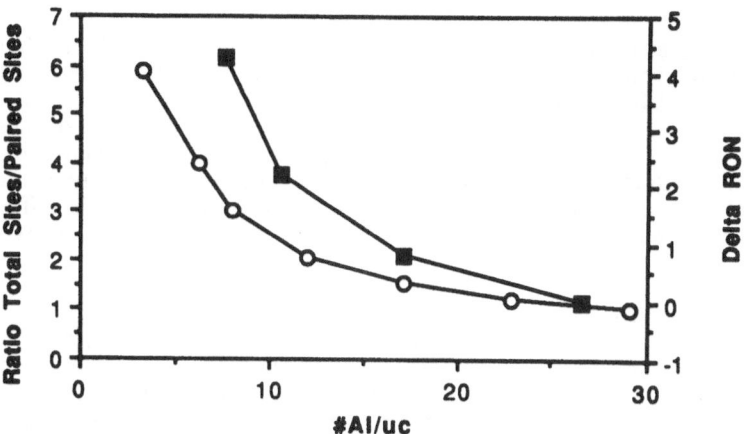

Figure 6. The experimental changes in the research octane number of the gasoline obtained as a product of cracking over a catalyst containing faujasite with varying aluminum content per unit cell, (■), are similar to changes in the ratio of the total number of sites (aluminum atoms) to the number of paired sites per unit cell, (O).

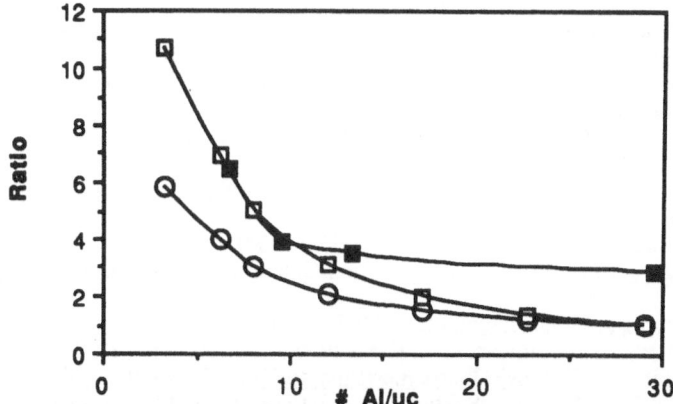

Figure 7. The behavior of the experimental ratio of the isomerization activity to the hydrogen transfer activity of faujasite at various dealumination levels, (■), and the corresponding behavior of the ratio of the number of total sites to the number of paired sites per unit cell, (O). Assuming the isolated sites to be twice as active as the paired sites, a ratio of total activity to paired site activity can also be obtained, (□).

A curve similar in shape to the octane curve has been obtained for the ratio of isomerization rate to the rate of hydrogen transfer in the cyclohexene system (6). This curve is also approximated by the ratio of total site density divided by the density of binary sites, Figure 7. Results recently published by Mirodatos and Barthomeuf (25) showed the existance of strong and weaker sites. If one identifies the strong sites with isolated sites and the weak sites with the binary sites, then total site activity is the number of isolated sites times two plus the number of binary sites. The total site activity defined in this way divided by the number of binary sites provides a somewhat better correlation with the isomerization/ hydrogen transfer ratio, also Figure 7.

These results indicate that isomerization and cracking reactions can occur on all sites, but the isolated sites are about two times stronger than the binary sites. Hydrogen transfer reactions require binary sites. This can be interpreted as a requirement for the formation of a bimolecular complex or for the formation of some rate determining transition state involving a double potential.

SUMMARY

The activity and selectivity of low unit cell faujasite is dominated by the framework aluminum. As the result of the operation of an avoidance rule, there are two types of framework sites in USY, isolated and binary. Selectivities for octane can be understood as a structure sensitivity derived from the relationship between these two types of sites.

Nonframework pentacoordnated and octahedral alumina can be removed by an acid treatment. This nonframework alumina does not appear to influence activity or selectivity. A tetrahedral form of nonframework alumina identified by A. Corma is different and cannot be as easily removed. The involvement of this species in selectivity effects has been suggested by A. Corma (28).

ACKNOWLEDGEMENTS

We wish to thank W. R. Grace for their support and for permission to publish this work. We also wish to thank John Mangano, Art Wadsworth, and Laura Peters for their assistance, the Colorado State Regional NMR Center, National Science Foundation Grant No. CHE-8616437 for high resolution masnmr measurements, and the National Museum of Natural History, Smithsonian Institution for donating a sample of andalusite.

REFERENCES

(1) D. H. Olson, J. Phys. Chem., **74**, 2758 (1970).
(2) R. Beaumont, and D. Barthomeuf, J. Catal. **27**, 45 (1972).
(3) E. Dempsey, J. of Catal. **39**, 155-157 (1975) and references therein.
(4) L.A. Pine, P.J. Maher, and W.A. Wachter, J. of Catal., **85**, 466 (1984).
(5) K. Rajagopalan and A. W. Peters, J. of Catal., **106**, 410-416 (1987).
(6) W. C. Cheng and K. Rajagopalan, J. of Catal., submitted.
(7) D.E.W. Vaughan, G. C. Edwards, and M. G. Barrett, U.S. Patent 4,178,352.
(8) G. R. Landolt, U. S. Patent 3702886 (1972).
(9) E. Lippmaa, M. Magi, A. Samoson, G. Engelhardt, and A.-R. Grimmer, J. Amer. Chem. Soc. **102**, 4889 (1980).
(10) E. Lippmaa, M. Magi, A. Samoson, M. Tarmak, and G. Engelhardt, J. Amer. Chem. Soc. **103**, 4992 (1981).
(11) M. T. Melchior, D. E. W. Vaughan, and A. J. Jacobson, J. Amer. Chem. Soc., **104**, 4859-4864 (1982).
(12) W. Loewenstein, Am. Mineral., **39**, 92-96, (1954).
(13) A. J. Vega, Intrazeolite Chemistry, Amer. Chem. Soc. Symposium 218, p. 217 (1983).
(14) A. W. Peters, J. Phys. Chem., **86**, 3489-3491 (1982).
(15) A. W. Peters, G. C. Edwards, M. P. Shatlock, T. G. Roberie, K. Rajagopalan, and E. T. Habib, XI Simposio Iberoamericano de Calalisis, 1988, p. 1417.
(16) C. A. Fyfe, C. G. Gobbi, J. S. Hartman, J. Klinowski, and M. Thomas, J. Phys. Chem., **86**, 1247 (1982).
(17) J.-B. Nagy, Z. Gabelica, G. Debras, E. G. Derouane, J.-P. Gilson, P. A. Jacobs, Zeolites, **4**, 133 (1984).
(18) J.-P. Gilson, G. C. Edwards, A. W. Peters, K. Rajagopalan, R. F. Wormsbecker, T. G. Roberie, and M. P. Shatlock, J. Chem. Soc., Chem. Commun., No. 2, 91-92 (1987).
(19) L. B. Alemany and G. W. Kirker, J. Amer. Chem. Soc. **108**, 6158-6162 (1986).
(20) A. Corma, V. Fornés, F. Melo, and J. Pérez-Pariente, Fluid Catalytic Cracking, ACS Symposium Series 375, M. Occelli ed., Amer. Chem. Soc., 1988, p.17.
(21) D. Breck, and E. Flanigan, Molecular Sieves, Society of the Chemical Industry, London, 1968, (22) p.47.
(22) J. Lunsford, Zeolites **6**, 225 (1986).
(23) B. Beagley, J. Dwyer, F. R. Fitch, R. Mann, and J. Walters, J. Phys Chem. **88**, 1744-1751 (1984).
(24) G. C. Edwards, K. Rajagopalan, A. W. Peters, G. W. Young, and J. E. Creighton, Fluid Catalytic Cracking, ACS Symposium Series 375, M. Occelli ed., Amer. Chem. Soc., 1988, p.101.
(25) C. Mirodatos, and D. Barthomeuf, J. Catal. **114**, 121 (1988).
(26) P. Magnoux, A. Gallet, and M. Guisnet, Bull. Soc. Chim. Fr., **22**, 810 (1987).
(28) G. Garralón, A. Corma and V. Fornés, J. Chem. Soc., Chem. Commun., 1256 (1988).

RELATIONSHIP BETWEEN THE PHYSICOCHEMICAL PROPERTIES OF ZEOLITIC SYSTEMS AND THEIR LOW DIMENSIONALITY

Eric G. Derouane

Facultés Universitaires N.-D. de la Paix, Laboratoire de Catalyse
Rue de Bruxelles, 61
B-5000 Namur. Belgium

INTRODUCTION

The ultimate goals of the NATO Workshop of which the proceedings are the object of this book were to delineate the relationships between the physicochemical properties of zeolitic systems and their low dimensionality, to assess our fundamental understanding of those, and to propose guidelines for the mastering of molecular sieve properties.

The timeliness of the meeting was demonstrated by both the amount and diversification of the new concepts and ideas which were advanced and the fierce but friendly discussions among the participants. Consequently, our evaluation and fundamental understanding of molecular sieve properties has indeed dramatically progressed, guidelines for future research and applications are proposed, and several basic questions which should be addressed in forthcoming work have been listed. These points are summarized in this general overview of the meeting with emphasis on:

1. The "intelligent" synthesis of zeolites and other molecular sieves.
2. The in-depth and careful characterization of zeolitic materials.
3. The understanding of molecular dynamics within the intracrystalline volume of microporous materials.
4. The definition of the acid function of zeolitic materials.
5. The delineation of ways to direct catalytic applications of such materials, in particular for the organic synthesis of fine and specialty chemicals.
6. The proposal of new concepts which need to be further assessed and amplified.

SYNTHESIS OF MOLECULAR SIEVES

Progresses

As demonstrated for the crystallization of zeolite Omega (F. FAJULA), the kinetic control of zeolite synthesis enables to direct crystal morphology and size and to master the distribution of the aluminum sites both in the framework (T-site type) and through the crystals (magnitude and orientation of the Al-concentration gradient). The control of morphology is possible if various crystal faces develop at different rates as a function of Al availability. Note that the latter factor also affects the degree of supersaturation of the liquid phase and the Al-siting at framework T-sites.

The analysis of synthesis kinetics, usually expressed in terms of (dm/dt), where m is a

Guidelines for Mastering the Properties of Molecular Sieves
Edited by D. Barthomeuf *et al.*
Plenum Press, New York, 1990

373

parameter proportional to the amount of crystalline material produced and t is time, is more rigorously achieved by consideration of the ratio (dl/dS), where l is a measure of crystal size and S is the degree of supersaturation of the solution from which crystal nucleation and growth occur. Supersaturation is affected by the Al concentration (F. FAJULA), the nature of the T-atoms mobilizing agent (OH^- or F^-) (J.-L. GUTH), and solvent effects (for example, NH_3 evolved in hydrothermal and basic conditions when performing a synthesis in the presence of NH_4^+ ions) (D. VAUGHAN).

One of the most important parameters differentiating the synthesis of aluminosilicates and silicoaluminophosphates (and other aluminophosphate-based materials) is pH: alkaline in the former case, acidic for the latter (E.G. DEROUANE, P.A. JACOBS). The effects of OH^- and its accompanying cations (organic or inorganic) can, and must in all cases, be carefully separated to understand the true molecular sieve crystallization mechanism.

Highlights

The use of fluoride media in zeolite synthesis (J.-L. GUTH) mobilizes T-species in solution at lower pH values. Larger and more perfect crystals can be obtained (note in particular very large crystals of ZSM-5). The method has been extended to the synthesis of aluminophosphate-based materials.

The use of neutral templates, e.g., crown-ethers, enables to control the synthesis of faujasitic (FAU)-type materials. FAU-polytypes can be prepared in a controlled manner as well as ZSM-20 (J.-L. GUTH). However, there are few such examples and this promising area deserves further attention.

With respect to the synthesis of silicoaluminophosphate (SAPO) molecular sieves, it was demonstrated that the availability of a monomeric Al-source (e.g., Al isopropoxide) may be a critical factor (P.A. JACOBS). The latter favors crystallization and the incorporation of Si. It should enable eventually the preparation of tailor-made SAPO materials. At high Si-content, SAPO's contain usually Si domains (substitution of both Al and P by Si). Aluminosilicate patches may even be formed, as observed in the case of SAPO-37. These considerations are of uttermost importance for the preparation of stable and acidic SAPO catalysts.

One of the most interesting challenges in the area of SAPO crystallization is the synthesis of siliceous or aluminosilicate materials isostructural to novel ALPO and SAPO structures. This apply in particular to the synthesis of aluminosilicate analogs of very large pore (18-membered rings) molecular sieves such as MCM-9 or VPI-5.

Questions

Monitoring and controlling during synthesis the Al concentration in both the solid (gel and crystals) and liquid phases is a prerequisite to the crystallization of predetermined aluminosilicate zeolite materials (F. FAJULA). Methods have to be developed for this purpose.

It has been rather well accepted (GENERAL DISCUSSION) that framework secondary building units (SBU) are probably not generic units in solution. However, little is known about the nature of the latter species and the way by which they are mobilized by the directing species (inorganic or organic cations, neutral template molecules,...) during crystal growth. Clearly, this is a research area which could benefit from the well-delineated principles of sol-gel chemistry.

Although it was agreed that zeolite nuclei and seeds were distinct species, and aside the classical definition of a crystallization nucleus, the general consensus was that only a poor knowledge of zeolite nucleation exists today and that this area is worth a great deal of attention. This probably relates to the need for a better understanding of the role(s) played by the inorganic and organic directing agents. Is it correct to believe that the latter species switch nucleation from "no go" to "go" for the formation of a given material whereas they act more "progressively" during crystal growth?

CHARACTERIZATION OF MOLECULAR SIEVES

Progresses

First hand information about molecular sieve structural features is now available by combining techniques (D. VAUGHAN) such as X-ray diffraction using synchrotron radiation (improved data collection, higher resolution, use of smaller crystals), physisorption measurements at very low pressure (ca. 10^{-4} torr) enabling the scaling of pore sizes below 1.5 nm, and NMR spectroscopy measurements using probe molecules (Xe, TMA$^+$, TMP$^+$) (E.G. DEROUANE). Note that all these measurements are affected by the hydration state of the material and the presence of cations. Thus, materials have to be compared and studied in a pre-determined (and achieved!) "equilibrated" state.

Isomorphous substitution induces novel physical and chemical properties which can be adequately monitored by a variety of techniques (J. VEDRINE): XRD, IR, UV, and ESR spectroscopies, calorimetry, etc... One of the major questions addresses the distribution of the added element, i.e., framework T-atom vs. extra-framework species and concentration gradient through the crystal. Checking framework isomorphous substitution appears feasible and rather clear-cut when a trivalent element replaces a tetravalent one (namely, one should not underestimate the potential of ion-exchange capacity measurements). It becomes a difficult problem when the substituting element is tetravalent. Little is known at this moment about isomorphous framework substitution by divalent elements. Clearly, careful material balances and analyses of both the solid, liquid, ... phases involved in the substitution process will enhance our understanding of isomorphous substitution (F.FAJULA).

Proving the latter from XRD measurements of unit cell expansion or contraction is, in most cases, not unambiguous. In particular, one should consider the side effects of the degree of hydration, of the cation(s) content and location(s), and of the presence of defects (GENERAL DISCUSSION). Equally difficult is the determination of Al-siting in zeolites on the basis of XRD data as it appears that the data rationalization needs the consideration of a unit larger than the TO_4 tetrahedron in which Al is sited (A. ALBERTI).

The potential of the solid state ion-exchange method (H. KARGE) has been demonstrated. In contrast to ion-exchange from aqueous solution, ionic motion and substitution involve non-hydrated species. The zeolite truly behaves in such conditions as a solid electrolyte (J.A. RABO).

Highlights

D. VAUGHAN described intergrowths of mazzite and mordenite (coined as "mazmorite") and showed the potential of medium-high resolution transmission electron microscopy and selected area electron diffraction for their characterization. When materials are layered, the observation of the layers stacking may not need supreme resolution equipment.

Effective pore size determination is of primary importance and can be achieved. Quantitative relationships are available which correlate the NMR chemical shift to the degree of confinement of probe molecules (E.G. DEROUANE). Thus 129-Xe enables the measurement of average pore sizes whereas 13-C NMR of TMA$^+$ and TMP$^+$ species quantifies the direct environment of the ion-exchange sites.

The catalytic probing of pore size and connectivity is now well-established. Constraint index (CI) measurements are presently recognized to be generally poor because of their dependence on temperature and other factors... However, the definition of CI was a first and important step in the right direction as it introduced the general concept of pore size catalytic probing. The n-decane and the spaciousness index tests (hydrocracking-hydroisomerization) are highly preferred and their use should be encouraged for the evaluation of 10-membered and 12-membered ring materials, respectively (P.A. JACOBS, J. WEITKAMP). Using n-heptadecane provides a way to probe even larger pores (18-membered rings and mesopores) (P.A. JACOBS). The m-xylene test has the advantage to be a simple and in principle monomolecular reaction. However, the occurence of xylene disproportionation should not be overlooked in large pore zeolite structures.

Low pressure physisorption isotherms have definite potential to scale micropores (D. VAUGHAN). At this moment, however, no satisfactory theory is available to evaluate quantitatively the results. Here is an area for future developments. It was also pointed out that pore size evaluations from nitrogen sorption could show artifacts (F. RAATZ) and suggested that argon might be a better sorption probe than nitrogen. This needs to be clarified in parallel to our definition of pore size.

The determination of the framework Al distribution and gradient in zeolite crystals is not straightforward by any technique. Because such distributions are directly relevant to the application of zeolites in selective sorption and catalysis, the development of methods for such determinations has become of primary importance (J. VEDRINE, A. ALBERTI).

The confinement (surface curvature effects) model (E.G. DEROUANE) has set the basis for a more rigorous description of sorption and diffusion effects in zeolites. However, important questions remain, namely the relative role, importance and/or complementarity of confinement, polarity, and electric field effects. Work is encouraged in these directions.

MOLECULAR DYNAMICS IN ZEOLITES

Progresses

The rationalization of surface curvature (confinement) effects acting on the behavior of molecules sorbed in zeolite frameworks provides a solid base for the understanding of their molecular dynamics (E.G. DEROUANE). Long chain molecules diffuse, when kT is low with respect to their sorption energy, in a "creeping" motion along the zeolite walls, the activation energy of which is about the heat of sorption of a CH_x (x = 2 or 3) group. "Floating" molecules, highly mobile, should be observed in tight pores. Finally, an intrinsic external surface energy barrier opposes, at the external pore rims, the penetration of the molecules in the intracrystalline free volume.

Because of this enhanced (by confinement) sorption energy and of the important activation energy barrier to desorption, the residence time of molecules in zeolites is expected to be rather long, which should favor catalytic events (J.A. RABO). In fact, considering a physisorption energy of about 30 kcal.mol^{-1} (which would be the case for decane in ZSM-5, for example), the residence time of a molecule on a sorption site can be estimated to be about one century at ambient temperature, 10^{-2} s at 600 K, and 10^{-7} s at 1000 K (E. COHEN DE LARA). The latter value is still rather long compared to the lifetime of transition or excited states. This, combined with the fact that sorption energies of reasonable size hydrocarbons in zeolites have values comparable to activation energies for chemical reactions, militates in favor of a synergy between physical and chemical effects for directing intracrystalline catalytic conversion events.

Specific interactions may affect the translation and reorientation dynamics of sorbed molecules. They arise from the presence of cations (E. COHEN DE LARA) and of structural Al (media effects) (J. RABO). When considering molecular dynamics, a distinction must be made between translation and reorientation mobilities. The latter property can be evaluated by the measurement of NMR relaxation times whereas the former necessitates techniques which can probe the self-diffusion of sorbed molecules (e.g., pulsed field gradient NMR) (H. LECHERT). Self-diffusion coefficients may also be obtained by the frequency response method. There is however a danger to measure "corrupted" diffusion coefficients because of bed-depth effects (L.V.C. REES).

Highlights

It appears that the evaluation of self-diffusion measurements at near reaction conditions should be possible using either the frequency response method (L.V.C. REES) or laser-heating pulsed field gradient NMR.

Heats of adsorption should preferably be evaluated by microcalorimetry, in particular by combining thermogravimetry and differential scanning calorimetry (TG-DSC), or from isosteric measurements. Note that the latter method has just been applied to sorption heat measurements on gas mixtures (L.V.C. REES).

Questions

Progresses in the general area of molecular dynamics of sorbed molecules are definitely dependent on a better definition of the effective pore size of molecular sieves and on the careful consideration and evaluation of confinement and media effects. Note in particular that, at this moment, a reliable and quantitative theory describing sorption in micropores (diameter < 1.5 nm) is still missing (GENERAL DISCUSSION).

The relevance of self-diffusion coefficients to describe the behavior of molecules in real catalytic conditions (non-equilibrium situation, presence of counterdiffusion) is not ascertained and needs clarification. This situation is further complicated by the existence of an external surface energy barrier at the pore rim, opposing the penetration of the molecules in the zeolite channels (E.G. DEROUANE).

"Old" and well-accepted observations, e.g., the "window effect", should be reproduced and evaluated in terms of novel concepts (such as surface curvature effects). Other areas which deserve attention are the identification of the role played by co-adsorbates, i.e., the study of competitive adsorption, and the comparison of liquid and gas phase adsorption kinetics and equilibria.

THE ACID FUNCTION OF ZEOLITES

Progresses

The acid activity of zeolite catalysts depends on the nature, concentration, density, and environment of the acid sites (E.G. DEROUANE). It is also effected by properties of the framework itself such as the ability to relax T-O bond lengths and T-O-T bond angles and to readjust globally upon proton "attack" (R. VAN SANTEN, J.DWYER, J.A. RABO). Topology factors also play a role (D. BARTHOMEUF).

Furthermore, the definition of acidity implies the consideration of the protic zeolite as an acid-base pair. The stronger the Brønsted acidity of the zeolite, the weaker the basicity of its conjugated anionic framework. The strength of the (reactant) base must also be accounted for (GENERAL DISCUSSION). Finally, the SHAB model can be used to obtain a better description of the acidic behavior of zeolites (A. CORMA) and intrinsic or cooperative effects may play a role in the stabilization of energetically favored adsorbate configurations.

Caution should be exerted when applying the Hammett H_0 concept to characterize zeolite acidity. This is not only due to steric factors which can perturb the approach and attachment of the base and of the indicator to the acid site. Indeed, zeolites possessing different structures will behave as different solvents in which the sorbed molecules will have different thermodynamic activities (E.G. DEROUANE). Strictly speaking, the H_0 concept can thus only be applied to the comparison of isostructural materials, e.g., series of zeolites dealuminated to different extents.

Highlights

Theoretical (quantum) calculations have decisively impacted on our understanding of the acidic function of zeolites. They have helped the quantification of effects arising from T-O bond length relaxation (R.VAN SANTEN), T-O-T bond angle variation (J. DWYER, J.A. RABO), and isomorphous substitution (J. DWYER). Global effects which maximize the stability of the crystal by resonance also play probably an important role (J.A. RABO).

Measuring acid strength remains a challenge, in particular for systems where the physical properties of the Brønsted sites measured by conventional techniques show little variation (e.g., no change in the O-H IR stretching frequency for zeolites with Si/Al greater than about 5-7). However, the perturbation of such O-H groups by organic base probe molecules leads to frequency shifts which can reveal different types of hydroxyls (J. DWYER, GENERAL DISCUSSION).

Questions

Clearly, extra-framework Al affects zeolite acidity. Investigations are encouraged to identify the nature of such species ((hydroxy)cations, (silica)alumina rings, chains, or debris,...), their location, and their role (J. DWYER, D. BARTHOMEUF).

The nature of reactions intermediates in acid-catalyzed reactions should be asessed. For example, the following species may exist in the reaction of olefins: free carbonium ions, carbonium ions complexed on the acid sites, and even surface alkoxides (J.A. RABO).

In partially ion-exchanged protic zeolites, the effect of the co-cation appears at a certain level of ion-exchange. This effect is not well understood. Ways should be found to investigate possible cooperative effects between cations and protons. Note that for multivalent cations (e.g., Mg^{2+}, La^{3+}), partial hydrolysis of the cation may generate Brønsted acidity (GENERAL DISCUSSION).

It is the consensus that many points relating to zeolite acidity are unresolved at this moment. As new concepts emerge (see above paragraphs and following section), they should be used to rationalize past and new observations and serve as guidelines to improve our fundamental and global understanding of zeolite acidity.

CATALYTIC APPLICATIONS OF ZEOLITES

At the fundamental level, the ultimate goal is to understand basic reaction mechanisms and to utilize this knowledge to characterize catalysts, improve their performance, and propose novel applications.

At this moment, only few zeolite-catalyzed reactions have been characterized in depth (over the last thirty years!...) such as paraffin cracking, hydrocracking and hydroisomerization, and to a lesser extent xylene isomerization (J. WEITKAMP). The ongoing trend is the application of zeolite catalysts to the organic synthesis of fine and specialty chemicals: a field that should be entered with caution and guidance in order to obtain significant results, in view of the great variety of available molecular sieve catalysts and reactions that will be considered.

Guidelines are needed; the following ones are proposed:

1. The thorough understanding of classical organic chemistry mechanisms which, a priori, should also hold for reactions conducted in the intracrystalline free volume of molecular sieves.
2. The evaluation of competitive adsorption effects which may potentially play a role in non-monomolecular reactions.
3. The recourse to molecular modeling, molecular graphics, and molecular dynamics as tools for catalyst selection and product distribution analysis or prediction (E.G. DEROUANE).

NOVEL CONCEPTS

Zeolites behave as solvents. Long range interactions manifest themselves in surface curvature and confinement effects (E.G. DEROUANE) whereas media effects (framework polarity and dielectric constant) are more localized (J.A. RABO). Both types of interactions affect the local (at the catalytic site) concentration of reactant(s) and molecular dynamics within the pores.

"Global" lattice effects are important. In particular, one needs to consider the readjustment of the framework bond structure upon proton attack or departure (J.A.RABO). Zeolite frameworks appear to be rather covalent in nature. Thus, bond length relaxation and conservation of bond order are important factors which impact on the understanding of their acid behavior (R. VAN SANTEN).

Long range effects can be described by theory. The quantum mechanical approach is suitable for the derivation of energy related parameters whereas properties relating to electron density are best evaluated by the density functional theory (W. MORTIER). Potential gradients are needed to observe molecular perturbations, which supports the proposal and observation of energy gradient selectivity (D. BARTHOMEUF). Physisorption energy gradients due to surface curvature effects should be comparable in magnitude to electric field energy gradients (E.G. DEROUANE).

Theory also enables the prediction of zeolite IR spectra as indicated by preliminary studies. It provides a better understanding of their lattice dynamics, local and global (R. VAN SANTEN).

CONCLUSIONS

Zeolite properties are defined by both short and long range interactions. A distinction can be made, although sometime with difficulty, between structure-dependent and structure-independent effects and structure-demanding vs. structure-non-demanding reactions. For this purpose, one has to consider both local and global effects.

Progress in our understanding of the relationship between the physicochemical properties of zeolites and their low dimensionality is dependent on both new concepts and careful evaluation and reconsideration of earlier data. As novel fundamental principles and techniques develop, with the help of new models and theories, guidelines will become available to the chemical designer interested by the mastering and the application of the unique properties of zeolites.

ORIENTATION OF CHEMICAL PROPERTIES BY DIRECT SYNTHESIS OF MOLECULAR SIEVES

D. E. W. Vaughan[+] and R. Szostak[*]

[+]Exxon Research and Engineering Company
Annandale, NJ, U.S.A.

[*]Georgia Institute of Technology
Atlanta, Ga., U.S.A.

SYNTHESIS VARIABLES

Nucleation and Growth in Zeolite

This subject was discussed on the basis of a broad collective experimental experience and not only that of the presented papers (1). The classical concept of a nuclei as a critically sized "species" allowing spontaneous growth in a system by virtue of lowering the free energy of the "solution" - was accepted. This may occur spontaneously in a supersaturated system or by the addition of such material promoters from an external source. The issue of whether the latter are nuclei or seeds was discussed at length, and depending on the specific zeolite or synthesis slurry in question, both mechanisms may be invoked. Nuclei in this case are structureless in the sense that such a source may induce crystal growth of many different structures in several different gel systems, whereas seeds are viewed as being structure specific. They direct the crystallization to only one structure - that of the seed.

Seeded systems, of the type discussed by Fajula for Omega/MAZ, included a discussion of crystal induced growth of X in Y (FAU) gels (2) and LTL in KL gels (3). The phenomenon of slurries of nuclei was discussed specifically with respect to highly basic aluminosilicate solutions originated by Elliot et al (4) which are capable of promoting zeolite growth in numerous kinds of gels. Dust, ash, insoluble residues and other micro-particulates were also noted to be effective nuclei in many chemical systems including zeolite crystallizations. In a thermodynamic sense the spontaneous growth of a crystal on a nucleus occurs in a supersaturated solution; the definition of

Guidelines for Mastering the Properties of Molecular Sieves
Edited by D. Barthomeuf *et al.*
Plenum Press, New York, 1990

381

supersaturation in the heterogeneous zeolite-gel media is difficult to both recognize and to measure. Its nature and necessity continues to be debatable, as does the mechanism of gel restructuring which results in "in situ" crystallization within the gel phase.

The issue of nucleus identification is of obvious interest and clearly will be a future topic of zeolite crystallization research. Various spectroscopies (NMR, IR and light scattering with laser and synchrotron sources) seem to be the most usefully applicable techniques. Analyses "in situ" seem to be necessary requirements.

Control of Crystal Size and Morphology

This is an area of growing interest in synthesis and catalysis research, particularly in those zeolites of catalytic interest without three dimensional pore systems (LTL, MOR, MAZ etc.). Fajula described several specific forms of MAZ, and Guth the specific morphologies of the high silica crown ether directed FAU and BSS (Breck structure formed by the faujasite sheets (5)). Methods of controlling crystal size include:

1. Seeding or nucleation level
2. Stirring or agitation (Collision breeding of nuclei) - rate and type
3. Gel or solution compositions (OH/Si; concentration gradients)
4. Surfactants or other surface specific additives

Crystal shape control is more complex, with gel composition, pressure and temperature having major roles, in addition to surface modifying agents. These effects are often zeolite specific. The extensive experience in several areas of crystallization chemistry is of interest here - such as in the crystallization of waxes, xylenes, urea, NaCl, etc. - and should be utilized.

The question of whether mechanisms of screw growth are observed in zeolite crystals was discussed without conclusions. Apparent screw forms have been observed in synthetic and natural zeolites but unequivocal identification is unconfirmed. The observation of growth steps at sheet edges is common, and sheet growth is well defined.

The role of fluoride as an effective mineralizer in geochemical experimentation is well known, but the results of Guth in zeolite syntheses are spectacular in some cases, particularly at low pH. These experiments are clearly a stimulus to future research, offering important compositional control under pH conditions where transition metal and ammonium cations are stable - an area of meager crystallization experience. The presence of less stable F⁻ (vs OH⁻) promotes the incorporation of hetero-atoms (Ga, Fe, Al) into MFI structures, and may increase crystal perfection. Such an approach overlaps the secondary synthesis modes presently in vogue, which have yielded major T atom incorporation into MFI and MEL structures.

The Nature of Templates

Experience with templates was discussed, with an observation of "primary" and "secondary" template roles. Alkali cations generally take on the former mode and alkyl ammonium or amines the latter role (6). Cations, molecules or salt pairs may behave as templates, and at this conference the use of crown ethers are a novel addition to the last. The Na^+ form of 15-crown-5 is specific for the crystallization of high silica FAU, and 18-crown-6 for the crystallization of the hexagonal ABAB stacked form of FAU sheets, often referred to as "Breck 6" or BSS - both at Si/Al ratios of 5. It was noted that other crown complexes yielded no products. The major activity in AlPO₄-SAPO-MeAPO syntheses field has stimulated more work on amines for syntheses in the aluminosilicate field, but with the exception of the crown-ether work, no new (organic) templates were reported.

The role of templates as structure directing agent is variable. They seem to have distinctive and possibly separate roles in both nucleation and crystal growth. In the described syntheses of MAZ, TMA is needed to form a nucleus but is not necessary for propagation of the MAZ structure. In other cases the reverse is true, and a third case may require the template for nucleation and propagation. There are several instances where a template was initially thought to be necessary (ZSM-5, VPI-5/HI) but later syntheses show it not to be the case. Whether template incorporation improves crystal perfection in such systems has not yet been established, although it has been demonstrated for GME using DABCO polymers (7).

ALUMINOPHOSPHATE BASED MATERIALS

Si Distribution in SAPOs

The addition of SiO_2 into the microporous aluminophosphate structures can be visualized by three substitutional mechanisms, the first two are methods of charge introduction and hence a potential way of introducing catalytic activity.

- o   SM1 is the substitution of Si for Al, generating a positive framework charge.

- o   SMII is the replacement of P by Si, giving rise to an anionic charged framework with the potential for a classic Bronsted acid site.

- o   SMIII represents the replacement of both Al and P by Si, leaving the neutral framework intact.

Acid catalytic behavior is known in the SAPO materials, but continuity between such activity and the bulk SiO_2 content of the material, in addition to an observed dependence on structure, still remains to be identified. Systematic examination of the SAPO synthesis parameters reported by Jacobs and coworkers primarily confirms the operation of the SMII and III mechanisms. Significant levels of SiO_2 substitution into framework structural sites is related to the degree of polymerization present in the starting materials as well as the pH of the crystallizing gel. The use of monomeric organosilicates and aluminates leads to greatly increased levels of Si incorporation compared to SAPOs prepared from the highly polymeric Ludox(R) and pseudobohemite starting materials originally used by the Union Carbide group.

The most revealing aspect of this study has been the identification of "silicate islands" within the SAPO crystals which develop with increasing Si incorporation. Acidic environments commonly used in $AlPO_4$ syntheses promote the incorporation of Si into $AlPO_4$ via SMII and SMIII. It was pointed out in the discussion that the minimum size of such Si patches must be five Si Td units based on the low probability of locating a Si adjoining a P as proposed in Flanigen's rules for molecular sieve structures (8). The generation of Bronsted acidity by back substitution of Al into the silicate islands was shown to be controlled by basic pH synthesis conditions akin the those found in conventional zeolite syntheses. Catalytic studies provided additional insight into the source of variable acid activity observed in the SAPO materials.

This is the first study to differentiate definitively the links between acidity and specific Si distributions and locations in the SAPOs. It illustrated the high degree of activity and selectivity attainable through control of synthesis variables and mode of Si incorporation into the structures.

18-Ring Structures

$AlPO_4$ syntheses by d'Yvoire (9) identified several new $AlPO_4$ structures based only on their unique X-ray diffraction patterns. These materials, prepared at low temperatures ($100^{\circ}C$) without amines, included two materials of recent interest, H1 and H3. Patents filed by workers at Mobil in 1985 (10) claimed the preparation of Si containing $AlPO_4$ with X-ray diffraction patterns related to those reported for these two species of d'Yvoire. The Mobil phases (MCM-1 and MCM-9) are compositionally unique as they contain Si, but appear structurally similar, based on the comparison of the diffraction patterns, to H3 and H1 respectively. More recently Davis and coworkers announced the preparation of the $AlPO_4$ called VPI-5 (11). This material exhibits a unique large molecule adsorption ability, equating it with a theoretical 18-membered ring structure. It too has an X-ray diffraction pattern similar to H1, although it was prepared in the presence of an organic amine. Possible structural inter-relationship between the materials called H1, VPI-5 and MCM-9, are unclear at the present time as at least 18 theoretical large channel structures are reported to have similar X-ray diffraction patterns to those for these materials. If all three are indeed identical materials topologically, then the H1 designation takes priority for the $AlPO_4$ forms, with possible DPA-H1 or TBA-H1 and DPA-(Si)H1 more appropriate names than VP1-5 and MCM-9, respectively, in the template containing structures (i.e. prior to removing the template). H3 would thus take structural priority over MCM-1. VPI-5 as well as MCM-1 and 9 would seem to be superfluous designations for these materials, if indeed they are structural similar and certainly in view of the obvious general ability of Si to substitute into $ALPO_4$ structures. Until such evidence is obtained, VPI-5 and MCM-9 will continue to be used.

The problem of zeolite misidentification and multiple names for related materials is a general one, reflecting inadequate generally available data bases for X-ray diffraction patterns, poor literature searching on the part of research workers, complexities and confusion in nomenclature in the patent literature, and the "cussedness" of zeolite structures, which do not follow any general trends for T atom substitutions because of the openness and flexibility of the frameworks. There is a clear need for the Zeolite community to work together either to upgrade the JCPDS zeolite files, or to publish a comprehensive independent zeolite database with frequent updates and promote guidelines for naming new materials.

REFERENCES

1. R.M. Barrer, "Hydrothermal Chemistry of Zeolites", Ch.4, Academic Press (1982)

2. H. Kacirek and H. Lechert, J. Phys. Chem., 79; 1589 (1975); ibid, 80; 1291 (1976)

3. D.E.W. Vaughan, Eur. Pat. Appl., 0132347 (1985)

4. C.H. Elliott, J. and C.V. McDaniel, U.S. Patent 3,639,099 (1972)

5. D.W. Breck "Zeolite Molecular Sieves", J. Wiley/Krieger, 58 (1984)

6. D.E.W. Vaughan, Mater. Res. Soc. Symp. Ser. 111: 89, Ed. M.M.J. Treacy, J.M. Thomas, and J.M. White (1988)

7. R.H. Daniels, G.T. Kerr and L.D. Rollmann, J. Amer. Chem. Soc., 100: 3097 (1978)

8. E.M. Flanigen, R.L. Patton and S.T. Wilson, "Innovation in Zeolite Materials Sciences", SSSC #37:13, Ed. P.J. Grobet, W.J. Mortier, E.F. Vansant and G. Schultz-Ekloff (1988)

9. F. d'Yvoire, Bull. Soc. Chim. France , 1762 (1961)

10. E.G.Derouane and R. von Ballmoos, Eur. Pat. Appl. 0146389 (1984)

11. M.E. Davis, J.M. Garces, C.H. Saldarriga and MdC. Montes de Correa, Intl. Pat. WO 89/01912 (1989)

CHARACTERIZATION OF STRUCTURAL AND PHYSICO–CHEMICAL PROPERTIES OF ZEOLITIC SYSTEMS

H.G. Karge[1] and J.B. Nagy[2]

[1] Fritz-Haber-Institut der Max-Planck-Gesellschaft
Faradayweg 4-6, 1000 Berlin 33 (West)
[2] Facultes Universitaires Notre-Dame de la Paix, Dept. de Chimie
Rue de Bruxelles 61, 8-5000 Namur, Belgium

INTRODUCTION

Zeolite characterization covers the fields of chemical composition, structural identification, the distribution of tetrahedral (T) atoms, the acidity and basicity linked to the framework, and extra-framework species, the post synthesis-modifications and the presence and nature of defect groups which can lead to a secondary pore structure.

CHEMICAL COMPOSITION

The determination of chemical composition is of paramount importance in understanding the physicochemical and catalytic properties of zeolites. Since, frequently, a concentration gradient is observed for the as-synthesized or hydrothermally treated microporous materials, both bulk and surface analysis techniques have to be used to characterize the samples.

Atomic absorption spectroscopy (AAS) and induced coupled plasma (ICP) are regarded as the best techniques, even for high Si/Al ratio (>120) materials [1-2]. In addition, proton-induced γ-ray emission [3] also yields accurate results for the same concentration range. Moreover, this technique offers the advantage of simultaneous multi-element analysis. For low Si/Al ratio samples, a variety of other techniques are available. The wet chemical technique has inherently larger errors in the Al determination [1].

X-ray fluorescence (XRF) measurements are less accurate in the Al and Si determinations. The high-resolution solid-state ^{29}Si-NMR data are reliable for low Si/Al samples, when distinct silicon NMR lines are detected for Si atoms surrounded by four, three, two, one or zero Al atoms in their second coordination sphere. However, crystallographic imperfections lead to line broadening, decreasing the resolution of the ^{29}Si-NMR spectra. In addition, in the case of multiple crystallographic sites, the assignment of the NMR lines has to be confirmed by measurements on highly siliceous samples. The solid state ^{27}Al-NMR technique requires the use of short pulses. Even so, some of the aluminum present in the

Guidelines for Mastering the Properties of Molecular Sieves
Edited by D. Barthomeuf *et al.*
Plenum Press, New York, 1990

sample (AAS results) still remains undetected. A method is proposed based on the complexation of extra-framework Al by pentadi-2,4-one. Accurate results have been obtained on Y and ZSM-5 zeolites [4].

The early surface analysis was made using XPS of powder samples. However, in this case, only average results are obtained and an apparent surface Al enrichment could stem from a mixture of highly aluminous and highly siliceous crystallites. Microprobe techniques are recommended for the determination of concentration gradients using combined bulk and surface techniques. Secondary ionization mass spectrometry (SIMS) was previously used, but it is complicated by the fluctuating charging effect.

Fast atom bombardment mass spectrometry (FABMS) avoids this difficulty and is recommended for future work. The use of laser microprobe mass spectrometry (LMMS) does not seem to have solved many problems in the field of zeolite analysis. A proposition is made to explore the possibilities of this technique, when a YAG laser is available.

STRUCTURAL DETERMINATION

The second section of the general discussion was concerned with structural determination. Suitability, advantages and drawbacks of techniques such as XRD, electron diffraction (ED), neutron diffraction (ND), tunneling electron microscopy, EXAFS and IR were dealt with. XRD was addressed as still being the most powerful tool for structural analysis. Use of synchrotron radiation, even though not yet employed extensively, is recommended as a means to improve experimental data due to its better intensity. However, there is complete agreement about the need to have, as a basis, carefully measured experimental data. These data strongly depend on experimental conditions, preparation of the samples, etc. On the basis of such good experimental data, refinement (Rietveld) will provide reliable structural information. It was pointed out that structural analysis carried out in this way leads to a "theoretical" pattern (Si form) but does not give information about structural modifications due to the presence of cations, water, etc.

Electron diffraction and neutron scattering are judged to be valuable complementary experimental tools in structure investigations, the latter being particularly relevant for determinations of H and D.

Progress was mentioned with respect to the problem of determination of structural changes upon adsorption of molecules into zeolite systems (e.g. ZSM 5 / ZSM II [5]). Also in this field NMR and neutron diffraction might be advantageously applied. Examples were mentioned where the presence of a template changes structural properties. Structural changes occur as a function of hydration as well. This in turn may depend on the crystal size. Similarly, it was mentioned that adsorption of molecules such as methane, acetylene, etc. changed the zeolite structure. This was revealed by careful analysis of neutron scattering data.

Since there is considerable interest in structural changes upon sorption of various species into the zeolite pore system, or as a function of temperature, the application of microcalorimetry and NMR were suggested to gain more insights into these phenomena.

Structural analysis of zeolites can be supported by IR investigations. IR spectroscopy provides information about structural building elements such as rings but is also able to help in identifying zeolite structures through the occurrence of specific bands. Observed bands can be compared with the calculated ones, which enables us to confirm or refute model assumptions for unknown structures. However, it was pointed out that problems frequently arise due to insufficient intensities of the respective bands. Provided that these IR structure investigations are carried out on self-supported wafers instead of KBr pellets, adsorption of probe molecules can yield additional information. EXAFS as a tool for structural analysis was considered to be less helpful. In particular, data analysis usually has to be achieved by Fourier transformation, which turns out to be critical. It was the experience of several experts that differences in methods of data evaluation gave significantly different results.

To date there seems to be no experience with the use of tunneling electron microscopy as a technique for evaluation of surface topology, despite the fact that it could be employed even for investigation of insulators such as zeolites via either rendering them surface-conductive (metal evaporation upon zeolite surfaces) or using force instead of current as the measuring variable.

With respect to pore size determinations, it should be kept in mind that the term "pore size" is by no means unequivocal. One must be aware of the fact that pore sizes usually do not coincide with those derived from XRD measurements, involving assumptions about the diameters of the pore-forming atoms, their rigidity, etc.

The use of inert-gas adsorption at very low pressures (Omicron technique) as a tool for pore-size investigation was extensively discussed; the question was raised as to whether N_2 should be a good adsorbate molecule since it would interact significantly with cations. Moreover, a review of the results suggests that they depend to a large extent on the model used for the evaluation of the pore-size distribution. Entropy effects may be important but are frequently disregarded.

Even though there was agreement about the problems in obtaining absolute figures for pore size and pore-size distribution, it was stressed that the technique is already useful as a means for ranking different materials with respect to their porous properties, in particular if those materials belong to homologous series of porous structures which were systematically modified. Moreover, N_2 might be used at least as a probe, e.g. to shed light on the effect of cations present in the adsorbent. It is hoped that confidence in the results obtained via low-pressure adsorption experiments will be enhanced when the method is more widely used and more data are collected. The method, at the present state of the art, is essentially a tool for characterization of the pore penetration but it does not differentiate between channels and cavities. Also it does not indicate absence or presence of intergrowth phases. Mesopores were either not detected or the results seem to depend on the nature of the probe (N_2 vs. O_2).

Reference was made to the successful application of [129] Xe NMR which provides data on pore and cavity size and their modifications, for instance as a result of introduction of metal particles or coke.

With respect to catalytic measurements [6], which were seen as methods of pore-size screening, most of the participants agreed that the "constraint index" has severe limitations. The introduction of this index had the merit of advancing a basically new idea of a way to characterize pore sizes of unknown zeolitic structures. However, the conditions under which the index has to be determined vary in an inacceptable manner from zeolite to zeolite, depending on the particular catalytic properties, e.g. acidity and accessibility of sites. Especially, the temperatures at which the measurements have to be carried out vary considerably, typically between 250 and 500°C.

In contrast, the estimation of pore sizes and their ranking from experiments with isomerisation and hydrocracking of long-chain paraffins (n-decane, n-undecane, n-tetradecane) seems to be much more reliable. The most appropriate chain length in the range between C_8H_{18} and $C_{17}H_{36}$ has to be chosen by the experimentalist according to his specific requirements. Today, this technique is considered to be most suitable for estimating medium pore sizes (10-membered ring zeolites).

The hydroconversion of n-heptadecane can be favourably used for estimating the size of mesopores. Also, the pore size of Si-VPI-5 has been characterized by this method.

The "spaciousness index", $Y(i-C_4) / Y(n-C_4)$, in hydrocracking of a C_{10}-naphthene can be elegantly and successfully employed in characterization of larger pores, i.e. 12-ring channels etc. [7, 8]. However, the isomerization of xylenes and their disproportionation was extended for the investigation of 12-membered ring structures as well. Disproportionation reactions, being bimolecular in character, have proven to be extremely valuable for pore-size characterization.

In general, one may conclude that a pronounced preference for one or the other probe reaction does not exist. Their successful application will ultimately depend on the data base available to the investigators.

The most promising trends in developing tools for characterization of structural properties of zeolites were addressed :

(i)   further improvements in IR spectroscopy, combined with theoretical calculations (in particular for high-silica zeolites);

(ii)   further elaboration of "indices" obtained by suitable catalytic reactions;

(iii)   adsorption of probes such as N_2 but also Ar, etc. at very low pressures, including refinement of adsorption models.

The distribution of Al atoms in zeolites greatly influences the direct environment of Al atoms in the framework and hence the Brønsted acidity linked to it.

The XRD techniques previously proposed for the determination of Al atoms should be abandoned. Indeed it is shown [9] that several factors influence the determination of Al-content in T-sites of the zeolites. These are :

a)   the variation of T-O distances on bonding forces;

b)   the variation of the T-O distances with the disorder in Si-Al distribution;

c)   the variation of the T-O distances on temperature and

d)   the apparent shortening of T-O distances when the apparent symmetry is higher than the true symmetry.

While the influence of the first three factors can be reduced, coping with the fourth effect is more difficult.

The neutron diffraction (ND) technique is recommended, by which an unambiguous distinction between Si and Al atoms can be made. It is also suggested that EDAX-STEM measurements can lead to information about the Al distribution in microcrystals.

The ^{29}Si-NMR data are always linked to modelling. Complications arise from the presence of several crystallographic sites, and the proposed models have to take into account the cross-terms for the Al distribution [10]. In special cases different ^{27}Al-NMR lines are also detected for the different crystallographic sites and the ^{27}Al-NMR data complement quite well the ^{29}Si-NMR [11]. It has to be emphasized that the NMR methods are only suitable for low Si/Al ratio samples.

The ^{29}Si-NMR analysis is also very helpful in the characterization of SAPOs. It is shown that at low Si content, Si replaces P in the $AlPO_4$-5, $AlPO_4$-11 and $AlPO_4$-37 structure (faujasite) to yield SAPO-5, SAPO-11 and SAPO-37. For higher Si content, silicon islands are formed, where the insertion of Al leads to zeolite material of SAPO structure [12].

ACIDITY - BASICITY

Because of the tremendous importance of zeolites as catalysts in acid-base catalysed reactions their characterization with respect to acidic and basic properties is a very substantial issue. Basically, one must distinguish between the nature, number and strength of the respective active sites. The techniques available for determination of acidity or basicity of zeolite catalysts have to be carefully selected and critically evaluated with respect to their applicability and reliability in characterising one or the other above-mentioned properties. Principal techniques, presently used, are

- titration in an aprotic solution using Hammett and/or aryl indicators
- thermochemical titration
- IR spectroscopy with and without probe molecules
- temperature-programmed desorption of suitable probe molecules
- microcalorimetry of specific heats of adsorption evolved upon adsorption of appropriate acidic or basic adsorbates
- esr spectroscopy of probes adsorbed on the acidic or basic catalyst under investigation
- high-resolution NMR of ^{1}H of the OH groups
- ^{13}C-, ^{15}N - or ^{31}P NMR of respective probes
- UV-visible spectroscopy
- test reactions
- shift of v (OH) of acidic hydroxyls

This list is by no means exhaustive and, because of the importance of the subject, modifications of the already existing techniques or even new ones are likely to appear.

While methods for identification of acidic or basic sites as well as those determining their density are well developed and widely accepted, there is still a need for techniques which enable us to characterize the strength of acidic or basic sites in a satisfactory way. Cases in which the classical titration method with indicators can be applied with some confidence are relatively rare. The situation is complicated by the fact that the term "acidity strength" is basically difficult to define. Strength of acidic (or basic) sites obviously may be affected by environmental effects such as electric fields, confinement etc. However, the pertinent techniques such as TPD, microcalorimetry (both advantageously combined with IR), and shift of OH bands of acidic hydroxyls upon adsorption of probes such as benzene should give at least a ranking of the acidic strength in homologous series of zeolite catalysts. Such series may be prepared, for instance, by systematic modification, via ion exchange, appropriate syntheses, dealumination etc.

It is widely agreed that, in using probe molecules, such species which exert a minimum of perturbation on the system are preferred. Better results might be obtained when the (Brønsted) sites are not deprotonated, but the protons exhibit weaker interaction with the probe. Suggestions were therefore made to use weak bases (H_2, paraffins, olefins) more extensively than strong ones (NH_3, pyridine) to probe the strength of acidic sites.

MODIFICATION OF ZEOLITES

The main modifications include dealumination, isomorphous substitution (using e.g. $SiCl_4$, BCl_3, $TiCl_4$...), ion exchange, impregnation and metal particle formation via reduction of incorporated cations.

Dealuminated samples have to be characterized by precise chemical composition, the structure, the T-distribution and the acid-base properties. Very often defect groups are found and the eventual secondary pore structure has to be characterized [see below].

The isomorphous substitution of T-atoms by B,Si,Ti, etc. leads to interesting catalytic properties. It is suggested that Ti is incorporated into silanol nests. For the Ti-atom identification as a framework tetrahedral T-atom, difficulties arise because the usual analytical techniques such as IR, unit cell determination via XRD, 29Si NMR do not lead to unambigious conclusions. The incorporation of B during impregnation and heating is also favoured by the presence of defect groups.

Ion exchange to introduce NH_4^+, Cs^+ or Ag^+ cations is emphasized to determine quantitatively the framework M(III)-T atoms.

The reduction of extra-framework cations leads to the formation of metal particles which can occupy, for instance, the supercages of faujasites or secondary pore structures. Care should be taken to determine the nature, size and location of the metal clusters. ^{129}Xe-NMR is proposed for the determination of the location of the particles. However, interpretation of the data proves to be difficult since a rather high additional chemical shift of the atom is observed due to its direct interaction with the metal particle. The extrapolated chemical shift to zero xenon pressure (δ_s) should be extracted to determine the size of the cavities occupied by the xenon atoms.

CHARACTERIZATION OF DEFECT GROUPS AND THE FORMATION OF SECONDARY PORE STRUCTURE

It has been observed on the as-made materials that a rather large number of defect groups are formed during the hydrothermal synthesis of zeolites under alkaline conditions. IR, ^{29}Si- and ^{1}H-NMR unambiguously show the presence of defect silanol groups. Silanol nests (four neighboring SiOH groups) are also found if one T-atom is missing in the structure. This model could account for the presence of some 32 defect groups per unit cell in the silicalite framework, where eight T-atoms are missing [13]. A similar explanation has been given for the large number of defects determined in the ZSM-48 structure, which was synthesized in the presence of hexamethylene-bis-trimethyl-ammonium ions, $((CH_3)_3N^+-(CH_2)_6-N^+(CH_3)_3)$ [14].

In most cases the defect groups are eliminated during calcination by healing the aluminosilicate framework. However, defect groups are also formed during dealumination by acid or hydrothermal treatments.

In addition, a secondary pore structure can be formed, the importance of which is recognized in catalysis. The pore distribution in dealuminated Y zeolites was studied using N_2 adsorption. It is proposed that ^{129}Xe-NMR of adsorbed xenon could adequately characterize the pores up to an average diameter of ca 50 Å. The formation of secondary pores introduces even larger changes in monodimensional (non interconnected) systems, where the interconnections of the channels lead to bi- and tridimensional networks [15].

REFERENCES

[1] D.R. Corbin, B.F. Burgess, Jr., A.J. Vega and R.D. Farlee, *Anal. Chem.* **59** (1987) 2722.

[2] K.J. Chao, S.H. Chen and M.H. Yang, *Fresenius Z. Anal.Chem.* **331** (1988) 418.

[3] G. Debras, E.G. Derouane, J.-P. Gilson, Z. Gabelica and G. Demortier, *Zeolites* **3** (1983) 37.

[4] P.J. Grobet, H. Geerts, J.A. Martens and P.A. Jacobs, *J. Chem. Soc., Chem. Commun.* **1987**, 1688.

[5] C.A. Fyfe, H. Strobl, G.T. Kokotailo, G.J. Kennedy and G.E. Barlow, *J. Am. Chem. Soc.* **110** (1988) 3373.

[6] P.A. Jacobs and J.A. Martens, in: "New Developments in Zeolite Science and Technology"; Proc. 7th Int. Zeolite Conference, Tokyo, August 17-22, 1986 (Y. Murakami, A. Iijima and J.W. Ward, Eds.) Kodansha, Tokyo; Elsevier, Amsterdam, **1986**, pp. 23-32.

[7] J. Weitkamp and St. Ernst, *Stud. Surf. Sci. Catal.* **38** (1987) 367 (Proc. 10th North American Meeting of the Catalysis Society, San Diego, CA, USA, May 17-22, 1987 (J.W. Ward, Ed.) Elsevier Amsterdam, 1987, pp. 367-382).

[8] J. Weitkamp, St. Ernst and C. Chen, in: Proc. 8th Int. Zeolite Conference, Amsterdam, The Netherlands, July 10-14, 1988, in press.

[9] A. Alberti and G. Gottardi, *Z. Kristallogr.* **184** (1988) 49.

[10] F. Raatz, J.C. Roussel, R. Cantiani, G. Ferré and J.B. Nagy, *Stud. Surf. Sci. Catal.* **37** (1988) 301.

[11] P. Massiani, F. Fajula, F. Figueras and J. Sanz, *Zeolites* **8** (1988) 332.

[12] J.A. Martens, M. Mertens, P.J. Grobet and P.A. Jacobs, *Stud. Surf. Sci. Catal.* **37** (1988) 97.

[13] J.B. Nagy, P. Bodart, H. Collette, J. El Hage-Al Asswad, Z. Gabelica, R. Aiello, A. Nastro and C. Pellegrino, *Zeolites* **8** (1988) 209.

[14] G. Giordano, J.B. Nagy, E.G. Derouane, N. Dewaele and Z. Gabelica, in: Proc. Symposium on Advances in Zeolite Synthesis, Los Angeles, **1988**, ACS Symposium Series, in press.

[15] J. Lynch, F. Raatz and P. Dufresne, *Zeolites* **7** (1987) 333.

STATIC AND DYNAMICS PARAMETERS IN ADSORPTION AND CATALYSIS IN ZEOLITES

L.V.C Rees[1] and Michael S. Spencer[2]

[1]Imperial Cellege of Science, Technology and Medicine
London, SW7 2AZ, England
[2]School of Chemistry and Applied Chemistry
University of Wales, Great Britain

INTRODUCTION

The discussion session was divided into four sections a) the techniques of measuring diffusion b) the significance of the results obtained by these various techniques and two corresponding sections on techniques for measuring adsorption phenomena and the significance of the results obtained.

DIFFUSION

Techniques: In the section devoted to the techniques of measuring diffusion the following methods were presented and discussed in turn.

(a) Pulsed field gradient NMR: This technique studies the motion of sorbate molecules located in the channels and cavities of zeolites under sorption equilibrium conditions[1,2]. The results obtained are not affected, therefore, by surface barrier effects. The method can only follow fast diffusing species, usually hydrocarbons because of their proton content. The lower limit of the diffusion coefficient which can be measured is $5 \times 10^{-14} m^2 s^{-1}$.

(b) Tracer desorption NMR: The rate of desorption of hydrocarbons from the sorbed phase into the gas phase is followed[2]. The diffusion coefficient is calculated from these desorption rates by the use of the appropriate solution of Fick's 2nd law of diffusion.

Guidelines for Mastering the Properties of Molecular Sieves
Edited by D. Barthomeuf *et al.*
Plenum Press, New York, 1990

(c) NMR Line Shape Analysis: The shape of the NMR signal is dependent
on the motion of the sorbate molecules within the zeolite channels and
cavities. Diffusion coefficients have been reported from the analysis of
these NMR spectra[3].

(d) Frequency-response[4]: In this technique sorption equilibrium is
established then the volume of the gas phase is subjected to a square-
wave modulation of ± 1%. The pressure response to such modulations is
followed. When the complete analysis is carried out square-waves over a
frequency range of 0.01 to 10 Hz are covered. However, the time constant
of the apparatus is ~ 25ms and single square-waves in the lower frequency
range can each be analysed using the appropriate solution of Fick's 2nd
diffusion law. In the complete analysis a Fourier transformation of the
volume and pressure "square-waves" is carried out and the phase angle
difference between the fundamental sine-waves is determined from which
the diffusion coefficient can be calculated[5]. Diffusion coefficients
for any type of sorbate species can be obtained over a wide range of
values at temperatures up to those which are required for catalytic
reactions. However large crystals may be required for these high
temperature studies.

(e) Sorption/Desorption rates: Analysis of these rates using the
appropriate solution of Fick's 2nd law for the boundary conditions which
exist in the system gives a diffusion coefficient which is an integral
value over the concentration range covered in the experiment. Time
constants as low as ~ 100ms are now obtainable by this method by the
application of valve corrections, etc[6]. This technique has shown that
heats of adsorption and bed-depth, mass transport effects (i.e. inter-
crystalline diffusion effects) are frequently the reasons for the large
discrepancies found between diffusion coefficients obtained by this method
and the self-diffusion coefficients obtained by pulsed field gradient NMR.

(f) Zero column length chromatography[7]: Sorption equilibrium is
established in a very shallow bed of zeolite crystals by flowing sorbate
over the sample in an inert carrier gas (eg. Ar). The desorption of the
sorbate is then followed by flowing a rapid stream of the pure carrier gas
over the zeolite crystals and analysing the sorbate concentration in the
gas stream using a flame-ionisation detector. The diffusion coefficient
is obtained from the application of the appropriate solution of Fick's
2nd law. This diffusion coefficient is an averaged value over the
concentration range followed in the desorption experiment.

(g) Membrane method[8]: A large single crystal of zeolite is embedded into a resin which is then microtomed on both sides to expose the zeolite crystal on both sides of the membrane. Sorbate is flowed across the ingoing surface of the activited crystal until steady-state conditions are obtained. The permeation rate of sorbate through the zeolite membrane is measured by analysing the sorbate concentration at the outgoing face as a function of time. The diffusion coefficient for the zeolite/ sorbate system under consideration can be determined from this permeation rate. The method assumes no permeation of sorbate around resin/zeolite crystal interface; no permanent sorption of solvent from the resin in the zeolite and, finally, no damage to the crystal surfaces on microtoming.

(h) In the lecture by Cohen De Lara at this meeting IR line shape analysis and Inelastic Neutron Scattering were used to determine the mobility of sorbate molecules in the zeolite while, in the lecture by LECHERT et al, NMR T_1 and T_2 relaxation times were shown to follow sorbate motions and to give energies of activation for transport of these sorbates in the zeolite channels. These techniques and results obtained are fully described in the respective papers.

Significance of results:- Several problems and comments were raised during the discussion which followed the presentation of these various techniques.

(i) How can one decide which technique is the best for a given sorbate/ sorbent system? All methods have limitations in their use. For example, the range of values of diffusion coefficient which can be determined; the range of temperature[5] which can be used in the apparatus; the type of diffusing sorbate which can be detected. Thus, careful analysis of these limitations is needed for each case.

(ii) Discrepancies (up to 5 orders of magnitude) in the values of the diffusion coefficients have been found in the use of different techniques with the same system (and in some cases with the same samples). There may be several reasons for these discrepancies of differing importance in different systems. Some techniques (eg (a) and (c)) follow movement of sorbate molecules within the zeolite channels and cavities only whereas, in others, (eg (b), (d), (e), (f), and (g),) the sorbate passes from the gas phase into and/or out of the zeolite crystal. A lack of knowledge of the exact structure of the zeolite surface prevents any detailed analysis of boundary effects. In some cases interference by inter-crystalline diffusion can seriously affect the diffusion coefficient determined. Non-equilibrum conditions may, also, give spurious values of D.

(iii) Diffusion data are frequently required under conditions where catalysis occurs. The main limitation here is that of temperature. In general, NMR methods are limited to temperatures not much above ambient by the sensitivity of the electronics in the NMR probe. A possible solution to this problem is the use of laser heating but control of sample temperature could be difficult. The frequency-response method has been shown to be potentially capable of use over a wide temperature range and it looks especially useful for measurements related to catalysis.

(iv) A further important problem, also, related to catalysis and some separation processes, is the measurement of diffusion in mixtures of sorbates and of counter-diffusion. Two techniques have potential (a) (initial work by H.G. Karge) time resolved FTIR of gas and sorbed phases; (b) proton NMR with mixtures of ^{1}H and ^{2}H substituted compounds.

(v) Diffusion from the liquid phase into the zeolite phase has been little studied. The few, reliable, existing results show some anomalies which have not been resolved. More work, both experimental and theoretical (eg on surface energy barriers) is needed.

One or two specific points were raised in the discussion. Lechert stated that NMR line shape analysis does not allow one to determine a diffusion coefficient (which disagrees with conclusions of ref.3). One is only observing motions of sorbate at a sorption site and not motions from one site to another. The mobility of Bronsted acid protons increases by three orders of magnitude with the introduction of only trace amounts of water. One needs to know all of the products which arise in a zeolite cavity during the course of a catalysed reaction in order to design or choose the zeolite which allows the required products to diffuse out into the gas phase.

ADSORPTION

In the Adsorption section only physical adsorption was considered and the following methods were presented for discussion.

Techniques:- (a) Gravimetric: Accurate isotherms can readily be measured using vacuum micro-balances which are now capable of weighing down to 10^{-7}g the mass of sorbate taken up by a zeolite at each equilibrium pressure. It is now possible with the latest instrumentation to measure heats of adsorption simultaneously with these gravimetric measurements using a combined differential scanning calorimeter/gravimetric cell.

(b) Volumetric: Isotherms can be measured accurately with simple, cheap vacuum systems by determining the quantity of gas admitted to the sorbent and the quantity of gas remaining in the gas phase after equilibrium. The difference gives the amount adsorbed at the final equilibrium pressure. More automated, commercial units based on this method are readily available mainly for determining surface areas. These units can be quite expensive, however.

(c) Isosteric method[9]: If a large mass of zeolite is used (~ 20g) with the minimum of free gas volume over the sorbent phase then on admitting a dose of sorbate some 97% of the dose is adsorbed. The method is ideal for studying binary or more complicated mixture isotherms. When equilibrium is established the equilibrium pressure and composition of the gas phase (if it is a mixture) is determined with an on-line mass-spectrometer. The temperature is then raised/lowered over (say) 10 increments of 5°C and the new equilibrium pressures and the gas phase compositions determined as above. It is assumed that the composition of the sorbed phase stays sensibly constant during these temperature changes at the composition of the original mixture. This assumption has been shown to be reasonable since, because of the design of the apparatus, some 97% of the original dose remains in the sorbed phase over the complete temperature range scanned. From these ten measurements an isostere can be obtained for the specific sorbate loading and composition. Isoteres are then determined at other sorbate loadings and sorbed phase compositions. From these isosteres the enthalpy of adsorption can be calculated from the slopes and isotherms can be calculated at any temperature within the range scanned by the isosteres and if the isosteres are well-behaved and linear, this temperature range can be extended by extrapolation of the isosteres.

(d) Liquid phase: Adsorption of pure liquid mixtures and mixtures dissolved in large excess of solvent can be measured. The standard procedure is to determine the amount and composition of the sorbed phase by analysis of the liquid phase before and after contact with the zeolite phase. Gas phase chromatography is the technique mainly used for the analysis of these liquid phases.

Significance of results: Adsorption measurements with zeolite substrates are carried out predominately for one or more of three reasons;

(i) As a means of characterising the internal structure of zeolites. This includes the topology of the internal surface (eg cage size) and its chemistry from the interaction energy with various sorbates.

(ii) For the investigation of the fundamentals of adsorption on well-defined micro-porous solids. Zeolites are, probably, the most regular and best characterised micro-porous solids for these studies. Zeolite surfaces can also, be readily modified by change in Si/Al ratios, ion-exchange, etc.

(iii) Generation of data for the design of industrial processes. Zeolites are used mainly in separation processes involving both liquid and gaseous feeds and in catalytic processes. Most separation processes are based on differences in the interaction energies between the components of the mixture and the zeolite channel and cage surfaces. Sometimes separations are based on different diffusion rates and in the extreme on complete molecular sieving effects, i.e. one or more components of the feed are too large to enter the zeolite channels. Adsorption data for mixtures are scarce at present. The isosteric technique is especially useful in generating this information. Further, once sufficient data for various mixtures have been accumulated it will be possible to have a better assessment of the various theoretical models used to calculate adsorption isotherms of mixtures from single component sorption data. If successful, this will greatly reduce the experimental work needed for the design of any given process, e.g. pressure swing separation. Little information has yet appeared in the open literature. Some gas and liquid phase separation processes have been well established for some years now. New developments are concerned with pressure swing separations of gas mixtures and liquid phase separations of sugar mixtures, e.g. glucose/ fructose mixtures. P.A. Jacobs emphasised the importance of the relation between the complex formation between sugars and divalent cations in homogeneous aqueous solutions. Calcium and magnesium exchanged zeolites should be excellent sorbents for the separation of sugars because of the differences in the properties of these sugar/divalent ion complexes.

REFERENCES

1. J. Karger and H. Pfeifer, "Perspectives in Molecular Sieve Science" (Ed W.H. Flank and T.E. Whyte Jr) ACS Symposium Series 368, ACS, Washington DC (1988), p 376.

2. J. Karger and H. Pfeifer, Zeolites 7:90 (1987).

3. D. Boddenberg, R. Burmeister and G. Spaeth, in "Zeolites as Catalysts", Sorbents and Detergent Builders" (Ed. H.G. Karge and J. Weitkamp) Studies in Surface Science and Catalysis 46, Elsevier, Amsterdam (1989) p 533.

4. N.G. Van-Den-Begin and L.V.C. Rees, in Proceedings 8th Int. Conf. Zeolites, Amsterdam, July 1989, Elsevier, Amsterdam in press.

5. Y. Yasuda, J. Phys. Chem. 86:1913 (1982).

6. M. Bulow, W. Mietk, P. Struve, and P. Lorenz, J. Chem.Soc. Faraday 1, 79:2457 (1983)

7. M. Eic and D.M. Ruthven, Zeolites 8:40 (1988).

8. A. Paravar and D.T. Hayhurst, in "Proceedings 6th Int. Zeolite Conf." (Ed D. Olson and A. Bisio) Butterworths, Guilford, UK (1984) p 217.

9. P. Graham, A.D. Hughes and L.V.C. Rees, Gas. Sep. Purf. 3:56 (1989)

LOCALIZED AND OVERALL PROPERTIES RELATED TO THE NATURE AND STRUCTURAL ORGANIZATION OF THE FRAMEWORK ATOMS

D Barthomeuf*, A. Corma**

* Laboratoire de Réactivité de Surface et Structure, URA 1106 CNRS
Université Pierre et Marie Curie
4 place Jussieu
75252 - Paris Cedex 05, France

** Instituto de Catalisis y Petroleoquimica, Calle Serrano 119
28006, Madrid, Spain

The understanding of zeolites properties, particularly the acidity, needs several simultaneous approaches. It has long been postulated that the properties at a given site may result from short interactions i.e. the local influence of the neighbours and also from interactions on a long range i.e. overall or global effects of the framework. The extent of the involvment of those two effects will be discussed in what follows. In addition the zeolite behaviour in acidic catalytic reactions requires to consider the system zeolite-reacting molecule.

STATE OF THE ART

Techniques

Infrared spectroscopy is a good tool to characterize the nature and strength of the acidity when the zeolite structure is known. Nevertheless it is not powerful enough in defining more precisely the acid site at the level required presently. Other approaches have been developped more recently. For instance XPS and X-ray emission spectroscopies show changes in the binding energy of electron on Si, Al and O with changes in Si/Al ratio (1). This shows a shift of global framework properties with the chemical composition. It has also to be realized that the exchange of a metal cation for a proton also involves a global shift in the energy level of the framework. XPS is representative when the composition of the external surface is the same as in the bulk.

Theoretical calculations

Theoretical studies have been more developped in the recent years, using either ab-initio or semi-empirical methods. Their contribution in solving physicochemical properties such as framework infrared spectra (2) or refractive index (3) is very helpful. They also give information on the framework energy. It is shown for instance that the cohesive energy of a zeolite crystal consists for approximately 10% of the long range electrostatic term and for 90% of the short range contribution (2).

In the field of acidity and catalysis the various parameters involved make it difficult presently to describe theoretically a real system. Those theoretical studies have then to be considered as not necessarily perfectly describing the reality but giving very helpful concepts in rationalizing observed experimental acidic and catalytic properties.

Guidelines for Mastering the Properties of Molecular Sieves
Edited by D. Barthomeuf *et al.*
Plenum Press, New York, 1990

The main recent ideas which can be pointed out concern the following points :

- The proton are mobile, jumping not only from oxygen on the same AlO_4 tetrahedra as it was already known but they may be located between two adjacent SiO_4 tetrahedra. They might move up to 0.6 nm from Al site. The concept of mobility depends on the technique used to look at. We have probably to envision the phenomena as a statistical distribution of positive charge in a given region.

- The proton strength increases as it is already known as the Al zeolite content decreases. A general agreement arises as to the fact that protons attached to oxygen belonging to large TOT angles have a high acid strength. This is in line with a structure influence on acid strength.

- The proton strength depends not only on the previous factors but also on the way the framework is able to relax upon protonation, changing bond lengths and bond angles. A small difference in computed charge on atoms may then be enhanced with regards to chemical properties by the framework relaxation.

- The difference between the charge (integration of the electron density distribution) and the electronegativity (compactness of the electron cloud) is pointed out (3)

- Electronegativity and covalence vary in the same way with the zeolite composition. The acid softness in zeolites varies in the same way than electronegativity. However, while electronegativity is an average value of the framework which can not take into account short range effects, the softness can have not only average but also local values. Therefore, it is a valuable concept to explain short range and long range effects on acidity and reactivity.

Acid-base properties

Due to the zeolites electroneutrality, the presence of positive charges related to acidity should not be disconnected from the associated negative charges on the oxygen atoms. This means that sites from acid-base pairs may both act in catalysis and adsorption. Their acid and base strengths are related.

Effect of substitution - New materials

Theoretical calculations and theoretical models do not take into account the presence of extraframework Al, which has an important implication not only on the total Brönsted acidity but also on acid strength and acidity type (Lewis). Therefore, one should not necessarily expect that the conclusion derived from theoretical models could directly be extrapolated to real cases, but they may serve as guidelines.

The understanding of acidic properties of zeolites with various substituants B, Ga, Ti... is making progress. For the new phosphorus based molecular sieves only few results appear. The correlation of acidic properties with chemical composition is complicated by the formation of phases mixtures (silica in SAPO-5 or zeolitic silica-alumina in SAPO-37)

Influence of the zeolite composition on the concentration and preactivation of reactants in the pores. Medium (solvent) effect.

In the case of zeolites the sorbates are always under the influence of pore walls. In other words the molecules in zeolites are solvated by the framework. The solvent effect produces a concentration enhancement of hydrocarbon reactants within the zeolite crystal relative to the vapour. Powerfull electrostatic fields in the zeolites are produced by charges on cations and framework atoms. This can polarize a sorbate leaving it susceptible to be attacked by strong acid centers. The extent of polarization of the sorbate increases when increasing the field gradient in the zeolite. Dipole or/and quadrupole may appear in sorbed reactants modifying the reactivity of C-C, C-H, C-O... bonds (1,4).

Zeolites with a high Al content have a high polarity which will favour the formation of free carbocations. On the other hand at high Si/Al ratio complexes will be formed. The solvent effect depends not only on the zeolite but also on the molecules in the cages. It rises a maximum for Al/Al+Si ratio between 0.25 and 0.33 (i.e. Si/Al from 2 to 3) (1).

There are non-specific van der waals interactions which are greater in zeolites than in amorphous aluminosilicates due to increased surface curvature (5).

The rates of reaction, activation energies and therefore selectivities are affected not only by the number and intrinsic activity of the active sites but also by the disturbances created in the sorbed reactant by the zeolite field and field gradient. This last effect changes the adsorption properties (adsorption constant and heat of adsorption) which control the concentration of the species inside the zeolite. This could be a contribution of long range effects to reactivity.

The number, quality and accesibility of sorption sites in zeolites are determined by the zeolite framework topology and chemical composition and by the sorbate-sorbent interactions.

The presence and type of extraframework Al species will modify the adsorption properties and short range interactions.

Simulation of catalytic reactions

Theoretical (MO) calculations on cluster compounds have been very effective in modelling reactive sites, indicating the important role of short range interactions on reactivity (6-8).

When the strong electrostatic fields existing inside the zeolite cavities are included in the calculations performed on the ab-initio level with full geometry optimization, this electrostatic perturbation stabilises both the ionic and the covalent complexes. The charge on the proton also increases after inclusion of these long range effects in the calculations (3).

During molecular interactions, the results obtained using semi-empirical methods (CNDO, MINDO, and EHT methods), should be considered carefully since they do not reproduce the correct polarization of the molecules (9-10). It would be better to estimate the interaction energies directly starting from ab-initio calculations of the interacting molecules.

Interactions and reactivity of molecules over acid zeolites can also be considered from the point of view of the frontier orbital theory. Following this, the distribution and energy of the frontier orbital of both, zeolite and the reactant molecule (HOMO and LUMO) should be considered. Using this approach, the influence of short range and long range interactions can be considered (11-12).

OPPORTUNITIES AND GOALS

Techniques

The improvement of techniques and of their accuracy (XPS, X-ray emission spectroscopies, NMR for atom distributions, XRD and neutron diffraction ...) open the access to the domain of experimental measurement of bond energies, bond angles, bond lengths in the framework in the absence or in the presence of adsorbed molecules. This is of major importance for a better understanding of acid-base properties and reactant-zeolite interactions.

Theoretical approaches

Despite the fact that it is not presently possible to introduce in such calculations all the real parameters defining the zeolite properties, thoses approaches have to be expanded. They give good ideas and new concepts which allow not only to rationalize known properties but which help in the prediction of the behaviour of new materials or eventually the prediction of new properties.

The importance of charge value which differs from electronegativity, of charge location or "delocalization", and of the correlation electronegativity - softness have to be kept in mind.

Acid-base properties

The idea of acid-base pairs should not be forgotten since the anion of zeolitic acid should have a role as important as that played in several usual inorganic or organic acids.

Effect of substitution, new materials

The fast expanding domain of those materials opens a field where very little is known or predictable on the properties induced by framework elements other than Si and Al. Much will be learned from a detailed study of selected solids. Local and global effects on zeolite properties could then be better understood and hence controlled.

Influence of the zeolite composition on the concentration and preactivation of reactants in the pores. Medium (solvent) effect

The concept of solvent effect is already used in usual silicium-aluminium zeolites (1). A good opportunity would be to extend it to zeolites substituted with other elements. Attempts could be done to reach a more quantitative approach in order to have accurate predictive tool. A way could be to use improved techniques such as I.R. and NMR looking into adsorbed molecules in conditions closer to chemical reactions.

It should be interesting to look into the adsorption-activation process not only from the point of view of a neutral molecule which is going to interact with a localized active site, but as an already "activated" molecule due to the electrostatic fields and field gradient, which is going to interact with a more particular active site.

The consideration of confinement effect and the effect of non-specific van der waals interactions together with more specific interactions due to strong localized effects can better model the interaction and reactivity of adsorbates in zeolites.

The possibility of modifications of the framework structure due to guest-host interaction in order to accomodate the molecular configuration of sorbates is an open field of both fundamental and applied interest.

It would help in our understanding of zeolite-sorbate interaction and catalysis to try to separate the influence of the different parameters (chemical composition, bond angle etc.) on the type and energetics of adsorption. This could be attacked by simulation techniques, as well as from experimental adsorption studies using well characterized zeolites.

Simulation of catalytic reactions

Studies on the interaction and reactivity of molecules with relatively small size zeolite clusters could be extended with the availability of more powerfull computers, to cluster approximating the actual channel or cage structure of zeolites. This will take into account the results from theoretical calculations, which show that in some structures the most negatively charged oxygens can be in Si-O-Si bonds (2,3).

Theoretical calculations will have to take into account when studying reactivity of zeolites that the zeolite crystals may respond to "proton attack" by a global readjustment of the bond structure of the whole zeolite matrix in order to prevent major local disturbance in bonding, leading to loss of long range symmetry, and thus, crystallinity.

In order to study the extent of crystal field effect it would be of interest to develop realistic simulations of amorphous materials.

RECOMMENDATIONS

- With regards to acid-base properties developp approaches which give information on the energy of the bonds and on the site geometry (angles, bond lengths) with presence or not of an adsorbed molecule (base, reactant).

- Find methods and models which establish unambiguously full or partial proton transfer from the zeolite to a base or a reactant.

- Developp methods to characterize the proton localization near or not far from AlO_4 tetrahedra.

- For high silica zeolites establish unambiguously whether Al sites are paired or not.

- Improve the characterization of extraframework Al (different types of Al in stabilized zeolites and its role on local changes in acidity (amount and strength) and on catalytic activity.

- Find ways to quantify the medium (solvent) effect.

- Establish reliable methods for basicity measurements.

- Look for experimental evidence on the stabilization of carbenium ions due to zeolite lattice walls.

- When comparing series of zeolites, perform activity measurements from kinetic rate constants in order to separate intrinsic activity and adsorption effects.

- Quantify electrostatic field and field gradient.

- Perform MO calculations considering clusters approximating the size of channels and cages.

- Develop models involving intermediate radical species.

- In general, develop procedures which allow to look into intermediate species and into zeolite structure while performing chemical reactions (I.R., Raman, NMR).

REFERENCES

1   J.A. Rabo, G.J. Gajda, this book, Chapter IV
2   R.A. Van Santen, B.M.H. Van Beest, A.J.M. de Man, this book, Chapter IV
3   W.J. Mortier, R. Vetrivel, this book, Chapter IV
4   C. Mirodatos, D. Barthomeuf, J. Catal. 114:121 (1988)
5   E.G. Derouane, this book, Chapter IV
6   A.G. Pelmenshchikov, V.I. Parlov, G.M. Zhidomirov, S. Beran, J. Phys. Chem.,
     91:3325 (1987)
7   P.J. O'Malley, J. Dwyer, Chem. Phys. Lett., 97:143 (1988)
     J. Dwyer, this book, Chapter IV
8   J. Sauer, Chem. Rev., 89:199 (1989)
9   W. Linert, V. Gutmann, P.G. Perkins, Inorg. Chim. Acta, 69:61 (1983)
10   Y. Wang, P. Nordlander, N.H. Tolk, J. Chem. Phys., 89:4163 (1988)
11   K.P. Wendlandt, H. Bremer, in Proc. 7th Intern. Cong. Catal., Berlin IV,
     Verlag Chemie, Weinheim 1984, p. 507
12   A. Corma, this book, Chapter IV

ORIENTATION OF THE PATH OF REACTIONS (CATALYSIS, ADSORPTION) BY

CHEMICAL OR OTHER GEOMETRIC OR NOW GEOMETRIC EFFECTS

A. Peters[1] and F. Raatz[2]

[1]W.R. Grace, Davison Chemical Division
Columbia, Maryland 21044 U.S.A.
[2]Institut Francais du Petrole (IFP)
1/4 rue de Bois Préau, BP 311
92506 Rueil Malmaison Cédex, France

STATE OF THE ART

Two types of factors control the selectivity of zeolites in catalytic applications : chemical and geometric. The chemical factors relate to the type of active functions present in the zeolite (strength and density of acid sites, metal, oxide or sulfide sites etc), while the geometric factors relate to the size and the connectivity of the channel system.

CHEMICAL FACTORS

Here it is necessary to distinguish between monofunctional and bifunctional catalysis.

Monofunctional catalysis.

Monofunctional zeolite catalysts will contain a single function, typically an acidic function but also a metal or an oxide function. Selectivity can be varied by controlling the strength and density or dispersion of the sites.

The transformation of hydrocarbons over large pore zeolites provides an illustration of this kind of control over selectivity. HY zeolites with a lower unit cell parameter and a higher silicon to aluminum ratios provide decreased activity for hydrogen transfer relative to isomerisation and cracking activities. The result is a more olefinic and higher octane gasoline. Initially it was believed that these changes in selectivity were a result of stronger acid sites. Current work by Corma in the previous section (Theme IV) and by A.Peters in this section suggests that these selectivity changes are more likely related to site density rather than acid site strength. Another control of selectivity by monitoring the acid function is the conversion of methanol over MFI. The selectivity for olefins or aromatics is strongly influenced by the silicon to aluminum ratio or site density of the zeolite (Wu and Keading, J. Catal, 88, (1984), p 478).

While acid site strength is known to be important, these results suggest that other factors such as the symmetry, local densities and

Guidelines for Mastering the Properties of Molecular Sieves
Edited by D. Barthomeuf *et al.*
Plenum Press, New York, 1990

409

topologies of the active sites may be important to selectivity. In this context A. Corma in the previous section discussed the application of frontier orbital theory and the concept of hard and soft acidity to zeolite catalytic selectivity.

Bifunctional catalysis

Bifunctional zeolite catalysts consist typically of an acidic zeolite containing a metallic function. The zeolite supplies the acidic site and the metal hydrogenation/dehydrogenation site. Catalyst selectivities over a large pore zeolite are controlled by the relative activities of the two catalytic functional sites. This defines the balance of the catalyst. Zeolite based commercial bifunctional catalysts include a variety of hydrocracking, isomerisation and more recently aromatization catalysts.

For instance, hydrocracking catalysts contain a modified HY zeolite combined with either a metallic or sulphided metallic function. Hydrocraking catalysts give three types of reactions : cracking, hydrogenolysis, and isomerisation. The selectivity of the catalyst for the different possible products is controlled by the relative activity of the two functions, ie the balance of the catalyst. In the specific case of hydrocracking the quantitative experimental criteria of a balanced catalyst in terms of the expected product selectivities are well defined, and clearly explained in the comprehensive review given by J. Weitkamp in this section.

Until recently a quantitative relationship between balance and measurements of the number and relative activities of the acid and metal sites has not been available. Recent work by Guisnet, Alvarez, Giannetto and coworkers has gone part way to correct this situation. They have shown that the relationships can be quantified, and that there can be a great deal of variation depending on the type, pore size and Si/Al ratio of the zeolite used.

Competitive adsorption and reversible poisoning, for example by H_2S, water and possibly by NH_3 can also have significant effects on the selectivity of hydrocraking catalysts and probably other bifunctional systems. Some of these effects are also discussed by J. Weitkamp in his review in this section. He points out that there is a lack of information on the effect of ammonia on selectivity, to that regard a recent work by P. Dufresne et al has been devoted to this subject (P. Dufresne et al, Catalyst in Petroleum Refining Conference, Kuwait, March 5-9, 1989, to be published).

GEOMETRIC FACTORS

Zeolites are molecular sieves. This property is of specific importance in catalysis since it allows the design of catalysts exhibiting shape selective properties. Shape selectivity can direct selectivity by controlling the diffusion of reactants or products through the pores of the zeolite or by restricting the formation of transition states. Shape selective properties of the medium pore size MFI type zeolites have lead to the design of selective industrial catalysts for applications such as dewaxing and the preparation of para-substituted aromatics. However, recent work suggests the existence of significant geometric factors other than conventional shape selectivity. For example, the surface may contain holes giving rise to a nest effect (E. Derouane, J. Catal, 88, (1986), p541) ; this type of effect is now a matter of controversy (M. Neuber and J. Weitkamp, 8th IZC, Amsterdam 10-14 july 1989, Recent Research Report n°198). The zeolite may also contain secondary pores induced by post-synthesis treatments which could affect the selectivity.

In practice, the combination of chemical and geometrical factors is very helpful to prepare catalysts with finely tuned selective properties. As an example the pore size of hydrocracking catalysts can be used to monitor the product distribution. This has led to a new method, known as the spaciousness index method (J. Weitkamp et al, Appl. Catal, 27,(1986), p207) for probing the actual pore size of new zeolite materials.

Chiral synthesis of optically active molecules is a so far unrealized goal of catalysis. The occurence of zeolites with channels containing predominantely a right or left handed screw symmetry or with a systematically asymmetric pore or surface geometry would suggest the possibility of chiral synthesis. The zeolite beta has a channel structure with the appropriate symmetry, but if polymorphs with both right and left handed symmetries occur with equal frequency, then there may be no chiral effect. While chiral synthesis is currently recognized as a possibility, there are no existing systems for such a synthesis.

OPPORTUNITIES AND GOALS

In many applications it would appear that the basic rules governing selectivities over either monofunctional or bifunctional zeolite catalysts are well established and well understood in terms of shape selectivity, site strength and density, and balance between functions. However there are applications for which there is less understanding than one might expect. This is true to an extent in the case of hydrocarbon transformations and is true to a greater extent for reactions involving the transformations of nonhydrocarbon organic reactants used in the preparation of fine chemicals. We will discuss hydrocarbon transformations first and secondly the preparation of fine chemicals.

HYDROCARBONS

Even though hydrocarbon chemistry over zeolites has been the subject of a great deal of work, opportunities in several areas can be identified including : i) bifunctional catalysis on systems other than the acid-metal sulfide or acid-noble metal systems with applications to hydrocracking, ii) shape selectivity of large pore zeolites, and iii) the role of secondary pore systems present in modified zeolites. Chiral synthesis would be an entirely new area, but would require a degree of control over site geometry not yet achievable.

In the last few years a lot of work has been devoted to bifunctional acid - oxide systems for the formation of aromatics from light hydrocarbon gas streams. These systems may contain gallium or zinc oxides. For Ga or Zn/MFI catalysts the required balance has not been clearly defined in terms of product selectivities or catalyst properties. It would seem that the concept of balance must survive, since for a given zeolite there will be a desired product selectivity and an optimum oxide activity, that in some sense will define the catalyst balance. Clearly, more work would be of value in this area.

Most often the term shape selectivity is applied to medium pore zeolites, for example the MFI system. However, for reactions involving sufficiently bulky molecules or transition states, large pore zeolites may show shape selective properties. This area has not been explored too much for hydrocarbon transformation yet, since it is perhaps more relevant to fine chemical transformations. As an example, refering again to hydrogen transfer reactions, it appears that in cyclohexene transformation and contrarily to Y zeolite, zeolites omega and beta exhibit a very low hydrogen transfer selectivity whatever their Si/Al ratio (E. Jacquinot et al, Stud. Surf. Sci. Catal, 46, (1989), p115). This could be tentitatively attributed to a restricted transition state selectivity in these zeolites which do not possess any large cages.

Hydrothermal dealumination has been observed to produce secondary pore systems in zeolites. The effect on selectivity in the conversion of heavy hydrocarbon feeds has been suggested but not clarified.

FINE CHEMICALS

Interest in the preparation and reactions of some the nonhydrocarbons used in the fine chemical industry has increased in the last few years. W. Holderich in his well documented review of this area addresses an important specific point : "what about the shape selective effect in fine chemicals transformations ?". His discussion covers a large variety of reactions including alkylation of phenols and nitrogen containing heteroaromatics, acylation of aromatics, halogenations, and a variety of rearrangements including the rearrangement of cyclohexanone oxime to e-caprolactam. He nicely shows that in many cases the observed selectivities are not easily rationalized within the usual scheme of zeolite shape selectivity. His discussion raises questions about the mechanism of some of these reactions including possible solvent effects as well as unexplained shape selectivities.

In hydrocarbon transformations on acidic zeolites, it is well accepted that carbocations act as intermediates. In the case of nonhydrocarbon transformations the case is not so clear. As an example, halogenation selectivities obtained for bulky compounds reacting on MFI, suggest a reaction mechanism involving a varying degree of radical character. There have been previous suggestions of a radical-like behavior over zeolites, however, there are no clear examples for this type of zeolite catalysis involving hydrocarbons. In the case of nonhydrocarbon chemistry on zeolite, do we continue to use the rules of carbocation chemistry when the reaction occurs in a zeolite media, or do other intermediates occur ? To what extend do the classical rules of organic chemistry, as they are known in homogeneous media, still apply ?

In many of the examples given by W.Holderich strong selectivities for specific isomers are obtained. These selectivities are not easily rationalized on the basis of the conventional shape selective models, and, in some cases, are the reverse of what one might expect solely on the basis of pore or channel size or framework structure. For example, in the alkylation of phenol, p-cresol is formed over a potassium faujasite while the MFI framework gives the ortho isomer. In another example, acylations with carboxylic acids give the unexpected result : para-selectivity over faujasite and ortho-selectivity over MFI zeolites. In these cases the reasons for the observed selectivities are not entirely understood. Site strength or site geometry either on the surface of zeolite particles or within the pore structure may be important. Acid site strength and site densities are frequently cited as important for controlling selectivity. Since in zeolites higher strength is associated with lower density, it is not always clear which is the real issue. Many of the rearrangements cited by W. Holderich appear to be sensitive mainly to acid stregth, requiring very weak acidity.

W. Holderich cites organic reactions occuring over zeolites that are very sensitive to the specific exchange cation ; for example the alkylation of pyridine and also the reduction of aldehydes and ketones to alcohols by the Meerwein-Ponndorf-Verley reduction. The existence of electrostatic effects associated with field gradients has been demonstrated by C. Mirodatos and Barthomeuf (J. Catal, 114, (1988), p121) in other contexts. These effects may also be involved in reactions where the selectivity is sensitive to the presence of specific exchange cations.

Some zeolites appear to be active and selective for the transformation of chemicals that are too bulky to easily react within the channel system ; for example the Beckmann rearrangements or halogenation reactions cited by W. Holderich using a MFI type of zeolite. In these cases the outer surface of the zeolite is likely to play an important catalytic role. In extreme cases, only some specific part of a bulky molecule may enter the pore and react at the entrance of the channel producing a surface shape selectivity (nest effect).

Another aspect of fine chemical transformations is the possible role of the solvent. The importance of the solvent effect is well-known in classical organic chemistry. The zeolite is often viewed as a solid solvent with specific electrostatic properties. However organic solvents are often present in the feed, and it may be possible to use a solvent to control selectivity.

RECOMMENDATIONS

When designing new catalyst systems, one attempts to control the reaction path and the product selectivity by a proper a priori choice of the catalytic function(s) and the structure of the zeolite. The choice is based on well understood rules involving estimates of acid site strength and density, the balance in the case of bifunctional systems, and shape considerations. These rules need to be extended to new catalytic systems and to new classes of nonhydrocarbon reactions. Without trying to be exhaustive and in agreement with the general discussion sessions organized during this NATO Worshop, we would propose the following areas for further work.

HYDROCARBONS

- Quantitative estimate of the meaning of balance in the case of new bifonctional catalyst systems (acid-oxide systems for aromatics production for instance),

- Shape selective effects in large pore zeolites and the consequence of secondary pore systems present in modified zeolites,

- Assessment of the effects of different types of acid sites (framework versus non framework) present in zeolites, and the effect of the density and topology of the site distribution,

- Chiral synthesis, a high risk project.

FINE CHEMICALS

- Mechanisms of reaction, does the same carbocation organic chemistry still apply ? Are the classical rules of organic reactions, as they are known in homogeneous media, still valid ?

- The relative importance of shape selectivity and the nature of the active site, especially in cases where the reaction is specific for very weak acid sites or for specific exchange cations,

- Organic solvent effect in zeolite catalysis,

- How does the zeolite outer surface affect the selectivity for the transformation of bulky compounds ?

PARTICIPANTS

Pr. A. ALBERTI

University of Sassari
Via G.M. Angioj 10
Sassari - Italy

Miss F. ARMANINI

Laboratoire de Réactivité de Surface
et Structure URA 1106 CNRS
Université Pierre et Marie Curie
4 place Jussieu
75252 - Paris- Cedex 05 - France

Dr. D. BARTHOMEUF

Laboratoire de Réactivité de Surface
et Structure
URA 1106 CNRS
Université P. et M. Curie
4 place Jussieu
75252-Paris Cedex 05 - France

Dr. E. COHEN DE LARA

Laboratoire de Recherches Physiques
Université Pierre et Marie Curie
4 place Jussieu Tour 22
75252 - Paris Cedex 05 - France

Pr. A. CORMA

.C.S.I.C
Institute de Catalisis y Petroleoquimic·
Calle Serrano 119
28006 - Madrid - Spain

Pr. E. DEROUANE

Facultés Universitaires Notre-Dame
de la Paix
Laboratoire de Catalyse
61 rue de Bruxelles
5000 - Namur - Belgium

Pr. J. DWYER

University of Manchester
Institute of Science and Technology
P.O. Box 88
Manchester M60 1QD - United Kingdom

Dr. F. FAJULA

Laboratoire Chimie Organique Physique
Appliquée
E.N.S.C.M - UA 418 CNRS
8 rue de l'Ecole Normale
34075 - Montpellier Cedex - France

Dr. G. GIANNETTO	Université de Poitiers Chimie 7 - UER Faculté des Sciences 40 avenue du Recteur Pineau 86022 - Poitiers - France
Dr. P.C. GRAVELLE	PIRSEM 4 rue Las Cases 75007 - Paris - France
Pr. J.L. GUTH	Université de Mulhouse Ecole Nationale Supérieure de Chimie 3 rue Alfred Werner 68093 - Mulhouse - France
Dr. W. HOELDERICH	BASF AG - Ludwigshafen ZAK/J-M300 6700 - Ludwigshafen - F.R.G
Pr. P.A. JACOBS	University of Leuven Centrum voor Oppervlaktescheikunde 42 De Croylaan 3030 - Leuven Heverlee - Belgium
Dr. H.G. KARGE	Fritz-Haber-Institut der Max-Plank- Gessellschaft Faradayweg 4-6 D-1000 Berlin 33 - F.R.G
Pr. M. KIZILYALLI	Department of Chemistry Middle East Technical University Ankara - Turkey
Pr. H. LECHERT	Universität Hamburg Institut für Physikalische Chemie Laufgraben 24 2000 - Hamburg 13 - F.R.G
Dr. P. MAGNOUX	ELF Centre de Recherche BP 22 69360 - Saint-Symphorien d'Ozon - France
Pr. J.B. MOFFAT	Department of Chemistry University of Waterloo Waterloo, Ontario N2L, 3G1 - Canada
Dr. W.J. MORTIER	Exxon Chemical Holland B.V. Basic Chemicals Technology P.O. Box 7335 3000 - HH Rotterdam - The Netherlands
Pr. J.B. NAGY	Facultés Universitaires Notre-Dame de la Paix Laboratory of Catalysis 61 rue de Bruxelles 5000 - Namur - Belgium
Dr. J. NIEMAN	Akzo chemie N.V. P.O. Box 15 1000 - Amsterdam - The Netherlands

Dr. A. PETERS	W.R. Grace et Company Davison Chemical Division Columbia Md 21044 - USA
Dr. F. RAATZ	Institut Français du Pétrole 1 avenue de Bois Préau BP 311 92502 - Rueil-Malmaison - France
Dr. J. RABO	UOP Molecular Sieve Department Tarrytown Technical Center Tarrytown, New York 10591 - USA
Pr. L.V.C. REES	Chemistry Department Imperial College of Science and Technology London SW7 2AY - Great Britain
Pr. F.R. RIBEIRO	Instituto Superior Tecnico Chem. Eng. Department Av. Rovisco Pais 1096 - Lisboa Codex - Portugal
Dr. E. ROLAND	Degussa AG Zweigniederiassung Wolfgang Rodenbacher Chaussee Postfach 1345 6450 - Hanau 1, F.R.G
Mr. A. SHIKOLESLAMI	Laboratoire de Réactivité de Surface et Structure URA 1106 CNRS Université Pierre et Marie Curie 4 place Jussieu 75252-Paris Cedex 05 - France
Dr. M.S. SPENCER	School of Chemistry and Applied Chemistry University of Wales P.O. Box 912, Cardif CF1 ATB-Great Britain
Dr. R. SZOSTAK	Georgia Tech Research Institute Zeolite Research Program Atlanta, Ga 30332 - USA
Dr. B. VAN BEEST	Koninklijke/Shell Laboratorium Postbus 3003 1003 AA Amsterdam-Noord-The Netherlands
Pr. P.H. VAN HOOFF	Eindhoven University of Technology Laboratory of Inorganic Chemistry and Catalysis P.O. Box 513 5600 MB Eindhoven - The Netherlands
Pr. R.A. VAN SANTEN	Technische Universiteit Eindhoven Den Dolech 2 Postbus 513 5600 - MB Eindhoven - The Netherlands
Dr. D.E. VAUGHAN	Exxon Research & Engineering Co. Corporate Research Science Lab. Route 22 East Clinton Township Annandale, NJ 08801 - USA

Dr. J. VEDRINE
Institut de Recherche sur la Catalyse
CNRS
2 avenue Albert Einstein
69626 - Villeurbanne Cedex - France

Pr. J. WEITKAMP
Institute of Chemical Technology
University of Stuttgart
Pfaffenweldring 55
7000 - Stuttgart 89 - F.R.G

CROWN-ETHERS
  As templates, for Si-rich
        faujasites, 79, 82
  Complexes with sodium, in zeolite
        micropores, 81, 84

CRYSTAL
  Habit : control of, 58
  Morphology, control of, 54
  Size : control of, 54

CRYSTALLIZATION
  Of ZSM-5 : factors influencing, 54

DIFFUSION
  Definition, 243
  In micropores of zeolites, 232
  Principles, 287
  Techniques, 395

DIFFUSIVITIES
  On heteropolyoxometallates, 198
  In zeolites, 287

FAUJASITE (SEE ALSO X and Y ZEOLITES)
  Activation energies of n-paraffins
        in, 187
  Benzene relaxation in dealuminated,
        190
  Benzene relaxation in coked, 191
  Correlation between structure and
        selectivity, 369
  Influence of structure on cracking
        selectivity, 365
  Investigation of framework
        aluminium by ^{29}Si NMR, 366
  Molecular motion of hydrocarbons in,
        by NMR, 183
  Polytypes, 82, 111
  Relaxation and self diffusion of
        aliphatic hydrocarbons in,
        186
  Relaxation of aromatic hydrocarbons
        in, 188
  Removal of non framework aluminium
        form, 368
  Structure imaging of polytypes, 111
  Use of crown ethers as templates
        in Si-rich, 79
  XRD of polytypes, 83

FINE CHEMICALS
  Preparation of, on zeolites, 412

FLUORIDE ANIONS
  As mobilizing agents in zeolite
        synthesis, 70, 73
  Solubility of complexes, in
        solution, 75
  Synthesis of MFI in presence of, 75
  Zeolites obtained in presence of, 74

FRAMEWORK ATOMS
  Nature of, in zeolites, 403
  Structural organization of, in
        zeolites, 403

GALLIUM
  Substitution in MFI zeolites, 76

GEL
  Cavities in, probed by ^{129}Xe NMR,
        89
  Interaction with TAA cations, 87
  Systems : ALPO and SAPO phase
        growth in, 95

GERMANIUM
  Incorporation in (Si) MFI, 76

HETEROPOLY OXOMETALLATES
  BET characterization of, 195
  Diffusivities on, 198
  Micropore average radii of, 197
  Sorption capacities of, 198
  Synthesis of 195

^{1}HNMR
  Of Silanol groups in zeolites, 131
  Studies of molecular motions of
        hydrocarbons in faujasites,
        183

HYDROCARBONS
  Aliphatic : relaxation in
        faujasites, 186
  Aromatic : relaxation in
        faujasite, 188
  Chemistry, over zeolites, 411
  Molecular motions of, in
        faujasites, 183

HYDROCRACKING
  Of alkanes, on bifunctional
        catalysts, 348
  Of alkanes, on PtY zeolites, 355
  Bifunctional : mechanisms of, 346
  Factors influencing, 343
  Of n-heptane on various Pt loaded
        zeolites, 355
  Selectivity and activity :
        influence of NH_3 and H_2S,
        352
  Selectivity of : influence of
        competitive adsorption, 350
  Types of, 343

INTERACTIONS (IN ZEOLITES)
  Long range, 263
  Molecular, in zeolitic channels,
        269
  In oxides, 268
  Short range, 263